Rhizosphere Biology

Series Editor

Anil Kumar Sharma, Biological Sciences, CBSH, G.B. Pant University of Agriculture & Technology, Pantnagar, Uttarakhand, India

The Series **Rhizosphere Biology**, emphasizes on the different aspects of Rhizosphere. Major increase in agricultural productivity, to meet growing food demands of human population is imperative, to survive in the future. Along with methods of crop improvement, an understanding of the rhizosphere biology, and the ways to manipulate it, could be an innovative strategy to deal with this demand of increasing productivity. This Series would provide comprehensive information for researchers, and encompass all aspects in field of rhizosphere biology. It would comprise of topics ranging from the classical studies to the most advanced application being done in the field. Rhizoshpere is a dynamic environment, and a series of processes take place to create a congenial environment for plant to grow and survive. There are factors which might hamper the growth of plants, resulting in productivity loss, but, the mechanisms are not very clear. Understanding the rhizosphere is needed, in order to create opportunities for researchers to come up with robust strategies to exploit the rhizosphere for sustainable agriculture.

There are titles already available in the market in the broad area of rhizosphere biology, but there is a major lack of information as to the functions and future applications of this field. These titles have not given all the up-to-date information required by the today's researchers and therefore, this Series aims to fill out those gaps.

Benjamin A. Horwitz • Prasun K. Mukherjee
Editors

Microbial Cross-talk in the Rhizosphere

Springer

Editors
Benjamin A. Horwitz
Faculty of Biology
Technion—Israel Institute of Technology
Haifa, Israel

Prasun K. Mukherjee
Nuclear Agriculture & Biotechnology Division
Bhabha Atomic Research Centre
Mumbai, Maharashtra, India

ISSN 2523-8442 ISSN 2523-8450 (electronic)
Rhizosphere Biology
ISBN 978-981-16-9506-3 ISBN 978-981-16-9507-0 (eBook)
https://doi.org/10.1007/978-981-16-9507-0

This Springer imprint is published by the registered company Springer Nature Singapore Pte Ltd.
The registered company address is: 152 Beach Road, #21-01/04 Gateway East, Singapore 189721, Singapore

Dedicated to our parents:
Lawrence P. Horwitz
Ruth Horwitz
Late Kalimohan Mukherjee
Nihar Mukherjee

Foreword

The soil is rich in microorganisms. In the rhizosphere, the soil volume is directly influenced by plant roots; it is an order of magnitude higher. The coexistence in the soil of roots with both beneficial and disease-causing microbes is an active and much reviewed research subject, including the earlier volumes in this book series. Here, the focus is on communication, including its molecular basis and consequences for the outcome and agricultural relevance of the interactions. The collection of chapters is, of necessity, a sampling of a very wide research spectrum, considering both methodologies and interaction types. Interactions can be studied one-to-one, but for many years, it has been clear that the multitude of microorganisms in the vicinity of roots, the rhizosphere microbiome, need to be considered as a whole. In the opening chapter, the editors extracted the essence of plant–microbe crosstalk in the rhizosphere, documenting the major highlights of the book which sets the stage for an exciting read for this rapidly evolving field of research (Chap. 1). Following that, there is an overview and update on studies of the various ways in which plants shape the composition of the microbiome (Chap. 2), underlining the need to consider microbial associations while improving the crop through breeding and selection. The most studied below-ground interactions that came to be known as biocontrol are only one type of interaction between soil microbes. These beneficial interactions drew attention because disease-suppressive soils, and later the most active members of their microbial populations, for example species belonging to the ascomycete genus *Trichoderma*, provided a way to fight soil-borne diseases with less fungicide load. Other beneficial microorganisms were not only well known to researchers, but even thought to be central to the evolution of life on land; for example, the mycorrhizae accompanied plants from the start. Rhizobia–root interactions permit fixation of atmospheric nitrogen and are almost synonymous with symbiosis. *Trichoderma* being at the centre stage of biological plant health management, plant–*Trichoderma* interactions are a model chosen to explain the principles of beneficial interactions (Chaps. 6 and 10). Quorum sensing (Chap. 5) within populations and between populations can be critical to their success within a particular soil or rhizosphere niche, as well as to the outcome of interactions with plants. The molecular modes of communication, which indeed are the connecting thread linking the subjects of this volume, are represented by plant hormones or microbial molecules with hormone-like activity and effector proteins (Chaps. 3, 4, 12). Each author or group of co-authors having chosen their model organism or interaction, many of the sections can be looked at from different angles. For example, the story of microbial effectors (Chap. 12) is also a view of fungal pathogens of insects, which have applications in biocontrol of pests. The narration across different chapters overlaps wherever similar mechanisms may reach different outcomes: positive or negative (for example Chaps. 7, 8, 11). Likewise, the methodologies are general, for example metabolomics (Chap. 6). The positive interactions are found in Chaps. 3, 4, 8 and 11, while the negative ones take over in Chap. 7. Thus, although each chapter is a review article in itself, the reader will find the comparisons equally rewarding, which is essentially the strength of this volume. While the major focus has been on unravelling the mechanisms, our understanding of how some of the plant–microbe interactions can make farming more sustainable is of utmost importance (Chaps. 8–10). It should be appreciated that the subject area is very vast with enormous amount of research data being added on a daily basis, and it is not possible to cover all the aspects under one umbrella. However, by integrating mechanisms with applications, the editors have succeeded to compile a volume that would stimulate research in this dynamic and rapidly evolving area of plant–microbe interactions leading to a better crop management for enhancing productivity.

The Robert H. Smith Faculty of Agriculture, Food and Environment,
The Hebrew University of Jerusalem,
Jerusalem, Israel

Ilan Chet

Preface

Rhizosphere, considered to be the "hidden half" of the hidden half (roots), has undergone transformative research with the advent of the omics era, especially the new high-throughput technologies applied to unravel the microbiome (collective microbial diversity). The amount of information on dynamicity of microbial interactions in the rhizosphere has been growing exponentially to the scale that could not have been imagined a few years ago, when the diversity study was limited to culturing of a miniscule (compared to the actual diversity) number of microbes. However, the challenge is to translate such huge information generated to application for improving plant productivity. To achieve this, there is a need to integrate the basic research with applications (research with a purpose). Such integration is mostly done in isolation, for a single beneficial organism or specific interacting pairs. Looking at a plant with its microbiome is an innovation mainly belonging to the past decade. The prime objective of compiling this book is based on this idea—to put together literature available on basic research done (understanding the rhizosphere) and a few chapters on how to apply the knowledge generated for improving crop yield. While talking about mechanisms of interactions (crosstalk) in the rhizosphere, the first thing that comes to our mind is the role of small molecules (small secreted proteins, hormones and other secondary metabolites). Consequently, majority of the chapters are devoted to this topic, on how the small molecules shape up the rhizosphere interactions. Some of these compounds have tremendous potential for applications in commercial agriculture. Rhizosphere microbes could be pathogenic or beneficial. The dominance of one partner will dictate whether the plant will be diseased or healthy. We have included chapters on pathogenic as well as plant-beneficial microbes, some of which like mycorrhizae, pseudomonads and *Trichoderma* spp. are widely used as biostimulants and biocontrol agents. Despite the current pandemic situation, the authors have worked hard to put together the information, and the editors are grateful to them; we also apologize for any inconveniences caused to them during the course of editing this volume. Despite very sincere efforts, we must admit that this book is far from complete in covering all topics relevant to the very broad topic of "plant–microbe crosstalk"; however, we hope that we will get an opportunity to compile a bigger volume on the related topic in near future.

Haifa, Israel Benjamin A. Horwitz
Mumbai, India Prasun K. Mukherjee

Acknowledgements

We take this opportunity to place on record our gratitude to Prof. Anil Sharma, Series Editor, for inviting us to edit this volume of topical importance. We are grateful to all the authors for spending their valuable time in putting together the current information available in the literature and sharing their thoughts through illustrations of outstanding quality. The editors also thank Mr. Suraj Kumar and the team at Springer for being so cooperative and especially flexible with regard to the timeline, at this time of global pandemic; without their help, it would not have been possible to put this volume together.

Contents

Editors and Contributors

About the Editors

Benjamin A. Horwitz is a Professor of Biology at the Technion, Israel Institute of Technology. Dr. Horwitz is well known internationally for his contributions to the field of signalling in filamentous fungi, studying the diverse ways that pathogenic and beneficial species co-opted conserved eukaryotic signalling pathways for their specific lifestyles. He has published over 90 research papers on photobiology and signalling, mostly on two fungi: *Trichoderma* and the agent of southern corn leaf blight; he has edited two books. He has mentored 28 graduate students and continues to communicate to students his enthusiasm for looking at complex biological processes through the simplifying lens of signalling input and output. Dr. Horwitz has served as an associate editor for Eukaryotic Cell, Microbiology and Phytoparasitica and co-chaired the 14th European Congress on Fungal Genetics, held at the Technion in Haifa. Dr. Horwitz holds the Joseph and Bessie Feinberg Academic Chair.

Prasun K. Mukherjee is a Scientific Officer "H", Bhabha Atomic Research Centre, and Professor and Head, Homi Bhabha National Institute, Mumbai. Dr. Mukherjee is an internationally acclaimed *Trichoderma* biologist who made immense contribution to the understanding of the mechanisms of *Trichoderma* action on plants and pathogenic fungi. He has developed *Trichoderma*-based formulations for applications in agriculture and in composting. Dr. Mukherjee has published several science articles in high-impact journals and has edited two books. Dr. Mukherjee is an associate editor of Fungal Genetics and Biology, 3 Biotech and Frontiers in Fungal Biology. He has received several awards and recognition in his research career of nearly 30 years: Homi Bhabha Science and Technology Award of the Department of Atomic Energy, Govt. of India; Fulbright Nehru Academic and Professional Excellence Fellowship; BOYSCAST Fellowship of the Department of Science and Technology, Govt. of India; and VASVIK Award, to mention a few. Dr. Mukherjee is an elected Fellow of the Indian National Science Academy; National Academy of Sciences, India; and National Academy of Agricultural Sciences. Dr. Mukherjee has commercialized two *Trichoderma*-based technologies.

Contributors

Ariella Alperovitch-Lavy Imagindairy LTD, Tel Aviv, Israel

Mariana Álvarez-Navarrete Departamento de Ingeniería Bioquímica, Instituto Tecnológico de Morelia, Morelia, Michoacán, Mexico

Anne J. Anderson Department of Biological Engineering, Utah State University, Logan, UT, USA

Michael J. Bidochka Department of Biological Sciences, Brock University, St. Catharines, ON, Canada

Yariv Brotman Department of Life Sciences, Ben-Gurion University of the Negev, Beersheva, Israel

Hexon Angel Contreras-Cornejo Instituto de Investigaciones Químico-Biológicas, Universidad Michoacana de San Nicolás de Hidalgo, Morelia, Michoacán, México

Monica Cortez Graduate Division of Biomedical and Biological Sciences program in Biochemistry, Cell, and Developmental Biology, Emory University, Atlanta, GA, USA

Valter Cruz-Magalhães Department of Phytopathology, Federal University of Lavras, Lavras, MG, Brazil

Kirstin Cutshaw Department of Ocean Engineering and Marine Sciences, Florida Institute of Technology, Melbourne, FL, USA

Alisdair R. Fernie Max-Planck-Institute of Molecular Plant Physiology, Potsdam, Germany

Marwa El Graoui AgroBiosciences Department (AgBS), Mohammed VI Polytechnic University (UM6P), Ben-Guerir, Morocco

Maria Lodovica Gullino Centre of Competence for the Innovation in the Agro-environmental Sector—AGROINNOVA, University of Turin, Grugliasco, TO, Italy

John Hampton Bio-Protection Research Centre, Lincoln University, Lincoln, New Zealand

Dave Handy Department of Ocean Engineering and Marine Sciences, Florida Institute of Technology, Melbourne, FL, USA

Alyssa Headlee Department of Mathematical Sciences, Florida Institute of Technology, Melbourne, FL, USA

Benjamin A. Horwitz Faculty of Biology, Technion—Israel Institute of Technology, Haifa, Israel

Ashutosh Joshi Plant Biotechnology Division, CSIR-Central Institute of Medicinal and Aromatic Plants, Lucknow, Uttar Pradesh, India

Annegret Kohler INRAE/Université de Lorraine, Interactions Arbres/Microorganismes, Labex ARBRE, Centre INRAE Grand Est Nancy, Champenoux, France

Lamfeddal Kouisni AgroBiosciences Department (AgBS), Mohammed VI Polytechnic University (UM6P), Ben-Guerir, Morocco
African Sustainable Agriculture Research Institute, Mohammed VI Polytechnic University (ASARI-UM6P), Laayoune, Morocco

Sunil Kumar Plant Biotechnology Division, CSIR-Central Institute of Medicinal and Aromatic Plants, Lucknow, Uttar Pradesh, India

Lourdes Macías-Rodríguez Instituto de Investigaciones Químico-Biológicas, Universidad Michoacana de San Nicolás de Hidalgo, Morelia, Michoacán, Mexico

Benedetto Marelli Department of Civil and Environmental Engineering, Massachusetts Institute of Technology, Cambridge, MA, USA

José Eduardo Marqués-Gálvez INRAE/Université de Lorraine, Interactions Arbres/Microorganismes, Labex ARBRE, Centre INRAE Grand Est Nancy, Champenoux, France

David Barbosa Medeiros Max-Planck-Institute of Molecular Plant Physiology, Potsdam, Germany

Artemio Mendoza-Mendoza Bio-Protection Research Centre, Lincoln University, Lincoln, New Zealand

Manal Mhada AgroBiosciences Department (AgBS), Mohammed VI Polytechnic University (UM6P), Ben-Guerir, Morocco

Cynthia Montanez Department of Aerospace, Physics and Space Sciences, Florida Institute of Technology, Melbourne, FL, USA

Prasun K. Mukherjee Nuclear Agriculture & Biotechnology Division, Bhabha Atomic Research Centre, Mumbai, Maharashtra, India

Fabiola Padilla-Arizmendi Bio-Protection Research Centre, Lincoln University, Lincoln, New Zealand

Israel Enrique Padilla-Guerrero Division of Natural and Exact Sciences, Department of Biology, University of Guanajuato, Campus Guanajuato, Guanajuato, Mexico

Andrew Palmer Department of Ocean Engineering and Marine Sciences, Florida Institute of Technology, Melbourne, FL, USA
Department of Biomedical and Chemical Engineering and Sciences, Florida Institute of Technology, Melbourne, FL, USA
Aldrin Space Institute, Florida Institute of Technology, Melbourne, FL, USA
Institute of Global Ecology, Florida Institute of Technology, Melbourne, FL, USA

Alejandro Perez Department of Biomedical and Chemical Engineering and Sciences, Florida Institute of Technology, Melbourne, FL, USA

Samantha Pryor Department of Aerospace, Physics and Space Sciences, Florida Institute of Technology, Melbourne, FL, USA

Arturo Ramírez-Ordorica Instituto de Investigaciones Químico-Biológicas, Universidad Michoacana de San Nicolás de Hidalgo, Morelia, Michoacán, Mexico

Rakesh Kumar Shukla Plant Biotechnology Division, CSIR-Central Institute of Medicinal and Aromatic Plants, Lucknow, Uttar Pradesh, India

Davide Spadaro Centre of Competence for the Innovation in the Agro-environmental Sector—AGROINNOVA, University of Turin, Grugliasco, TO, Italy
Department of Agricultural, Forestry and Food Sciences (DiSAFA), University of Torino, Grugliasco, TO, Italy

Lauren B. L. Stone Department of Biological Sciences, Brock University, St. Catharines, ON, Canada

Katherine Vanselow Department of Biomedical and Chemical Engineering and Sciences, Florida Institute of Technology, Melbourne, FL, USA

Claire Veneault-Fourrey INRAE/Université de Lorraine, Interactions Arbres/Microorganismes, Labex ARBRE, Centre INRAE Grand Est Nancy, Champenoux, France

Joshua Weissman Department of Aerospace, Physics and Space Sciences, Florida Institute of Technology, Melbourne, FL, USA

Brooke Wheeler College of Aeronautics, Florida Institute of Technology, Melbourne, FL, USA
Aldrin Space Institute, Florida Institute of Technology, Melbourne, FL, USA

Oumaima Zaher AgroBiosciences Department (AgBS), Mohammed VI Polytechnic University (UM6P), Ben-Guerir, Morocco

Eric Ziegler Department of Biomedical and Chemical Engineering and Sciences, Florida Institute of Technology, Melbourne, FL, USA

Augustine T. Zvinavashe Department of Civil and Environmental Engineering, Massachusetts Institute of Technology, Cambridge, MA, USA

Chapter 1
Plant-Microbe Cross Talk in the Rhizosphere: Introductory Remarks

Benjamin A. Horwitz and Prasun K. Mukherjee

Abstract Rhizosphere microorganisms populate the soil near and under the influence of plant roots. In agriculture, the microbes interact with a particular crop plant, while in nature, the plant population is diverse. This leads to an almost un-countable number of possible interactions, but research has focused on some that are significant because of their contribution to plant nutrition and crop yield, as models for basic research, or agents of soilborne diseases. This includes symbioses with nitrogen-fixing rhizobia, and mycorrhizae. To introduce some of the concepts discussed in this volume, we emphasize these two examples, because the molecular signals from the plant and from the microbial symbiont are both known. The second focus of this chapter is to chart the types of interactions between plant and microbes and among the soil microbial populations. Finally, it is important to note the outlook for applications in agriculture, in particular where the microbes and their capacity for interaction can be engineered or selected to improve crop yield and suppress diseases.

Keywords Plant-microbe interactions · Plant growth · Soil-borne pathogens · Beneficial soil microbes · Inter-kingdom communication · Rhizobia · Mycorrhizae · Lipo-chitooligosaccharide

1.1 Plant-Microbe Cross Talk

Rhizosphere, the soil under the influence of roots, is the stage on which microbial populations act out their effects on plant growth and crop yield. These populations, collectively, make up the rhizosphere microbiome. The composition and activity of

B. A. Horwitz (✉)
Faculty of Biology, Technion—Israel Institute of Technology, Haifa, Israel
e-mail: horwitz@technion.ac.il

P. K. Mukherjee
Nuclear Agriculture & Biotechnology Division, Bhabha Atomic Research Centre, Mumbai, Maharashtra, India

B. A. Horwitz, P. K. Mukherjee (eds.), *Microbial Cross-talk in the Rhizosphere*, Rhizosphere Biology, https://doi.org/10.1007/978-981-16-9507-0_1

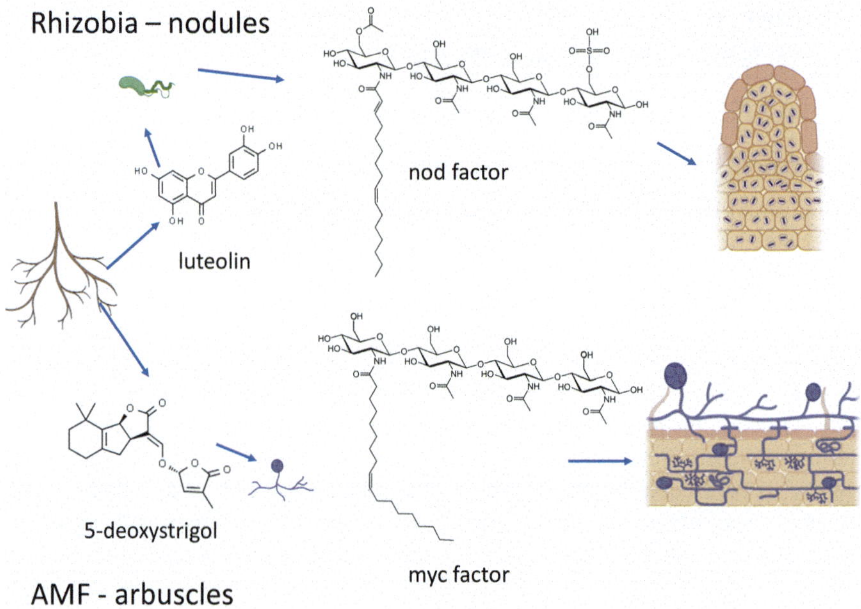

Fig. 1.1 Plant-microbe cross talk in rhizobial and AMF symbioses. Fungal and bacterial (lipo-chitooligosaccharide) and plant (luteolin, strigolactone) signals (Peters et al. 1986; Lerouge et al. 1990; Akiyama et al. 2005; Maillet et al. 2011) are illustrated. AMF arbuscular mycorrhizal fungi. (The scheme is designed following Oldroyd (2013). The small diagrams are from biorender.com or redrawn)

the members of the rhizosphere microbiome determine whether they will promote plant growth, cause disease, or induce systemic resistance. Interaction between plants and microbes implies inter-kingdom recognition of signals, which relies on a molecular language (Bonfante and Genre 2015).

To illustrate this, consider four words from the molecular languages used in two of the best-studied root symbioses: rhizobacteria and arbuscular mycorrhizal fungi (AMF). The importance of these to plants cannot be overestimated: rhizobia-containing nodules of legumes fix atmospheric nitrogen, and AMF colonize most land plants, providing mineral nutrients and promoting growth. Given their importance, the four words (luteolin, 5-deoxystrigol, nod factor and myc factor) in Fig. 1.1 were discovered before microbiome took center stage. Though produced by a prokaryote and eukaryote, respectively, the microbial signals are strikingly similar. The plant signals, in contrast, are different, the only similarity being that they are small diffusible organic molecules. Thus, the microbes have followed parallel evolutionary paths to "design" their signals, while the two plant signals fulfill similar functions despite being the products of carotenoid and flavonoid biosynthetic pathways, respectively. The evolutionary logic, clearly, could be that the microbes are divergent, while the plant host in the two symbioses is similar (or even identical, for example, a legume hosting either or both AMF and rhizobia). A further level of

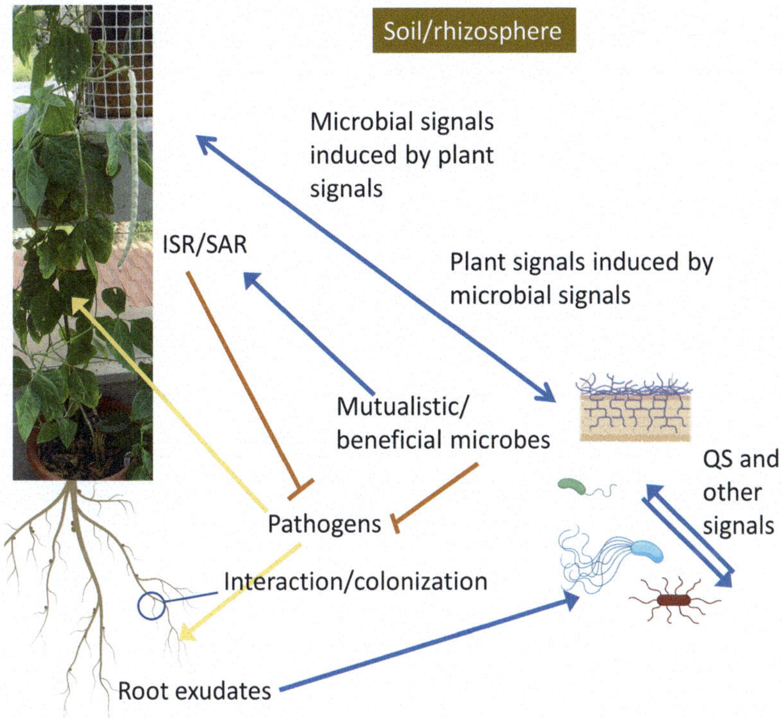

Fig. 1.2 Diagram of relevant cross talks in the rhizosphere. (For a scheme like this one, see, for example, Phour et al. (2020). Image credits: schematic diagrams, Biorender.com; *Vigna unguiculata*, B.A.H and Joseph Mouyal)

complexity is that microbial signals like the nod and myc factors are not unique to symbiosis, as shown recently for lipo-chitooligosaccharides and chitooligosaccharides (Rush et al. 2020; Khokhani et al. 2021).

One challenge for current and future research is to deepen the understanding of important one-to-one plant-microbe interactions in the rhizosphere. Another is to reach a similar level of understanding, but microbiome-wide. Basically, there are three relevant types of cross talk: the plant shapes its rhizosphere microbiome, members of the microbiome interact with each other, and signals from the microbiome modulate plant functions (Fig. 1.2).

1.2 Microbiome

When generalizing plant-microbe interactions to the microbiome, the questions are orders of magnitude larger than for one-to-one pairs. Plants interact with complex sets of microbiome populations (Chialva et al. 2022). Plants produce chemical signals and metabolites carried by root exudates. Primary metabolites can diffuse from the root, as can compounds that are not essential for plant metabolism. The latter are defined as "secondary" metabolites; the term "specialized metabolites" has seen increasing use; both are often abbreviated SM. The root provides a niche for microbes that are associated to varying degrees: from influence at a distance to superficial, and endophytic root colonization. Plants thus modulate their rhizosphere microbiome; plant-driven control of the microbiome depends on genotype, and also on the developmental stage. A recent model centers on the realization that the plant phenotype (rather than the genotype itself) directly drives the composition of the microbiome (Wagner 2021). Defense against pathogens adds another layer; indeed the plant recruits beneficial microbes ("microbiome to the rescue") (Zamioudis and Pieterse 2012). Microbe-driven control should be equally important. Indeed, the microbiome can program plant functions, in particular some that are expressed in the signals the plant sends back to the microbiome (Chap. 2). Thus, plant- and microbe-driven controls actually form a set of feedback loops.

1.3 Microbial Effectors

In the example in Fig. 1.1, the "words" are small molecules specific to the two symbiotic dialogs shown. Perhaps more typical though, microbial effector molecules are often small polypeptides, whose hallmarks are lack of obvious enzyme activity, and multiple cysteine residues. The term effector points to their ability to manipulate the plant's metabolism or gene expression in way that favors the pathogen or symbiont (Chap. 12). Phytohormones are involved in every phase of plant development, and microbes can manipulate phytohormone signaling (Chap. 4). Fungi in particular have evolved to produce effectors that impact phytohormone signaling (Shen et al. 2018).

1.4 Communication Between Microbes (QS)

In the rhizosphere, microbes communicate with the plant and with each other (Fig. 1.2). Quorum sensing by microbial populations is the subject of Chap. 5. Microbes not only communicate but compete, and antibiosis is a fundamental part of this composition, driving evolution of fungal and bacterial secondary metabolite production (Khalid and Keller 2021). In a recent study, a β-lactamase from *Fusarium oxysporum* alters the rhizosphere microbiota of soybean, showing that even under the plant's supervision, a beneficial rhizobacterium might not always win the competition (Chang et al. 2021a). AMF mycorrhizae also have multiple levels of interaction, where AMF modulates the bacterial microbiome. AM symbiosis was recently shown to promote rhizobia accumulation in the rhizosphere of *Medicago truncatula* and hence rhizobial symbiosis (Wang et al. 2021). Conversely, "helper" bacteria promote AMF (Sangwan and Prasanna 2021) and indeed there is a specific AMF bacterial microbiome (Emmett et al. 2021).

1.5 Spectrum of Interactions from Mutualist to Pathogen

Many individual points could be chosen on the continuous spectrum of plant-microbe interactions that range from beneficial to clearly deleterious. Beneficial interactions include nitrogen-fixing rhizobia, mycorrhizal fungi, antibiotic producing bacteria, antagonistic and plant-beneficial fungi and entomopathogens. Deleterious ones include diseases caused by plant pathogenic wilt-causing fungi (Chap. 7) and bacteria and root-rot pathogens. The points on the spectrum chosen here reflect timeliness and the choices made by the authors who joined this volume. Critical to the outcome of a given interaction, as mentioned above, is that the plant "decides" and "recruits" beneficial interacting microbes. The molecular details of how this happens are fascinating (see Chiu and Paszkowski 2021), because the same microbe-associated molecular pattern (MAMP) could be recognized to trigger immunity or symbiosis. Pathogens can desensitize MAMP sensing to overcome plant immunity, for example (Lammertz et al. 2019). How, though, can the plant balance immunity and symbiosis? One newly discovered mechanism is chitotetraose receptor competition at the plant membrane (Zhang et al. 2021). Despite significant progress made in understanding plant-microbe interactions, in many cases, the most intriguing question remains how the outcome (pathogenic or symbiosis) is decided, for example, when and how it is decided that a *Fusarium* will be pathogenic, while the root colonizing *Trichoderma* will establish itself as a root symbiont, despite being able to invade roots, and thus overcoming the first line of plant defense. Interplay of plant hormones especially the salicylic acid and the jasmonic acid is crucial, but again the factors that regulate the temporal and spatial expression of the hormonal signaling is not very well understood. In case of the ectomycorrhiza *Laccaria bicolor*, several mycorrhiza-induced secreted proteins (MiSSPs) play crucial roles in regulation of host invasion and formation of symbiotic relationship (Plett et al. 2011, 2014a, b; Kang et al. 2020). In case of *Trichoderma*, the plant SA restricts the systemic development inside roots (Alonso-Ramírez et al. 2014; Martínez-Medina et al. 2017), while in case of AM fungus *Glomus intraradices*, it's the JA signaling which restricts the fungus (Herrera-Medina et al. 2008). *Laccaria bicolor* secretes MiSSP7 to block the JA signaling via stabilization of the JAZ6 protein, in order to be able to colonize roots of Poplar. Direct evidences on the involvement of small secreted proteins in root invasion and stabilization of symbiotic association are lacking in many plant-fungal symbiotic associations and this should be a research priority for the future (Chap. 11).

1.6 Engineering Rhizosphere Cross Talk for Agriculture

Practical importance of the topics discussed here cannot be overemphasized. Innovations for sustainable agriculture are often rooted in the rhizosphere (Trivedi et al. 2021; Hakim et al. 2021; Phour et al. 2020). A time-tested yet still evolving strategy is biocontrol of soilborne disease, and induction of systemic resistance, by beneficial bacteria (Chap. 8) or fungi (Chap. 10). From the engineering side, it is essential to design and optimize ways to manipulate the rhizosphere composition, for example, by seed coating (Chap. 9). Looking for the most effective microbes to coat the seeds with, a direct approach involves selection and engineering, starting with known beneficial microbes (Mukherjee et al. 2019). Overall, the choice of rhizosphere microbes and communities must draw on evolutionary understanding, and the abundance of new data on how the plant and its microbiome influence each other (Chap. 2). Recent evidences suggest cultivar-specific heritability of the rhizosphere microbiome in various crops, and the rhizosphere microbiome composition directly influences the yield (Liu et al. 2021; Chang et al. 2021b; Deng et al. 2021; Wagner 2021; Gaete et al. 2021). Thus, the crop improvement programs should take into account the influence of the "invisible" partners in the rhizosphere while selecting for improved plant types.

References

Akiyama K, Matsuzaki KI, Hayashi H (2005) Plant sesquiterpenes induce hyphal branching in arbuscular mycorrhizal fungi. Nature 435:824–827. https://doi.org/10.1038/nature03608

Alonso-Ramírez A, Poveda J, Martín I et al (2014) Salicylic acid prevents *Trichoderma harzianum* from entering the vascular system of roots. Mol Plant Pathol 15:823–831. https://doi.org/10.1111/mpp.12141

Bonfante P, Genre A (2015) Arbuscular mycorrhizal dialogues: do you speak "plantish" or "fungish"? Trends Plant Sci 20:150–154

Chang HX, Noel ZA, Chilvers MI (2021a) A β-lactamase gene of *Fusarium oxysporum* alters the rhizosphere microbiota of soybean. Plant J 106:1588–1604. https://doi.org/10.1111/tpj.15257

Chang J, Sun Y, Tian L et al (2021b) The structure of rhizosphere fungal communities of wild and domesticated rice: changes in diversity and co-occurrence patterns. Front Microbiol 12:610823. https://doi.org/10.3389/fmicb.2021.610823

Chialva M, Lanfranco L, Bonfante P (2022) The plant microbiota: composition, functions, and engineering. Curr Opin Biotechnol 73:135–142

Chiu CH, Paszkowski U (2021) How membrane receptors tread the fine balance between symbiosis and immunity signaling. Proc Natl Acad Sci U S A 118(24):e2106567118

Deng S, Caddell DF, Xu G et al (2021) Genome wide association study reveals plant loci controlling heritability of the rhizosphere microbiome. ISME J. https://doi.org/10.1038/s41396-021-00993-z

Emmett BD, Lévesque-Tremblay V, Harrison MJ (2021) Conserved and reproducible bacterial communities associate with extraradical hyphae of arbuscular mycorrhizal fungi. ISME J 15: 2276–2288. https://doi.org/10.1038/s41396-021-00920-2

Gaete A, Pulgar R, Hodar C et al (2021) Tomato cultivars with variable tolerances to water deficit differentially modulate the composition and interaction patterns of their rhizosphere microbial communities. Front Plant Sci 12:688533. https://doi.org/10.3389/fpls.2021.688533

Hakim S, Naqqash T, Nawaz MS et al (2021) Rhizosphere engineering with plant growth-promoting microorganisms for agriculture and ecological sustainability. Front Sustain Food Syst 5:16

Herrera-Medina MJ, Tamayo MI, Vierheilig H et al (2008) The jasmonic acid signalling pathway restricts the development of the arbuscular mycorrhizal association in tomato. J Plant Growth Regul 27:221–230. https://doi.org/10.1007/s00344-008-9049-4

Kang H, Chen X, Kemppainen M et al (2020) The small secreted effector protein MiSSP7.6 of Laccaria bicolor is required for the establishment of ectomycorrhizal symbiosis. Environ Microbiol 22:1435–1446. https://doi.org/10.1111/1462-2920.14959

Khalid S, Keller NP (2021) Chemical signals driving bacterial–fungal interactions. Environ Microbiol 23:1334–1347

Khokhani D, Carrera Carriel C, Vayla S et al (2021) Deciphering the chitin code in plant symbiosis, defense, and microbial networks. Annu Rev Microbiol 75:583–607. https://doi.org/10.1146/annurev-micro-051921-114809

Lammertz M, Kuhn H, Pfeilmeier S et al (2019) Widely conserved attenuation of plant MAMP-induced calcium influx by bacteria depends on multiple virulence factors and may involve desensitization of host pattern recognition receptors. Mol Plant-Microbe Interact 32:608–621. https://doi.org/10.1094/MPMI-10-18-0291-R

Lerouge P, Roche P, Faucher C et al (1990) Symbiotic host-specificity of Rhizobium meliloti is determined by a sulphated and acylated glucosamine oligosaccharide signal. Nature 344:781–784. https://doi.org/10.1038/344781a0

Liu Y, Yang H, Liu Q et al (2021) Effect of two different sugarcane cultivars on rhizosphere bacterial communities of sugarcane and soybean upon intercropping. Front Microbiol 11: 596472. https://doi.org/10.3389/fmicb.2020.596472

Maillet F, Poinsot V, André O et al (2011) Fungal lipochitooligosaccharide symbiotic signals in arbuscular mycorrhiza. Nature 469:58–64. https://doi.org/10.1038/nature09622

Martínez-Medina A, Appels FVW, van Wees SCM (2017) Impact of salicylic acid- and jasmonic acid-regulated defences on root colonization by Trichoderma harzianum T-78. Plant Signal Behav 12:e1345404. https://doi.org/10.1080/15592324.2017.1345404

Mukherjee PK, Mehetre ST, Sherkhane PD et al (2019) A novel seed-dressing formulation based on an improved mutant strain of Trichoderma virens, and its field evaluation. Front Microbiol 10: 1910. https://doi.org/10.3389/fmicb.2019.01910

Oldroyd GED (2013) Speak, friend, and enter: signalling systems that promote beneficial symbiotic associations in plants. Nat Rev Microbiol 11:252–263

Peters NK, Frost JW, Long SR (1986) A plant flavone, luteolin, induces expression of Rhizobium meliloti nodulation genes. Science 233:977–980. https://doi.org/10.1126/science.3738520

Phour M, Sehrawat A, Sindhu SS, Glick BR (2020) Interkingdom signaling in plant-rhizomicrobiome interactions for sustainable agriculture. Microbiol Res 241:126589

Plett JM, Kemppainen M, Kale SD et al (2011) A secreted effector protein of Laccaria bicolor is required for symbiosis development. Curr Biol 21:1197–1203. https://doi.org/10.1016/j.cub.2011.05.033

Plett JM, Daguerre Y, Wittulsky S et al (2014a) Effector MiSSP7 of the mutualistic fungus Laccaria bicolor stabilizes the Populus JAZ6 protein and represses jasmonic acid (JA) responsive genes. Proc Natl Acad Sci U S A 111:8299–8304. https://doi.org/10.1073/pnas.1322671111

Plett JM, Khachane A, Ouassou M et al (2014b) Ethylene and jasmonic acid act as negative modulators during mutualistic symbiosis between Laccaria bicolor and Populus roots. New Phytol 202:270–286. https://doi.org/10.1111/NPH.12655

Rush TA, Puech-Pagès V, Bascaules A et al (2020) Lipo-chitooligosaccharides as regulatory signals of fungal growth and development. Nat Commun 11:3897. https://doi.org/10.1038/s41467-020-17615-5

Sangwan S, Prasanna R (2021) Mycorrhizae helper bacteria: unlocking their potential as bioenhancers of plant–arbuscular mycorrhizal fungal associations. Microb Ecol. https://doi.org/10.1007/s00248-021-01831-7
Shen Q, Liu Y, Naqvi NI (2018) Fungal effectors at the crossroads of phytohormone signaling. Curr Opin Microbiol 46:1–6
Trivedi P, Mattupalli C, Eversole K, Leach JE (2021) Enabling sustainable agriculture through understanding and enhancement of microbiomes. New Phytol 230:2129–2147
Wagner MR (2021) Prioritizing host phenotype to understand microbiome heritability in plants. New Phytol 232(2):502–509
Wang X, Feng H, Wang Y et al (2021) Mycorrhizal symbiosis modulates the rhizosphere microbiota to promote rhizobia–legume symbiosis. Mol Plant 14:503–516. https://doi.org/10.1016/j.molp.2020.12.002
Zamioudis C, Pieterse CMJ (2012) Modulation of host immunity by beneficial microbes. Mol Plant-Microbe Interact 25:139–150
Zhang C, He J, Dai H et al (2021) Discriminating symbiosis and immunity signals by receptor competition in rice. Proc Natl Acad Sci U S A 118:e2023738118. https://doi.org/10.1073/PNAS.2023738118

Chapter 2
How Plants Modulate Their Rhizosphere Microbiome

Ariella Alperovitch-Lavy

Abstract Rhizosphere, the soil zone in close proximity with plant roots, is a very special environment for soil microorganisms. The plant secretes a variety of nutrients and bioactive compounds. These have a direct influence on any microbe, and in recent years it has become clear that plant signals shape the rhizosphere microbiome. Plant microbe coevolution has, no doubt, produced interactions that benefit all the partners. The complexity of the microbiome, though, would preclude a simple positive-negative interaction, so that to reach a complete understanding, a full network of interactions needs to be studied. This would be difficult, but recent studies have extracted some of the principles. Thus, the microbiome is shaped in a specific way by the phytochemical composition of the exudates, which in turn reflect the developmental stage of the plants that are present in a given soil neighborhood, their physiological state, as well as the consequences of interactions in the soil. Soil microbes, as assemblies of microbiome populations, can, indeed program the composition of the phytochemicals released by the plant. There is, therefore, a continuous exchange of signals between plants and their associated bacteria and fungi.

Keywords Root exudates · Exometabolites · Plant phenolics · Flavonoids · Systemic signals

2.1 Can Roots Manage Their Microbiomes?

The rhizosphere, the layer of soil close to and influenced by a plant's root, was actually defined as a way to give a spatial meaning to the influence of plant root exudates on soil microorganisms (see Hirsch and Mauchline 2012). The rhizoplane, analogous to the leaf phylloplane, is the region on, or closest to, the root surface (Fig. 2.1). Rhizosphere microorganisms, to the extent that the population differs from the composition of the surrounding soil, are recruited and selected for by the

---

A. Alperovitch-Lavy (✉)
Imagindairy LTD, Yokneam, Israel
e-mail: Ariella@Imagindairy.com

B. A. Horwitz, P. K. Mukherjee (eds.), *Microbial Cross-talk in the Rhizosphere*, Rhizosphere Biology, https://doi.org/10.1007/978-981-16-9507-0_2

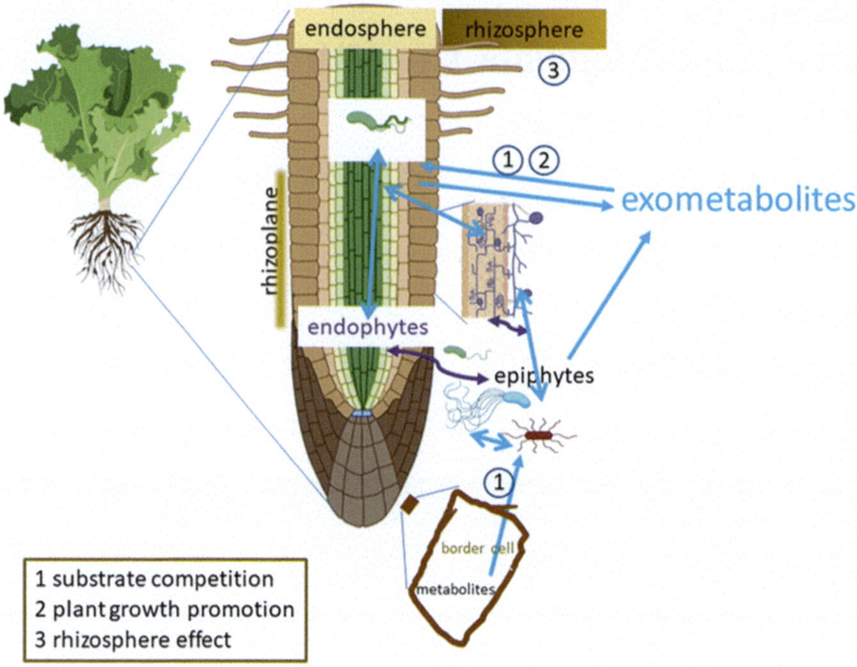

Fig. 2.1 Scheme emphasizing how metabolites carry information between plant and microbiome. The diagram is a simplified and adapted version of Fig. 2.1 from Sasse et al. (2018), focusing on the routes of communication. Root, microbe, and lettuce images are from Biorender.com. Exometabolites include root exudates and diffusible factors secreted by microbes belonging to the rhizosphere microbiome. Light blue arrows show communication via exometabolites, including within the root where endophytes can communicate with plant cells and, in principle, with each other. Purple arrows indicate mobility of microbes from the rhizosphere (epiphytes) to the root interior (endophytes). In the case of fungi, like the AMF symbolized to the right of the root, "mobility" means growth into the root tissues and colonization, intra- as in the diagram shown, or extracellularly. Rhizoplane and rhizosphere are shown with fuzzy or graded color, to emphasize that the boundaries are not sharp

plant. Microorganisms that antagonize colonization by pathogens are beneficial. Members of the rhizosphere and rhizoplane microbiome provide multiple benefits to the plant. It is no surprise, then, that plants and microbiome species have coevolved; here I address how plant factors influence microbiome assembly and persistence. Plants and the members of the rhizosphere microbiome communicate with each other. Microorganisms that establish close interactions can communicate even before physical contact. The main drivers of this cross talk are soluble metabolites (Fig. 2.1). In 2011 Mendes et al. reported a metagenomic study of the microbiome characteristic of disease-suppressive soils (Mendes et al. 2011). The authors noted that as early as 1995 (Cook et al. 1995) raised the possibility that the plant is an active participant rather than just a bystander in beneficial interactions. The hypothesis that plants can recruit beneficial members of the microbiome rapidly gained momentum (Zamioudis and Pieterse 2012). Root-associated microbiomes vary by soil and host genotype (Ofek-Lalzar et al. 2014) and can be affected by environmental variables as well as by the host physiological conditions. Furthermore it has become clear that interactions between microbes are of no less importance than plant-microbe interactions, for example, in biocontrol of oomycete pathogens by bacterial microbiota (Durán et al. 2018). In this chapter, I will review some of the multitude of interactions and mechanisms by which plants shape their microbiome, and indeed how the microbiome can modulate plant metabolism.

The influence of age and developmental stage of Arabidopsis plants on the root microbiome was assessed by Chaparro et al. (2014). The core microbiome of *Arabidopsis* was established at the seedling stage, with bacterial phyla representation of Acidobacteria, Actinobacteria, and less abundant Bacteroidetes and Cyanobacteria. While the abundance of Acidobacteria exhibited no change between the seedling and the flowering stages, Actinobacteria, Bacteroidetes, and Cyanobacteria exhibited significant changes. The plant's ability to influence bacterial assemblage is suggested by the secretion of specific root exudate compounds correlated with microbiome functional analysis of chemotaxis and antibiosis activity (Chaparro et al. 2014). Environmental conditions, such as biotic and abiotic stresses, exert a strong selective pressure on the composition of the rhizosphere microbiome. Re-composition of bacterial and fungal communities of *Arabidopsis* was demonstrated across a Pi gradient by Finkel et al. (2019). A profound effect on the plant microbiome due to the induction of the Phosphate Starvation Response (PSR) was noted. This suggested the effect of the suspected cross talk between the PSR and the plant defense system, together with the changes in the composition of the root exudates resulting in a new order of the bacterial and fungal representation (Finkel et al. 2019). Re-composition of fungal communities was also observed under field conditions, in a two-factor comparison looking at maize root type (axial or lateral) and at high or low phosphate levels. At low P concentration, high beta diversity was demonstrated on the lateral roots compared to the axial roots, with a similar beta diversity in high and low P levels. The authors hypothesized based on work with Arabidopsis (Hiruma et al. 2016; Hacquard et al. 2016) that physiological conditions of the axial roots versus the lateral roots in correlation with P determine the extent of defense responses, which can explain the differences between the fungal composition (Yu et al. 2018).

The effect of biotic stress on the defense system has a role in plant microbiome assembly. Lee et al. (2012) show how biotic stress can trigger elicitation of the plant defense response in pepper (*Capsicum annuum*) upon aphid infestation, which caused recruitment of the beneficial bacteria *B. subtilis* GB03 and a reduction in the pathogenic strain SL1931 *R. solanacearum*. This beneficial recruitment of the bacterial strain is suspected as a helper factor against pathogenic attack (Lee et al. 2012).

The plant immune system has a crucial role as the gatekeeper for beneficial microbes to exist on or in the root, while preventing colonization by pathogenic microbes. The plant hormones, jasmonic acid (JA) and ethylene, are part of the Induced Systemic Resistance (ISR) pathway, antagonizing necrotrophic pathogens.

A separate and in some ways complementary pathway is mediated by salicylic acid (SA) belonging to the Systemic Acquired Resistance (SAR) response to biotrophic microbes.

Lebeis et al. (2015) described the role of SA in the regulation of the composition of *Arabidopsis* root endophyte families. Alternation in bacterial composition was assessed in the hyperimmunity strain mutant *cpr5* and the immunocompromised *pad4* mutant. Under constitutive expression of SA by the *cpr5* mutant, some of the bacterial endophyte growth was depleted, indicating the strains as SA sensitive. In contrast, under a deficient immune system, some bacterial isolates from Synthetic Microbial Communities (SynCom) were more abundant compared to the wild type, which was defined as "sporadic or non-colonizer" microbes (Lebeis et al. 2015).

Here we overview some of the environmental conditions that can influence the assembly of the root microbiome. Other forces such as nutrients availability, drought, soil salinity (Trivedi et al. 2020; Pascale et al. 2020), and the composition of the soil microorganisms as the main force of the Plant-Soil-Feedback (PSF) (Fitzpatrick et al. 2018) can trigger the elicitation of the plant signaling pathways that will lead to reorganization in the root microbiome.

2.2 Root Exudates

The rhizosphere is a rich soil-plant environment due to the release of variety of primary and secondary metabolites by the plant's roots, known as root exudates (Sasse et al. 2018). Primary metabolites are estimated to be released in larger quantities compared to the secondary metabolites, and consist of carbohydrates, amino acids, and organic acids. Up to 50% of the plant's fixed carbon can be secreted in the exudates, depending on the plant physiological condition (Massalha et al. 2017; Jaitz et al. 2011; van Dam and Bouwmeester 2016). Thus, the amount and relative composition of the exudates, even when looking only at primary metabolites, does not fit into any fixed pattern. Nevertheless, a core pattern might be extracted from data by looking across a large number of studies, but it seems that this had never been done on a large scale. An example is presented in Fig. 2.2. Already it is apparent, comparing the pie chart to the left-hand side of Fig. 2.2a, that some metabolites, even primary ones, make up a small fraction of the exudate composition. Low concentrations of a particular metabolite, obviously, do not imply that this metabolite lacks importance in plant-microbe communication. This principle is even more striking when looking at secondary metabolites, which may be very active at low levels.

The root exudates also include secondary metabolites such as flavonoids, phytohormones, phenolic compound, and VOCs (volatile organic compounds). These secondary metabolites are commonly abundant among different plant species including maize, Arabidopsis, rice, common bean and soybean, and others (van Dam and Bouwmeester 2016). Plant secondary metabolites (PSM), now increasingly referred to as specialized metabolites (Erb and Kliebenstein 2020), are low molecular weight

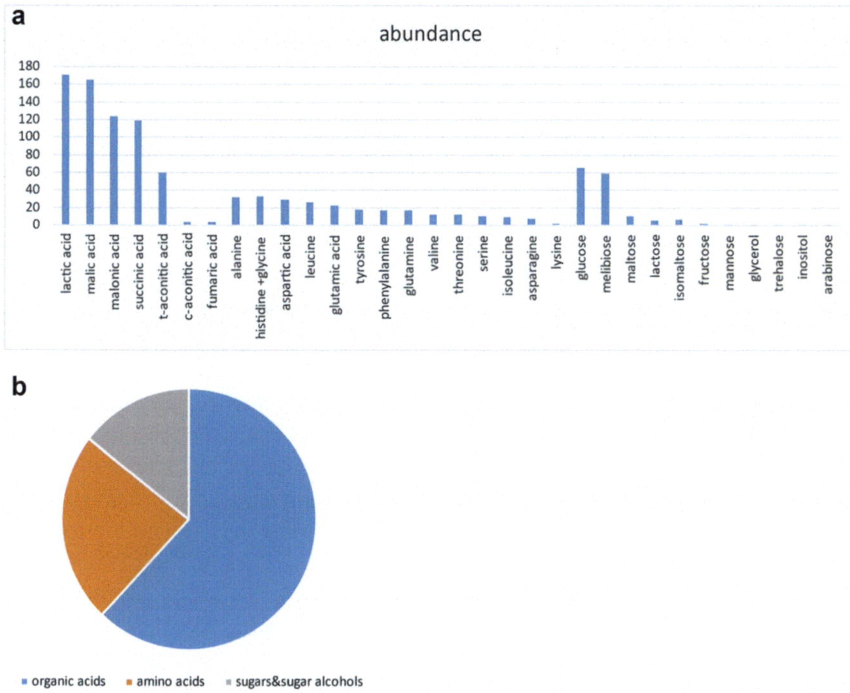

Fig. 2.2 Example of root exudate composition. In the example shown, data from Fan et al. (2012) are shown in (**a**). Maize root exudates from axenic hydroponic culture were analyzed for major primary metabolites. In (**b**), the data are replotted as a pie chart to illustrate that it might be possible to define a core exudate primary metabolite composition

compounds and usually are considered as nonessential for plant development and reproduction. Generally, they are defined by their biosynthesis pathways generated from primary metabolites or as their intermediates from these pathways. A wide range of compounds are assigned to major structural classes: phenolics, flavonoids, alkaloids, steroids, and terpenes (Kessler and Kalske 2018). Their function is linked with environmental stress response, defense response, growth, and development processes; these are also associated with plant-microbe interactions and influence the microbiome assembly (Stringlis et al. 2019; Pascale et al. 2020; Pang et al. 2021). The effect of representative PSM on the modulation of the root microbiome is discussed here.

2.2.1 Coumarin, Benzoxazinoids, and Terpenes

2.2.1.1 Coumarins

Coumarins are secondary metabolites produced via the phenylpropanoid pathway and present at different plant organs including leaves and roots. Coumarin biosynthesis was observed under Fe starvation conditions to participate in Fe^{3+} reduction and to improve its transport to the root cells, in Arabidopsis. The main production of coumarins upon Fe deficiency response in Arabidopsis are scopoletin, esculin, asculetin, fraxetine, and sideretin, and their presence is dependent on the pH level of their growth environment (Tsai and Schmidt 2017; Fourcroy et al. 2016; Schmid et al. 2014; Schmidt et al. 2014; Stringlis et al. 2019). The influence of coumarins on the root microbiome composition was shown in recent studies; for example, the coumarin-deficient mutant *f6′h1* of Arabidopsis had an increase in the abundance of Proteobacteria compared to the wild-type strain, and a decrease of Firmicutes around the roots (Schmidt et al. 2014; Pang et al. 2021). Voges et al. (2019) demonstrated the ability of coumarins to inhibit the growth of *Pseudomonas* around the roots of Arabidopsis WT plants compared to the knockout *f6′h1* lines, under Fe limitation. The reshaping of the SynCom in the WT plants was proposed to be due to the production of reactive oxygen species as antimicrobial agents (Voges et al. 2019). Another study showed the selective antimicrobial effect of the coumarin scopoletin on selected soil-borne fungal pathogens *Verticillium dahliae* and *Fusarium oxysporum*, by inhibition of their growth. The plant-growth-promoting rhizobacteria *Pseudomonas simiea* and *Pseudomonas capeferrum* exhibited tolerance to the antimicrobial activity of scopoletin. Moreover, a correlation between scopoletin function under iron deficiency and ISR priming was indicated through the regulation of the transcription factor MYB72 and the BGLU42 importer regulator (Stringlis et al. 2018).

2.2.1.2 Triterpenes and Camalexin

Camalexin, a defense antimicrobial compound in plants roots, was found to have a role in determining the Arabidopsis root microbiome. Camalexin-deficient strain *cyp71A27* modified the ability of three growth-promoting bacteria: *Pseudomonas* sp. CH267, *Pseudomonas simiae* WCS417r, and *Paraburkholderia phyofirmas* PsJN, to induce a growth promotion effect, which was observed in cocultivation with WT plants. The importance of camalexin was also reflected in the ability of the endophytic fungus *Serendipita indica* to colonize the *cyp71A27* mutant plants. These results suggest the importance of camalexin in plant-microbiome interactions (Koprivova et al. 2019).

Another large and structurally diverse group of natural plant metabolites is the triterpenes. They are produced via the mevalonate pathway. They possess antimicrobial activity and function in signaling. In the study of Huang et al. (2019), the impairment in the production of arabidin and thalianin pathways exhibited different bacterial communities profiling compared to the WT Arabidopsis. Moreover, the mutant strain revealed similarity in the bacterial composition, with an enrichment of *Bacteriodetes* and depletion of *Deltaproteobacteria* representation compared with the WT (Huang et al. 2019).

2.2.1.3 Benzoxazinoids (BXs)

Benzoxazinoids (BXs) are a common secondary metabolite class produced in plant defense responses. They are abundant in roots of plants such as wheat and maize and were documented to alter the roots' microbiome composition. Hu et al. (2018) investigated the effect of BXs on the bacterial and fungal communities in maize root compared with the deficient plant bx−. Differential bacterial and fungal composition was detected in the mutant plant root compared to the WT, with no specification of the bacterial or fungal OTUs (operational taxonomic units). The ability of the BXs to prime the defense system of the next generation of plants against insects was tested on WT plants growing with and without BX, in the soil. Different bacterial and fungal population profiling was observed in the root and the rhizosphere of +BX versus −BX. The OTUs of the fungal phyla Ascomycota and Glomeromycota and the bacterial Actinobacteria were differential between root and rhizosphere. The bacterial variability was clear with higher abundance of OTUs from Chloroflexi under BX− and Proteobacteria in BX+. On the other hand, fungal community composition exhibited less differences of OTUs between the roots and rhizosphere, +/−BX. OTUs of Ascomycota were detected in all types of soil and cannot indicate any effect on this phylum. In contrast, a negative effect was observed on the Glomeromycota with less OTUs representation under +BX growth conditions. The authors suggested that the ability of the next-generation plants to cope with insects is due to the effect of benzoxazinoids on the bacterial communities, with less contribution by the fungal composition (Hu et al. 2018). In another study by Cotton et al. (2019), the correlation between the metabolome profiling and the rhizobiome composition was assessed, in maize roots. A significant effect of BXs was documented on the bacterial OTUs profiling, while a weak effect was demonstrated on the fungal OTUs composition. The impairment in the biosynthesis of BX in two mutants at the beginning of the pathway, *bx1* and *bx2*, had a profound effect on the root metabolome composition compared with the downstream mutant bx6 and the WT. This can indicate that BX acts as a regulator of other metabolites production in addition to its activity as a defense compound. Positive correlation was observed with the enrichment of Methylophilaceae bacterial OTUs in WT roots, which was linked with the most abundant group of metabolites, flavonoids (Cotton et al. 2019), which are another class of SMs that possesses an antimicrobial activity and signaling ability to recruit beneficial soil bacteria, for example, rhizobia, and regulate the interaction between plants and the beneficial root colonizer fungus, arbuscular mycorrhizae (Hassan and Mathesius 2012; Pang et al. 2021).

2.2.2 *Phytohormones*

Plant hormones (phytohormones) are small molecules with a role in plant defense response as well as in morphological signaling. Phytohormones can modify the plant microbiome by inhibiting the proliferation of microorganisms, while at the same time allowing colonization. Furthermore they are metabolized by microbes and are thus utilized as nutrients, for example, as a carbon source. Salicylic acid (SA), jasmonic acid (JA), abscisic acid (ABA), ethylene (ET), and strigolactones (SL) are among the phytohormones that were shown to affect the plant microbiome, and are overviewed in brief. The effect of SA on reassembly of the root microbiome of Arabidopsis was mentioned previously (Sect. 2.1). Similarly, the effect of the defense hormone JA was investigated on *Azoarcus olearius*, an abundant nitrogen-fixing rice endophyte, by Chen et al. (2020). Coculturing of the JA-deficient rice plants (cpm2) with *A. olearius* led to an increase in the ability of the endophyte to colonize the plant roots (Chen et al. 2020). For detailed discussion of phytohormones in rhizosphere communication, see Chaps. 3 and 4.

2.3 Influence of Plant Species and Genotypes on the Microbiome

2.3.1 *Domestication*

Plant domestication has been developed (and is developing) to sustain the food supply to the local community, and it's thought of as one of the most meaningful accomplishments in human civilization. However, the rising demand for high crop yield led to a reduction in the genetic diversity of cultivated plants, as was revealed in rice (Ganesh Ram et al. 2007), maize (Yamasaki et al. 2005), and wheat (Haudry et al. 2007). The loss of diversity has led to an increase in the use of chemical pesticides and fertilizers, and a reduction in the microbiome diversity. Ramirez et al. (2012) showed that high levels of nitrogen in the soil led to a decrease in the microbial biomass and re-composition of the communities with increased representation of *Actinobacteria* and *Firmicutes*, and a decrease in the abundance of *Acidobateria* and *Verrucomicrobia*. This finding correlates with other observations and supports the *copiotrophic* hypothesis that microbial groups with a fast growth rate under high nutrient conditions will be more abundant compared to slow growing microbial groups, with lower nutrient conditions (Ramirez et al. 2012). Weese et al. (2015) assessed the influence of long-term nitrogen fertilization on the mutualism interaction between legumes and the N-fixation bacteria belonging to the genus *Rhizobium*. The study highlighted the importance of rhizobia as a natural fertilizer and its ability to induce plant growth promotion, in a non-fertilized environment (Weese et al. 2015). Rhizobia can supply nitrogen to the host, such as legumes, by forming unique root-nodule structures as a microenvironment to convert nitrogen into ammonia. The host provides carbohydrates as carbon source, and together this mutualistic relationship is part of the nitrogen and carbon cycle in the soil, with a significant role to maintain sustainable agriculture (Pérez-Jaramillo et al. 2016; Liu et al. 2020). The compatibility between rhizobial species with specific legume genotypes is modulated mainly by the host immune response to the bacterial effectors, with involvement of a variety of genes. The formation of the nodule occurs upon bacterial infection which triggers reprograming of the root tissues cells, to develop the nodule primordium. The recruitment of the rhizobia by the host involves the secretion of the secondary metabolites, flavonoids, that were mentioned in Sect. 2.2.1.3. Secretion of flavonoids can activate the bacterial nodulation genes also known as Nod factors, as they possess a pivotal role in the host nodulogenesis (Geurts and Bisseling 2002).

The influence of human selection on legumes was shown to affect the interaction with rhizobia, with a reduction in the ability to form nodulation compared with the wild species. Mutch and Young (2004) reported a reduction in the ability of *Rhizobium leguminosarum* strains to nodulate broad pea (*Vicia faba*), compared to wild species. They suggested that crop domestication impaired the interaction of legumes and rhizobia, as their interaction with the rhizo-biome is less promiscuous compared to the wild strains (Mutch and Young 2004). The ability to generate promiscuous interactions between legumes and bacteria was also shown in the study of Chang et al. (2019). The ability of wild soybeans, *Glycine soja*, to recruit beneficial bacteria (*Bradyrhizobium* and *Pseudomonas*) was higher compared to cultivated *Glycine max*. Similarly, reduction in the genetic diversity of cultivated chickpeas, *Cicer arietinum*, led to the reduction in the diversity of symbionts colonizing root nodules, as was demonstrated in the study of Kim et al. (2014).

The effect of plant domestication on the assemblage of the root-rhizosphere microbiome reveals differences between the wild plants and the cultivars, as was reported in a few studies. For example, Leff et al. (2017) investigated the root and rhizosphere composition on wild and modern strains of sunflower, with evidence for differences in hosting fungal community (Leff et al. 2017). The authors suggested that a strong impact on the fungal community is linked to the genetic and physiological background of the plants. The influence of the plant genotype and root length on the domestication effect was also assessed by Pérez-Jaramillo et al. (2017), with *Phaseolus vulgaris* (common bean) as the host. Enrichment of the bacterial phylum *Bacteriodetes* in the rhizosphere of wild bean accessions was observed. In contrast, dominant representation of *Proteobacteria* and *Actinobacteria* was associated with the modern relatives (Pérez-Jaramillo et al. 2017). Interestingly, consistent with other observations, Bacteriodetes was found as the dominant phylum in and on roots of wild plants (bean, barley, lettuce). As mentioned above, changes in the concentration of secondary metabolites have been suggested to impact the structure of the microbial community surrounding the host (Pérez-Jaramillo et al. 2018).

2.3.1.1 Seed Domestication

Plant seeds are the embryonic plants protected by the seed coat, and they are responsible for the reproduction of gymnosperm and angiosperm plants. Seeds are susceptible to targeted selection in the plant domestication process, as their phenotype is correlated with changes in traits, usually related to size/weight as was shown for maize seed by Liu et al. (2016).

Plant microbiomes have an important influence on the plant growth, health, and microbial composition in the rhizosphere as mentioned above. Hence, the seed's microbiome has a crucial role in establishing the emerging plant. Plants can transmit microbes to the progeny vertically, through seeds (Robinson et al. 2016; Shahzad et al. 2018; Ridout et al. 2019). Endophytes, generally considered as beneficial microbes in close association with the host tissues, were detected in the seed layers (endosperm and embryo) after germination (Berg and Raaijmakers 2018; Soldan et al. 2021).

The effect of genotype and domestication of rice on the fungal and bacterial communities was investigated by Kim et al. (2020). In their study, the results are consistent with other reports demonstrating enrichment of the bacterial phyla OTUs, *Bacteriodetes*, in wild rice seeds, whereas enrichment in *Proteobacteria* and *Actinobacteria* phyla were observed in domesticated seeds. Concerning the fungal community, changes were observed in the composition of *Ascomycota* membership of wild seeds compared to domesticated seeds, with additional representation of *Basidiomycota*. Correlated with these results, assessment of the fungal network structures shows decrease in the putative associated connections with domesticated rice compared to the wild rice strains. The authors suggested that this might be because of gene loss in the host, which could impair the symbiotic association with fungal species.

The ability to create microbial interactions in cereal seeds of wild strains vs. cultivars was explored in the study by Abdullaeva et al. (2021). A strong correlation between host genotype and microbiome diversity was observed in cultivated seeds, with higher bacterial diversification, and less microbial interactions. In the wild-type seeds, in contrast, *Actinobacter* and *Pseudomonas* were predominant. The authors suggest that this representation of bacterial phyla can support wild plants under stress conditions. The study also identifies a common core microbiome of *Pseudomonas*, *Actinobater*, *Pantoea*, *Sternotrophomonas*, and *Burkholderiaceae* which can emphasize the strength of the association during evolution, as a result of direct selection. The host genotype has a profound influence on the plant microbiome structure and will be reviewed in the next section.

2.3.2 Ecotypes

As discussed in the previous section, plant domestication has led to a loss of genes compared to the wild type, with correlation of reduction in the connections with microbes. The need to improve the plant's traits is usually not in correlation with the plant's microbiome and a plant's ability to attract beneficial microbes enhance tolerance to biotic and abiotic stresses should be considered (Compant et al. 2019). The influence of the host genotype on the microbial structure was mentioned in the Sects. 2.3.1.1 and 2.2.1, and supported by additional studies (Schlaeppi et al. 2014; Naylor et al. 2017; Wei and Jousset 2017 among others).

A recent study published by Xiong et al. (2021) demonstrates the strong role of the soil/planting location on the establishment of the rhizosphere microbial structure, on rice cultivars and hybrid line. Microbial consortia were shown to be associated in host-genotype specific manner. In the case of hybrid rice, higher representation of bacteria with functional genes participating in N, K, and P bioavailability to the host, compared to the Japonica cultivars, was noticed. Among the bacterial phyla are the *Anaeromyxobacter* spp. and *Sideroxydans* spp. with their contribution to the N-fixation ability. Compared to cultivars, wild-type plants supported *Nakamurella* spp. possessing enzymes involved in phosphate and carbon cycles; however, low abundance of methanotrophs was noticed (Xiong et al. 2021).

Differences in the microbiome composition with a correlation to the host genetic background was demonstrated by Kwak et al. (2018). In their study, they cultivated the flavobacterium—TRM1 from the rhizosphere of the resistant tomato strain, Hawaii 7996 to *R. solanacearum*, and revealed its ability to suppress disease development by *R. solanacearum*, in the sensitive Moneymaker stain. The precision recruitment of the beneficial bacterium TRM1 by the resistant strain Hawaii 7996 emphasizes the intimacy of the relationship between the host and its microbiome population.

Environmental conditions have a strong influence on the composition of the host microbiome, as mentioned before. In recent research conducted by Liu et al. (2021), the effect of two switchgrass ecotypes Alamo and Kanlow under drought conditions was evaluated. Bacterial OTUs analysis from the rhizo-compartment Rhizosheath (the contact layer of root hairs, soil, and aggregated mucilage) revealed variation mostly at the genera level. This evidence was explained by the authors as a reflection of a direct effect of the root exudates respectively to the plant ecotype. In other words, the root exudates are the outcome of the plant genotypic background and the consequences of the plant-environmental feedback.

To link between the host genotype and the associated microbes, Deng et al. (2021) used GWAS-Genome Wide Association Studies. By sorting 200 rhizosphere populations of sorghum genotypes, suspected loci were identified as the bridge for the colonization of the bacterial subset. However, further investigation with genetic approaches is needed for evaluation of these indications.

2.4 Can the Rhizosphere Microbiome Modulate Root Exudation?

The preceding sections include numerous examples of how plants can manage their rhizosphere microbiomes. Less was known, though, about how the microbiome reprograms root exudation. If this possible that the microbiome would be remodeling the most central signals responsible for its own composition, as hinted by the arrows in schemes like the one in Fig. 2.1; experimental evidence comes from a recent study (Korenblum et al. 2020). Using a hydroponic split root-system assay in tomato, the authors defined a novel regulatory loop: systemically induced root exudation of metabolites (SIREM). The split root-system was set up to see how interaction of the "local" side with different soil-derived microbiomes (high, medium, and low diversity) influenced the composition of exudates from the "systemic" side in which the roots were uncolonized, incubated only in axenic hydroponic medium. Metabolome profiling of the root exudates from the systemic root system compartment identified metabolites that were modulated in a differential way by the microbiomes in the local compartment. Thus, long-distance signaling triggered a systemic, specific root exudate composition. These include acyl sugars, hydroxy-cinnamic acid conjugates, oxylipins, and azelaic acid. Notably, azelaic acid hexose glycoside induced systemic changes in metabolite profile. The steroidal glycoalkaloid α-tomatine, a known antimicrobial compound from tomato, was secreted from the systemic root when the azelaic acid glycoside was applied to the local side. The study went on to map the spatial distribution of SIREM metabolites and gene expression. These processes are proposed to have a major role in chemical diversity in the soil, in particular in the rhizosphere, where local exposure to microbes provides a systemic signal to reprogram plant exudate composition, in turn altering the rhizosphere microbiome.

References

Abdullaeva Y, Ambika Manirajan B, Honermeier B et al (2021) Domestication affects the composition, diversity, and co-occurrence of the cereal seed microbiota. J Adv Res 31:75–86. https://doi.org/10.1016/j.jare.2020.12.008

Berg G, Raaijmakers JM (2018) Saving seed microbiomes. ISME J 12:1167–1170. https://doi.org/10.1038/s41396-017-0028-2

Chang C, Chen W, Luo S et al (2019) Rhizosphere microbiota assemblage associated with wild and cultivated soybeans grown in three types of soil suspensions. Arch Agron Soil Sci 65:74–87. https://doi.org/10.1080/03650340.2018.1485147

Chaparro JM, Badri DV, Vivanco JM (2014) Rhizosphere microbiome assemblage is affected by plant development. ISME J 8:790–803. https://doi.org/10.1038/ismej.2013.196

Chen X, Marszałkowska M, Reinhold-Hurek B (2020) Jasmonic acid, not salicyclic acid restricts endophytic root colonization of rice. Front Plant Sci 10:1758. https://doi.org/10.3389/fpls.2019.01758

Compant S, Samad A, Faist H, Sessitsch A (2019) A review on the plant microbiome: ecology, functions, and emerging trends in microbial application. J Adv Res 19:29–37

Cook RJ, Thomashow LS, Weller DM et al (1995) Molecular mechanisms of defense by rhizobacteria against root disease. Proc Natl Acad Sci U S A 92:4197–4201. https://doi.org/10.1073/pnas.92.10.4197

Cotton TEA, Pétriacq P, Cameron DD et al (2019) Metabolic regulation of the maize rhizobiome by benzoxazinoids. ISME J 13:1647–1658. https://doi.org/10.1038/s41396-019-0375-2

Deng S, Caddell DF, Xu G et al (2021) Genome wide association study reveals plant loci controlling heritability of the rhizosphere microbiome. ISME J. https://doi.org/10.1038/s41396-021-00993-z

Durán P, Thiergart T, Garrido-Oter R et al (2018) Microbial interkingdom interactions in roots promote Arabidopsis survival. Cell 175:973–983.e14. https://doi.org/10.1016/j.cell.2018.10.020

Erb M, Kliebenstein DJ (2020) Plant secondary metabolites as defenses, regulators, and primary metabolites: the blurred functional Trichotomy1[OPEN]. Plant Physiol 184:39–52

Fan B, Carvalhais LC, Becker A et al (2012) Transcriptomic profiling of Bacillus amyloliquefaciens FZB42 in response to maize root exudates. BMC Microbiol 12:1–13. https://doi.org/10.1186/1471-2180-12-116

Finkel OM, Salas-González I, Castrillo G et al (2019) The effects of soil phosphorus content on plant microbiota are driven by the plant phosphate starvation response. PLoS Biol 17:1–34. https://doi.org/10.1371/journal.pbio.3000534

Fitzpatrick CR, Copeland J, Wang PW et al (2018) Assembly and ecological function of the root microbiome across angiosperm plant species. Proc Natl Acad Sci U S A 115:E1157–E1165. https://doi.org/10.1073/pnas.1717617115

Fourcroy P, Tissot N, Gaymard F et al (2016) Facilitated Fe nutrition by phenolic compounds excreted by the Arabidopsis ABCG37/PDR9 transporter requires the IRT1/FRO2 high-affinity root Fe^{2+} transport system. Mol Plant 9:485–488

Ganesh Ram S, Thiruvengadam V, Vinod KK (2007) Genetic diversity among cultivars, landraces and wild relatives of rice as revealed by microsatellite markers. J Appl Genet 48:337–345. https://doi.org/10.1007/bf03195230

Geurts R, Bisseling T (2002) Rhizobium nod factor perception and signalling. Plant Cell 14:S239–S249. https://doi.org/10.1105/tpc.002451

Hacquard S, Kracher B, Hiruma K et al (2016) Survival trade-offs in plant roots during colonization by closely related beneficial and pathogenic fungi. Nat Commun 7:11362. https://doi.org/10.1038/ncomms11362

Hassan S, Mathesius U (2012) The role of flavonoids in root-rhizosphere signalling: opportunities and challenges for improving plant-microbe interactions. J Exp Bot 63:3429–3444

Haudry A, Cenci A, Ravel C et al (2007) Grinding up wheat: a massive loss of nucleotide diversity since domestication. Mol Biol Evol 24:1506–1517. https://doi.org/10.1093/molbev/msm077

Hirsch PR, Mauchline TH (2012) Who's who in the plant root microbiome? Nat Biotechnol 30:961–962. https://doi.org/10.1038/nbt.2387

Hiruma K, Gerlach N, Sacristán S et al (2016) Root endophyte Colletotrichum tofieldiae confers plant fitness benefits that are phosphate status dependent. Cell 165:464–474. https://doi.org/10.1016/j.cell.2016.02.028

Hu L, Robert CAM, Cadot S et al (2018) Root exudate metabolites drive plant-soil feedbacks on growth and defense by shaping the rhizosphere microbiota. Nat Commun 9:1–13. https://doi.org/10.1038/s41467-018-05122-7

Huang AC, Jiang T, Liu YX et al (2019) A specialized metabolic network selectively modulates Arabidopsis root microbiota. Science 364:eaau6389. https://doi.org/10.1126/science.aau6389

Jaitz L, Mueller B, Koellensperger G et al (2011) LC-MS analysis of low molecular weight organic acids derived from root exudation. Anal Bioanal Chem 400:2587–2596. https://doi.org/10.1007/s00216-010-4090-0

Kessler A, Kalske A (2018) Plant secondary metabolite diversity and species interactions. Annu Rev Ecol Evol Syst 49:115–138

Kim DH, Kaashyap M, Rathore A et al (2014) Phylogenetic diversity of Mesorhizobium in chickpea. J Biosci 39:513–517. https://doi.org/10.1007/s12038-014-9429-9

Kim H, Lee KK, Jeon J et al (2020) Domestication of Oryza species eco-evolutionarily shapes bacterial and fungal communities in rice seed. Microbiome 8:20. https://doi.org/10.1186/s40168-020-00805-0

Koprivova A, Schuck S, Jacoby RP et al (2019) Root-specific camalexin biosynthesis controls the plant growth-promoting effects of multiple bacterial strains. Proc Natl Acad Sci U S A 116: 15735–15744. https://doi.org/10.1073/pnas.1818604116

Korenblum E, Dong Y, Szymanski J et al (2020) Rhizosphere microbiome mediates systemic root metabolite exudation by root-to-root signaling. Proc Natl Acad Sci U S A 117:3874–3883. https://doi.org/10.1073/pnas.1912130117

Kwak MJ, Kong HG, Choi K et al (2018) Rhizosphere microbiome structure alters to enable wilt resistance in tomato. Nat Biotechnol 36:1100–1116. https://doi.org/10.1038/nbt.4232

Lebeis SL, Paredes SH, Lundberg DS et al (2015) Salicylic acid modulates colonization of the root microbiome by specific bacterial taxa. Science 349:860–864. https://doi.org/10.1126/science.aaa8764

Lee B, Lee S, Ryu CM (2012) Foliar aphid feeding recruits rhizosphere bacteria and primes plant immunity against pathogenic and non-pathogenic bacteria in pepper. Ann Bot 110:281–290. https://doi.org/10.1093/aob/mcs055

Leff JW, Lynch RC, Kane NC, Fierer N (2017) Plant domestication and the assembly of bacterial and fungal communities associated with strains of the common sunflower, Helianthus annuus. New Phytol 214:412–423. https://doi.org/10.1111/nph.14323

Liu Z, Garcia A, McMullen MD, Flint-Garcia SA (2016) Genetic analysis of kernel traits in maize-teosinte introgression populations. G3 6:2523–2530. https://doi.org/10.1534/g3.116.030155

Liu J, Yu X, Qin Q et al (2020) The impacts of domestication and breeding on nitrogen fixation symbiosis in legumes. Front Genet 11:1–9

Liu TY, Ye N, Wang X et al (2021) Drought stress and plant ecotype drive microbiome recruitment in switchgrass rhizosheath. J Integr Plant Biol 63:1753–1774. https://doi.org/10.1111/jipb.13154

Massalha H, Korenblum E, Tholl D, Aharoni A (2017) Small molecules below-ground: the role of specialized metabolites in the rhizosphere. Plant J 90:788–807. https://doi.org/10.1111/tpj.13543

Mendes R, Kruijt M, De Bruijn I et al (2011) Deciphering the rhizosphere microbiome for disease-suppressive bacteria. Science 332:1097–1100. https://doi.org/10.1126/science.1203980

Mutch LA, Young JPW (2004) Diversity and specificity of Rhizobium leguminosarum biovar viciae on wild and cultivated legumes. Mol Ecol 13:2435–2444. https://doi.org/10.1111/j.1365-294X.2004.02259.x

Naylor D, Degraaf S, Purdom E, Coleman-Derr D (2017) Drought and host selection influence bacterial community dynamics in the grass root microbiome. ISME J 11:2691–2704. https://doi.org/10.1038/ismej.2017.118

Ofek-Lalzar M, Sela N, Goldman-Voronov M et al (2014) Niche and host-associated functional signatures of the root surface microbiome. Nat Commun 5:4950. https://doi.org/10.1038/ncomms5950

Pang Z, Chen J, Wang T et al (2021) Linking plant secondary metabolites and plant microbiomes: a review. Front Plant Sci 12:621276

Pascale A, Proietti S, Pantelides IS, Stringlis IA (2020) Modulation of the root microbiome by plant molecules: the basis for targeted disease suppression and plant growth promotion. Front Plant Sci 10:1–23

Pérez-Jaramillo JE, Mendes R, Raaijmakers JM (2016) Impact of plant domestication on rhizosphere microbiome assembly and functions. Plant Mol Biol 90:635–644. https://doi.org/10.1007/s11103-015-0337-7

Pérez-Jaramillo JE, Carrión VJ, Bosse M et al (2017) Linking rhizosphere microbiome composition of wild and domesticated Phaseolus vulgaris to genotypic and root phenotypic traits. ISME J 11: 2244–2257. https://doi.org/10.1038/ismej.2017.85

Pérez-Jaramillo JE, Carrión VJ, de Hollander M, Raaijmakers JM (2018) The wild side of plant microbiomes. Microbiome 6:1–6. https://doi.org/10.1186/s40168-018-0519-z

Ramirez KS, Craine JM, Fierer N (2012) Consistent effects of nitrogen amendments on soil microbial communities and processes across biomes. Glob Chang Biol 18:1918–1927. https://doi.org/10.1111/j.1365-2486.2012.02639.x

Ridout ME, Schroeder KL, Hunter SS et al (2019) Priority effects of wheat seed endophytes on a rhizosphere symbiosis. Symbiosis 78:19–31. https://doi.org/10.1007/s13199-019-00606-6

Robinson RJ, Fraaije BA, Clark IM et al (2016) Wheat seed embryo excision enables the creation of axenic seedlings and Koch's postulates testing of putative bacterial endophytes. Sci Rep 6: 25581. https://doi.org/10.1038/srep25581

Sasse J, Martinoia E, Northen T (2018) Feed your friends: do plant exudates shape the root microbiome? Trends Plant Sci 23:25–41. https://doi.org/10.1016/j.tplants.2017.09.003

Schlaeppi K, Dombrowski N, Oter RG et al (2014) Quantitative divergence of the bacterial root microbiota in Arabidopsis thaliana relatives. Proc Natl Acad Sci U S A 111:585–592. https://doi.org/10.1073/pnas.1321597111

Schmid NB, Giehl RFH, Döll S et al (2014) Feruloyl-CoA 6′-Hydroxylase1-dependent coumarins mediate iron acquisition from alkaline substrates in Arabidopsis. Plant Physiol 164:160–172. https://doi.org/10.1104/pp.113.228544

Schmidt H, Günther C, Weber M et al (2014) Metabolome analysis of Arabidopsis thaliana roots identifies a key metabolic pathway for iron acquisition. PLoS One 9:1–11. https://doi.org/10.1371/journal.pone.0102444

Shahzad R, Khan AL, Bilal S et al (2018) What is there in seeds? Vertically transmitted endophytic resources for sustainable improvement in plant growth. Front Plant Sci 9:24. https://doi.org/10.3389/fpls.2018.00024

Soldan R, Fusi M, Cardinale M et al (2021) The effect of plant domestication on host control of the microbiota. Commun Biol 4:1–9

Stringlis IA, Yu K, Feussner K et al (2018) MYB72-dependent coumarin exudation shapes root microbiome assembly to promote plant health. Proc Natl Acad Sci U S A 115:E5213–E5222. https://doi.org/10.1073/pnas.1722335115

Stringlis IA, De Jonge R, Pieterse CMJ (2019) The age of coumarins in plant-microbe interactions. Plant Cell Physiol 60:1405–1419

Trivedi P, Leach JE, Tringe SG et al (2020) Plant–microbiome interactions: from community assembly to plant health. Nat Rev Microbiol 18:607–621

Tsai HH, Schmidt W (2017) Mobilization of iron by plant-borne coumarins. Trends Plant Sci 22: 538–548

van Dam NM, Bouwmeester HJ (2016) Metabolomics in the rhizosphere: tapping into belowground chemical communication. Trends Plant Sci 21:256–265

Voges MJEEE, Bai Y, Schulze-Lefert P, Sattely ES (2019) Plant-derived coumarins shape the composition of an Arabidopsis synthetic root microbiome. Proc Natl Acad Sci U S A 116: 12558–12565. https://doi.org/10.1073/pnas.1820691116

Weese DJ, Heath KD, Dentinger BTM, Lau JA (2015) Long-term nitrogen addition causes the evolution of less-cooperative mutualists. Evolution (N Y) 69:631–642. https://doi.org/10.1111/evo.12594

Wei Z, Jousset A (2017) Plant breeding goes microbial. Trends Plant Sci 22:555–558

Xiong J, Lu J, Li X et al (2021) Effect of rice (Oryza sativa L.) genotype on yield: evidence from recruiting spatially consistent rhizosphere microbiome. Soil Biol Biochem 161:108395. https://doi.org/10.1016/J.SOILBIO.2021.108395

Yamasaki M, Tenaillon MI, Bi IV et al (2005) A large-scale screen for artificial selection in maize identifies candidate agronomic loci for domestication and crop improvement. Plant Cell 17: 2859–2872. https://doi.org/10.1105/tpc.105.037242

Yu P, Wang C, Baldauf JA et al (2018) Root type and soil phosphate determine the taxonomic landscape of colonizing fungi and the transcriptome of field-grown maize roots. New Phytol 217:1240–1253. https://doi.org/10.1111/nph.14893

Zamioudis C, Pieterse CMJ (2012) Modulation of host immunity by beneficial microbes. Mol Plant-Microbe Interact 25:139–150

Chapter 3
Strigolactone Signalling and Plant-Microbe Communications

Sunil Kumar, Ashutosh Joshi, and Rakesh Kumar Shukla

Abstract The widespread role of strigolactones makes them compounds of significance in the context of plant hormones, as well as in adaptive processes. Strigolactones are isoprenoid derivatives having a characteristic butenolide ring (D ring) connected to the lactone (ABC ring). However, the non-canonical strigolactones could lack the ABC ring while only retaining the D-ring moiety. The bioassay studies on the widely adopted bioactive synthetic analog GR24 led to appreciable findings thus enhancing our knowledge about the molecules. So far only a few strigolactones have been purified due to the extremely unstable nature of the compound and the fact that they are produced in very low quantities by the plants. Therefore, we reflect upon the diversity of the strigolactones (SLs) isolated so far and tried to understand the structural diversity assigned on the basis of the revelation by the physical data. We also summarize the important advances regarding biosynthesis, perception, regulation, roots symbiotic interactions with AM fungus, role of SLs as a kind of quorum-sensing molecule, and the biological functions associated with the strigolactones.

Keywords Strigolactones · Phytohormone · Strigol · Orobranchol · GR24 · AMF colonization · Auxin efflux transporters

Sunil Kumar and Ashutosh Joshi contributed equally with all other contributors.

S. Kumar · A. Joshi · R. K. Shukla (✉)
Plant Biotechnology Division, CSIR-Central Institute of Medicinal and Aromatic Plants, Lucknow, Uttar Pradesh, India
e-mail: rk.shukla@cimap.res.in

B. A. Horwitz, P. K. Mukherjee (eds.), *Microbial Cross-talk in the Rhizosphere*, Rhizosphere Biology, https://doi.org/10.1007/978-981-16-9507-0_3

3.1 Introduction

3.1.1 Strigolactone as an Endogenous Plant Hormone

Strigolactone (SLs) are a class of terpenoids initially identified as a signalling compound present in root exudates that stimulates the germination of the parasitic plant *Striga* (Cook et al. 1966) which lends the name to this compound. The presence of lactone in the chemical structure justifies the other part of the name. The involvement of these compounds in plant development as well designated them as endogenous plant hormones (Ravazzolo et al. 2021; Gomez-Roldan et al. 2008). SLs endogenously regulate various aspects of plant development such as root architecture (Koltai and Kapulnik 2011; Villaécija-Aguilar et al. 2021), shoot branching (Umehara et al. 2008; Okazaki et al. 2021), leaf senescence (Bennett et al. 2016), and regulation of secondary growth (Cheng et al. 2013). SLs are also known to exogenously establish a symbiosis by promoting hyphal branching in arbuscular mycorrhizal fungi, ensuring the supply of fixed carbon to the fungi in exchange for mineral and water supplies (Akiyama et al. 2005; Fernández et al. 2019).

3.1.2 Structural Diversity and Classification

SLs are carotenoid derivatives which in turn are synthesized from the terpenes or isoprenoid units (Matusova et al. 2005). Diverse SL are reported from different plant species. The general structure suggests that SLs are composed of a tricyclic ABC part connected to a butenolide ring (D ring) in the $2'R$ configuration and connected to a variable second moiety via an enol ether bridge. D ring and the enol ether bridge are considered to be crucial to its bioactivity (Scaffidi et al. 2014; Flematti et al. 2016). Such SLs which have conserved $2'R$ configuration in between the C and D ring are designated as non-natural SLs while those having $2'S$ configuration instead are called natural SLs (Flematti et al. 2016). SLs as per the variable moiety are classified into those having a tricyclic lactone (ABC) ring as canonical SLs (Cook et al. 1966). On the other hand, those which lack the A, B, or C ring but have somehow retained the enol-ether D ring are classified as non-canonical SLs (Abe et al. 2014; Ueno et al. 2014). Owing to the difference in the stereochemistry of the B and C ring, canonical SLs are further divided into the strigol- (Cook et al. 1966) and orobanchol-like strigolactones (Fig. 3.1) (Akiyama 2007). Strigol-like SLs are identified as having the C ring in α orientation while the C ring is in β-orientation in orobanchol-like strigolactones (Jia et al. 2018).

The identification and classification of SL molecules are based on the analyses of the exudates from different species containing secreted SLs (Table 3.1). However, it is not to be concluded that the endogenous SLs necessarily have a similar structural configuration. 5-deoxy Strigol (Fig. 3.1) was collected from root exudates of *Gossypium hirsutum* L. and was the first SL to be extracted. The discovery was

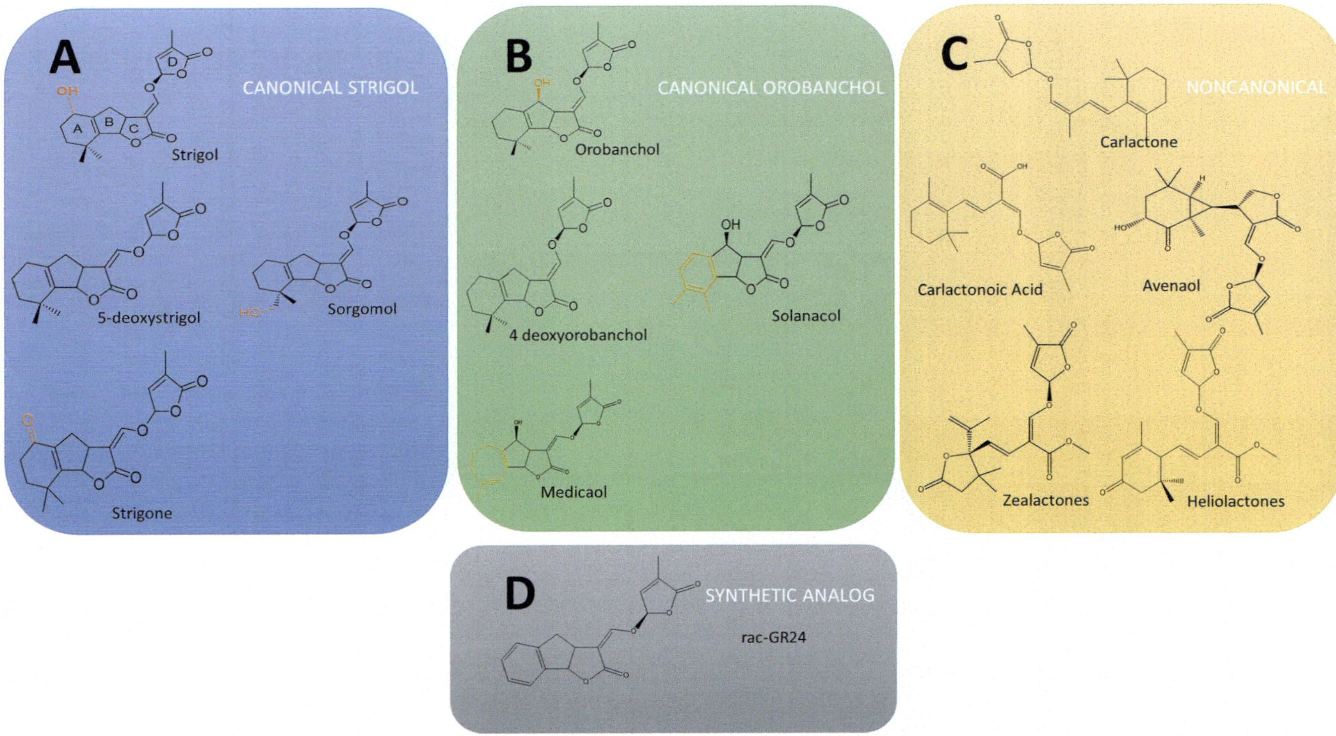

Fig. 3.1 A Schematic view of the structure of strigolactones. (**a**) Canonical SLs are defined by complete A, B, C, and D ring structures. (**b**) Canonical SLs are divided into a separate family based on different conformation between B and C ring. (**c**) Non-canonical SLs have an intact D ring like that of canonical SLs while the other rings can have different chemistry. (**d**) The synthetic analog GR24

Table 3.1 Strigolactone-like compounds isolated from the root exudates of different plants

S. No.	Strigolactones (SLs)	Plant species	References
1.	7-Oxo-, 7-α-hydroxyorobanchol, 7β-hydroxyorobanchol	*Linum usitatissimum* L., *Cucumis sativus* L.	Yoneyama et al. (2008)
2.	Fabacyl acetate	*Pisum sativum* L.	Xie et al. (2009)
3.	Strigol	*Gossypium hirsutum*; *Houttuynia cordata*; *Menispermum dauricum*; *Panicum miliaceum*; *Sorghum bicolor*; *Trifolium pratense*; *Vigna unguiculata*; *Zea mays*	Cook et al. (1966), Kisugi et al. (2013), Yasuda et al. (2003), Siame et al. (1993), Awad et al. (2006), Yokota et al. (1998), Sato et al. (2003), Xie et al. (2008)
4.	Sorgomol	*Astragalus sinicus*; *Cosmos bipinnatus*; *Lupinus albus*; *Sorghum bicolor*	Yoneyama et al. (2008, 2011), Xie et al. (2008), Jamil et al. (2011)
5.	Orobanchol	*Actium lappa*; *Arabidopsis thaliana*; *Arachis hypogaea*; *Astragalus sinicus*; *Carthamus tinctorius*; *Glycine max*; *Hedypnois rhagodioloides*; *Lactuca sativa*; *Linum usitatissimum*; *Lupinus albus*; *Medicago sativa*; *Nicotiana tabacum*; *Oryza japonica*; *Phaseolus vulgaris*; *Pisum sativum*; *Solanum lycopersicum*; *Sorghum bicolor*; *Tagetes erecta*; *Trifolium incarnatum*	Yoneyama et al. (2008, 2011), Xie et al. (2008, 2009), Jamil et al. (2011), Goldwasser et al. (2008), López-Ráez et al. (2010), Koltai et al. (2010), Sato et al. (2003), Ueno et al. (2011)

elemental as it was capable to stimulate seed germination of the parasitic plant *Striga*, (Cook et al. 1966; Cook et al. 1972). Strigol was isolated in 1996 but not until 1985 was its full structure determined (Fig. 3.1). Orobanchol (Fig. 3.1) from the other parasitic family *Orobanchaceae* (broomrapes) was isolated in 1998 (Akiyama 2007). It first had an incorrect structure assigned, which was later corrected. Along with Orobanchol, the compound alcetrol was also sequestered from the sorghum and cowpea root exudates to be later identified as the SL orobanchyl acetate (Goldwasser et al. 2008). In the years to come, Strigol was also identified in the root exudates from maize and proso millet (*Pennisetum glaucum* R.Br.) (Siame et al. 1993). The detection of 5-deoxystrigol in both monocots (Awad et al. 2006) and dicots (Yoneyama et al. 2008) leads to the idea that other SLs are likely to be derived from 5-deoxystrigol by hydroxylation (Matusova et al. 2005; Bouwmeester et al. 2007; Rani et al. 2008). Allylic hydroxylation of 5-deoxystrigol yields strigol or orobanchol, while homoallylic hydroxylation yields sorgomol (Fig. 3.1) (Yoneyama et al. 2008). The difference in the B, C ring configurations led to the strigol types and orobanchol types (Ueno et al. 2011; Xie et al. 2013) with *Oryza* and *Solanum* strictly producing orobanchol type SLs while *Nicotiana* has both kinds of SLs (Xie et al.

Table 3.2 Synthetic strigolactone analogs which mimic SL and have high potent bioactivity

S. No.	Strigolactones (SLs)	Reference
1.	GR24	Mangnus and Zwanenburg (1992)
2.	Nijmegen-1	Nefkens et al. (1997)
3.	T-010	Samejima et al. (2016)
4.	SPL7	Uraguchi et al. (2018)
5.	2NOD	Li et al. (2021)

2013; Yoneyama et al. 2018). Orobanchol appears to be commonly distributed in the plant kingdom, as shown by the isolation of 7-oxoorobanchyl acetate from flax (*Linum usitatissimum* L.) (Zwanenburg et al. 2009) and 7-hydroxyorobanchol acetate from cucumber (*Cucumis sativus* L.) (Xie et al. 2009) root exudates. Since then, more than 25 SLs have been identified across the plant kingdom and categorized based on structural variations such as differences in stereochemistry and the existence or absence of the ABC ring system (canonical and non-canonical SLs) (Wang and Bouwmeester 2018; Yoneyama et al. 2018). Non-canonical SLs such as zeapyranolactone (Charnikhova et al. 2018), heliolactone (Yoshimura et al. 2019), and aveanol (Kim et al. 2014) have been isolated from maize, sunflower, and the black oat (Fig. 3.1). Most recently, lotuslactone has been isolated from *Lotus japonicus* (Xie et al. 2019). A similar butenolide structure is present in the chemical contained in burnt plant smoke known as karrikin (Flematti et al. 2004), which triggers the germination of dormant seeds after the burning. Non-natural SLs bind to the karrikin receptor KAI2 (Guo et al. 2013) though at high concentrations, suggesting that KAI2 is a homolog of *D14*, the receptor of SLs (Scaffidi et al. 2014). Thus, karrikins mimic the SLs as the endogenous signal and share a similar signalling pathway, which is conserved in the liverwort *Marchantia polymorpha* where *MpSMXL* is crucial for MpKAI2A-dependent signalling (Mizuno et al. 2021).

The isolation of Strigol from the cotton plant led us to believe that plants do not have a single side to the story of the production of strigolactones as cotton plants are not usually compromised by Striga species. As a result, the assumption that SLs may play other roles apart from host-parasite interactions became much more entrenched. It took nearly 50 years, however, to discover that strigol is needed as a branching factor to assist AMF in interacting with plant roots (Akiyama et al. 2005). The tracing of the biosynthesis of SLs to the precursor molecule being isoprenoid also overlaps with the pathway of synthesis of some plant hormones such as ABA, highlighting the role of SLs in plant development too. The elucidation of structures and isolation of SLs have been always very tricky due to the varied stereochemistry. In fact, SLs despite having such enormous potential received very scarce attention from the plant science community as the total synthesis of SLs has been a tedious task to perform (Zwanenburg et al. 2015). Synthetic SLs are a versatile tool for deciphering the underlying mechanism of action of such signalling molecules (Table 3.2). Chemical synthesis, on the other hand, is a difficult process and the yield is very low as it involves the synthesis of the ABC scaffold followed by the selective oxidation and coupling with the D ring along with the installation of the correct stereochemistry (Reizelman et al. 2000). Chemical SL analogs should have a simpler structure while retaining their bioactivity. The analog GR24 prepared by Johnson et al. (1981) from 1-indanone is the most commonly used in experiments. GR24 is a solanocol modification that lacks methyls in the A ring and OH in the B ring and has a half-life of 100 h at pH 7 (Kannan and Zwanenburg 2014) (Fig. 3.1). The biological activities of SL analogs differ between plant species (Umehara et al. 2015). The bioactivity of GR24 was effective enough to germinate seeds of parasitic weeds Striga and Orobanche, but the synthesis of GR24 has also created the temptation to study the other aspects of the fascinating molecule, i.e., strigolactone. 2-Nitrodebranone (2NOD) outperformed rac-GR24 in several aspects of *Arabidopsis* plant growth and development, including inhibition of hypocotyl elongation, promotion of root hair development, and senescence (Li et al. 2021).

3.2 Biosynthesis, Perception, and Regulation

3.2.1 Biosynthesis of Strigolactones

Natural SLs are derivatives of the carotenoid pathway (Matusova et al. 2005). Natural SLs are made up of a tricyclic lactone (ABC ring) and a butenolid ring (D ring), and it is thought that these two moieties evolved independently (Matusova et al. 2005; Crawford et al. 2010) only to later couple into a common biosynthetic step. Carlactone (CL) (Fig. 3.2) with an SL-like skeleton is considered a putative precursor for strigolactones, but not detected as an endogenous compound from plant tissues (Alder et al. 2012). Strigolactones are known to inhibit tillering and shoot branching in plants, and higher tillering was observed in plants with *Carotenoid Cleavage Dioxygenase 7* (*CCD7*) and *CCD8* mutations in *Arabidopsis* and rice (Umehara et al. 2015) and in *DWARF27*, a small polypeptide identified in a SL-deficient mutant in rice (Lin et al. 2009). Thus, CL is the end result of concurrent reactions involving three biosynthetic enzymes: *D27*, *CCD7* (Schwartz et al. 2004), and *CCD8* utilizing *trans*-β-carotene as a substrate (Alder et al. 2008). *DWARF27* (*D27*) is a β-carotene isomerase that catalyzes the reversible conversion of all *trans*-β-carotene to 9-*cis*-β-carotene (Alder et al. 2012; Bruno et al. 2016; Harrison et al. 2015). The stereospecific enzyme *CCD7* (*MAX3*) cleaves only 9-*cis*-β-carotene and generates β-ionone (C13) and 9-*cis*-configured β-apo-10-carotenal (C27) from C40 carotenoid (Schwartz et al. 2004). Furthermore, the secondary cleavage of apocarotenoid is catalyzed by *CCD8*. *CCD8* in vitro cleaves all *trans*-β-apo-10-carotenal into β-apo-13-carotenone (Schwartz et al. 2004), carlactone (C19) which subsequently leads to the product with an aldehyde and alcohol group, i.e., ω-OH-(4-CH_3)-hepta-2,4,6-trien-al (C8) (Bruno et al. 2017; Alder et al. 2012). Carlactone, being identical to 4-deoxyorobanchol and 5-deoxystrigol, is the central player and the parent compound to all canonical and non-canonical SLs (Alder et al. 2012; Seto et al. 2014). *CCD8* is an unusual carotenoid cleavage enzyme that can catalyze both *cis* and *trans* isomers depending on the substrate's stereochemistry.

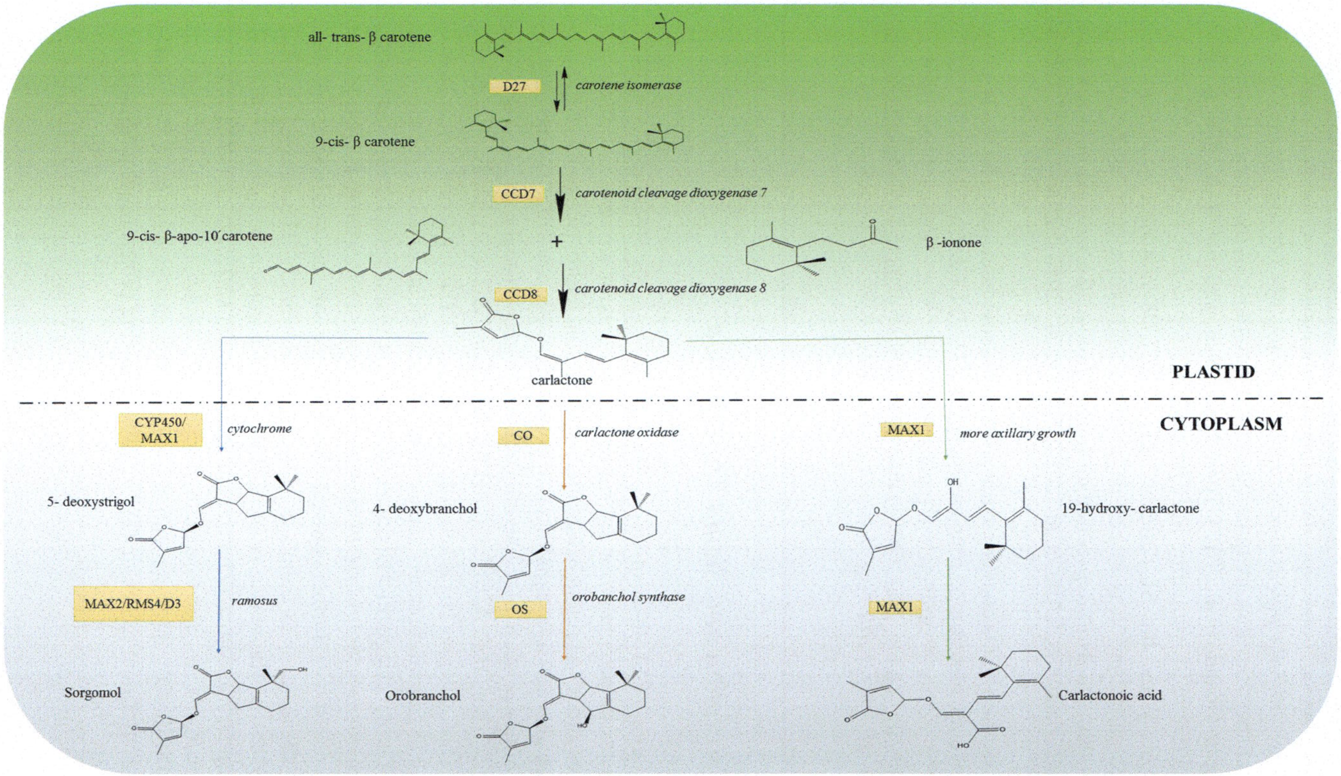

Fig. 3.2 Synthesis of strigolactones (SLs). The carotene isomerase D27 converts all-*trans*-β carotene into 9-*cis*-β carotene, which undergoes cleavage by carotenoid cleavage dioxygenase (CCD7) to yield 9-*cis*-β-apo-10′ carotene (C27) and β-ionone (C13). CCD8 converts 9-*cis*-β-apo-10′ carotene into carlactone (C19) which is considered as the central metabolite of the pathway. MAX1 in Arabidopsis leads to the oxygenation of carlactone to produce carlactonoic acid while its homolog carlactone oxidase (CO) catalyzes oxygenation of carlactone to form 4-deoxybranchol. Hydroxylation of 4-deoxybranchol by orobanchol synthase (OS) further converts 4-deoxybranchol into orobanchol. Carlactone is presumed to be the parent molecule of all the SLs

Decker et al. (2017) expressed and identified the enzymatic activity of *PpCCD7* and *PpCCD8* in *E. coli*. *PpCCD7* was a stereospecific 9-*cis*-CCD. *PpCCD8* was called CL synthase. *ppccd7* and *ppccd8* mutants demonstrated substantial caulomela growth. *Physcomitrella patens* has a single homolog *MAX2*; the *Ppmax2* mutant exhibits characteristics that are quite dissimilar from those of the moss SL-deficient mutant (Lopez-Obando et al. 2018). Their data also showed that *MAX2* does not play a conserved role in SL signalling in *P. patens* and *Arabidopsis* as both are functionally different. When the SL-deficient rice *d10/CCD8* mutant was administered ^{13}C-labelled carlactone, it produced deoxystrigol and orobanchol, while the same experiment with the *Arabidopsis* mutant (*max4*) produced a labelled non-canonical SL methyl carlactonate (Abe et al. 2014; Seto et al. 2014). Zhang et al. (2014) determined the role of the four rice *MAX1* homologs and demonstrated that carlactone oxidase (CO) converts carlactone to 4-deoxybranchol. The study delineated the overall function of four enzymes, i.e., *D27*, *CCD7*, *CCD8*, and *CO*, as capable of generating orobanchol parent molecules like strigolactones from all *trans*-β-carotene. Zhang et al. (2014) also described orobanchol synthase, a second *MAX1* homolog *Os1400* catalyzing 4-deoxyorobanchol hydroxylation to yield orobanchol.

3.2.2 Perception and Signal Transduction

Characterization of SL-insensitive rice *D3* and *D14* mutants identified the products of these genes as perception or signalling components of strigolactone pathways. The *D14* gene codes a protein belonging to the α/β-fold hydrolase family (Arite et al. 2009) which also acts as a signalling component (GA receptor GID1) in the pathway of other hormones such as gibberellin (GA) (Ueguchi-Tanaka et al. 2005). The catalytic triad (Ser, His, Asp), required for the hydrolysis reaction, is preserved in the case of *D14*, whereas His is substituted by Val in GID1, rendering it incapable of hydrolyzing even *p*-nitrophenyl acetate used as an artificial substrate for the protein hydrolase family (Ueguchi-Tanaka et al. 2005). Crystallographic studies of the homolog of *Petunia hybrida DECREASED APICAL DOMINANCE* (*DAD2*) and *Arabidopsis* (*AtD14*) further proved the presence of the catalytic triad, thus confirming these genes as members of the hydrolase family (Hamiaux et al. 2012; Zhao et al. 2013; Kagiyama et al. 2013). Other groups established *D14* as SL receptors in *Oryza sativa* (Zhao et al. 2013) and *RAMOSUS 3* (*RMS3*) as SL receptors in *Pisum sativum* (Germain et al. 2013). Within 15–30 min of exposure with SL, *D14* directly interacts with an F-box protein encoded by *MAX2/D3/RMS4* and the transcriptional repressor *D53*, thereby inducing proteasomal mediated degradation, thus abrogating the activity of *D53* on the SL signalling pathway (Zhou et al. 2013; Jiang et al. 2013; Soundappan et al. 2015; Wang et al. 2015; Khosla et al. 2020).

The biochemical studies suggest that the SL binding proteins (*D14/AtD14/DAD2*) catalyze chemical synthetic analog hydrolysis, GR24 into D ring structure

(Hydroxymethyl Butenolide; HMB) and ABC ring structure (Formyllactone) (Nakamura et al. 2013). According to yeast-two hybrid (Y2H) studies, *DAD2* interacts with a petunia *MAX2* ortholog (*PhMAX2*) (Hamiaux et al. 2012). The catalytic triad residues confer ligand binding stability as if replacing the Ser residue in the catalytic triad with Ala (*DAD2:S96A*), the resulting *DAD2:S96A* had lost the GR24 hydrolyzing activity in vitro, and the ability to prevent shoot branching. In particular, *DAD2:S96A* did not interact with *PhMAX2* in the Y2H system. The petunia *DAD2* mutant phenotype was also not restored by GR24 hydrolyzed products, indicating the function of *DAD2* in SL perception and that hydrolyzed products are required for SL signal transduction (Hamiaux et al. 2012).

Using differential scanning fluorimetry (DSF) (Niesen et al. 2007) in further interaction studies between *DAD2* and GR24 showed GR24 mediated *DAD2* protein destabilization (Hamiaux et al. 2012). Nakamura et al. (2013) showed similar results using trypsin digestion assay. This implies that the GR24 can induce a conformation change in *DAD2* and *D14* protein structures, which further plays an important role in the interaction of *D14* with *D53* resulting in the degradation of *D53* via the proteasome pathway (Jiang et al. 2013; Niesen et al. 2007). The crystal structure of *DAD2* revealed a four-helix lid-like domain and a cavity for the accommodation of the strigolactone molecule. The hydrolase proteins show slow substrate kinetics when examined using the isothermal titration calorimetry (ITC) method (Kagiyama et al. 2013). The authors hypothesized that the active form for shoot branching inhibition is the D ring (HMB) above D14's head, resulting in altered protein structure (Nakamura et al. 2013). Therefore, this adapted structure was assumed to be appropriate for *D14* interaction with F-box protein *D3*. Instead, the other group interpreted electron density concentrated at the active serine site as an intermediate produced by the serine's nucleophilic attack on GR24 molecule (Zhao et al. 2013). Both molecules were active in SL hydrolysis, but no apparent conformational change occurred in protein organization upon binding with SL. Zhao et al. 2015 interpreted the intermediate hydrolysis at the serine then proposed the concept that the intact GR24 molecule but no D ring is responsible for the altered *D14* protein surface. SL's crystallographic analysis *AtD14-OsD3-AtASK1*suggested that *AtD14* covalently binds to the D ring on SL hydrolysis to undergo a conformational shift from open to closed, thus promoting interaction with *D3*. The latter study also concluded that the electron density extended from the active site's serine to the histidine, adding a ligand as a "covalently connected intermediate molecule" (CLIM) (Yao et al. 2016). This intermediate which is attached simultaneously to both of these amino acids possibly triggers the change in conformation observed in *D14* when bound to the *D3*. The mechanism of allosteric activation and stabilization of *D14* remains elusive. CLIM is part of the *AtD14-OsD3-AtASK1* complex and it is important to study the conformational change which occurs in *D14* when it is bound by F-box protein *D3*. The change in structure in helix αT1 in *D14* is a prerequisite for binding between *AtD14* and *D3*. On binding of *D14* with *MAX2*, helix αT1in *D14* extends and terminates at residue G158 which otherwise is located in helix αT2 in a free *D14* structure. *D14* cannot bind to *D3* but has the capability to hydrolyze the strigolactone when the glycine is replaced by glutamate in *AtD14* G158E raising a certain question whether the position strictly demands glycine or it could do with certain other charges (Yao et al. 2016). In the end, the authors concluded that the pliability in G158 residue is plausible to form a TT-turn structure at the end of the alpha-helix (Yao et al. 2016). Recently the idea of hydrolysis mediated conformational change has been challenged (Seto et al. 2019) and it was proposed instead that the hydrolysis of the strigolactone molecule is not necessary for signalling but rather happens after the degradation of *D53/SMXLs*. So, the hydrolysis of SL happens just to destroy it so that it could not get subsequently used further in signalling.

D14, a member of the α/β-fold hydrolase family, may have its hydrolase activity regulated. *D3*, the rice homolog of *AtMAX2*, could exist in two different conformational states due to the presence of α-helix at the C-terminal (Shabek et al. 2018). The C-terminal is therefore highly mobile and can eject from the remaining LRR domain. This ejected helix before hydrolysis has the ability to bind to the *D14* present in open conformation and arrests *D14* to prevent premature hydrolysis of SL before the *D53* is polyubiquitinated. The binding of *D53* results in the reactivation of the *D3-D14* pair allowing it to conduct SL hydrolysis which then gives rise to the intermediate. The intermediate then stabilizes the closed conformation of *D14*, thus causing the C-terminal helix of *D3* to coincide with the rest of the leucine-rich repeats (LRR) which leads to the ubiquitination of D14. The C-terminal helix's SL-dependent tethering to D14 has yet to be concluded. Numerous proteins have been hypothesized as *D14-SCF*^{MAX2} signalling complex proteolytic targets; however, there is no genetic or physiological proof for any of the proposed protein targets. Rather, all existing research shows that *D14-SCF*^{MAX2}*'s* primary, if not exclusive, target is the protein family SMAX1-LIKE (SMXL). Physical interactions were found in vitro and *in planta* between *SMXL7/D53* and *D14* and *MAX2/D3* (Machin et al. 2020). *D53* has been demonstrated to interact strigolactone-independently with *D3* in vitro (Jiang et al. 2013; Wang et al. 2015), but no direct contact between *SMXL7* and *MAX2* has been observed *in planta*, even with strigolactone (Liang et al. 2016). In any instance, the *SCFD3* complex requires active *D14* to polyubiquitinate and degrade *D53*.

Two other hydrolases similar to *D14* exist in *Arabidopsis*: *D14-like 2* (*DLK2*) and *HYPOSENSITIVE TO LIGHT/KARRIKIN INSENSITIVE 2* (*KAI2*) (Aquino et al. 2021; Nelson et al. 2011; Waters et al. 2015). *Physcomitrella patens*, unlike vascular plants, has a large number of α/β-fold hydrolase receptors (PpKAI2) that are linked to KAI2 from angiosperms (Lopez-Obando et al. 2016).

3.2.3 Regulation of SL Biosynthesis

Strigolactones have an intriguing level of complexity and many tiers of regulation, but knowledge is still limited. The transcriptional regulation of SL biosynthesis is determined by a negative feedback mechanism that determines hormone homeostasis. In nitrogen-deficient roots of *Medicago truncatula*, the expression of *D27* and a *MAX1* ortholog was *NSP1* and *NSP2* (NODULATION SIGNALING PATHWAY) dependent (Liu et al. 2011). *D27* expression is induced by inorganic phosphate availability in wild-type plants but not in *nsp1* mutants, with a non-significant increase in *nsp2* mutants, indicating that *NSP1* and *NSP2* act as transcription factors regulating strigolactone biosynthetic genes. In addition, under phosphate starvation conditions, transcription levels of SL transporter genes and biosynthetic genes are also regulated, as in petunia *DAD1* (Breuillin et al. 2010) and *PDR1* (Kretzschmar et al. 2012), rice *D10*, *D27*, and *D17/HDT1* (Sun et al. 2014), *DgCCD7*, *DgCCD8*, *DgD27*, and *Chrysanthemum* (Bonneau et al. 2013). Similarly, in *Medicago truncatula*, the SL transcripts such as *MtCCD7*, *MtCCD8*, *MtD27*, and *MtMAX1* are regulated by *NSP1* and *NSP2* transcription factors under phosphate deficiency (Wen et al. 2016; Liu et al. 2011; van Zeijl et al. 2015). In fact, the promoter region of *MtD27* contains nodulation-responsive elements similar to those found in the promoter regions of early nodulin responsive genes *ENOD11*, and it binds to the *NSP1* transcription factor, implying that these genes are regulated under nitrogen-deficient conditions as well (Hirsch et al. 2009). *MdIAA24* overexpression in apple (*Malus domestica*) resulted in increased expression of SL biosynthetic genes, including *MdD27*, *MdCCD7*, *MdCCD8a*, *MdCCD8b*, and *MdMAXa*, resulting in increased strigolactone levels and thus AMF colonization in OE lines (Huang et al. 2021). The regulation of SL biosynthesis under nutrient-deficient conditions in moss (*Physcomitrella patens*) (Yamada et al. 2014) indicates the conserved function of the SLs which helps plants to cope up such adverse conditions. A recent study has concluded that Zaxinone inhibits SL production in rice roots under normal and phosphorus-deficient conditions. Zaxinone was demonstrated to inhibit the transcription of D27, D17, D10, and CYP711A2 in phosphate-depleted roots independently of SL signalling (Wang et al. 2019). In contrast, Zaxinone promoted MAX3 and MAX4 expression in Arabidopsis roots under normal and phosphate-starved conditions (Ablazov et al. 2020).

Multiple factors help to regulate strigolactone levels in plants. The SL content of the root exudates and tissues decreased in *Lotus* under drought conditions. However, when treated with SLs exogenously, the drought stress-induced ABA increase was inhibited. This indicates that the SLs might interact with the hormones such as ABA to combat response to abiotic stress (Liu et al. 2015). Exogenous treatment of synthetic analog GR24 on *Vitis vinifera* induced the transcription of *VvMybA1* and *VvUFGT*, thus reportedly inhibiting the ABA-induced accumulation of anthocyanins and sugars (Ferrero et al. 2018). The use of AbaminSG, an inhibitor of NCED, which is a key enzyme in ABA biosynthesis, reduced the amount of SLs in tomatoes, implying that ABA is most likely a positive regulator of SL biosynthesis (López-Ráez et al. 2010). Furthermore, ABA-deficient mutants (*notabilis*, *siliens*, *and flaccca*) displayed a decreased level of SL, which is associated with the downregulation of *LeCCD7* and *LeCCD8* in tomatoes. This was previously known to occur in the root exudates of ABA-deficient mutant *Viviparous14* in maize (López-Ráez et al. 2010; Schwartz et al. 1997).

BdCYP711A29 is a gene that encodes a Cytochrome P450 in *Brachypodium distachyon* which regulates the biosynthesis of orobanchol, as suggested by the studies of the root exudates of the *BdCYP711A29* OE lines (Changenet et al.

2021). The protein *Bd*CYP711A29 belongs to the same clade as Os900 and Os1400 protein. Both Os900 and Os1400 protein converts CL to orobanchol in two step reaction. Firstly, Os900 which is a CL oxidase enzyme catalyzes the conversion of CL into ent-22-epi-5-deoxystrigol. In the second step ent-22-epi-5-deoxystrigol acts as substrate and *Os1400* converts it into orobanchol (Zhang et al. 2014). Our knowledge of strigolactone regulation is minimal, and the coming years will be instrumental in this particular direction, as researchers are fascinated by SLs with quite diverse and exceptional features.

3.3 Strigolactone Signalling in Plant Development

Years of study have shown that strigolactones are active in other facets of plant growth (Fig. 3.3), including the control of root system development by promoting primary roots (PRs) elongation and inhibiting the creation of adventitious roots (ARs). The impact of the strigolactone on the growth of lateral roots (LRs) was discovered to be dependent on the accessibility of nutrients, specifically phosphorus (P) and nitrogen (N). Strigolactones prevent the elongation of LRs under ideal growth conditions, but they promote LR growth when plants are starved (Waters et al. 2017). This result suggests that strigolactones perform a critical role during nutrient stress in the plant. The biosynthesis of strigolactones gets boosted under both P and N deficiency, which leads to the secretion of this hormone in greater concentrations in the rhizosphere, most likely to facilitate symbiotic relationships with *Rhizobium* spp. and Arbuscular mycorrhizal fungi (AMF). Increased strigolactone concentrations affect plant morphology by restricting the growth of

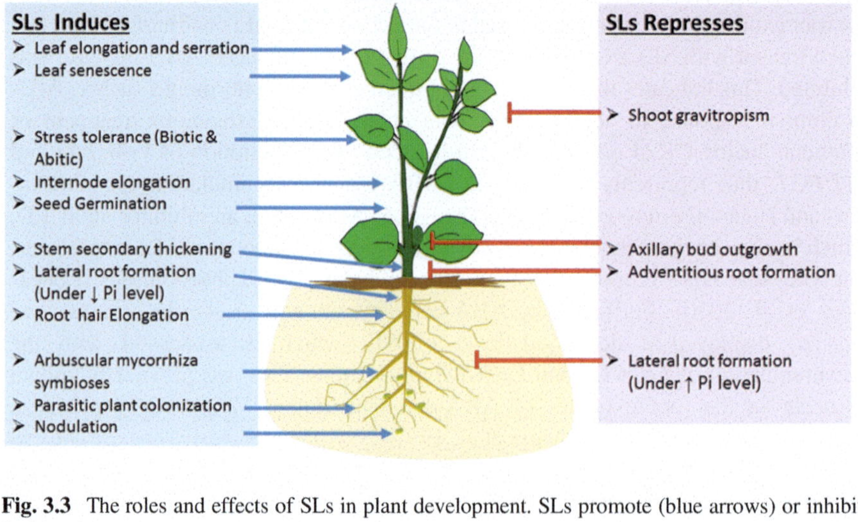

Fig. 3.3 The roles and effects of SLs in plant development. SLs promote (blue arrows) or inhibit (red bars)

shoots and inducing the growth of lateral root, thereby allowing the plant to adapt to various abiotic stresses (Marzec et al. 2013; Ito et al. 2016). It has been reported that strigolactones also control the senescence of sepals (Xu et al. 2021). Although the importance of strigolactones has been identified during various abiotic stresses, more research is needed to confirm it as a stress-related hormone (Mishra et al. 2017). Additional research on *Lotus japonicus* L. (Liu et al. 2015) and *Solanum lycopersicum* L. (Ruiz-Lozano et al. 2016; Visentin et al. 2016) indicated that strigolactones along with abscisic acid (ABA), make the plant tolerant to water scarcity. Tomato plants exhibit reduced strigolactone concentration in the roots as compared to the shoots in response to drought (Visentin et al. 2016). These findings support the notion that strigolactones have a wide range of physiological and morphological impacts on the plant (Mostofa et al. 2018) (Fig. 3.3). According to an in silico study of the genes involved in the strigolactone biosynthesis pathway in *A. thaliana* and rice, it was reported that strigolactone can provide resistance to plants against various biotic and abiotic stresses (Marzec and Muszynska 2015).

3.3.1 Regulation of Shoot Branching

The modulation of shoot budding is the most well-studied function of strigolactones in the growth and development of the plant. There are two major models that have been suggested (Fig. 3.4). The first proposes that strigolactones control a transcription factor that belongs to the TCP domain TF family, namely *BRANCHED1* (*BRC1*), while the second proposes that strigolactones regulate PIN1 protein which is an auxin efflux carrier protein and is localized at the plasma membrane (Dun et al. 2012; Crawford et al. 2010; Shinohara et al. 2013). In general, these strigolactones interaction models are comparable to the two major shoot branching regulation models, i.e., the first one is the direct-action model and the second is the canalization model (Domagalska and Leyser 2011).

The direct-action model suggests that hormones such as strigolactone and cytokinin, which are synthesized mainly in the root, are transferred by the xylem vessels to the shoot and specifically influence shoot branching (Brewer et al. 2015). Auxin released by active shoot apices often inhibits dormant bud outgrowth (apical dominance), but this is an indirect effect (Domagalska and Leyser 2011). Various past studies provided evidence that in auxin signalling there are some secondary messengers that transmit auxin signals in buds. Since auxin controls the biosynthesis of strigolactone, as well as cytokinin, and both these hormones are thought to function directly in the bud, these two hormones are likely to be auxin's second messengers (Domagalska and Leyser 2011). Since recent research suggests that cytokinin isn't an important target of auxin in apical dominance (Mori et al. 2016), it's still conceivable that strigolactones play a role. In the direct-action paradigm, both strigolactones and cytokinins control transcription of *BRC1* in buds (Dun et al. 2012), a possible regulatory axis for the control of shoot branching (Fig. 3.4).

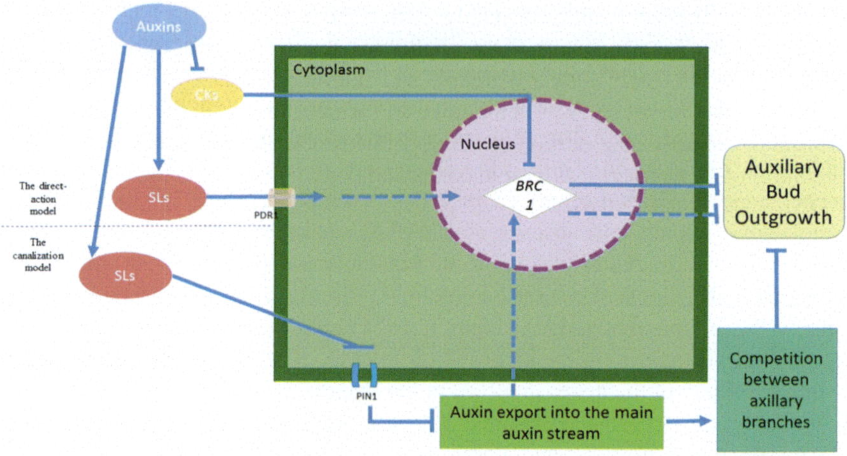

Fig. 3.4 Two models suggested for the strigolactone signalling which regulates shoot branching. (a) The direct-action model. In this model, strigolactones and cytokinins promote and inhibit, respectively, the expression of the transcription factor that belongs to the TCP domain TF family, namely *BRANCHED1* (*BRC1*) which regulates bud growth. (b) The canalization models. In this model, bud outgrowth is based on the bud's ability to export auxin from a high concentration (the source) to a lower concentration (the sink). Such coordinated export requires the subcellular and supracellular mobilization of the PIN1 transporter. In some species, SL may also affect bud outgrowth via the promotion of *BRC1*expression (shown by dashed lines)

The direct-action paradigm is a simple and compelling justification for strigolactone's impact on branching, but it is defied by several observations. In *Arabidopsis* and pea, for example, *brc1* and strigolactone mutants have significant variations in branching, and so these mutations have cumulative consequences on branching in Arabidopsis (Chevalier et al. 2014). As a result, the lack of *BRC1* cannot account for all of the increased branching in SL mutants. According to these findings, BRC1 is unlikely to be a transcriptional target of strigolactone signalling. There are some other active bud-regulatory genes that may be controlled by strigolactones and can support the direct-action model, but such genes are yet to be discovered. As a result, the direct-action model does not yet provide a full description of how flowering plants regulate shoot branching.

The canalization model (Bennett and Leyser 2014), which is derived from the eponymous vascular patterning model, offers an alternate outline for shoot branching regulation that involves a method for organized development of branches in the plant (Domagalska and Leyser 2011). As per this model, buds are said to be the auxin sources, and they will only develop when they transport the auxin to the main stem (Prusinkiewicz et al. 2009) (Fig. 3.4). The amount of auxin which can be exported from the buds is dependent on the vitality of the auxin sink, which is calculated primarily by the capacity of the stem to move auxin toward the root. As a result, buds compete for auxin export, and the ones with the best auxin sources canalize auxin export to the stem and hence grow out.

Canalization is well supported by both experimental findings and statistical models in the case of vascular patterning, although to some extent it supports shoot branching (Prusinkiewicz et al. 2009). Canalization's mechanism is not well known, but it is believed to be fueled by the self-organizing action of auxin efflux carriers belonging to the PIN family (Bennett et al. 2014).

In the canalization model, by facilitating the elimination of PIN1 transporter protein from stem and bud, strigolactone improves competitiveness between buds. This potentially reduces the stem's sink strength, limiting the growth and development of buds consequently making the export of auxin more difficult (Prusinkiewicz et al. 2009). The significance of the strigolactone impact on PIN1 for shot branching is, however, debatable. Physiological studies have supported and refuted the significance of this effect (Brewer et al. 2009, 2015; Shinohara et al. 2013).

Experiments to validate the canalization model have been plagued by a lack of mechanistic comprehension, resulting in contradictory evidence that supports as well as contradicts it (Brewer et al. 2015; Shinohara et al. 2013). Over the last decade, this has resulted in a polarizing debate about the role of strigolactones in the branching of the shoot. It is necessary to note that these models are not exclusive. A hybrid model has been proposed that incorporates both canalizations and the direct model which controls branching. The next move in resolving this problem would be to figure out how strigolactone controls *BRC1* and *PIN1* in eudicots. Shoot branching is a complicated operation, so it is not surprising that many interconnected regulatory processes have accumulated over time. Understanding the underlying mechanism of how strigolactone controls branching is still a subject of research.

3.3.2 Regulation of Root Architecture

As in the shoots, evidence suggests that strigolactones play a function in the control of PIN protein production in the roots. The first evidence for this was reported after the exogenous supplementation of a synthetic auxin, i.e., 2,4-D, which reversed the impact of synthetic strigolactone GR24 in tomato root (*Solanum lycopersicum*), suggesting the participation of GR24 in the export of auxin into the root (Koltai et al. 2010). Second evidence emerged after studying *Arabidopsis* where reduced PIN1-GFP strength was found in the lateral root of GR24-treated *Arabidopsis* seedlings, implying that auxin flux is controlled by PIN1, which is, in turn, controlled by strigolactone GR24 (Ruyter-Spira et al. 2011). Strigolactones thus modify the auxin optima required for the development of lateral root. Kumar et al. reported that, in Arabidopsis, PIN2 polarization was elevated on the root epidermis' plasma membrane in the wild-type plant compared to the *max2* mutant after GR24 treatment, a synthetic strigolactone which causes extension of the root hair. Furthermore, in a *MAX2*-dependent mechanism, GR24 administration caused a spike in PIN2 endocytosis, which in turn resulted in elevation of endosomal mobility in the epidermal cells, which is responsible for alterations in actin filament structure and

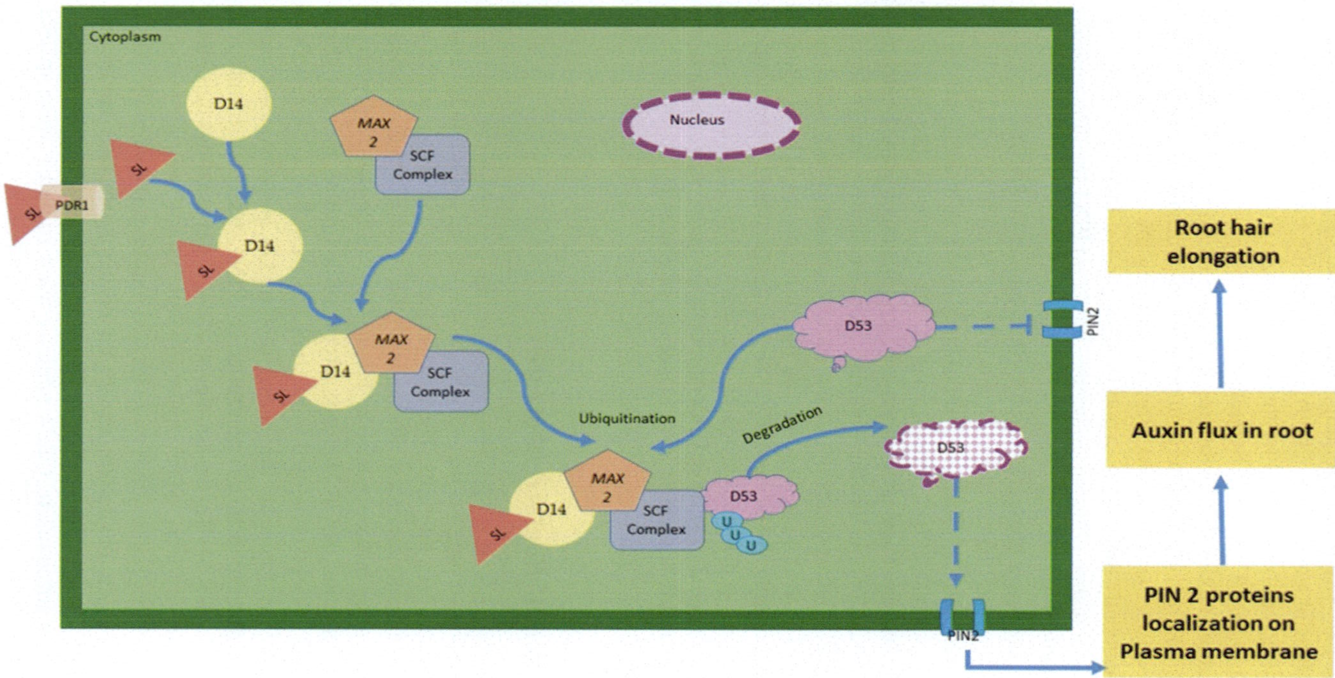

Fig. 3.5 A model of the signalling pathway of strigolactones in determining roots architecture. An α/β-fold hydrolase protein (D14) serves as the strigolactone receptor. It interacts with the SCF complex (Skp, Cullin, F box) and, as a result, leads to degradation of D53(a class I Chloroplast adenosine 59-triphosphate protein). This degradation leads to changes in the architecture and dynamics of actin filaments and PIN endocytosis, which is important for PIN2 polarization. As a result, PIN2 protein polarization is affected, which leads to changes in auxin flux and execution of strigolactone-associated root effects, i.e., root hair elongation

dynamics. These findings indicate that strigolactones influence PIN protein localization in the plasma membrane (Fig. 3.5; Kumar et al. 2014).

Strigolactone signalling which is responsible for regulating the root architecture act through the proteasomal degradation of *D53* (class I Chloroplast adenosine 5′-triphosphate) through an *Skp*, *Cullin*, *F box-con*taining (*MAX2*) complex (SCF complex) (Moon et al. 2004) in a *D14*-dependent manner. *D53* acts as a repressor of the strigolactone signalling system since it represses the PIN endocytosis, which is crucial for the polarization of PIN2 on the root epidermis' plasma membrane, which is, in turn, responsible for auxin flux in the root, resulting in root hair elongation. The D14 protein belongs to the α/β-fold hydrolase superfamily, i.e., is the upstream receptor of strigolactone (Arite et al. 2009). Experimental data in rice support the binding of D14 with GR24 which is a type of synthetic strigolactone (Nakamura et al. 2013; Kagiyama et al. 2013). In a yeast two-hybrid experiment, petunia *DECREASED APICAL DOMINANCE2* (*DAD2*), a homolog of *D14*, was found to associate with the petunia *MAX2* only in the presence of GR24 (Hamiaux et al. 2012). This complex then ubiquitinates the D53 and this degradation by strigolactones promotes the polarization of PIN2 (Kapulnik and Koltai 2014; Zhou et al. 2013; Jiang et al. 2013). Moreover, it was also reported that *D14* in Arabidopsis was degraded vis the proteasome. Degradation was found to be triggered by strigolactone itself in a *MAX2*-dependent manner (Chevalier et al. 2014). This observation signifies that strigolactone also has a negative regulatory signalling pathway.

3.4 Strigolactone in the Symbiotic Interaction

Root symbiotic associations with AM fungi and *Rhizobium* spp. increase plant nutrient absorption and strigolactones are essential signals in these interactions (Bouwmeester et al. 2007). While AM symbioses are very common in plants, some plant families (e.g., *Brassicaceae*) have lost the potential to form these molecules (Bravo et al. 2016). As a result, the function of strigolactones in the symbiotic association has already been well explored in host plants of AMF such as rice, tomato, etc., as well as various legumes which form a symbiotic relationship with *Rhizobium* spp. resulting in nodulation (De Cuyper and Goormachtig 2017).

3.4.1 With Arbuscular Mycorrhiza

The arbuscular mycorrhizal (AM) symbiosis, which involves the interaction of higher plant roots with soil AMF, is the most common symbiosis on the planet. The AMF belongs to the class *Glomeromycota* and is found to establish symbiotic relationships with almost all plants that bloom (Redecker and Raab 2006), particularly under unideal growing circumstances. Arbuscular Mycorrhizal Fungi hyphae that spread via the ground have a wider root surface region, allowing the plant to make use of a greater percentage of soil and thus raise the consumption of nutrients (Rausch and Bucher 2002). The Arbuscular Mycorrhizal Fungi obtains, in exchange, Sucrose, or hexoses (Solaiman and Saito 1997) including glucose (Douds Jr et al. 2000) derived from the host plant.

The presymbiotic and the symbiotic periods are two separate functional phases of the AM symbiosis. Fungal spore germination, together with fungal hyphae development, are induced due to the existence of the host root during the presymbiotic phase (Requena et al. 2007). Fungal spore germination together with hyphae development, and the hyphal branching reaction, may indicate a specific rhizosphere interaction aimed at enhancing active mycorrhization on the host (Koske and Gemma 1992).

In several AMF strains, artificial strigolactones or strigolactones filtered from the host rhizosphere promoted the branching of AMF hyphae, germination of spores, and transcriptional reformatting in AMF (Lanfranco et al. 2018; Akiyama et al. 2005). The presence of these molecules in the rhizosphere is measured in nanograms (Akiyama and Hayashi 2006). In less fertile soil, a plant is found to synthesize and flux much larger levels of strigolactones (Schlemper et al. 2017). Additionally, at a very low concentration, the synthetic strigolactone GR24 significantly enhanced the branching of AMF hyphae (Gomez-Roldan et al. 2008). It was also found that branching of AMF hyphae was reduced in the rhizosphere of strigolactone-deficient mutant plants as compared to the wild type (Koltai et al. 2010).

From further thorough investigation, it became clear that, in AMF, strigolactones trigger rapid shifts in energy metabolism much earlier than its gene expression mechanism. It was also reported that strigolactone causes rapid modification in the structure of the AMF hyphae (like shape and mass) and mortality (within 60 min) of hyphal mitochondria. Within a minute after the introduction of a synthetic strigolactone like GR24 into the rhizosphere, it has been observed that in AMF like *Gigaspora rosea* the concentrations ATP and NADH get significantly elevated (Besserer et al. 2006, 2008).

The significant role of strigolactones in AM fungi symbiosis association was strengthened when a tomato strigolactone biosynthesis mutant was found to have a lower colonization rate as compared to a wild-type plant when only AMF spores were used as inoculum. When a complete inoculum containing fungal spores together with fungal hyphae was introduced into the rhizosphere of these plants, these variations were less prominent (Koltai et al. 2010).

However, seed germination in *Striga* spp. was found to be induced to some degree when treated with root exudates of the mycorrhitic plant when compared with root exudates of the non-mycorrhitic plant (Fernández-Aparicio et al. 2010; Lendzemo et al. 2009). In addition, Lopez-Raez and his group (2011) found that strigolactone synthesis in the roots of mycorrhized tomato plants was substantially reduced. As a result, strigolactones could be negatively influenced by AMF through a feedback mechanism (Fernández et al. 2019). Colonization by Arbuscular mycorrhizal fungi can have a major effect on AM symbiosis enhancement and, as a result, procurement of phosphate increases, resulting in increased phosphate content, which further suppresses the strigolactone biosynthesis. Limiting strigolactone synthesis would have a detrimental impact on AM colonization, combining strigolactone agonists with AM fungus spores immediately before crop sowing might hasten AM colonization and ultimately result in increased tolerance to stress. Another means of improving resistance to various stresses is by SAR (systemic acquired resistance) initiation, which would be augmented by SAR-inducing agents like 1,2-benzisothiazol-3(2*H*)-one1,1-dioxide (BIT) (Yoneyama et al. 2019).

It is still unknown strigolactones play any part in the morphogenesis of the fungus in the cortical cells of the host or during the symbiosis periods. Furthermore, whether strigolactones are needed for AMF interaction is unknown and could be explored by chemotaxis studies to disclose possible strigolactone receptors in the fungus (Carvalhais et al. 2019).

3.4.2 In Nodulation

On the roots of legumes plants like alfalfa, pea, etc., *Rhizobium* spp., a genus of nitrogen-fixing bacteria, play a significant role in nodule formation and beneficial symbiotic associations. The symbiotic relationship entails the exchange of signals between the host plant and *Rhizobium*, which contributes to mutual recognition and result in nodulation. In general, the microbe detects flavonoids released by the root of the host plant during the preinfection period. These flavonoid compounds are unique to each legume-rhizobium relationship, cause the bacteria to produce and secrete lipochitooligosaccharides known as Nod factors, which the host plant recognizes. In certain instances, Nod-factor perception triggers a series of signalling activities, including root hair distortion and the development of infection threads, which requires alteration of the cell division orientation in the root cortex. This contributes to the development of nodule organs (Fig. 3.6). Many rhizobium bacteria produce Nod factors and exopolysaccharides, which are responsible for the formation of an infection thread through which the bacteria penetrate and reproduce. Bacterial cells in the intracellular space then divide into bacteroides (the nitrogen-fixing stage of the symbiosis). During the process of nodulation, plants and bacteria undergo various morphological and cellular changes. However, many of the molecular pathways by which they participate in the differentiation phase remain unknown. Rhizobia bacteroides supply fixed nitrogen to the plant in the form of ammonia which the plant utilizes for making amino acids and nucleotides, and in turn, it obtains organic acids like malate and succinate as an energy source in the nodule meristem after cell differentiation (Kapulnik and Koltai 2014; Markmann and Parniske 2009).

Foo and Davies (2011) reported that a pea *rms1* mutant (a strigolactone-deficient mutant) which lacks the ability to create a strigolactone ring, was able to inhibit nodulation by 40% as compared to the wild-type plants, which possess strigolactones. This finding demonstrated that endogenous strigolactones are promotors of nodulation in plants. When the strigolactone mimic GR24 was applied to the *rms1* mutant plant, the number of nodules on the root of the mutant plant

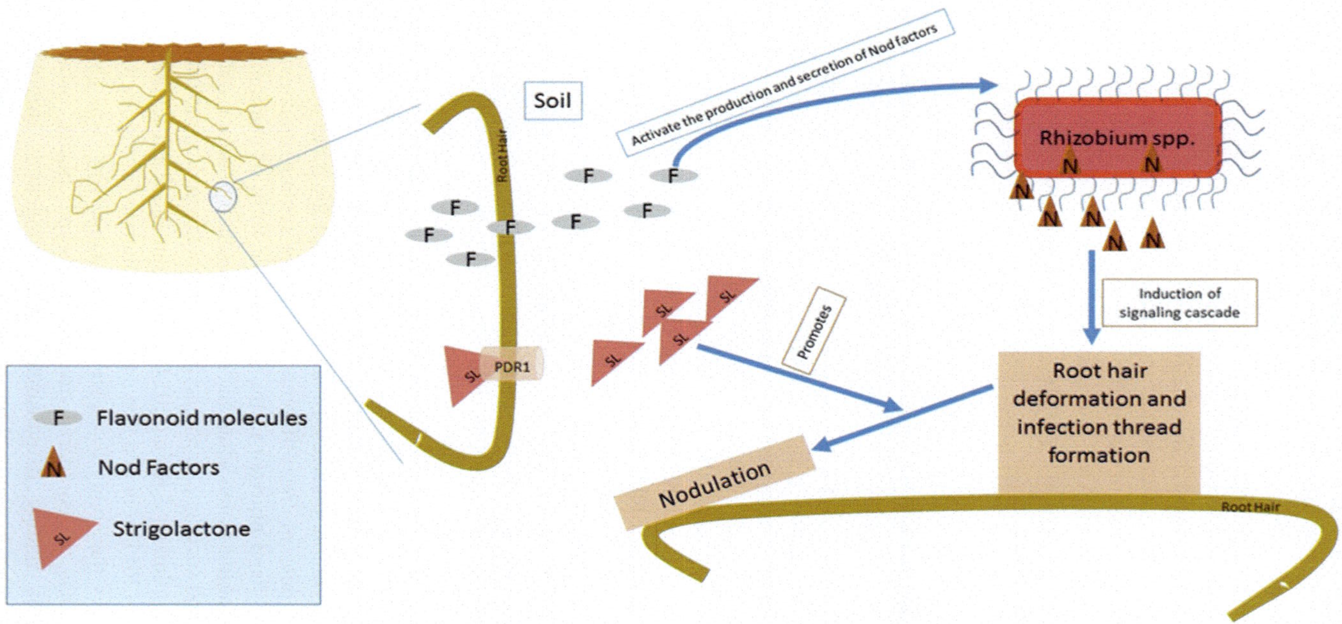

Fig. 3.6 A model for the symbiotic relationship with *Rhizobia* spp. Plant roots secrete the bacterial-specific flavonoids which are sensed by nodulating microbes and starts the secretion of Nod factors, which induces a signalling cascade resulting in infection thread formation. Thereafter SLs promote the formation of nodules

matched that of the wild type (without GR24). The use of GR24 also increases the number of nodules in wild-type pea, *Lotus japonicus*, etc. Genetic research revealed that strigolactones may operate much earlier during nodule development instead of during nodule organogenesis. However, strigolactones were found to have no effect on nodulation either by direct action on the *Rhizobium* (Soto et al. 2010) or the Ca^{2+} signalling which occurs after flavonoids sensing (Moscatiello et al. 2010). It has been proposed that SLs are not required for the production of a functional nodule, rather they influence nodule quantity (Foo et al. 2014). The pathway controlling strigolactone-induced nodulation is currently not well explored. One possible theory is that strigolactones interfere with auxin transport during the cell division phase which contributes to nodule formation (Kapulnik and Koltai 2014).

3.5 Strigolactones as Quorum-Sensing Signal Molecule

Strigolactones are produced by plants and used by them as a signal molecule to interact with other species and to react to an external stimulant. Arbuscular mycorrhizal fungi, lichens, and *Rhizobia* are perhaps the most conspicuous responders to this hormone and so develop a symbiotic or parasitic relationship with the host plant. Strigolactones have a comparable mode of operation to that of quorum-sensing signal molecules (Kalia et al. 2021). In contrast to the autoinducers usually thought of as quorum-sensing signal molecules, Strigolactone-mediated communication involves interactions between distinct species. To designate strigolactone as a true quorum-sensing signal molecule, it should fulfill some more criteria like (a) should be constantly excreted into the rhizosphere (Kretzschmar et al. 2012); (b) should be found outside the cell and can be recognized (Wang and Bouwmeester 2018); (c) should have a targeted response rather than a generalized stress response (Al-Babili and Bouwmeester 2015); (d) should be responsive at physiological levels (Kretzschmar et al. 2012); (e) should benefit the community in terms of overall fitness; (f) should have a strong correlation between the concentration of quorum-sensing signal molecule and density of cells; and (g) should have a threshold concentration level to activate the quorum-sensing signal molecule response. To some extent some of the criteria to designate strigolactone as a quorum-sensing signal molecule are fulfilled. Since our knowledge is limited, further study of strigolactone is needed to decipher all the criteria to conclusively identify the strigolactones as quorum-sensing signal molecules (Aquino et al. 2021).

3.6 Potential Agronomical Application of Strigolactones

Following World War II, the "Green Revolution" was accompanied by unsustainable soil degradation and the overuse and exploitation of various agrochemicals like chemical fertilizers, insecticides, fungicides, and different types of chemical weedkillers. These days, as civic awareness grows regarding the adverse consequences of these chemical substances, there is growing interest in developing more environmentally sustainable agricultural practices. AMF symbiotic relationship typically benefits the host plant's growth and development by stimulating water and mineral nutrient acquisition, especially during times of stress. However, its adverse effects are also being reported, particularly in crop plants (Grace et al. 2009). Additionally, AMF is widespread and is able to colonize a number of plant species. Undoubtedly, AMF is often used in agricultural practices as biofertilizers to boost plant growth and development and hence enhancing its productivity (Barea et al. 2005; Gianinazzi et al. 2010; Duhamel and Vandenkoornhuyse 2013). Since AM symbiosis enhances the plant's resistance against various environmental as well as pathogenic stresses apart from enhancing the plant nutrient uptake, it could be used as a biocontrol tool toward various environmental and biotic stresses.

Strigolactones are required for establishing AMF symbiosis (Foo et al. 2013; Kohlen et al. 2012). As a result, selecting cultivars with elevated SL output may be used as a tool for enhancing mycorrhizal colonization under agronomic conditions. This may also be accomplished by exogenous administration of natural strigolactones or their synthetic derivatives. Stress factors such as nutritional scarcity, drought, and salinity, on the other side, have been shown to affect SL biosynthesis. Another approach to promote AM symbiosis is to add regulated stress conditions and controlling various defense responsive hormones like methyl jasmonate that do not have a significant negative impact on the plant (Lahari et al. 2019). Moreover, it's important to take into consideration that strigolactones are often germination agonists for various root parasitic plants and are engaged in a wide range of biological functions inside the plant. Furthermore, all species of flora produce a unique combination of SLs, which can vary depending on the stage of development and environmental factors (Ćavar et al. 2014). Indeed, little is known regarding their uniqueness in different species. As a consequence, a greater knowledge of their structure-function relationship, as well as its mode of action, is needed prior to their utilization. Some advancements have been made, with evidence of the impact of structural variations among strigolactones on AMF symbiosis, root and shoot architecture, and parasitic weed seed germination (Boyer et al. 2012, 2014). Kohlen and colleagues discovered that several strigolactones are exudated mostly in the rhizosphere, whereas the remaining are loaded and transferred to the shoot via xylem. The identification of SLs that are theoretically exclusive to the host plant-AM fungus relationship would undoubtedly aid in the better implementation of AM symbiosis in the agroecosystems.

Strigolactones have the capacity to be very useful in farming. The structures of these natural strigolactones, though, are very complicated so it is infeasible to scale up its industrial production. However, a large variety of strigolactone analogs and mimics are available which have a much simpler structure as compared to their natural counterparts but have the same biobehavior and bioactivity (Zwanenburg and Blanco-Ania 2018). Since the structure of these analogs and mimics are simple, hey can be synthesized easily, resulting in scaling up their production and reducing production cost. These analogs and mimics can be then used in agriculture practices with a wide range of application range from their use as hormones to alter and/or monitor plant architecture, to their use as biostimulants to stimulate seed germination of parasitic weeds to act as a biocontrol agent, to their use as "biostimulants" of plant root colonization by AMF, enhancing plant nutrition, and have yet unknown effects on soil microbial populations (Vurro et al. 2016).

3.7 Conclusion

Strigolactones are plant secondary metabolites produced from carotenoids and are active in both endogenous and exogenous signalling responses. After their classification as phytohormones, scientists have made significant strides in understanding their signal transduction pathways along with finding the key components which are involved in its signalling cascade in both monocots and dicot plants. Several facets of SL signalling cascades have been discovered to be peculiar among plant hormones and there is a necessity for further study to fully comprehend their mechanism. Exogenously, strigolactones were found to stimulate AMF symbiosis, leguminous plant nodulation, and parasitic weed seed germination. It has been also reported that it also controls the architecture of the shoots and roots, secondary development, apoptosis, and fruit ripening on an endogenous level. The endogenous as well as exogenous signalling cascade is influenced by environmental factors such as illumination, temperature, food supply, and abiotic and biotic stress.

Acknowledgments The authors acknowledge the Director, CSIR-Central Institute of Medicinal & Aromatic Plants, Lucknow, India, for providing the necessary facilities. SK acknowledges UGC for fellowship and AJ acknowledges CSIR for fellowship.

References

Abe S, Sado A, Tanaka K, Kisugi T, Asami K, Ota S, Kim HI, Yoneyama K, Xie X, Ohnishi T, Seto Y (2014) Carlactone is converted to carlactonoic acid by MAX1 in Arabidopsis and its methyl ester can directly interact with AtD14 in vitro. Proc Natl Acad Sci U S A 111(50):18084–18089. https://doi.org/10.1073/pnas.1410801111

Ablazov A, Mi J, Jamil M, Jia KP, Wang JY, Feng Q, Al-Babili S (2020) The apocarotenoid zaxinone is a positive regulator of strigolactone and abscisic acid biosynthesis in Arabidopsis roots. Front Plant Sci 11:578. https://doi.org/10.3389/fpls.2020.00578

Akiyama K (2007) Chemical identification and functional analysis of apocarotenoids involved in the development of arbuscular mycorrhizal symbiosis. Biosci Biotechnol Biochem 2007: 0705070393–0705070393. https://doi.org/10.1271/bbb.70023

Akiyama K, Hayashi H (2006) Strigolactones: chemical signals for fungal symbionts and parasitic weeds in plant roots. Ann Bot (Lond) 97:925–931. https://doi.org/10.1093/aob/mcl063

Akiyama K, Matsuzaki K, Hayashi H (2005) Plant sesquiterpenes induce hyphal branching in arbuscular mycorrhizal fungi. Nature 435:824–827. https://doi.org/10.1038/nature03608

Al-Babili S, Bouwmeester HJ (2015) Strigolactones, a novel carotenoid-derived plant hormone. Annu Rev Plant Biol 66:161–186. https://doi.org/10.1146/annurev-arplant-043014-114759

Alder A, Holdermann I, Beyer P, Al-Babili S (2008) Carotenoid oxygenases involved in plant branching catalyse a highly specific conserved apocarotenoid cleavage reaction. Biochem J 416 (2):289–296. https://doi.org/10.1042/BJ20080568

Alder A, Jamil M, Marzorati M, Bruno M, Vermathen M, Bigler P, Ghisla S, Bouwmeester H, Beyer P, Al-Babili S (2012) The path from β-carotene to carlactone, a strigolactone-like plant hormone. Science 335(6074):1348–1351. https://doi.org/10.1126/science.1218094

Aquino B, Bradley JM, Lumba S (2021) On the outside looking in: roles of endogenous and exogenous strigolactones. Plant J 105(2):322–334. https://doi.org/10.1111/tpj.15087

Arite T, Umehara M, Ishikawa S, Hanada A, Maekawa M, Yamaguchi S, Kyozuka J (2009) d14, a strigolactone-insensitive mutant of rice, shows an accelerated outgrowth of tillers. Plant Cell Physiol 50(8):1416–1424. https://doi.org/10.1093/pcp/pcp091

Awad AA, Sato D, Kusumoto D, Kamioka H, Takeuchi Y, Yoneyama K (2006) Characterization of strigolactones, germination stimulants for the root parasitic plants Striga and Orobanche, produced by maize, millet and sorghum. Plant Growth Regul 48:221–227. https://doi.org/10.1007/s10725-006-0009-3

Barea JM, Pozo MJ, Azcon R, Azcon-Aguilar C (2005) Microbial co-operation in the rhizosphere. J Exp Bot 56:1761–1778. https://doi.org/10.1093/jxb/eri197

Bennett T, Leyser O (2014) Strigolactone signalling: standing on the shoulders of DWARFs. Curr Opin Plant Biol 22:7–13. https://doi.org/10.1016/j.pbi.2014.08.001

Bennett T, Hines G, Leyser O (2014) Canalization: what the flux? Trends Genet 30:41–48. https://doi.org/10.1016/j.tig.2013.11.001

Bennett T, Liang Y, Seale M, Ward S, Müller D, Leyser O (2016) Strigolactone regulates shoot development through a core signalling pathway. Biol Open 5(12):1806–1820. https://doi.org/10.1242/bio.021402

Besserer A, Puech-Pages V, Kiefer P, Gomez-Roldan V, Jauneau A, Roy S, Portais JC, Roux C, Becard G, Sejalon-Delmas N (2006) Strigolactones stimulate arbuscular mycorrhizal fungi by activating mitochondria. PLoS Biol 4:1239–1247. https://doi.org/10.1371/journal.pbio.0040226

Besserer A, Bécard G, Jauneau A, Roux C, Séjalon-Delmas N (2008) GR24, a synthetic analog of strigolactones, stimulates the mitosis and growth of the arbuscular mycorrhizal fungus *Gigaspora rosea* by boosting its energy metabolism. Plant Physiol 148:402–413. https://doi.org/10.1104/pp.108.121400

Bonneau L, Huguet S, Wipf D, Pauly N, Truong HN (2013) Combined phosphate and nitrogen limitation generates a nutrient stress transcriptome favorable for arbuscular mycorrhizal symbiosis in Medicago truncatula. New Phytol 199(1):188–202. https://doi.org/10.1111/nph.12234

Bouwmeester HJ, Roux C, Lopez-Raez JA, Becard G (2007) Rhizosphere communication of plants, parasitic plants and AM fungi. Trends Plant Sci 12:224–230. https://doi.org/10.1016/j.tplants.2007.03.009

Boyer FD, Germain ADS, Pillot J-P, Pouvreau J-B, Chen VX, Ramos S, Stevenin A, Simier P, Delavault P, Beau J-M, Rameau C (2012) Structure-activity relationship studies of strigolactonerelated molecules for branching inhibition in garden pea: molecule design for shoot branching. Plant Physiol 159:1524–1544. https://doi.org/10.1104/pp.112.195826

Boyer FD, De Saint GA, Pouvreau JB, Clavé G, Pillot JP, Roux A, Rasmussen A, Depuydt S, Lauressergues D, Frei Dit Frey N, Heugebaert TSA, Stevens CV, Geelen D, Goormachtig S, Rameau C (2014) New strigolactone analogs as plant hormones with low activities in the rhizosphere. Mol Plant 7:675–690. https://doi.org/10.1093/mp/sst163

Bravo A, York T, Pumplin N, Mueller LA, Harrison MJ (2016) Genes conserved for arbuscular mycorrhizal symbiosis identified through phylogenomics. Nat Plants 2:15208. https://doi.org/10.1038/nplants.2015.208

Breuillin F, Schramm J, Hajirezaei M, Ahkami A, Favre P, Druege U, Hause B, Bucher M, Kretzschmar T, Bossolini E, Kuhlemeier C (2010) Phosphate systemically inhibits development of arbuscular mycorrhiza in Petunia hybrida and represses genes involved in mycorrhizal functioning. Plant J 64(6):1002–1017. https://doi.org/10.1111/j.1365-313X.2010.04385.x

Brewer PB, Dun EA, Ferguson BJ, Rameau C, Beveridge CA (2009) Strigolactone acts downstream of auxin to regulate bud outgrowth in pea and Arabidopsis. Plant Physiol 150:482–493. https://doi.org/10.1104/pp.108.134783

Brewer PB, Dun EA, Gui R, Mason MG, Beveridge CA (2015) Strigolactone inhibition of branching independent of polar auxin transport. Plant Physiol 168:1820–1829. https://doi.org/10.1104/pp.15.00014

Bruno M, Koschmieder J, Wuest F, Schaub P, Fehling-Kaschek M, Timmer J, Beyer P, Al-Babili S (2016) Enzymatic study on AtCCD4 and AtCCD7 and their potential to form acyclic regulatory metabolites. J Exp Bot 67(21):5993–6005. https://doi.org/10.1093/jxb/erw356

Bruno M, Vermathen M, Alder A, Wüst F, Schaub P, van der Steen R, Beyer P, Ghisla S, Al-Babili S (2017) Insights into the formation of carlactone from in-depth analysis of the CCD8-catalyzed reactions. FEBS Lett 591(5):792–800. https://doi.org/10.1002/1873-3468.12593

Carvalhais LC, Rincon-Florez VA, Brewer PB, Beveridge CA, Dennis PG, Schenk PM (2019) The ability of plants to produce strigolactones affects rhizosphere community composition of fungi but not bacteria. Rhizosphere 9:18–26. https://doi.org/10.1016/j.rhisph.2018.10.002

Ćavar S, Zwanenburg B, Tarkowski P (2014) Strigolactones: occurrence, structure, and biological activity in the rhizosphere. Phytochem Rev 14:691–711. https://doi.org/10.1007/s11101-014-9370-4

Changenet V, Macadré C, Boutet-Mercey S, Magne K, Januario M, Dalmais M, Bendahmane A, Mouille G, Dufresne M (2021) Overexpression of a cytochrome P450 monooxygenase involved in orobanchol biosynthesis increases susceptibility to Fusarium head blight. Front Plant Sci 12: 546. https://doi.org/10.3389/fpls.2021.662025

Charnikhova TV, Gaus K, Lumbroso A, Sanders M, Vincken JP, De Mesmaeker A, Ruyter-Spira CP, Screpanti C, Bouwmeester HJ (2018) Zeapyranolactone – a novel strigolactone from maize. Phytochem Lett 24:172–178. https://doi.org/10.1016/j.phytol.2018.01.003

Cheng X, Ruyter-Spira C, Bouwmeester H (2013) The interaction between strigolactones and other plant hormones in the regulation of plant development. Front Plant Sci 4:199. https://doi.org/10.3389/fpls.2013.00199

Chevalier F, Nieminen K, Sanchez-Ferrero JC, Rodnguez ML, Chagoyen M (2014) Strigolactone promotes degradation of DWARF14, an α/βhydrolase essential for strigolactone signaling in *Arabidopsis*. Plant Cell 26:1134–1150. https://doi.org/10.1105/tpc.114.122903

Cook CE, Whichard LP, Turner B, Wall ME, Egley GH (1966) Germination of witchweed (*Striga lutea* Lour.): isolation and properties of a potent stimulant. Science 154:1189–1190. https://doi.org/10.1126/science.154.3753.1189

Cook CE, Whichard LP, Wall ME, Egley GH, Coggon P, Luhan PA (1972) Germination stimulants. II. Structure of strigol, a potent seed germination stimulant for witchweed (Striga lutea). J Am Chem Soc 94:6198–6199. https://10.1021/ja00772a048

Crawford S, Shinohara N, Sieberer T, Williamson L, George G (2010) Strigolactones enhance competition between shoot branches by dampening auxin transport. Development 137:2905–2913. https://doi.org/10.1242/dev.051987

De Cuyper C, Goormachtig S (2017) Strigolactones in the rhizosphere: friend or foe? Mol Plant-Microbe Interact 30:683–690. https://doi.org/10.1094/MPMI-02-17-0051-CR

Decker EL, Alder A, Hunn S, Ferguson J, Lehtonen MT, Scheler B, Kerres KL, Wiedemann G, Safavi-Rizi V, Nordzieke S, Balakrishna A, Baz L, Avalos J, Valkonen JPT, Reski R, Al-Babili S (2017) Strigolactone biosynthesis is evolutionarily conserved, regulated by phosphate starvation and contributes to resistance against phytopathogenic fungi in a moss. Physcomitrella patens New Phytol 216(2):455–468. https://doi.org/10.1111/nph.14506

Domagalska MA, Leyser O (2011) Signal integration in the control of shoot branching. Nat Rev Mol Cell Biol 12:211–221. https://doi.org/10.1038/nrm3088

Douds DD Jr, Pfeffer PE, Shachar-Hill Y (2000) Carbon partitioning, cost, and metabolism of arbuscular mycorrhizas. In: Kapulnik Y, Douds DD Jr (eds) Arbuscular mycorrhizas: physiology and function. Kluwer Academic Publishers, Dordrecht, The Netherlands, pp 107–130. https://doi.org/10.1007/978-94-017-0776-3_6

Duhamel M, Vandenkoornhuyse P (2013) Sustainable agriculture: possible trajectories from mutualistic symbiosis and plant neodomestication. Trends Plant Sci 18:597–600. https://doi.org/10.1016/j.tplants.2013.08.010

Dun EA, de Saint Germain A, Rameau C, Beveridge CA (2012) Antagonistic action of strigolactone and cytokinin in bud outgrowth control. Plant Physiol 158:487–498. https://doi.org/10.1104/pp.111.186783

Fernández I, Cosme M, Stringlis IA, Yu K, de Jonge R, van Wees S, Pozo MJ, Pieterse CMJ, van der Heijden MGA (2019) Molecular dialogue between arbuscular mycorrhizal fungi and the nonhost plant *Arabidopsis thaliana* switches from initial detection to antagonism. New Phytol 223:867–881. https://doi.org/10.1111/nph.15798

Fernández-Aparicio M, García-Garrido JM, Ocampo JA, Rubiales D (2010) Colonisation of field pea roots by arbuscular mycorrhizal fungi reduces *Orobanche* and *Phelipanche* species seed germination. Weed Res 50:262–268. https://doi.org/10.1111/j.1365-3180.2010.00771.x

Ferrero M, Pagliarani C, Novák O, Ferrandino A, Cardinale F, Visentin I, Schubert A (2018) Exogenous strigolactone interacts with abscisic acid-mediated accumulation of anthocyanins in grapevine berries. J Exp Bot 69(9):2391–2401. https://doi.org/10.1093/jxb/ery033

Flematti GR, Ghisalberti EL, Dixon KW, Trengove RD (2004) A compound from smoke that promotes seed germination. Science 305(5686):977. https://doi.org/10.1126/science.1099944

Flematti GR, Scaffidi A, Waters MT, Smith SM (2016) Stereospecificity in strigolactone biosynthesis and perception. Planta 243(6):1361–1373. https://doi.org/10.1007/s00425-016-2523-5

Foo E, Davies NW (2011) Strigolactones promote nodulation in pea. Planta 234:1073–1081. https://doi.org/10.1007/s00425-011-1516-7

Foo E, Yoneyama K, Hugill CJ, Quittenden LJ, Reid JB (2013) Strigolactones and the regulation of pea symbioses in response to nitrate and phosphate deficiency. Mol Plant 6:76–87. https://doi.org/10.1093/mp/sss115

Foo E, Ferguson BJ, Reid JB (2014) The potential roles of strigolactones and brassinosteroids in the autoregulation of nodulation pathway. Ann Bot (Lond) 113:1037–1045. https://doi.org/10.1093/aob/mcu030

Germain ADS, Ligerot Y, Dun EA, Pillot JP, Ross JJ, Beveridge CA, Rameau C (2013) Strigolactones stimulate internode elongation independently of gibberellins. Plant Physiol 163(2):1012–1025. https://doi.org/10.1104/pp.113.220541

Gianinazzi S, Gollotte A, Binet MN, van Tuinen D, Redecker D, Wipf D (2010) Agroecology: the key role of arbuscular mycorrhizas in ecosystem services. Mycorrhiza 20:519–530. https://doi.org/10.1007/s00572-010-0333-3

Goldwasser Y, Yoneyama K, Xie X, Yoneyama K (2008) Production of strigolactones by Arabidopsis thaliana responsible for Orobanche aegyptiaca seed germination. Plant Growth Regul 55:21–28. https://doi.org/10.1007/s10725-008-9253-z

Gomez-Roldan V, Fermas S, Brewer PB, Puech-Pagès V, Dun EA, Pillot JP, Letisse F, Matusova R, Danoun S, Portais JC, Bouwmeester H (2008) Strigolactone inhibition of shoot branching. Nature 455(7210):189–194. https://doi.org/10.1038/nature07271

Grace EJ, Cotsaftis O, Tester M, Smith FA, Smith SE (2009) Arbuscular mycorrhizal inhibition of growth in barley cannot be attributed to extent of colonization, fungal phosphorus uptake or effects on expression of plant phosphate transporter genes. New Phytol 181:938–949. https://doi.org/10.1111/j.1469-8137.2008.02720.x

Guo Y, Zheng Z, La Clair JJ, Chory J, Noel JP (2013) Smoke-derived karrikin perception by the α/β-hydrolase KAI2 from Arabidopsis. Proc Natl Acad Sci U S A 110(20):8284–8289. https://doi.org/10.1073/pnas.1306265110

Hamiaux C, Drummond RS, Janssen BJ, Ledger SE, Cooney JM, Newcomb RD, Snowden KC (2012) DAD2 is an α/β hydrolase likely to be involved in the perception of the plant branching hormone, strigolactone. Curr Biol 22(21):2032–2036. https://doi.org/10.1016/j.cub.2012.08.007

Harrison PJ, Newgas SA, Descombes F, Shepherd SA, Thompson AJ, Bugg TD (2015) Biochemical characterization and selective inhibition of β-carotene cis–trans isomerase D27 and carotenoid cleavage dioxygenase CCD 8 on the strigolactone biosynthetic pathway. FEBS J 282(20):3986–4000. https://doi.org/10.1111/febs.13400

Hirsch S, Kim J, Muñoz A, Heckmann AB, Downie JA, Oldroyd GE (2009) GRAS proteins form a DNA binding complex to induce gene expression during nodulation signaling in Medicago truncatula. Plant Cell 21(2):545–557. https://doi.org/10.1105/tpc.108.064501

Huang D, Wang Q, Jing G, Ma M, Li C, Ma F (2021) Overexpression of MdIAA24 improves apple drought resistance by positively regulating strigolactone biosynthesis and mycorrhization. Tree Physiol 41(1):134–146. https://doi.org/10.1093/treephys/tpaa109

Ito S, Ito K, Abeta N, Takahashi R, Sasaki Y, Yajima S (2016) Effects of strigolactone signaling on Arabidopsis growth under nitrogen deficient stress condition. Plant Signal Behav 11(1): e1126031. https://doi.org/10.1080/15592324.2015.1126031

Jamil M, Charnikhova T, Cardoso C, Jamil T, Ueno K, Verstappen F, Asami T, Bouwmeester HJ (2011) Quantification of the relationship between strigolactones and Striga hermonthica infection in rice under varying levels of nitrogen and phosphorus. Weed Res 51:373–385. https://doi.org/10.1111/j.1365-3180.2011.00847.x

Jia K-P, Baz L, Al-Babili S (2018) From carotenoids to strigolactones. J Exp Bot 69(9):2189–2204. https://doi.org/10.1093/jxb/erx476

Jiang L, Liu X, Xiong G, Liu H, Chen F, Wang L, Meng X, Liu G, Yu H, Yuan Y, Yi W (2013) DWARF 53 acts as a repressor of strigolactone signalling in rice. Nature 504(7480):401–405. https://doi.org/10.1038/nature12870

Johnson AW, Gowada G, Hassanali A, Knox J, Monaco S, Razavi Z, Rosebery G (1981) The preparation of synthetic analogues of strigol. J Chem Soc Perkin Trans I:1734–1743. https://doi.org/10.1039/P19810001734

Kagiyama M, Hirano Y, Mori T, Kim SY, Kyozuka J, Seto Y, Yamaguchi S, Hakoshima T (2013) Structures of D14 and D14L in the strigolactone and karrikin signaling pathways. Genes Cells 18(2):147–160. https://doi.org/10.1111/gtc.12025

Kalia VC, Gong C, Patel S, Lee JK (2021) Regulation of plant mineral nutrition by signal molecules. Microorganisms 9(4):774. https://doi.org/10.3390/microorganisms9040774

Kannan C, Zwanenburg B (2014) A novel concept for the control of parasitic weeds by decomposing germination stimulants prior to action. J Crop Prot 61:11–15. https://doi.org/10.1016/j.cropro.2014.03.008

Kapulnik Y, Koltai H (2014) Strigolactone involvement in root development, response to abiotic stress, and interactions with the biotic soil environment. Plant Physiol 166(2):560–569. https://doi.org/10.1104/pp.114.244939

Khosla A, Morffy N, Li Q, Faure L, Chang SH, Yao J, Zheng J, Cai ML, Stanga J, Flematti GR, Waters MT (2020) Structure–function analysis of SMAX1 reveals domains that mediate its karrikin-induced proteolysis and interaction with the receptor KAI2. Plant Cell 32(8): 2639–2659. https://doi.org/10.1105/tpc.19.00752

Kim HI, Kisugi T, Khetkam P, Xie X, Yoneyama K, Uchida K, Yokota T, Nomura T, McErlean CS, Yoneyama K (2014) Avenaol, a germination stimulant for root parasitic plants from Avenastrigosa. Phytochemistry 103:85–88. https://doi.org/10.1016/j.phytochem.2014.03.030

Kisugi T, Xie X, Kim HI, Yoneyama K, Sado A, Akiyama K, Hayashi H, Uchida K, Yokota T, Nomura T, Yoneyama K (2013) Strigone, isolation and identification as a natural strigolactone from Houttuynia cordata. Phytochemistry 87:60–64. https://doi.org/10.1016/j.phytochem.2012.11.013

Kohlen W, Charnikhova T, Lammers M, Pollina T, Toth P, Haider I, Pozo MJ, de Maagd RA, Ruyter-Spira C, Bouwmeester HJ, López-Ráez JA (2012) The tomato CAROTENOID CLEAVAGE DIOXYGENASE8 (SlCCD8) regulates rhizosphere signaling, plant architecture and affects reproductive development through strigolactone biosynthesis. New Phytol 196:535–547. https://doi.org/10.1111/j.1469-8137.2012.04265.x

Koltai H, Kapulnik Y (2011) Strigolactones as mediators of plant growth responses to environmental conditions. Plant Signal Behav 6(1):37–41. https://doi.org/10.4161/psb.6.1.13245

Koltai H, LekKala SP, Bhattacharya C, Mayzlish-Gati E, Resnick N, Wininger S, Dor E, Yoneyama K, Yoneyama K, Hershenhorn J et al (2010) A tomato strigolactone-impaired mutant displays aberrant shoot morphology and plant interactions. J Exp Bot 61:1739–1749. https://doi.org/10.1093/jxb/erq041

Koske RE, Gemma JN (1992) Fungal reactions to plants prior to mycorrhizal formation. In: Allen MF (ed) Mycorrhizal functioning: an integrative plant-fungal process. Chapman & Hall, New York, pp 3–36

Kretzschmar T, Kohlen W, Sasse J, Borghi L, Schlegel M, Bachelier JB, Reinhardt D, Bours R, Bouwmeester HJ, Martinoia E (2012) A petunia ABC protein controls strigolactone-dependent symbiotic signalling and branching. Nature 483(7389):341–344. https://doi.org/10.1038/nature10873

Kumar N, Shema R, Kumar M, Mayzlish-Gati E, Levy D, Zemach H, Belausov E, Wininger S, Abu-Abied M, Kapulnik Y, Koltai H (2014) Strigolactone analog GR24 triggers changes in PIN2 polarity, vesicle trafficking and actin filament architecture. New Phytol 202(4):1184–1196. https://doi.org/10.1111/nph.12744

Lahari Z, Ullah C, Kyndt T, Gershenzon J, Gheysen G (2019) Strigolactones enhance root-knot nematode (*Meloidogyne graminicola*) infection in rice by antagonizing the jasmonate pathway. New Phytol 224:454–465. https://doi.org/10.1111/nph.15953

Lanfranco L, Fiorilli V, Venice F, Bonfante P (2018) Strigolactones cross the kingdoms: plants, fungi, and bacteria in the arbuscular mycorrhizal symbiosis. J Exp Bot 69:2175–2188. https://doi.org/10.1093/jxb/erx432

Lendzemo V, Kuyper TW, Vierheilig H (2009) *Striga* seed-germination activity of root exudates and compounds present in stems of *Striga* host and nonhost (trap crop) plants is reduced due to root colonization by arbuscular mycorrhizal fungi. Mycorrhiza 19:287–294. https://doi.org/10.1007/s00572-009-0235-4

Li S, Li Y, Chen L, Zhang C, Wang F, Li H, Wang M, Wang Y, Nan F, Xie D, Yan J (2021) Strigolactone mimic 2-nitrodebranone is highly active in Arabidopsis growth and development. Plant J. https://doi.org/10.1111/tpj.15274

Liang Y, Ward S, Li P, Bennett T, Leyser O (2016) SMAX1-LIKE7 signals from the nucleus to regulate shoot development in Arabidopsis via partially EAR motif-independent mechanisms. Plant Cell 28(7):1581–1601. https://doi.org/10.1105/tpc.16.00286

Lin H, Wang R, Qian Q, Yan M, Meng X, Fu Z, Yan C, Jiang B, Su Z, Li J, Wang Y (2009) DWARF27, an iron-containing protein required for the biosynthesis of strigolactones, regulates rice tiller bud outgrowth. Plant Cell 21(5):1512–1525. https://doi.org/10.1105/tpc.109.065987

Liu W, Kohlen W, Lillo A, den Camp RO, Ivanov S, Hartog M, Limpens E, Jamil M, Smaczniak C, Kaufmann K, Yang WC (2011) Strigolactone biosynthesis in Medicago truncatula and rice requires the symbiotic GRAS-type transcription factors NSP1 and NSP2. Plant Cell 23(10): 3853–3865. https://doi.org/10.1105/tpc.111.089771

Liu J, He H, Vitali M, Visentin I, Charnikhova T, Haider I, Schubert A, Ruyter-Spira C, Bouwmeester HJ, Lovisolo C, Cardinale F (2015) Osmotic stress represses strigolactone biosynthesis in Lotus japonicus roots: exploring the interaction between strigolactones and ABA under abiotic stress. Planta 41(6):1435–1451. https://doi.org/10.1007/s00425-015-2266-8

Lopez-Obando M, Conn CE, Hoffmann B, Bythell-Douglas R, Nelson DC, Rameau C, Bonhomme S (2016) Structural modelling and transcriptional responses highlight a clade of PpKAI2-LIKE genes as candidate receptors for strigolactones in Physcomitrella patens. Planta 243(6): 1441–1453. https://doi.org/10.1007/s00425-016-2481-y

Lopez-Obando M, de Villiers R, Hoffmann B, Ma L, de Saint Germain A, Kossmann J, Coudert Y, Harrison CJ, Rameau C, Hills P, Bonhomme S (2018) Physcomitrella patens MAX2 characterization suggests an ancient role for this F-box protein in photomorphogenesis rather than strigolactone signalling. New Phytol 219(2):743–756. https://doi.org/10.1111/nph.15214

López-Ráez JA, Kohlen W, Charnikhova T, Mulder P, Undas AK, Sergeant MJ, Verstappen F, Bugg TD, Thompson AJ, Ruyter-Spira C, Bouwmeester H (2010) Does abscisic acid affect strigolactone biosynthesis? New Phytol 187(2):343–354. https://doi.org/10.1111/j.1469-8137.2010.03291.x

López-Ráez JA, Charnikhova T, Fernández I, Bouwmeester H, Pozo MJ (2011) Arbuscular mycorrhizal symbiosis decreases strigolactone production in tomato. J Plant Physiol 168:294–297. https://doi.org/10.1016/j.jplph.2010.08.011

Machin DC, Hamon-Josse M, Bennett T (2020) Fellowship of the rings: a saga of strigolactones and other small signals. New Phytol 225(2):621–636. https://doi.org/10.1111/nph.16135

Mangnus EM, Zwanenburg B (1992) Tentative molecular mechanism for germination stimulation of Striga and Orobanche seeds by strigol and its synthetic analogs. J Agric Food Chem 40(6): 1066–1070. https://doi.org/10.1021/jf00018a032

Markmann K, Parniske M (2009) Evolution of root endosymbiosis with bacteria: how novel are nodules? Trends Plant Sci 14:77–86. https://doi.org/10.1016/j.tplants.2008.11.009

Marzec M, Muszynska A (2015) In silico analysis of the genes encoding proteins that are involved in the biosynthesis of the RMS/MAX/D pathway revealed new roles of strigolactones in plants. Int J Mol Sci 16(4):6757–6782. https://doi.org/10.3390/ijms16046757

Marzec M, Muszynska A, Gruszka D (2013) The role of strigolactones in nutrient-stress responses in plants. Int J Mol Sci 14(5):9286–9304. https://doi.org/10.3390/ijms14059286

Matusova R, Rani K, Verstappen FW, Franssen MC, Beale MH, Bouwmeester HJ (2005) The strigolactone germination stimulants of the plant-parasitic Striga and Orobanche spp. are derived from the carotenoid pathway. Plant Physiol 139:920–934. https://doi.org/10.1104/pp.105.061382

Mishra S, Upadhyay S, Shukla RK (2017) The role of strigolactones and their potential cross-talk under hostile ecological conditions in plants. Front Physiol 7:691. https://doi.org/10.3389/fphys.2016.00691

Mizuno Y, Komatsu A, Shimazaki S, Naramoto S, Inoue K, Xie X, Ishizaki K, Kohchi T, Kyozuka J (2021) Major components of the KARRIKIN INSENSITIVE2-dependent signaling pathway are conserved in the liverwort *Marchantia polymorpha*. Plant Cell. https://doi.org/10.1093/plcell/koab106

Moon J, Parry G, Estelle M (2004) The ubiquitin-proteasome pathway and plant development. Plant Cell 16(12):3181–3195. https://doi.org/10.1105/tpc.104.161220

Mori N, Nishiuma K, Sugiyama T, Hayashi H, Akiyama K (2016) Carlactone-type strigolactones and their synthetic analogues as inducers of hyphal branching in arbuscular mycorrhizal fungi. Phytochemistry 130:90–98. https://doi.org/10.1016/j.phytochem.2016.05.012

Moscatiello R, Squartini A, Mariani P, Navazio L (2010) Flavonoid induced calcium signalling in *Rhizobium leguminosarum* bv. viciae. New Phytol 188:814–823. https://doi.org/10.1111/j.1469-8137.2010.03411.x

Mostofa MG, Li W, Nguyen KH, Fujita M, Tran L-SP (2018) Strigolactones in plant adaptation to abiotic stresses: an emerging avenue of plant research. Plant Cell Environ 41:2227–2243. https://doi.org/10.1111/pce.13364

Nakamura H, Xue YL, Miyakawa T, Hou F, Qin HM, Fukui K, Shi X, Ito E, Ito S, Park SH, Miyauchi Y (2013) Molecular mechanism of strigolactone perception by DWARF14. Nat Commun 4(1):1–10. https://doi.org/10.1038/ncomms3613

Nefkens GH, Thuring JW, Beenakkers MF, Zwanenburg B (1997) Synthesis of a phthaloylglycine-derived strigol analogue and its germination stimulatory activity toward seeds of the parasitic weeds Striga hermonthica and Orobanche crenata. J Agric Food Chem 45(6):2273–2277. https://doi.org/10.1021/jf9604504

Nelson DC, Scaffidi A, Dun EA, Waters MT, Flematti GR, Dixon KW, Beveridge CA, Ghisalberti EL, Smith SM (2011) F-box protein MAX2 has dual roles in karrikin and strigolactone signaling in Arabidopsis thaliana. Proc Natl Acad Sci U S A 108(21):8897–8902. https://doi.org/10.1073/pnas.1100987108

Niesen FH, Berglund H, Vedadi M (2007) The use of differential scanning fluorimetry to detect ligand interactions that promote protein stability. Nat Protoc 2(9):2212. https://doi.org/10.1038/nprot.2007.321

Okazaki K, Watanabe S, Koike I, Kawada K, Ito S, Nakamura H, Asami T, Shimomura K, Umehara M (2021) Strigolactone signaling inhibition increases adventitious shoot formation on internodal segments of ipecac. Planta 253(6):1–14. https://doi.org/10.1007/s00425-021-03640-1

Prusinkiewicz P, Crawford S, Smith RS, Ljung K, Bennett T (2009) Control of bud activation by an auxin transport switch. Proc Natl Acad Sci U S A 106:17431–17436. https://doi.org/10.1073/pnas.0906696106

Rani K, Zwanenburg B, Sugimoto Y, Yoneyama K, Bouwmeester HJ (2008) Biosynthetic considerations could assist the structure elucidation of host plant produced rhizosphere signalling compounds (strigolactones) for arbuscular mycorrhizal fungi and parasitic plants. Plant Physiol Biochem 46(7):617–626. https://doi.org/10.1016/j.plaphy.2008.04.012

Rausch C, Bucher M (2002) Molecular mechanisms of phosphate transport in plants. Planta 216: 23–37. https://doi.org/10.1007/s00425-002-0921-3

Ravazzolo L, Boutet-Mercey S, Perreau F, Forestan C, Varotto S, Ruperti B, Quaggiotti S (2021) Strigolactones and auxin cooperate to regulate maize root development and response to nitrate. Plant Cell Physiol. https://doi.org/10.1093/pcp/pcab014

Redecker D, Raab P (2006) Phylogeny of the glomeromycota (arbuscular mycorrhizal fungi): recent developments and new gene markers. Mycologia 98:885–895. https://doi.org/10.3852/mycologia.98.6.885

Reizelman A, Scheren M, Nefkens G, Zwanenburg B (2000) Synthesis of all eight stereoisomers of the germination stimulant Strigol. Synthesis-Stuttgart 2000:1944–1951. https://doi.org/10.1055/s-2000-8230

Requena N, Serrano E, Ocon A, Breuninger M (2007) Plant signals and fungal perception during arbuscular mycorrhiza establishment. Phytochemistry 68:33–40. https://doi.org/10.1016/j.phytochem.2006.09.036

Ruiz-Lozano JM, Aroca R, Zamarreño ÁM, Molina S, Andreo-Jiménez B, Porcel R (2016) Arbuscular mycorrhizal symbiosis induces strigolactone biosynthesis under drought and improves drought tolerance in lettuce and tomato. Plant Cell Environ 39(2):441–452. https://doi.org/10.1111/pce.12631

Ruyter-Spira C, Kohlen W, Charnikhova T, van Zeijl A, van Bezouwen L, de Ruijter N, Cardoso C, Lopez-Raez JA, Matusova R, Bours R, Verstappen F, Bouwmeester H (2011) Physiological effects of the synthetic strigolactone analog GR24 on root system architecture in Arabidopsis: another belowground role for strigolactones? Plant Physiol 155(2):721–734. https://doi.org/10.1104/pp.110.166645

Samejima H, Babiker AG, Takikawa H, Sasaki M, Sugimoto Y (2016) Practicality of the suicidal germination approach for controlling Striga hermonthica. Pest Manag Sci 72(11):2035–2042. https://doi.org/10.1002/ps.4215

Sato D, Awad AA, Chae SH, Yokota T, Sugimoto Y, Takeuchi Y, Yoneyama K (2003) Analysis of strigolactones, germination stimulants for Striga and Orobanche, by high-performance liquid chromatography/tandem mass spectrometry. J Agric Food Chem 51(5):1162–1168. https://doi.org/10.1021/jf025997z

Scaffidi A, Waters MT, Sun YK, Skelton BW, Dixon KW, Ghisalberti EL, Flematti GR, Smith SM (2014) Strigolactone hormones and their stereoisomers signal through two related receptor proteins to induce different physiological responses in Arabidopsis. Plant Physiol 165(3): 1221–1232. https://doi.org/10.1104/pp.114.240036

Schlemper TR, Leite MFA, Lucheta AR, Shimels M, Bouwmeester HJ, Veen JAV, Kuramae EE (2017) Rhizobacterial community structure differences among sorghum cultivars in different growth stages and soils. FEMS Microbiol Ecol 93. https://doi.org/10.1093/femsec/fix096

Schwartz SH, Tan BC, Gage DA, Zeevaart JA, McCarty DR (1997) Specific oxidative cleavage of carotenoids by VP14 of maize. Science 276(5320):1872–1874. https://doi.org/10.1126/science.276.5320.1872

Schwartz SH, Qin X, Loewen MC (2004) The biochemical characterization of two carotenoid cleavage enzymes from Arabidopsis indicates that a carotenoid-derived compound inhibits lateral branching. J Biol Chem 279:46940–46945. https://doi.org/10.1074/jbc.M409004200

Seto Y, Sado A, Asami K, Hanada A, Umehara M, Akiyama K, Yamaguchi S (2014) Carlactone is an endogenous biosynthetic precursor for strigolactones. Proc Natl Acad Sci U S A 111 (4):1640–1645. https://doi.org/10.1073/pnas.1314805111

Seto Y, Yasui R, Kameoka H, Tamiru M, Cao M, Terauchi R, Sakurada A, Hirano R, Kisugi T, Hanada A, Umehara M (2019) Strigolactone perception and deactivation by a hydrolase receptor DWARF14. Nat Commun 10(1):1–10. https://doi.org/10.1038/s41467-018-08124-7

Shabek N, Ticchiarelli F, Mao H, Hinds TR, Leyser O, Zheng N (2018) Structural plasticity of D3–D14 ubiquitin ligase in strigolactone signalling. Nature 563(7733):652–656. https://doi.org/10.1038/s41586-018-0743-5

Shinohara N, Taylor C, Leyser O (2013) Strigolactone can promote or inhibit shoot branching by triggering rapid depletion of the auxin efflux protein PIN1 from the plasma membrane. PLoS Biol 11:e1001474. https://doi.org/10.1371/journal.pbio.1001474

Siame BP, Weerasuriya Y, Wood K, Ejeta G, Butler LG (1993) Isolation of strigol, a germination stimulant for Striga asiatica, from host plants. J Agric Food Chem 41:1486–1491. https://doi.org/10.1021/jf00033a025

Solaiman MZ, Saito M (1997) Use of sugars by intraradical hyphae of arbuscular mycorrhizal fungi revealed by radiorespirometry. New Phytol 136:533–538. https://doi.org/10.1046/j.1469-8137.1997.00757.x

Soto MJ, Fernandez-Aparicio MN, Castellanos-Morales V, Garcia Garrido JM, Ocampo JA, Delgado MJ, Vierheilig H (2010) First indications for the involvement of strigolactones on nodule formation in alfalfa (*Medicago sativa*). Soil Biol Biochem 42:383–385. https://doi.org/10.1016/j.soilbio.2009.11.007

Soundappan I, Bennett T, Morffy N, Liang Y, Stanga JP, Abbas A, Leyser O, Nelson DC (2015) SMAX1-LIKE/D53 family members enable distinct MAX2-dependent responses to Strigolactones and Karrikins in Arabidopsis. Plant Cell 27(11):3143–3159. https://doi.org/10.1105/tpc.15.00562

Sun H, Tao J, Liu S, Huang S, Chen S, Xie X, Yoneyama K, Zhang Y, Xu G (2014) Strigolactones are involved in phosphate- and nitrate-deficiency-induced root development and auxin transport in rice. J Exp Bot 65(22):6735–6746. https://doi.org/10.1093/jxb/eru029

Ueguchi-Tanaka M, Ashikari M, Nakajima M, Itoh H, Katoh E, Kobayashi M, Chow TY, Yue-ie CH, Kitano H, Yamaguchi I, Matsuoka M (2005) GIBBERELLIN INSENSITIVE DWARF1 encodes a soluble receptor for gibberellin. Nature 437(7059):693–698. https://doi.org/10.1038/nature04028

Ueno K, Nomura S, Muranaka S, Mizutani M, Takikawa H, Sugimoto Y (2011) Ent-2-′-epi-orobanchol and its acetate, as germination stimulants for Striga gesnerioides seeds isolated from cowpea and red clover. J Agric Food Chem 59(19):10485–10490. https://doi.org/10.1021/jf2024193

Ueno K, Furumoto T, Umeda S, Mizutani M, Takikawa H, Batchvarova R, Sugimoto Y (2014) Heliolactone, a non-sesquiterpene lactone germination stimulant for root parasitic weeds from sunflower. Phytochemistry 108:122–128. https://doi.org/10.1016/j.phytochem.2014.09.018

Umehara M, Hanada A, Yoshida S, Akiyama K, Arite T, Takeda-Kamiya N, Magome H, Kamiya Y, Shirasu K, Yoneyama K, Kyozuka J (2008) Inhibition of shoot branching by new terpenoid plant hormones. Nature 455(7210):195–200. https://doi.org/10.1038/nature07272

Umehara M, Cao MM, Akiyama K, Akatsu T, Seto Y, Hanada A, Li WQ, Takeda-Kamiya N, Morimoto Y, Yamaguchi S (2015) Structural requirements of strigolactones for shoot branching inhibition in rice and Arabidopsis. Plant Cell Physiol 56:1059–1072. https://doi.org/10.1093/pcp/pcv028

Uraguchi D, Kuwata K, Hijikata Y, Yamaguchi R, Imaizumi H, Sathiyanarayanan AM, Rakers C, Mori N, Akiyama K, Irle S, McCourt P (2018) A femtomolar-range suicide germination stimulant for the parasitic plant Striga hermonthica. Science 362(6420):1301–1305. https://doi.org/10.1126/science.aau5445

van Zeijl A, Liu W, Xiao TT, Kohlen W, Yang WC, Bisseling T, Geurts R (2015) The strigolactone biosynthesis gene DWARF27 is co-opted in rhizobium symbiosis. BMC Plant Biol 15(1):1–15. https://doi.org/10.1186/s12870-015-0651-x

Villaécija-Aguilar JA, Struk S, Goormachtig S, Gutjahr C (2021) Bioassays for the effects of strigolactones and other small molecules on root and root hair development. In: Strigolactones. Humana, New York, pp 129–142. https://doi.org/10.1007/978-1-0716-1429-7_1

Visentin I, Vitali M, Ferrero M, Zhang Y, Ruyter-Spira C, Novák O, Cardinale F (2016) Low levels of strigolactones in roots as a component of the systemic signal of drought stress in tomato. New Phytol 212(4):954–963. https://doi.org/10.1111/nph.14190

Vurro M, Prandi C, Baroccio F (2016) Strigolactones: how far is their commercial use for agricultural purposes? Pest Manag Sci 72:2026–2034. https://doi.org/10.1002/ps.4254

Wang Y, Bouwmeester HJ (2018) Structural diversity in the strigolactones. J Exp Bot 69(9): 2219–2230. https://doi.org/10.1093/jxb/ery091

Wang L, Wang B, Jiang L, Liu X, Li X, Lu Z, Meng X, Wang Y, Smith SM, Li J (2015) Strigolactone Signaling in Arabidopsis regulates shoot development by targeting D53-like SMXL repressor proteins for ubiquitination and degradation. Plant Cell 27(11):3128–3142. https://doi.org/10.1105/tpc.15.00605

Wang JY, Haider I, Jamil M, Fiorilli V, Saito Y, Mi J, Baz L, Kountche BA, Jia KP, Guo X, Balakrishna A (2019) The apocarotenoid metabolite zaxinone regulates growth and strigolactone biosynthesis in rice. Nat Commun 10(1):1–9. https://doi.org/10.1038/s41467-019-08461-1

Waters MT, Scaffidi A, Moulin SL, Sun YK, Flematti GR, Smith SM (2015) A Selaginella moellendorffii ortholog of KARRIKIN INSENSITIVE2 functions in Arabidopsis development but cannot mediate responses to karrikins or strigolactones. Plant Cell 27(7):1925–1944. https://doi.org/10.1105/tpc.15.00146

Waters MT, Gutjahr C, Bennett T, Nelson DC (2017) Strigolactone signaling and evolution. Annu Rev Plant Biol 68(8):1–31. https://doi.org/10.1146/annurev-arplant-042916-040925

Wen C, Zhao Q, Nie J, Liu G, Shen L, Cheng C, Xi L, Ma N, Zhao L (2016) Physiological controls of chrysanthemum DgD27 gene expression in regulation of shoot branching. Plant Cell Rep 35(5):1053–1070. https://doi.org/10.1007/s00299-016-1938-6

Xie X, Yoneyama K, Kusumoto D, Yamada Y, Yokota T, Takeuchi Y, Yoneyama K (2008) Isolation and identification of alectrol as (+)-orobanchyl acetate, a germination stimulant for root parasitic plants. Phytochemistry 69(2):427–431. https://doi.org/10.1016/j.phytochem.2007.07.017

Xie X, Yoneyama K, Kurita JY, Harada Y, Yamada Y, Takeuchi Y, Yoneyama K (2009) 7-Oxoorobanchyl acetate and 7-oxoorobanchol as germination stimulants for root parasitic plants from flax (Linum usitatissimum). Biosci Biotechnol Biochem 73(6):1367–1370. https://doi.org/10.1271/bbb.90021

Xie X, Yoneyama K, Kisugi T, Uchida K, Ito S, Akiyama K, Hayashi H, Yokota T, Nomura T, Yoneyama K (2013) Confirming stereochemical structures of strigolactones produced by rice and tobacco. Mol Plant 6(1):153–163. https://doi.org/10.1093/mp/sss139

Xie X, Mori N, Yoneyama K, Nomura T, Uchida K, Yoneyama K, Akiyama K (2019) Lotuslactone, a non-canonical strigolactone from Lotus japonicus. Phytochemistry 157:200–205. https://doi.org/10.1016/j.phytochem.2018.10.034

Xu X, Jibran R, Wang Y, Dong L, Flokova K, Esfandiari A, McLachlan ARG, Heiser A, Sutherland-Smith AJ, Brummell DA, Bouwmeester HJ, Dijkwel PP, Hunter DA (2021) Strigolactones regulates sepal senescence in Arabidopsis. J Exp Bot 199. https://doi.org/10.1093/jxb/erab199

Yamada Y, Furusawa S, Nagasaka S, Shimomura K, Yamaguchi S, Umehara M (2014) Strigolactone signaling regulates rice leaf senescence in response to a phosphate deficiency. Planta 240(2):399–408. https://doi.org/10.1007/s00425-014-2096-0

Yao R, Ming Z, Yan L, Li S, Wang F, Ma S, Yu C, Yang M, Chen L, Chen L, Li Y (2016) DWARF14 is a non-canonical hormone receptor for strigolactone. Nature 536(7617):469–473. https://doi.org/10.1038/nature19073

Yasuda N, Sugimoto Y, Kato M, Inanaga S, Yoneyama K (2003) (+)-Strigol, a witchweed seed germination stimulant, from *Menispermum dauricum* root culture. Phytochemistry 62(7): 1115–1119. https://doi.org/10.1016/S0031-9422(02)00679-9

Yokota T, Sakai H, Okuno K, Yoneyama K, Takeuchi Y (1998) Alectrol and orobanchol, germination stimulants for Orobanche minor, from its host red clover. Phytochemistry 49(7): 1967–1973. https://doi.org/10.1016/S0031-9422(98)00419-1

Yoneyama K, Xiaonan X, Hitoshi S, Yasutomo T, Shin O, Kohki A, Hideo H, Koichi Y (2008) Strigolactones, host recognition signals for root parasitic plants and arbuscular mycorrhizal fungi, from Fabaceae plants. New Phytol 179:484–494. https://doi.org/10.1111/j.1469-8137.2008.02462.x

Yoneyama K, Xie X, Kisugi T, Nomura T, Sekimoto H, Yokota T, Yoneyama K (2011) Characterization of strigolactones exuded by Asteraceae plants. Plant Growth Regul 65(3):495–504. https://doi.org/10.1007/s10725-011-9620-z

Yoneyama K, Xie X, Yoneyama K, Kisugi T, Nomura T, Nakatani Y, Akiyama K, McErlean CS (2018) Which are the major players, canonical or non-canonical strigolactones? J Exp Bot 69(9): 2231–2239. https://doi.org/10.1093/jxb/ery090

Yoneyama K, Xie X, Yoneyama K, Nomura T, Takahashi I, Asami T, Mori N, Akiyama K, Kusajima M, Nakashita H (2019) Regulation of biosynthesis, perception, and functions of strigolactones for promoting arbuscular mycorrhizal symbiosis and managing root parasitic weeds. Pest Manag Sci 75:2353–2359. https://doi.org/10.1002/ps.5401

Yoshimura M, Fonné-Pfister R, Screpanti C, Hermann K, Rendine S, Dieckmann M, Quinodoz P, De Mesmaeker A (2019) Total synthesis and biological evaluation of heliolactone. Helvetica Chimica Acta 102(11):1900211. https://doi.org/10.1002/hlca.201900211

Zhang Y, van Dijk AD, Scaffidi A, Flematti GR, Hofmann M, Charnikhova T, Verstappen F, Hepworth J, van der Krol S, Leyser O, Smith SM, Zwanenburg B, Al-Babili S, Ruyter-Spira C, Bouwmeester HJ (2014) Rice cytochrome P450 MAX1 homologs catalyze distinct steps in strigolactone biosynthesis. Nat Chem Biol 10(12):1028–1033. https://doi.org/10.1038/nchembio

Zhao LH, Zhou XE, Wu ZS, Yi W, Xu Y, Li S, Xu TH, Liu Y, Chen RZ, Kovach A, Kang Y (2013) Crystal structures of two phytohormone signal-transducing α/β hydrolases: karrikin-signaling KAI2 and strigolactone-signaling DWARF14. Cell Res 23(3):436–439. https://doi.org/10.1038/cr.2013.19

Zhao LH, Zhou XE, Yi W, Wu Z, Liu Y, Kang Y, Hou L, De Waal PW, Li S, Jiang Y, Scaffidi A (2015) Destabilization of strigolactone receptor DWARF14 by binding of ligand and E3 ligase signaling effector DWARF3. Cell Res 25(11):1219–1236. https://doi.org/10.1038/cr.2015.122

Zhou F, Lin Q, Zhu L, Ren Y, Zhou K, Shabek N, Wu F, Mao H, Dong W, Gan L, Ma W, Gao H, Chen J, Yang C, Wang D, Tan J, Zhang X, Guo X, Wang J, Jiang L, Liu X, Chen W, Chu J, Yan C, Ueno K, Ito S, Asami T, Cheng Z, Wang J, Lei C, Zhai H, Wu C, Wang H, Zheng N, Wan J (2013) D14-SCF(D3)-dependent degradation of D53 regulates strigolactone signalling. Nature 504(7480):406–410. https://doi.org/10.1038/nature12878

Zwanenburg B, Blanco-Ania D (2018) Strigolactones: new plant hormones in the spotlight. J Exp Bot 69(9):2205–2218. https://doi.org/10.1093/jxb/erx487

Zwanenburg B, Mwakaboko AS, Reizelman A, Anilkumar G, Sethumadhavan D (2009) Structure and function of natural and synthetic signalling molecules in parasitic weed germination. Pest Manag Sci 65(5):478–491. https://doi.org/10.1002/ps.1706

Zwanenburg B, Cavar Zeljkovic S, Pospisil T (2015) Synthesis of strigolactones, a strategic account. Pest Manag Sci 72(3):637. https://doi.org/10.1002/ps.4229

Chapter 4
The Role of Phytohormones in Cross-communication Between Plants and Rhizo-Microbes

Hexon Angel Contreras-Cornejo, Arturo Ramírez-Ordorica, Mariana Álvarez-Navarrete, and Lourdes Macías-Rodríguez

Abstract Plants in their ecological niches establish multiple interactions with arthropods and rhizosphere microorganisms. Plant growth-promoting rhizobacteria (PGPR) and soil-borne fungi can establish molecular dialogues with plants by producing powerful molecules such as ethylene (ET), auxins (IAA), cytokinins (CKs), and gibberellins (GAs) that activate specific molecular mechanisms that subsequently modulate specific physiological processes such as cell division, expansion, or cellular differentiation, whereas abscisic acid (ABA) or enzymatic components as the 1-aminocyclopropane-1-carboxylate deaminase (ACCase) have the ability to induce resistance to different kinds of abiotic stresses as salinity and drought. The microbial root interaction might activate defense responses mediated by the phytohormones salicylic acid (SA) or jasmonic acid (JA) that result effectives against the attack of plant pathogenic microorganisms (hemi- and biotrophic or necrotrophic pathogens) or chewing and piercing sucking insects. This chapter highlights the role of several microbial metabolites that impact on the molecular mechanisms modulated by phytohormones that regulate defense responses and the growth and development of plants.

Keywords Phytohormones · Plant-microbe interactions · Rhizobacteria · Secondary metabolites · Soil microbes

H. A. Contreras-Cornejo (✉) · A. Ramírez-Ordorica · L. Macías-Rodríguez
Instituto de Investigaciones Químico-Biológicas, Universidad Michoacana de San Nicolás de Hidalgo, Morelia, Michoacán, Mexico

M. Álvarez-Navarrete
Departamento de Ingeniería Bioquímica, Instituto Tecnológico de Morelia, Morelia, Michoacán, Mexico

B. A. Horwitz, P. K. Mukherjee (eds.), *Microbial Cross-talk in the Rhizosphere*, Rhizosphere Biology, https://doi.org/10.1007/978-981-16-9507-0_4

4.1 Introduction

Plants and microorganisms interact in practically all the ecosystems and their presence is essential to maintain life on earth due to the ecosystemic services that both provide (Herrera-Paredes and Lebeis 2016). Because plants are anchored to the soil through the roots, it is through this system that plants interact with myriads of rhizosphere microorganisms, such as bacteria, ectomycorrhizal, arbuscular mycorrhizal fungi (AMF), and free-living fungi (Reboutier et al. 2002; Splivallo et al. 2009). Some microorganisms can be found associated with the seeds, stems, leaves, and fruits of plants (Romero et al. 2014; Coleman-Derr et al. 2016; Ramírez-Ordorica et al. 2020). These microorganisms induce on plants a broad range of effects that can be negative, neutral, or beneficial for the host (Newton et al. 2010; Zúniga et al. 2017). Beneficial interactions involve communication among the two organisms through signaling metabolites that are exuded by plant roots and microorganisms. These compounds subsequently are perceived by the epidermal cells of the host plant, and in turn impact on the endogenous signaling programs of the cells (Ortíz-Castro et al. 2008; Baetz and Martinoia 2014). Interactions between plants and microorganisms can be through physical contact such as root colonization or interactions without physical contact that involve the exchange of low molecular weight compounds that function as specific signals to modulate some physiological, biochemical, or molecular processes in the recipient organism (Ryu et al. 2003; Contreras-Cornejo et al. 2009; Rasmann and Turlings 2016). Rhizosphere bacteria and fungi produce a number of metabolites that when perceived in the target cell can function as phytohormones in plants (Splivallo et al. 2009; Shi et al. 2017). To date, the number of microbial metabolites identified is broad and several of them have plant growth modulatory activity that resembles the function of the canonical phytohormones but such compounds have very different chemical structure, thus the molecular mechanisms through which the plants perceive and respond to them are hot topics at this moment (Contreras-Cornejo et al. 2015a, b, c; Garnica-Vergara et al. 2016).

On the other hand, some compounds of microbial origin can also activate defense responses. Under starvation, rhizosphere microorganisms can release low molecular weight compounds or peptidic compounds that stimulate plant immunity. Such defensive priming has been found to be effective to resist the attack of plant pathogen microorganisms or even herbivorous insects (Battaglia et al. 2013; Coppola et al. 2019a, b; Contreras-Cornejo et al. 2020a, b). In order to illustrate the biological, ecological, and agricultural importance of the microorganisms in the rhizosphere, this chapter describes the most recent scientific advances that highlight some secondary metabolites of microbial origin that are perceived by plants and modulate different aspects of growth, development, and defense under different environmental conditions.

4.2 Microorganisms in the Rhizosphere

Soil microorganisms include various genera and species of bacteria, fungi, oomycetes, nematodes, and amoebae (Newton et al. 2010; van Dam and Bouwmeester 2016). The composition of the microbial communities in the rhizosphere differs between plant species (Hardoim et al. 2008; Redford et al. 2010). For example, *Arabidopsis thaliana* rhizosphere community is mainly associated with bacterial phyla Acidobacteria, Actinobacteria, Bacteroidetes, Cyanobacteria, Firmicutes, Gemmatimonadetes, and Proteobacteria (Lundberg et al. 2012). Plant root exudates, soil pH, nutrients availability, soil humidity, and other abiotic factors participate in the modulation of the microorganism populations (Kim and Lee 2020; Juan-Ovejero et al. 2020). Microorganisms have important roles for the optimal functioning of the ecosystem, for example: mobilization and stabilization of carbon, organic matter decomposition, nutrient mineralization and biocontrol of pathogenic microorganisms, and the different types of interactions with plants (Mitchell et al. 2003).

Microbial community structure is dependent on temperature variations (Manzoni et al. 2012). The depth in the soil can affect the location of microorganisms; the microbial population is found to be more in aerated soils (Liu et al. 2016). Additionally, the edaphic properties of the soil, for example, the concentrations of electron donors or acceptors in bulk soil, agricultural practices as the utilization of organic nitrogen fertilizers, and plant vegetation cover species, can largely modulate impact in microbial communities assemblage (Caradonia et al. 2019; Yuan et al. 2019). In a recent report it was shown that the microbiota of tomato (*Solanum lycopersicum*) is constituted mainly by members of Actinobacteria, Bacteroidetes, Alpha-, Beta-, Gamma-, and Deltaproteobacteria and Verrucomicrobia (Caradonia et al. 2019). However, in the peatland rhizosphere, the phyla Ascomycota and Basidiomycota are fungal communities present at 46% and 40%, respectively (Thormann et al. 2007). Otherwise, the lifestyles of the microorganisms determine if they can live freely in the rhizosphere, or if they form endophytic associations with plants (Akiyama et al. 2005; Bae et al. 2011).

4.3 Plant-Microorganism Interaction

The microorganisms in the rhizosphere may have no apparent effect on plants, be harmful if they cause disease, or can be beneficial for plants. Beneficial microorganisms can stimulate plant growth through: (1) direct mechanisms, by the production of growth regulating molecules, and processes that improve the acquisition of macronutrients (N, K, Ca, Mg, P, and S) and micronutrients (Fe, B, Mn, Zn, Cu, Mo, and Ni), and (2) indirect mechanisms, mainly by the biocontrol of phytopathogen microorganisms (Contreras-Cornejo et al. 2016; Ramakrishna et al. 2019). It has often been observed that beneficial soil microorganisms stimulate plant immunity and promote resistance to different types of abiotic stress such as those caused by water deficit, salinity, or soil contamination by heavy metals (Martínez-Medina et al. 2019; Kim and Lee 2020).

Root exudates are an important source of compounds to enrich the rhizosphere near to the plants with nutrients, and as a consequence the microorganisms in close association with the root increase their population density and association with the plants (Zúniga et al. 2017; Sasse et al. 2018). Plant metabolites or those of microbial origin, depending on their physicochemical characteristics, can be diffused through the rhizosphere or towards the atmosphere (Tyc et al. 2017). Plants release the root compounds that are exuded through passive processes such as diffusion, channels, vesicular transport, or more complex processes as ATP-binding cassette (ABC) transporters (Baetz and Martinoia 2014). Some compounds derived from plants function as signaling molecules to attract the beneficial microorganisms towards the root, and later root colonization takes place (Akiyama et al. 2005).

In the case of associations established by AMF, host plants initiate the interaction through the production and release of stringolactones that are carotenoid-derived compounds of low molecular weight less than 500 Da (Rasmann and Turlings 2016). These metabolites, when perceived by the fungal spores, promote the growth of the hyphae towards the host plant. Specifically, root exudates from *Lotus japonicus* plants release 5-deoxy-stringol, which at concentrations of 30 pg to 100 ng is a branching factor for *Gigaspora margarita* hyphae (Akiyama et al. 2005). It is known that plants growing in phosphorus limitation are forced to increase their ratio of root exudation, which favors the exudation of agents that promote the branching of AMF hyphae (Nagahashi and Douds 2000). In the rhizosphere, there are some species of fungi that also establish associations with plants and colonize the roots without apparent participation of signaling molecules that favor interaction; such is the case of the endophytic fungus *Piriformospora indica* that colonizes the root tissue of barley (*Hordeum vulgare* L.) (Waller et al. 2008).

Soil microorganisms can perceive the components present in root exudates (De Weert et al. 2002). More recently it was shown that during plant-*Trichoderma* interactions, plants release oxylipins and carbohydrates through root exudates that are perceived by the fungus and function as chemotactic agents for the attraction of the microorganism towards the root (Lombardi et al. 2018; Macías-Rodríguez et al. 2018). In the particular case of tomato plants inoculated with *Trichoderma atroviride* IMI 206040, it was observed that before physical contact between the two organisms, the plants mainly exude the monosaccharides arabinose, xylose, glucose, myo-inositol, and fructose. Later, when *T. atroviride* grew towards the root, the levels of carbohydrate exudation were slightly reduced and the exudation of sucrose that served as a nutritional and as a carbon source for the fungus increased its levels (Macías-Rodríguez et al. 2018).

In regard with the associations of rhizobia with roots, it is known that plants release flavonoids which are chemotactic agents for nitrogen-fixing bacteria and in response such bacteria release molecules named "nod factors," which are substances structurally related to acylated chitin oligomers (Bisseling and Geurts 2020). It has been assumed that plants in the cell membrane possess a receptor for nod factors, which is a heterodimeric receptor that in *Medicago truncatula* is composed of the kinases LYSM DOMAIN-CONTAINING RECEPTOR LIKE KINASE 3 (MTLYK3) and NOD FACTOR PERCEPTION (MtNFP) (Zipfel and Oldroyd 2017). Then, in order to activate this mechanism, bacteria approach the proximity of the root hairs, and such structures cover the bacterial cell, thus forming the nodule where the fixation of atmospheric nitrogen will occur (Oldroyd et al. 2011; Podlešáková et al. 2013).

4.3.1 Root Perception of Microbial Signals

Changes in the shape and structure of the root are often the result of the perception of microbial signals by the host plant cells (Gutiérrez-Luna et al. 2010). Table 4.1 shows the physicochemical properties of the canonical phytohormones. Some of them can be produced by rhizosphere microorganisms. Depending on the type of signal, microbial molecules can induce changes in the plasticity of the root system, which implies an increase in the formation of secondary roots and root hairs (Contreras-Cornejo et al. 2009, 2015b; Garnica-Vergara et al. 2016). These changes in the root system allow the plants a more vigorous anchoring to the soil, improved uptake of water and nutrients and a greater contact surface with the microorganisms of the rhizosphere (Ortíz-Castro et al. 2009). Figure 4.1 shows the impact of the inoculation of *Bacillus amyloliquefaciens* M496 on the growth of maize and *Medicago truncatula* plants grown in MS medium. Bacterial inoculation enhances both shoot length growth and root branching, which suggests that such strain uses an

Table 4.1 Physicochemical properties of phytohormones

Name	Molecular formula	Molecular mass (g/mol)	Density (g/cm^3)	Appearance	Class of compound
Indole-3-acetic acid (IAA)	$C_{10}H_9NO_2$	175.19	1.4 ± 0.1	White solid	Indolic compound
Ethylene (ET)	C_2H_2	28.05	1.138[a]	Colorless gas	Hydrocarbon
Abscisic acid (ABA)	$C_{15}H_{20}O_4$	264.32	1.2 ± 0.1	Colorless solid crystals	Isoprenoid-derived compound
Zeatin (CK)	$C_{10}H_{13}N_5O$	219.24	1.4 ± 0.1	Yellowish to off-white crystals	Adenine-derived compound
Jasmonic acid (JA)	$C_{12}H_{18}O_3$	210.27	1.1 ± 0.1	Off-white oil	Oxylipin-derived compound
Salicylic acid (SA)	$C_7H_6O_3$	138.12	1.4 ± 0.1	White solid	Phenolic compound

Partial information in this table was taken from Contreras-Cornejo et al. (2009, 2015a, b)
[a] This property is in kg/m^3 at 25 °C

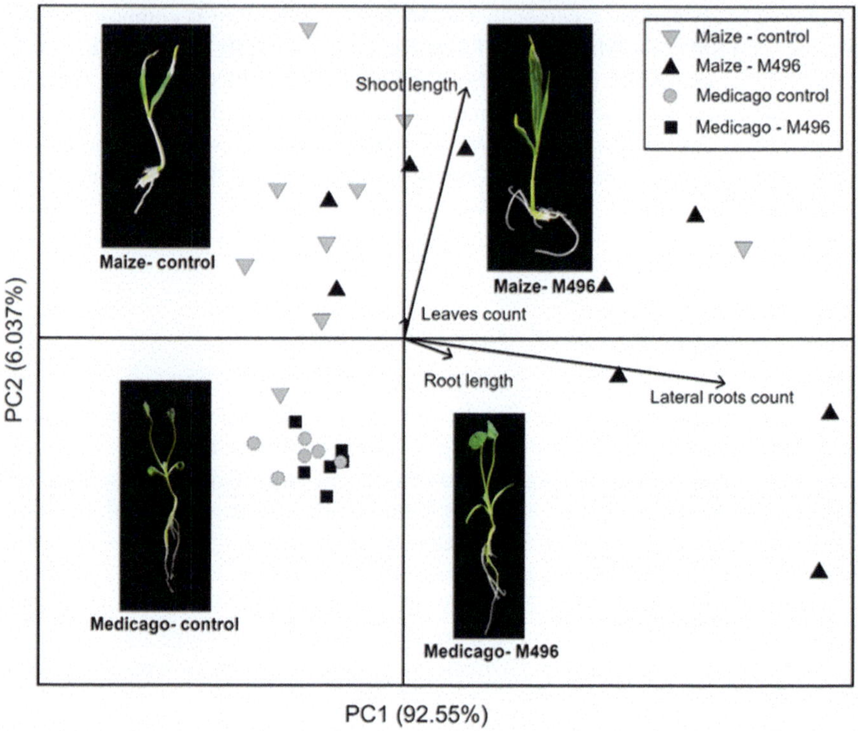

Fig. 4.1 Principal component analysis (PCA) that highlights the effects of *Bacillus amyloliquefaciens* M496 on maize and *M. truncatula* plants. Notice that the main effect of this microorganism is promoting shoot growth and lateral root formation. This pattern suggests a plant species-independent response to the same bacterial strain

efficient mechanism to induce beneficial effects on mono- and dicotyledonous plants.

Some bacteria and fungi in the rhizosphere can produce metabolites that directly impact some signaling pathways, modulated by canonical phytohormones, thus producing the typical hormonal effects at the physiological level (Splivallo et al. 2009; Contreras-Cornejo et al. 2014a, b; Garnica-Vergara et al. 2016; Gouda et al. 2018). Figure 4.2 shows the chemical structure of some phytohormones. However, there are some microbial compounds that can antagonize the effect of phytohormones (Reboutier et al. 2002). For example, the basidiomycetous fungus *Pisolithus tinctorius* that establishes ectomycorrhizas with *Eucalyptus globulus* releases an indole alkaloid metabolite called hypaphorine that reduces the root hair elongation and counteracts the activity of the auxin IAA (Ditengou and Lapeyrie 2000; Reboutier et al. 2002). Below are described some rhizospheric microorganisms, microbial metabolites, and the molecular mechanisms that are modulated in plants when both kind of organisms cointeract.

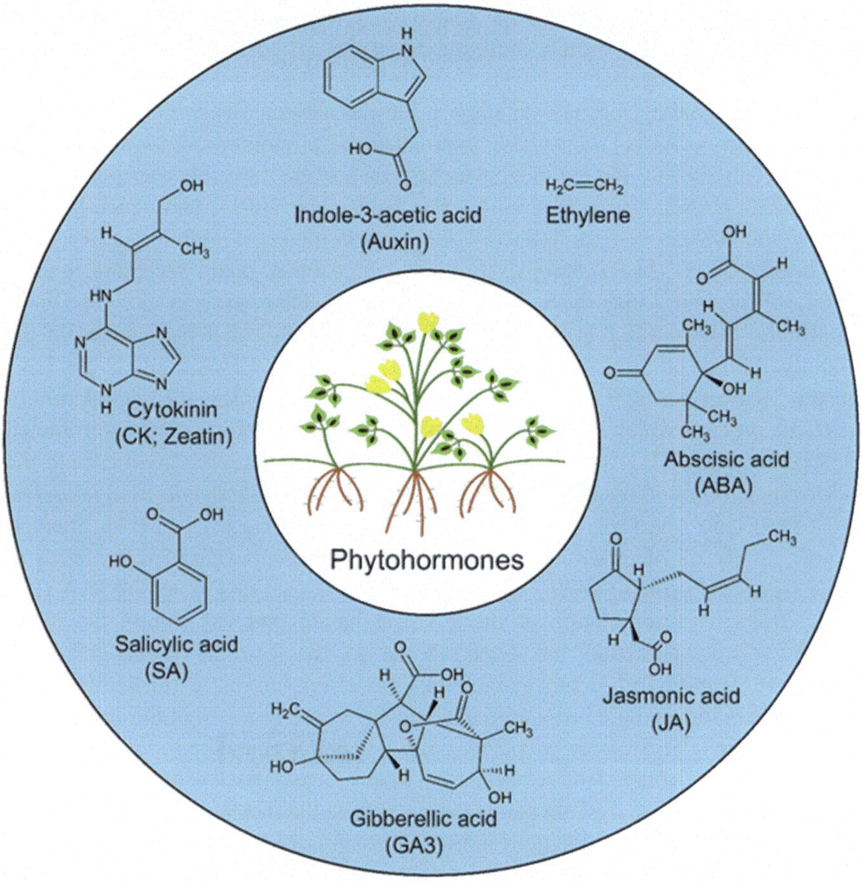

Fig. 4.2 Molecular structures of the seven canonical phytohormones. Rhizosphere microorganisms as bacteria or fungi can produce this class of compounds. To the best of our knowledge JA has not been identified in microorganisms but beneficial plant microorganisms activate its signaling mechanism. Notice differences among chemical structures

4.3.1.1 Auxins

Auxins are a group of compounds whose main chemical characteristic is that they have the indole molecule as structural base. It is known that a large number of pathogenic and beneficial soil bacteria and fungi produce auxins (Contreras-Cornejo et al. 2020b). The main auxin in plants that participates in multiple development processes is indole-3-acetic acid (IAA). The IAA biosynthesis pathway starts from chorismate from which anthranilate results, and then the amino acid tryptophan (Spaepen et al. 2007). In bacteria, several routes have been proposed for IAA biosynthesis that use the following metabolites as precursors: (1) indole-3-acetamide, (2) indole-3-pyruvate, (3) tryptamine, and/or (4) indole-3-acetonitrile

(Spaepen et al. 2007). It has been speculated that IAA released to the environment is present in its protonated form because the rhizosphere is considered a weakly acidic environment (Hinsinger et al. 2003).

The IAA signaling mechanism starts when the molecule is perceived by influx transporters (i.e., AUX1) and these internalize the phytohormone within the cell (Swarup et al. 2001). Also involved in auxin transport are the efflux transporters PIN FORMED 2 (PIN2, also named EIR1) (De Billy et al. 2001). Subsequently, IAA binds to the nuclear receptor TIR1 (a F-box protein transport inhibitor 1) (Kepinski and Leyser 2005). In *Arabidopsis thaliana*, three other auxin receptors, named AFB1, AFB2, and AFB3, have also been identified (Dharmasiri et al. 2005; Parry and Estelle 2006). TIR1 is part of the SCF^{TIR1} complex that participates in the ubiquitination of proteins that will later be degraded. When IAA binds to the TIR1 receptor, part of the SCF^{TIR1} ubiquitination molecular complex, the degradation of Aux/IAA proteins which repress the expression of auxin response genes is promoted (Spaepen et al. 2007). Auxins modulate different genes that are classified into the following families: *Gretchen Hagen 3* (*GH3*), *auxin/indole-acetic acid-inducible* (*AUX/IAA*), and *small auxin up RNA* (*SAUR*) genes (Zhao et al. 2014). There is another family of genes different from those described above and it is named as *auxin-repressed proteins* (*ARP*) whose expression is also modulated by IAA (Lee et al. 2013). The modulation of the auxin signaling pathway during the plant-microorganism interaction has been shown in the case of poplar (*Populus tremula* × *Populus alba*) inoculated with the ectomycorrhizal fungus *Laccaria bicolor* which induced the formation of lateral roots and increases of the auxin levels at the root tips. Treatments of plants with 1-naphthylphthalamic acid, an auxin transport blocker, affected the accumulation of auxins and the development of lateral roots. An oligoarray analysis in plants that perceived the metabolites released by the fungus revealed important changes in the expression of several genes belonging to the molecular mechanism of auxins as *PtaPIN* and *PtaAUX* of the auxin transport, *PtaGH3* involved in auxin conjugation, and *PtaIAA* genes involved in the signaling of auxins (Felten et al. 2009). In vitro assays between *Arabidopsis thaliana* and *Trichoderma virens* Gv29-8 revealed that the fungus promotes plant growth. The beneficial effect on the accumulation of foliar biomass, induction of lateral roots, and formation of root hairs was the result of the production of IAA (13.48 ± 0.97 μg/L), indole-3-acetaldehyde (59.40 ± 4.47 μg/L), and indole-3-ethanol (72.33 ± 1.41 μg/L). Genetic confirmation for the participation of IAA in the foliar growth promotion of *Arabidopsis thaliana* by *T. virens* Gv29-8 was supported by the phenotype observed in the plant due to the reduced response on growth promotion in the mutants *aux1-7*, *eir1-1*, and *axr1-3* inoculated with the fungus (Contreras-Cornejo et al. 2009).

IAA can also interact with the ET signaling pathway and modulate plant development as evidenced by the production of this volatile hormone and its release from the truffle mycelium and due to its hormonal activity on the host plant *Cistus incanus* and non-host *Arabidopsis thaliana* (Splivallo et al. 2009). In the case of the in vitro interaction between *Trichoderma atroviride* IMI206040 and *Arabidopsis thaliana*, it has also been observed that the fungus modulates plant growth by activating the IAA and ET pathways, with cross talk between both hormonal response pathways. This complex signaling mechanism involves the activity of the MAP kinase MPK6, which is induced after 15 min by 1 μM of IAA and by high concentrations of indole-3-acetaldehyde and indole-3-ethanol produced by *T. atroviride* IMI206040. In this interaction, MPK6 possibly functions as a repressor element, which results in the activation of molecular events, leading to the negative regulation of the formation of lateral roots, root hairs, and the growth of the primary root (Contreras-Cornejo et al. 2015b). In the same work, it was found that the fungus activates the ET pathway through the production of this phytohormone; the developmental processes that ET regulates on the root system for the formation of lateral roots and root hairs involve the components of the ET pathway (ET RESPONSE 1) ETR1, ETHYLENE INSENSITIVE 2 (EIN2), and EIN3, probably using MPK6 and CTR1 as modulators of fungal signals between IAA and ET pathways (Contreras-Cornejo et al. 2015b). In the case of the plant-pathogen interaction a key role for *OsGH3.1* in the induction of rice resistance against *Magnaporthe grisea* was evidenced in experiments with the transformant plant overexpressing *OsGH3.1* because in such lines the resistance to the pathogenic fungus was higher (Domingo et al. 2009).

4.3.1.2 ACCase/Ethylene (ET)

Some rhizospheric microorganisms such as the bacterium *Burkholderia phytofirmans* PsJN can promote plant growth through the action of the enzyme called 1-aminocyclopropane-1-carboxylic acid deaminase [ACCase] (Sessitsch et al. 2005). This mechanism begins when the microorganisms that possess ACCase perceive the ACC exuded by the roots of the plants, and this organic acid is cleaved into α-ketobutyrate and ammonia (Santoyo et al. 2016). ACCase encoding gene has been identified in several microorganisms that include the yeast *Hansenula saturnus*, *Trichoderma asperellum* T203, and *Penicillium citrinum* (Minami et al. 1998; Jia et al. 1999; Viterbo et al. 2010).

It has been proposed that the amino acid L-methionine is used as a precursor for the production of α-keto-γ-(methylthio) butyric acid, and due to its photosensitive properties it releases ET (Splivallo et al. 2009). In microorganisms, there are at least two ET biosynthetic pathways involving the compounds 2-oxo-4-methylthiobutyrate (OMTB) or 2-oxoglutarate (Cristescu et al. 2002). The OMTB pathway has been found to exist in the fungi *Botrytis cinerea*, *Penicillium digitatum*, and the yeast *Saccharomyces cerevisiae*, and in the bacterium *Pseudomonas syringae* (Nagahama et al. 1994; Weingart et al. 1999; Cristescu et al. 2002).

ET is a gaseous molecule of very low molecular weight that coordinates different aspects of plant physiology such as development and defense (Jaroszuk-Ściseł et al. 2019). In *Arabidopsis thaliana*, at least five receptors of ET have been detected called: ET RESPONSE 1 (ETR1), ETR2, ET RESPONSE SENSOR1 (ERS1), ERS2, and ET INSENSITIVE 4 (EIN4) (Kazan 2015). The activation of these receptors causes repression of CONSTITUTIVE TRIPLE RESPONSE 1 (CTR1), that permits ETHYLENE INSENSITIVE 2 (EIN2) to relay the ET signal to the transcription factors EIN3 and ETHYLENE-INSENSITIVE3-LIKE (EIL). Then, EIN3 activates ETHYLENE RESPONSE FACTOR 1 (ERF1), inducing the expression of ET-responsive genes (Huang et al. 2003; Kieber et al. 1993; Sánchez-Rodríguez et al. 2010). Nine genes encoding ACC synthetases (ACS) have been detected in *Arabidopsis thaliana* (Tsuchisaka and Theologis 2004). It is known that *Piriformospora indica* induces changes in the expression of the *ACS1* and *ACS8* genes in the tips of the roots of reporter plants that express the activity of β-glucuronidase (GUS) (Khatabi et al. 2012).

4.3.1.3 Cytokinins (CKs)

These types of compounds are derivatives of N_6-substituted adenine compounds (Wang et al. 2017). Some plant-mutualistic bacteria such as *Sinorhizobium meliloti* and *Mesorhizobium loti* have been detected to produce CKs (Lohar et al. 2004; Frugier et al. 2008). In *Streptomyces turgidiscabies*, the causal agent of potato scab, a biosynthetic pathway for this phytohormone has also been detected (Joshi and Loria 2007). For several decades, it has been known that *Agrobacterium tumefaciens* transfers to its host plants a specific fragment of DNA (transfer DNA [T-DNA]) that comes from a tumor-inducing (Ti) plasmid and among the genes that are encoded in such plasmid, a *trans*-zeatin synthesizing gene is found, which promotes the production of *trans*-zeatin in bacteria (Hwang et al. 2013).

In plants, CKs promote cell division but also participate in various processes associated with active growth, metabolism, and in modulating defense responses (Giron et al. 2013). CKs are active at the site of their synthesis, but also at distal sites where they are concentrated after being transported from other very distant parts (Zhao 2008). It has been detected in *Arabidopsis thaliana* that the transporter ABCG14 (an ATP-binding cassette) regulates the translocation of ^{14}C-*tZ* type CK from roots to foliages through the xylem (Ko et al. 2014).

In *Medicago truncatula*, the CK signaling mechanism is triggered when the phytohormone is perceived by the extracellular cyclase/histidine kinase-associated sensing extracellular domain in the CK response 1 (CRE1) transmembrane receptor (Gonzalez-Rizzo et al. 2006; Kundu and DasGupta 2018). On the other hand, the endophytic fungus *Piriformospora indica* can promote the growth of the model plant *Arabidopsis thaliana* by activating CKs signaling; importantly, increases in the content of *cis*-CKs were detected in the roots of the plant and there is evidence that shows that the combination of the CRE1/AHK2 receptors is necessary during the signaling mechanism (Vadassery et al. 2008).

It has been identified in rice plants (*Oryza sativa*) infected with the pathogen *Magnaporthe oryzae*, that there is a space-dependent hormonal modulation between components of the signaling pathway mediated by CKs and ABA. In the early stages of the infection by *M. oryzae*, a suppression of the immune system occurs in the plant, which is partly related to the activation of the signaling pathway mediated by ABA, which leads to activation of the signaling pathway modulated by CKs and the expression of genes that code for sugar transporters (Cao et al. 2016).

In *Arabidopsis thaliana* ecotype Columbia-0, an interesting effect on the modulation of the components of the CKs and auxins signaling pathways has been detected during the clubroot disease caused by the obligate biotrophic *Plasmodiophora brassicae*. During this disease, genes belonging to the *GH3* family and members involved in the auxin homeostasis were upregulated. In contrast, some components involved in the CK homeostasis were downregulated (Siemens et al. 2006). It is known that in early stages of the interaction, *Plasmodiophora* provides CKs to the host plant, which causes a re-initiation of cell division in the cortex, and consequently several signaling and metabolic events occur that will allow to maintain the plant pathogen and later the gall formation (Devos et al. 2006).

4.3.1.4 Gibberellins (GAs)

These compounds act in multiple processes of plant growth such as modulation of cell division in growing tissues, stem elongation, flowering, seed germination, fruit formation, and senescence (Yamaguchi 2008; Hamayun et al. 2009). GAs are tetracyclic diterpenoid molecules (Qin et al. 2013). These types of compounds are produced by various bacteria and fungi and a little more than 130 molecules structurally related to GAs have been identified (Contreras-Cornejo et al. 2015c).

Initially, these molecules were identified in the phytopathogenic fungus *Fusarium fujikuroi* (teleomorph *Gibberella fujikuroi*) which attacks rice plants (*Oryza sativa*), but also in other microorganisms such as the phytopathogenic bacterium *Xanthomonas oryzae* pv. *oryzicola* and *Mesorhizobium loti* MAFF303099 (Nett et al. 2017). These compounds have also been identified in *Sphaceloma manihoticola*, strains of *Phaeosphaeria* sp. (MacMillan 2002; Bömke and Tudzynski 2009). The main active molecule in plants of this group of compounds is gibberellic acid 3 (GA3), although there is evidence showing that GA4 is the most active in *Arabidopsis thaliana* (Contreras-Cornejo et al. 2015c). In the signaling mechanism, bioactive GA in the plant promotes the interaction of the GA-Insensitive Dwarf 1 receptor (GID1) with DELLA proteins (Hirano et al. 2008). The signaling mechanism of GAs is finely modulated through the inactivation of bioactive GAs by modifying the molecular structure through the epoxidation of the 16, 17 double bond of GAs by the enzymatic activity of a cytochrome P450 monooxygenase, or alternatively by the methylation of the molecule by the action of GA methyl transferases (Zhu et al. 2006; Varbanova et al. 2007). In bacteria the production of GAs is modulated by a cytochrome P450 (CYP)-rich operon. This group of enzymes in turn produces *ent*-kaurene which is the predecessor intermediate product of (*E*,*E*,*E*)-geranylgeranyl diphosphate, which will be the substrate of the *ent*-copalyl diphosphate synthase and the resulting product of this reaction will be the target of the *ent*-kaurene synthase (Morrone et al. 2009; Hershey et al. 2014). The operon that regulates the production of GAs has been identified in nitrogen-fixing rhizobacteria and in phytopathogenic microorganisms. There is evidence to suggest that GA9, which is inactive in plants and which is produced by rhizobia, is converted by the plant to its active form GA4, this probably in order to block further nodulation

(Tatsukami and Ueda 2016). The signaling mechanism of these phytohormones is also key in establishing symbiotic interactions. There is evidence showing that GAs and the GID1 receptor are components involved during the interaction of *P. indica* with roots (Schäfer et al. 2009). Similarly, gibberellin DELLA proteins are components that modulate arbuscular mycorrhizal symbioses (Foo et al. 2013).

4.3.1.5 Abscisic Acid (ABA)

This phytohormone is involved in modulating various physiological processes in plants such as abscission of leaves and stomatal opening to control transpiration and gas exchange that implies the entry of CO_2 that will be used during photosynthesis (Brodribb and McAdam 2011). ABA is also involved in the modulation of lateral root development and in the activation of mechanisms of resistance or tolerance to abiotic stress caused by water deficit or salinity (Achard et al. 2006). The ABA signaling mechanism occurs in the nucleus of cells. The phytohormone has a heterodimeric receptor made up of the RCAR protein (REGULATORY COMPONENT OF ABA RECEPTOR) and the PP2C protein (PROTEIN PHOSPHATASE TYPE 2C) that includes ABI1 (ABSCISIC ACID INSENSITIVE 1 [ABI1]). PP2Cs phosphatases have been proposed to regulate the ABA pathway by repressing the activity of the SUCROSE-NONFERMENTING KINASE1-RELATED PROTEIN KINASES (SnRKs) protein kinases. However, in the presence of ABA the activity of PP2Cs is inhibited. As a consequence, the repressed protein kinases are released, remaining active to phosphorylate the respective components of the ABA pathway, such as bZIP (basic leucine zipper) transcription factors such as ABI5 (Raghavendra et al. 2010). Two mutants insensitive to ABA, *abi1* and *abi2* have been isolated in *Arabidopsis thaliana*; such mutants have the characteristic that they do not close their stomata in response to the exogenous application of ABA (Allen et al. 1999). On the other hand, ABI4 is a transcriptional factor of the APETALA 2 (AP2) type that is modulated in response to ABA during plant development or in response to salt and sugar (Arroyo et al. 2003; Finkelstein et al. 2011). In vitro experiments to study the interaction between *Arabidopsis thaliana* and *T. virens* Gv29-8 and *T. atroviride* IMI206040 showed that both fungal species modulate the opening of stomata in leaves. This effect was correlated with the modulation of leaf transpiration. Interestingly, it was found that the *abi1-1* and *abi2-1* mutants did not close their stomata in response to the fungal inoculation, suggesting that modulation of stomatal activity by *Trichoderma* involves the ABA pathway. Chemical analysis revealed that *T. virens* Gv29-8 and *T. atroviride* IMI 206040 produce ABA, and that both species modulate the expression of the reporter gene *abi4:uidA* (Contreras-Cornejo et al. 2015a).

However, other types of metabolites of bacterial and fungal origin have been identified and they are not structurally related with the typical phytohormones but can also activate growth and development processes in plants. Below are described some of such compounds and the mechanisms that they activate when communicating with plants.

4.3.1.6 Homoserine Lactones

Different Gram-negative bacteria have a self-modulation system known as "Quorum-Sensing" (QS) through a group of metabolites identified as homoserine-lactones (HSL), and themselves regulate their population density, motility, biofilm formation, and biosynthesis of both exopolysaccharides and siderophores (Chalupowicz et al. 2008; Zúniga et al. 2017). The QS system of bacteria is key in the establishment and colonization of these microorganisms on their host plant as revealed in *Arabidopsis thaliana* and maize (*Zea mays*) (Coutinho et al. 2013; Zúñiga et al. 2013). *Paraburkholderia phytofirmans* PsJN is known to have two different QS systems named Bpl.1./R.1 and Bpl.2/R2 (BraI/R-like QS system) that modulate HSL production (Zúniga et al. 2017).

HSL have an acyl chain and when they are perceived by the roots of the plants they can be captured towards the interior of the root and modulate the root growth (Götz et al. 2007; Ortíz-Castro et al. 2008; von Rad et al. 2008). For example, it has been observed that the modulation of the growth of rice (*Oryza sativa*) plants by the *Pseudomonas aeruginosa* PUPa3 system is due in part to the regulation of QS by acyl-homoserine lactones (Steindler et al. 2009). The QS system has also been reported in the host-specific tumorigenic pathogen bacterium *Pantoea agglomerans* pv. *gypsophilae*. This microorganism mainly biosynthesizes *N*-butanoyl-L-homoserine lactone (C_4-HSL) and *N*-hexanoyl-L-homoserine lactone (C_6-HSL) in lower concentrations (Chalupowicz et al. 2009).

Interestingly, it has been observed that the modulation of galls induced by *P. agglomerans* is due to the modulation of the QS system, auxin levels, and CK (Chalupowicz et al. 2009). It is known that HSL *N*-decanoyl-homoserine lactone (C_{10}-HLS) can regulate the growth of *Arabidopsis thaliana* (Ortíz-Castro et al. 2008). Interestingly, *N*-3-oxo-hexanoyl-homoserine lactone ($3OC_6$-HSL) stimulates the elongation of the primary root of *Arabidopsis thaliana*. This effect on the roots was correlated with the increase in the expression of the *AtMYB44* gene that codes for a transcriptional factor involved in plant defense and saline stress. The effect of $3OC_6$-HLS on root elongation was affected in the *atmyb44* mutant. It was detected that the modulation of root growth by $3OC_6$-HSL was due in part to the regulation of the expression of genes *ARR15* and *ARR4* in response to CK and *IAA7*, *IAA14*, and *MAX2* in response to auxins, respectively (Zhao et al. 2016).

4.3.1.7 Polyamines

Polyamines are a class of low molecular weight compounds found in higher organisms, microorganisms, and plants. The most common are cadaverine, putrescine, spermine, and spermidine. It is known that *Azospirillum brasilense* Az30 can promote the root growth of rice plants in a mechanism partially mediated by the production of cadaverine (Cassán et al. 2009). Recently, Xie et al. (2014) reported that *Bacillus subtilis* OKB105 has the ability to promote the growth of *Nicotiana tabacum* plants and spermidine was identified as one of the molecules responsible for regulating this phenotype in plants. The growth promotion of *N. tabacum* by *B. subtilis* involved the induction of the expression of cell expansion genes *Nt-EXPA1* and *Nt-EXP2* and the inhibition of the *Nt-ACO1* gene that encodes a 1-aminocyclopropane-1-carboxylic acid oxidase that participates in the biosynthesis of ethylene.

4.3.1.8 Volatile Organic Compounds

Soil microorganisms emit volatile metabolites that fulfill multiple functions in ecosystems and among the activities is the communication between the members of the microbial community and with the associated plants (Camarena-Pozoz et al. 2018). Although some rhizobacteria and rhizosphere fungi are grouped in the same genus, there may be differences at the metabolic level, and as a consequence they impact differently on the physiology of plants (Gutiérrez-Luna et al. 2010; Contreras-Cornejo et al. 2014a, b; Garnica-Vergara et al. 2016; Guo et al. 2020). Table 4.2 shows some particular cases of plant growth-promoting rhizobacteria and their chemicals that are bioactive on plants. It has been observed that volatile metabolites that alter plant growth have molecular masses of less than 300 Da and are of lipidic nature (Fincheira and Quiroz 2018; García-Gómez et al. 2019). However, there is still too much work to do on the role of microbial volatile metabolites in different interactions because ~2000 volatile compounds have been identified from 1000 microbial strains (Lemfack et al. 2018). It has been observed that the soil bacteria *Enterobacter cloacae*, *B. amyloliquefaciens*, and *B. subtilis* can promote plant growth through the emission of volatile compounds (Lugtenberg and Kamilova 2009).

In a study carried out by Gutiérrez-Luna et al. (2010) with several isolates of rhizobacteria to evaluate their effect on the growth promotion of *Arabidopsis thaliana* through the emission of volatile organic compounds, it was found that the isolates L254 and L272a, although both belong to the genus *Bacillus*, had a very different impact on the growth and development of the root system as evidenced by the different pattern of lateral root and root hair formation. When chemical analyses were performed to determine the profile of the volatiles emitted by the bacterial isolates L254, L255, L265a, L266, L270, and L272a, the following metabolites were found: 1-butanol, 6-methyl 2-heptanol, 2-nonenal, 1-octen-3-ol, benzaldehyde, butyrolactone, acetophenone, tridecanal, tetradecanal, 4-decanone, 6-undecanone, 5-tridecanone, cyclodecane, 3-tetradecanone, 2-pentadecanone, 1-tridecanol, 6,10,14-trimethyl 2-pentadecanone, 2-pentadecanol, 9-octadecanone, 4-octadecyl morpholine, cyclododecane, and 2-morpholinomethyl-1,3-diphenyl-2-propanol.

It has been determined that *T. virens* Gv29-8 releases a mixture of VOCs constituted in part by mono (C_{10})- and sesquiterpenes (C_{15}) that included camphene, 3-carene, β-myrcene, β-phellandrene, eucalyptol, *trans*-β-ocimene, β-terpinene, β-caryophyllene, τ-cadinene, δ-cadinene, α-amorphene, and τ-selinene. In experiments carried out in vitro in Petri dishes to test the effects of the *T. virens* VOCs on

Table 4.2 Microbial metabolites and their effect on plants

Strain	Chemicals	Plant	Effect	References
Arthrobacter agilis UMCV2	Dimethylhexadecylamine	*Medicago truncatula*	Growth promotion	Orozco-Mosqueda et al. (2013)
	Dimethylhexadecylamine	*Sorghum bicolor*	Iron uptake stimulation	Castulo-Rubio et al. (2015)
	Dimethylhexadecylamine	*Medicago sativa*	Growth promotion	Velázquez-Becerra et al. (2011)
	Pool of VOCs	*Medicago truncatula*	Iron deficiency stress response, brassinosteroid content increased	Flores-Cortez et al. (2019)
Azospirillum brasilense SP7	Pool of VOCs	*Mentha* × *piperita*	Growth promotion, monoterpene biosynthesis stimulated	Santoro et al. (2011)
Bacillus amyloliquefaciens FZB42	Pool of VOCs	*Arabidopsis thaliana*	Growth promotion, multiple transcriptional modifications related to stress responses	Hao et al. (2016)
Bacillus amyloliquefaciens strains	Pool of VOCs	*Arabidopsis thaliana*	Growth promotion and phytopathogens growth inhibition	Asari et al. (2016)
Bacillus megaterium XTBG34	2-Pentylfuran	*Arabidopsis thaliana*	Growth promotion	Zou et al. (2010)
Bacillus methylotrophicus M496	3-Hydroxy-2-butanone	*Zea mays*	Growth promotion	Pérez-Flores et al. (2017)
Bacillus methylotrophicus M4-96	3-Hydroxy-2-butanone	*Fragaria* × *ananassa*	Growth promotion	Vicente-Hernández et al. (2019)
Bacillus sp. B55	Dimethyl disulfide	*Nicotiana attenuata*	Sulfur assimilation	Meldau et al. (2013)

(continued)

Table 4.2 (continued)

Strain	Chemicals	Plant	Effect	References
Bacillus strains	Pool of VOCs	*Arabidopsis thaliana*	Growth promotion, root architecture modifications	Gutiérrez-Luna et al. (2010)
Bacillus subtilis SYST2	1,3-Propanediol	*Solanum lycopersicum*	Growth promotion	Tahir et al. (2017)
Bacillus subtilis GB03	2,3-Butanediol	*Arabidopsis thaliana*	Growth promotion	Ryu et al. (2003)
	Pool of VOCs	*Arabidopsis thaliana*	Growth promotion, inflorescence development, chlorophyll content	Xie et al. (2009)
	Pool of VOCs	*Ocimum basilicum*	Growth promotion, essential oil accumulation	Banchio et al. (2009)
	Pool of VOCs	*Arabidopsis thaliana*	Iron acquisition induction	Zhang et al. (2009)
	Pool of VOCs	*Mentha × piperita*	Growth promotion, monoterpene biosynthesis stimulated	Santoro et al. (2011)
Bacillus vallismortis EXT-1	3-Hydroxy-2-Butanone	*Nicotiana tabacum*	Growth promotion	Ann et al. (2013)
Burkholderia ambifaria	Dimethyl trisulfide, acetophenone, 3-hexanone	*Arabidopsis thaliana*	Growth promotion, fungi growth inhibition	Groenhagen et al. (2013)
Burkholderia pyrrocinia LMG 21822	Indole	*Arabidopsis thaliana*	Growth promotion	Blom et al. (2011)
Chromobacterium violaceum CV0	Indole	*Arabidopsis thaliana*	Growth promotion	Blom et al. (2011)
Escherichia coli	Indole	*Arabidopsis thaliana*	Growth promotion	Bailly et al. (2014)
Paenibacillus polymyxa E681	n-Tridecane	*Arabidopsis thaliana*	Induced systemic response (ISR)	Lee et al. (2012)
Proteus vulgaris JBLS202	Indole	*Arabidopsis thaliana*	Growth promotion	Bhattacharyya et al. (2015)

Pseudomonas fluorescens SS101	13-Tetradecadien-1-ol, 2-butanone, 2-Methyl-n-1-tridecene	*Nicotiana tabacum*	Growth promotion	Park et al. (2015)
Pseudomonas fluorescens WCS417r	Pool of VOCs	*Mentha × piperita*	Growth promotion, monoterpene biosynthesis stimulated	Santoro et al. (2011)
Pseudomonas simiae AU	Pool of VOCs	*Glycine max*	Growth promotion and induced systemic tolerance (IST) to salt stress	Vaishnav et al. (2015)
Streptomyces coelicolor	3-Octanone	*Arabidopsis thaliana*	Growth promotion	Dotson et al. (2020)

the growth of *Arabidopsis thaliana*, it was found that the blend of VOCs promoted both the shoot growth and lateral root formation (Contreras-Cornejo et al. 2014b). In other hand, although species of the phytopathogenic fungus *Verticillium* cause rot in different crops, it has been observed that natural mixtures of volatile compounds constituted in part by 2-methyl-1-propanol, 3-methyl-1-butanol, 2-methyl-1-butanol, 3-pentenol, 3-methyl-3-buten-1-ol, 3-octanone, acetoin, 1-hexanol, 4-methyl-5-hexen-2-ol, 3-octanol, 4-methyl-6-hepten-3-ol, 1-octen-3-ol, 3,5,5-trimethyl-2-hexene, 2,3-butanediol, 3-ethyl-4-methyl-3-penten-2-one, himachala-2,4-diene, β-caryophyllene, α-amorphene, azulene, and phenylethyl alcohol but their effects on plants have not been fully elucidated. Such compounds from fungi can promote the *Arabidopsis thaliana* growth and such effect is related in part with the manipulation of the auxin-mediated signaling pathway (Li et al. 2018).

However, few studies have shown that certain pure metabolites are biologically active to promote growth. Among the group of the sesquiterpenoid compounds, it is known that β-caryophyllene is a metabolite whose range of biological activity in *Lactuca sativa* plants is 25–100 μM (Minerdi et al. 2011). Among the different volatile metabolites produced by *Laccaria bicolor* the sesquiterpenoid compound (−)-thujopsene is present, which is responsible for stimulating the formation of lateral roots in *Arabidopsis thaliana* (Ditengou et al. 2015). In experiments performed in vitro, it has been found that some species of *Trichoderma* emit the lactone 6-pentyl-2*H*-pyran-2-one. Applied exogenously on *Arabidopsis thaliana* at concentrations of 50–175 μM, such metabolites are responsible for promoting the plant growth through the modulation of the auxin transporters expression (Garnica-Vergara et al. 2016).

The volatile metabolites 2,3-butanediol and 3-hydroxy-2-butanone (acetoin) are compounds produced by *Bacillus subtilis* GB03 and *Bacillus amyloliquefaciens* IN937a released in the soil and triggered towards the atmosphere by the plant growth-promoting bacteria (Ryu et al. 2003). In pharmacological tests with the pure compound 2,3-butanediol and *Arabidopsis thaliana* it was found that such volatile metabolite is active in the range of 1–100 μg as revealed by the increased growth of the leaf area (Fincheira and Quiroz 2018).

4.3.2 Root Colonization

Microorganisms of the rhizosphere can colonize root tissues and establish eventual, facultative, or obligate associations with their host (Santoyo et al. 2016). The processes of root colonization involve the production of hydrolytic enzymes of microbial origin that are used as tools to break the layers of root cells and penetrate the root tissue (Contreras-Cornejo et al. 2016). Cellulases and xylanases are fungal enzymes that degrade specific plant tissue constituents, and most likely are components involved in the root colonization process (Henrissat et al. 1985; Payne et al. 2015; Estrada-Rivera et al. 2019).

There are other types of non-hydrolytic enzymes that favor the binding or adhesion of fungal hyphae to root cells. For example, swollenin (Swo) produced by *T. asperellum* and other fungal strains (Brotman et al. 2008; Meng et al. 2019). Swo possesses a cellulose-binding module, and in turn disrupts the structure of the crystalline cellulose present in the plant cell wall. It has been demonstrated that during the plant-*Trichoderma* interaction Swo facilitates the expansion of the plant cell wall in roots and root hairs (Brotman et al. 2008). Once the physical interaction between the root and the microorganism occurs, a series of physiological, biochemical, and molecular events are triggered in the host plant.

4.3.3 Plant Growth Promotion

The promotion of plant growth and development regulated by rhizosphere microorganisms can occur at different stages of the interaction (Sevilla et al. 2001; Ryu et al. 2003; Real-Santillán et al. 2019). In nature, some plants and microorganisms establish specific associations, while there are microorganisms that have the ability to promote growth without host specificity (Ray et al. 2018). Figure 4.3 shows the dynamics among phytohormones and other microbial molecules that modulate multiple process of the plant physiology. It has often been observed that soil bacteria and fungi can modulate the growth and plasticity of the root system (Contreras-Cornejo et al. 2014a, b). These effects are the result of inducing the differentiation of epidermal cells for the formation of root hairs, the activation of the cells of the primary root pericycle to induce the formation of lateral roots, and by activating the cell cycle in the cells of the tip of the primary root (López-Bucio et al. 2007).

It is important to note that the modulation of the plant growth and development by microorganisms can vary because not all microorganisms impact in the same way (Reboutier et al. 2002; Gutiérrez-Luna et al. 2010; García-Gómez et al. 2019). This effect is because there are microorganisms that produce more than one growth regulating metabolite. For example, *Burkholderia phytofirmans* PsJN and *Pseudomonas putida* W619 produce IAA and ACCase, and *Enterobacter* sp. 638 produces siderophore, IAA, acetoin, and 2,3-butanediol (Santoyo et al. 2016). Figure 4.4 shows the kind of VOCs produced by some soil fungi and the roles of such metabolites in plant physiology remain to be determined.

It has been observed that bacteria that have the same 16S rRNA gene sequence may impact differently on the phenotype of the plants with which they interact (Blakney and Patten 2011; Haney et al. 2015; Timm et al. 2015; Herrera-Paredes and Lebeis 2016). While some microorganisms can impact on the formation of lateral roots, others can do so on the formation of root hairs or the growth of the primary root (Reboutier et al. 2002; Gutiérrez-Luna et al. 2010; Garnica-Vergara et al. 2016). Even a certain microorganism can modulate the development of the host plant depending on the type of interaction that can occur without physical contact or by establishing the colonization of the root tissue (Macías-Rodríguez et al. 2018). Frequently, the aerial zone of the plants is also altered after the microbial stimulus.

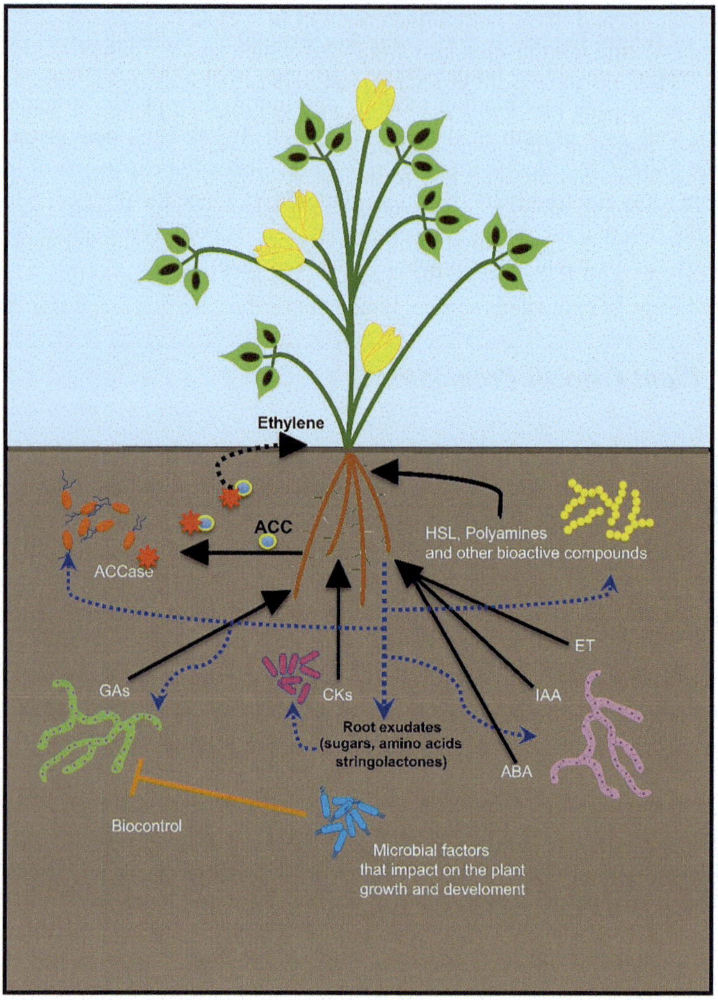

Fig. 4.3 A simplified model for a plant-microorganism interaction. In the rhizosphere detrimental or plant beneficial microorganisms produce several metabolites (black lines) with capacity to modulate the plant growth or induce defense responses. In addition, root exudates (blue lines) impact on the microbial communities and key plant-derived compounds are signaling molecules to coordinate specific interactions with rhizobia or mycorrhizal arbuscular fungi. Furthermore, sugar derived from roots can serve as nutritional source for soil microorganisms. *ABA* abscisic acid, *IAA* indole-3-acetic acid, *ET* ethylene, *CKs* cytokinins, *GAs* gibberellins, *ACCase* 1-aminocyclopropane-1-carboxylate deaminase, *ACC* 1-aminocyclopropane-1-carboxylic acid, *HSL* homoserine-lactones

Petioles, leaf size, and accumulation of foliar biomass generally increase (Flores-Cortez et al. 2019).

The opening of foliar stomata can also be modulated, which will have an impact on the transpiration of the leaves and the exchange of gases towards the atmosphere

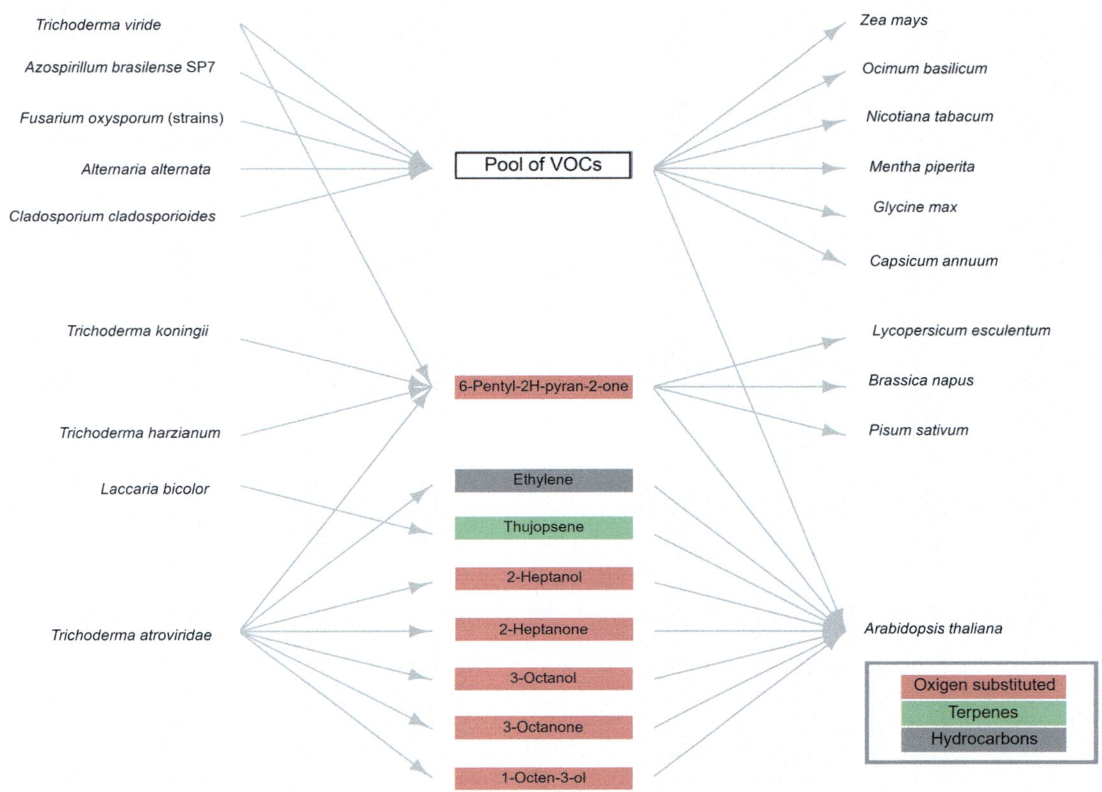

Fig. 4.4 Schematic representation of volatile compounds produced by soil fungi and plants in which have been investigated their effects

(Contreras-Cornejo et al. 2015b). The modulation of plant growth and development has been observed in different plants of agricultural interest and under different growing conditions. It has been observed that the rhizobacterium *Arthrobacter agilis* UMCV2 modulates in an organ-specific manner the expression of *FERRIC REDUCTION OXIDADSE (FRO)* genes *MtFRO1*, *MtFRO2*, *MtFRO3*, *MtFRO4*, and *MtFRO5* of *Medicago truncatula* involved in iron uptake in both conditions of sufficiency and deficiency (Montejano-Ramírez et al. 2018).

Another beneficial effect for sugarcane plants cultivar SP70-1143 induced by the colonization of the diazotrophic bacterium *Acetobacter diazotrophicus* (syn. *Gluconacetobacter diazotrophicus*) is the promotion of growth and the incorporation of nitrogen as revealed by a $^{15}N_2$ incorporation bioassay (Sevilla et al. 2001). *Azospirillum brasilense* is another diazotrophic bacterium whose effects on growth promotion are mainly attributed to the production of IAA (Barbieri et al. 1991; Okon and Labandera-Gonzalez 1994). Rhizosphere microorganisms can also promote the plant growth through the production and emission of volatile organic compounds. For example, in vitro experiments with bean plants (*Phaseolus vulgaris*) and *Arabidopsis thaliana* inoculated with the rhizobacteria *Bacillus megaterium*, it was found that compared with non-inoculated plants, the microorganism promoted the growth of both types of plant as shown by the accumulation of fresh biomass and the induction of lateral roots (López-Bucio et al. 2007).

4.4 Plant Defense Responses

One of the first events that occur during the plant-microorganism interactions is the activation of local or systemic defense responses, and it will depend on the type of microorganism detected (Stringlis et al. 2018). After the perception of the microbial signals by the host plant, there are increases in the cytoplasmic Ca^{2+} content, changes in the endogenous content of reactive oxygen species (i.e., H_2O_2, NO), and primary metabolites that include the phytohormones salicylic acid, jasmonic acid, and ethylene (Farag et al. 2013).

Defense responses can be induced after the perception of bacterial components such as low molecular compounds (AHLs), flagella, siderophores, and lipopolysaccharides (Lugtenberg and Kamilova 2009). The tomato mottle virus (ToMoV) which is transmitted by natural dynamics of viruliferous silverleaf whitefly *Bemisia argentifolii* adults. An experiment carried out in the spring of 1998 under field conditions to evaluate the effect of the inoculation of various plant growth-promoting bacteria on tomato plants infected with *Bemisia argentifolii* revealed that treatments with *Bacillus subtilis* 937b and *Bacillus pumilus* SE34 significantly reduced the rate of symptom severity (Murphy et al. 2000). Some microorganisms have the potential to control multiple plant diseases.

Liu et al. (2017) observed that the strains AP196, AP197, AP203, AP208, and AP298 of *Bacillus velezensis* showed some effectiveness in reducing postemergence damping off in pepper caused by *Rhizoctonia solani*, in cucumber caused by

Pythium ultimum, and in tomato caused by *Xanthomonas axonopodis* and *Pseudomonas syringae* pv. tomato in plants grown in controlled conditions. The rhizobacterium *Pseudomonas fluorescens* FPT9601-T5 isolated from tomato plants also induces defense responses that partially suppress the infection caused by *Pseudomonas syringae* pv. tomato DC3000 in *Arabidopsis thaliana*; these effects were related to increases in the expression of genes involved in signal transduction and metabolic processes. Among the genes that were downregulated, some are members of the family of transcriptional factors *ETHYLENE RESPONSE FACTOR* (*ERF*) and *MyB* belonging to the signaling pathway mediated by ET (Wang et al. 2005). Microorganisms that establish physical interactions with roots also often emit volatile organic compounds that can activate defense responses in plants (Farag et al. 2013).

For example, it was observed that some rhizobacteria can stimulate defense responses through the emission of VOCs (Farag et al. 2013). In an elegant work by Ryu et al. (2004) with *Arabidopsis thaliana* plants exposed to the VOCs of the bacteria *Bacillus subtilis* GB03, *Bacillus amyloliquefaciens* IN937a, and *Escherichia coli* DH5α and infected with the pathogenic plant bacterium *Erwinia carotova*, it was found that the strains GB05 and IN937a significantly reduced the foliar attack of the pathogen. This effect was due in part to the activity of 2,3-butanediol in the concentration range of 0.2 pg to 20 μg. In the same work, it was found that the VOCs of the GB05 strain induced changes in the expression of the reporter line *PDF1.2::GUS* in response to JA/ET, observed as increased activity of β-glucuronidase.

It was found that VOCs released by *T. virens* also induced changes in the expression of the transgenic line *LOX2::GUS*, a JA-responsive gene. Such priming on plant defense was correlated with accumulations in the contents of JA and H_2O_2. *T. virens* Gv29-8-elicited plant immunity was effective to restrict the attack caused by the necrotrophic fungus *Botrytis cinerea* on the shoot (Contreras-Cornejo et al. 2014b).

These data suggested that bacterial or fungal VOCs can modulate defense responses and confer protection against the attack of pathogenic microorganisms. On the other hand, the complex mixtures of VOCs emitted by plants in interaction with soil microorganisms generally have a strong impact on the third trophic level because they act as alarm signals, inducing indirect defense responses to attract natural enemies of herbivorous insects (Dicke et al. 2009; Schausberger et al. 2012). For example, maize plants constantly suffer the attack of the fall armyworm *Spodoptera frugiperda*, but in field conditions the insect is endoparasitized by female wasps of *Campoletis sonorensis* and during the maize root-*T. atroviride* association, the fungus releases in the soil 6-PP, which attracts *C. sonorensis* towards plants that are suffering herbivory. Endoparasitism is then stimulated, providing biocontrol of *S. frugiperda* at least in controlled conditions (Contreras-Cornejo et al. 2018a).

4.4.1 JA/ET Mediated Immunity

The induction of systemic resistance is a characteristic of plant beneficial microorganisms. This phenomenon was first detected in plant growth-promoting rhizobacteria and later identified in biocontrol and mycorrhizal fungi (Jaroszuk-Ściseł et al. 2019). Such microorganisms can induce increases in the content of JA, a metabolite derived from oxylipins (Wang et al. 2020). The production of JA occurs within the cell, in the chloroplast; there in the organelle linoleic acid (18:3) and hexadecatrienoic acid (16:3) through the corresponding catalytic activity of the enzymes lipoxigenase, allene oxidase synthase, and allene oxide cyclase are converted to 12-oxophytodienoic acid (OPDA) and dinor-OPDA. Subsequently, OPDA reductase 3 (OPR3) acts on OPDA and converts it to 3-oxo-2 (2-′-pentenyl)-cyclopentane-1-octanoic acid (OPC-8:0). This last compound is subjected to three cycles of β-oxidation resulting then in the production of JA (Al-Zahrani et al. 2020).

JA activates effective defense responses against necrotrophic pathogens or herbivorous chewing insects (Browse 2009). Recently, it was reported that *T. virens* Gv29-8 when associated with the roots of maize plants increases the content of 2-oxo-phytodienoic acid in the xylem (12-OPDA) and α-ketol of octadecadienoic acid (KODA); these oxylipins were shown to be key in the induction of ISR because the mutant line *lox10* of maize that is susceptible to the attack of the pathogenic fungus *Colletotrichum graminicola* and that was later transfused with sap that contained high concentrations of 12-OPDA restored the ISR in such mutant (Wang et al. 2020).

In multiple plant-microorganism interactions a cross talk can occur between phytohormones that simultaneously modulate plant growth and defense processes (Villalobos-Escobedo et al. 2020). In a more recent work, it was reported that during the interaction of *T. atroviride* with *Arabidopsis thaliana*, the fungal NADPH oxidase plays a key role in both plant growth promotion and defense responses enhanced by the fungus. In regard with the modulation of the root-system architecture by the fungus, important information was revealed by the Δ*noxR* mutant of *T. atroviride*. Such strain failed in modulating the developmental progress of lateral roots formation as observed with the wild-type strain. In contrast, Δ*noxR* efficiently activated JA-mediated plant defense, which was effective against the attack of the necrotrophic fungus *B. cinerea* (Villalobos-Escobedo et al. 2020). Modulations of the endogenous levels of JA in maize plants by *T. atroviride* IMI 206040 after 5 days of interaction with roots also resulted in an effective priming of plant defense to reduce the attack of the chewing insect *S. frugiperda* (Contreras-Cornejo et al. 2018b).

On the other hand, auxin has been shown to alter SA production and JA signaling, and ABA can act as a hormone antagonistic to ET/JA-mediated signaling (Ponzio et al. 2013). In a study carried out in tobacco plants (*Nicotiana tabacum*) infected with *Pseudomonas syringae*, it was found that resistance to the infection caused by this pathogen was due to the modulation of the ABA and kinetin (a CK) levels. In this condition, CK decreased the ABA levels in tobacco plants. On the other hand, the decrease in ABA concentrations allowed the participation of CK to activate the resistance mechanism against *P. syringae* (Großkinsky et al. 2014).

4.4.2 SA-Mediated Immunity

It has been observed that some pathogenic bacteria and soil fungi can activate defense responses mediated by SA (Contreras-Cornejo et al. 2011). This type of defense is known as systemic acquired resistance (SAR) and is effective against biotrophic microorganisms (Vlot et al. 2009). The SA-mediated signaling mechanism results in the induction of the expression of *PATHOGENESIS-RELATED* (*PR*) genes and their subsequent translation into their respective defensin proteins (Barriuso et al. 2008). There is evidence, though, that the signaling routes of the JA and SA are antagonistic (Derksen et al. 2013). The activation of both routes simultaneously by some beneficial microorganisms has been reported (Contreras-Cornejo et al. 2011; Niu et al. 2011). For example, the rhizobacterium *Bacillus cereus* AR156 when inoculated in *Arabidopsis thaliana* ecotype Col-0 activates effective defense responses against the pathogenic bacterium *Pseudomonas syringae* pv. tomato DC3000. AR156 induced the expression of the *PR-1*, *PR-2*, and *PR-5* genes in response to SA and *PDF1.2* that encodes plant defensin in response to JA/ET, which suggests that the SA, JA, and ET pathways are simultaneously activated, probably involving the NPR1 component (Niu et al. 2011).

A similar effect in the activation of the JA and SA signaling pathways was found in *Arabidopsis thaliana* after the inoculation with *T. atroviride* IMI 206040 and *T. virens* Gv.29-8 because both markers *PR-1::GUS* of response to SA and *LOX2::GUS* of response to JA were activated after 6 days of inoculation when the root colonization was established. In such interactions, although both signaling pathways were activated, they were slightly primed one on the other as revealed by the activity of β-glucuronidase after GUS staining and the levels of SA and JA accumulated in shoot tissues (Contreras-Cornejo et al. 2011).

4.5 Conclusion

In the environment, plants interact with myriads of microorganisms. In the case of plant beneficial rhizobacteria or fungi, they impact on physiological responses modulating the plant growth, adaptation to resist abiotic stress caused by salinity, drought or soil contamination and the activation of defense responses. A plant-microorganism interaction can initiate with the emission of a certain metabolite derived from one of the involved organisms. Then such signaling molecule encodes a key message to coordinate and form a specific interaction; for example, arbuscular mycorrhizal associations are established after the rhizodeposition by the plant and perception of strigolactones by the spores of the target fungus, which in turn will grow tropically towards the host root. In the case of nitrogen-fixing bacteria, plant roots release chemotactic agents to attract the rhizobia and after a complex signaling mechanism the nodule is formed.

In other interactions, microorganisms release volatile or soil diffusible signaling molecules that coordinate both plant growth and defense responses. In those cases, such plant responses are regulated after the activation of endogenous mechanisms modulated by phytohormones. For example, plant growth and development are coordinated by auxins, gibberellins, CKs, ET, and ABA. Cross talk among phytohormones can occur to efficiently coordinate some plant responses. For example, lateral root formation and root hair induction is a process that requires the participation of the MPK6, auxins, and ET. Depending on the kind of rhizosphere microorganism, plant immunity is primed by the canonical defense mediators SA, JA, and ET. This can occur due to the perception of microbial defense elicitors as peptides or flagellin or when the root colonization is established. In some interactions, plant defense involves the synergistic activity of JA and ET, but frequently antagonistic roles among JA and SA, SA and IAA, and CKs and ABA have been observed. Currently, there is a broad body of scientific literature that describes the role of rhizosphere microorganisms in plant performance. However, there is a need to investigate and generate more information about the role of unknown microbial signals that are released in the atmosphere and in the soil, and how they impact on plant performance under different changing environmental conditions or when plants are challenged simultaneously with more than one type of aggressor (i.e., pathogens and herbivores). This kind of information will be relevant to generate knowledge with potential application to manage crops in cultured lands with serious troubles of pests or disturbances caused by the climatic global change.

Acknowledgments We thank Prasun K. Mukherjee (Nuclear Agriculture and Biotechnology Division, Bhabha Atomic Research Centre, Trombay, Mumbai, India) and Benjamin A. Horwitz (Faculty of Biology Technion—Israel Institute of Technology) for kindly inviting us to contribute this book chapter. We apologize to colleagues whose relevant work we were unable to cite owing to space limitations. The authors declare that they have no conflict of interest.

References

Achard P, Cheng H, De Grauwe L, Decat J, Schoutteten H, Moritz T, Van Der Straeten D, Peng J, Harberd NP (2006) Integration of plant responses to environmentally activated phytohormonal signals. Science 311:91–94. https://doi.org/10.1126/science.1118642

Akiyama K, Matsuzaki K, Hayashi H (2005) Plant sesquiterpenes induce hyphal branching in arbuscular mycorrhizal fungi. Nature 435:824–827. https://doi.org/10.1038/nature03608

Allen GJ, Kuchitsu K, Chu SP, Murata Y, Schroeder JI (1999) *Arabidopsis abi1-1* and *abi2-1* phosphatase mutations reduce abscisic acid-induced cytoplasmic calcium rises in guard cells. Plant Cell 11:1785–1798. https://doi.org/10.1105/tpc.11.9.1785

Al-Zahrani W, Bafeel SO, El-Zohri M (2020) Jasmonates mediate plant defense responses to *Spodoptera exigua* herbivory in tomato and maize foliage. Plant Signal Behav 15:1746898. https://doi.org/10.1080/15592324.2020.1746898

Ann M, Cho Y, Ryu H, Kim H, Park K (2013) Growth promotion of tobacco plant by 3-hydroxy-2-butanone from Bacillus vallismortis EXTN-1. Korean J Pestic Sci 17:388–393. https://doi.org/10.7585/kjps.2013.17.4.388

Arroyo A, Bossi F, Finkelstein RR, León P (2003) Three genes that affects sugar sensing (Abscisic acid insensitive 4, abscisic acid insensitive 5, and constitutive triple response 1) are differentially regulated by glucose in *Arabidopsis*. Plant Physiol 133:231–242. https://doi.org/10.1104/pp.103.021089

Asari S, Matzén S, Petersen MA, Bejai S, Meijer J (2016) Multiple effects of *Bacillus amyloliquefaciens* volatile compounds: plant growth promotion and growth inhibition of phytopathogens. FEMS Microbiol Ecol 92:fiw070. https://doi.org/10.1093/femsec/fiw070

Bae H, Roberts DP, Lim HS, Strem MD, Park SC, Ryu CM, Melnick RL, Bailey BA (2011) Endophytic *Trichoderma* isolates from tropical environments delay disease onset and induce resistance against *Phytophthora capsici* in hot pepper using multiple mechanisms. Mol Plant-Microbe Interact 24:336–351. https://doi.org/10.1094/MPMI-09-10-0221

Baetz U, Martinoia E (2014) Root exudates: the hidden part of plant defense. Trends Plant Sci 19:90–98. https://doi.org/10.5167/uzh-106507

Bailly A, Groenhagen U, Schulz S, Geisler M, Eberl L, Weisskopf L (2014) The inter-kingdom volatile signal indole promotes root development by interfering with auxin signalling. Plant J 80:758–771. https://doi.org/10.1111/tpj.12666

Banchio E, Xie X, Zhang H, Paré PW (2009) Soil bacteria elevate essential oil accumulation and emissions in sweet basil. J Agric Food Chem 57:653–657. https://doi.org/10.1021/jf8020305

Barbieri P, Baggio C, Bazzicalupo M, Galli E, Zanetti G, Nuti MP (1991) *Azospirillum*-graminae interaction: effect of indole-3-acetic acid. In: Nitrogen fixation, vol 48. Springer, Dordrecht, pp 161–168. https://doi.org/10.1007/978-94-011-3486-6_29

Barriuso J, Ramos B, Gutiérrez FJ (2008) Protection against pathogen and salt stress by four plant growth-promoting rhizobacteria isolated from *Pinus* sp. on *Arabidopsis thaliana*. Phytopathology 98:666–672. https://doi.org/10.1094/PHYTO-98-6-0666

Battaglia D, Bossi S, Cascone P, Digilio MC, Duran Prieto J, Fanti P, Guerrieri E, Iodice L, Lingua G, Lorito M, Maffei ME, Massa N, Ruocco M, Sasso R, Trotta V (2013) Tomato below ground-above ground interactions: *Trichoderma longibrachiatum* affects the performance of *Macrosiphum euphorbiae* and its natural antagonists. Mol Plant-Microbe Interact 26:1249–1256. https://doi.org/10.1094/MPMI-02-13-0059-R

Bhattacharyya D, Garladinne M, Lee YH (2015) Volatile indole produced by rhizobacterium *Proteus vulgaris* JBLS202 stimulates growth of *Arabidopsis thaliana* through auxin, cytokinin, and brassinosteroid pathways. J Plant Growth Regul 34:158–168. https://doi.org/10.1007/s00344-014-9453-x

Bisseling T, Geurts R (2020) Specificity in legume nodule symbiosis. Science 369:620–621. https://doi.org/10.1126/science.abd3857

Blakney AJ, Patten CL (2011) A plant growth-promoting pseudomonad is closely related to the *Pseudomonas syringae* complex of plant pathogens. FEMS Microbiol Ecol 77:546–557. https://doi.org/10.1111/j.1574-6941.2011.01136.x

Blom D, Fabbri C, Connor E, Schiestl F, Klauser D, Boller T, Eberl L, Weisskopf L (2011) Production of plant growth modulating volatiles is widespread among rhizosphere bacteria and strongly depends on culture conditions. Environ Microbiol 13:3047–3058. https://doi.org/10.1111/j.1462-2920.2011.02582.x

Bömke C, Tudzynski B (2009) Diversity, regulation, and evolution of the gibberellin biosynthetic pathway in fungi compared to plants and bacteria. Phytochemistry 70:1876–1893. https://doi.org/10.1016/j.phytochem.2009.05.020

Brodribb TJ, McAdam SAM (2011) Passive origins of stomatal control in vascular plants. Science 331:582–585. https://doi.org/10.1126/science.1197985

Brotman Y, Briff E, Viterbo A, Chet I (2008) Role of Swollenin, an expansin-like protein from *Trichoderma*, in plant root colonization. Plant Physiol 147:779–789. https://doi.org/10.1104/pp.108.116293

Browse J (2009) Jasmonate passes muster: a receptor and targets for the defense hormone. Annu Rev Plant Biol 60:183–205. https://doi.org/10.1146/annurev.arplant.043008.092007

Camarena-Pozoz DA, Flores-Nuñez VM, López MG, López-Bucio J, Partida-Martínez LP (2018) Smells from the desert: microbial volatiles that affect plant growth and development of native and non-native plant species. Plant Cell Environ 42:1368–1380. https://doi.org/10.1111/pce.13476

Cao J, Yang C, Li L, Jiang L, Wu Y, Wu C, Bu Q, Xia G, Liu X, Luo Y, Liu J (2016) Rice plasma membrane proteomics reveals *Magnaporthe oryzae* promotes susceptibility by sequential activation of host hormone signaling pathways. Mol Plant-Microbe Interact 29:902–913. https://doi.org/10.1094/MPMI-08-16-0165-R

Caradonia F, Ronga D, Catellani M, Giaretta Azevedo CV, Alegria Terrazas R, Robertson-Albertyn S, Francia E, Bulgarell D (2019) Nitrogen fertilizers shape the composition and predicted functions of the microbiota of field-grown tomato plants. Phytobiomes J 3:315–325. https://doi.org/10.1094/PBIOMES-06-19-0028-R

Cassán F, Maiale S, Masciarelli O, Vidal A, Luna V, Ruiz O (2009) Cadaverine production by *Azospirillum brasilense* and its possible role in plant growth promotion and osmotic stress mitigation. Eur J Soil Biol 45:12–19. https://doi.org/10.1016/j.ejsobi.2008.08.003

Castulo-Rubio DY, Alejandre-Ramírez NA, Orozco-Mosqueda MC, Santoyo G, Macías-Rodríguez L, Valencia-Cantero E (2015) Volatile organic compounds produced by the rhizobacterium *Arthrobacter agilis* UMCV2 modulate *Sorghum bicolor* (strategy II plant) morphogenesis and SbFRO1 transcription in vitro. J Plant Growth Regul 34:611–623. https://doi.org/10.1007/s00344-015-9495-8

Chalupowicz L, Manulis-Sasson M, Itkin M, Sacher A, Sessa G, Barash I (2008) Quorum sensing system affects gall development incited by *Pantoea agglomerans* pv. *gypsophilae*. Mol Plant-Microbe Interact 21:1094–1105. https://doi.org/10.1094/MPMI-21-8-1094

Chalupowicz L, Barash I, Panijel M, Sessa G, Manulis-Sasson S (2009) Regulatory interactions between quorum-sensing, auxin, cytokinin, and the Hrp regulon in relation to gall formation and epiphytic fitness of *Pantoea agglomerans* pv. *gypsophilae*. Mol Plant-Microbe Interact 22:849–856. https://doi.org/10.1094/MPMI-22-7-0849

Coleman-Derr D, Desgarennes D, Fonseca-García C, Gross S, Clingenpeel G, Woyke T, North G, Visel A, Partida-Martinez LP, Tringe SG (2016) Plant compartment and biogeography affect microbiome composition in cultivated and native *Agave* species. New Phytol 209:798–811. https://doi.org/10.1111/nph.13697

Contreras-Cornejo HA, Macías-Rodríguez L, Cortés-Penagos C, López-Bucio J (2009) *Trichoderma virens*, a plant beneficial fungus, enhances biomass production and promotes lateral root growth through an auxin-dependent mechanism in *Arabidopsis*. Plant Physiol 149:1579–1592. https://doi.org/10.1104/pp.108.130369

Contreras-Cornejo HA, Macías-Rodríguez L, Beltrán-Peña E, Herrera-Estrella A, López-Bucio J (2011) *Trichoderma*-induced plant immunity likely involves both hormonal and camalexin dependent mechanisms in *Arabidopsis thaliana* and confers resistance against necrotrophic fungi *Botrytis cinerea*. Plant Signal Behav 6:1554–1563. https://doi.org/10.4161/psb.6.10.17443

Contreras-Cornejo HA, Macías-Rodríguez LI, Alfaro-Cuevas R, López-Bucio J (2014a) *Trichoderma* improves growth of *Arabidopsis* seedlings under salt stress through enhanced root development, osmolite production and Na^+ elimination through root exudates. Mol Plant-Microbe Interact 27:503–514. https://doi.org/10.1094/MPMI-09-13-0265-R

Contreras-Cornejo HA, Macías-Rodríguez L, Herrera-Estrella A, López-Bucio J (2014b) The 4-phosphopantetheinyl transferase of *Trichoderma virens* plays a role in plant protection against *Botrytis cinerea* through volatile organic compound emission. Plant Soil 379:261–274. https://doi.org/10.1007/s11104-014-2069-x

Contreras-Cornejo HA, Macías-Rodríguez L, Garnica-Vergara A, López-Bucio J (2015a) *Trichoderma* modulates stomatal aperture and leaf transpiration through an abscisic acid-dependent mechanism. J Plant Growth Regul 34:425–432. https://doi.org/10.1007/s00344-014-9471-8

Contreras-Cornejo HA, López-Bucio JS, Méndez-Bravo A, Macías-Rodríguez L, Ramos-Vega M, Guevara-García AA, López-Bucio J (2015b) Mitogen-activated protein kinase 6 and ethylene and auxin signaling pathways are involved in *Arabidopsis* root-system architecture alterations by *Trichoderma atroviride.* Mol Plant-Microbe Interact 28:701–710. https://doi.org/10.1094/MPMI-01-15-0005-R

Contreras-Cornejo HA, Macías-Rodríguez L, López-Bucio J (2015c) Fungal biomolecules in plant growth promotion. In: Gupta VK, Mach RL, Sreenivasaprasad S (eds) Fungal biomolecules: sources, applications and recent developments. Wiley, Hoboken, NJ. https://doi.org/10.1002/9781118958308.ch23

Contreras-Cornejo HA, Macías-Rodríguez L, del Val E, Larsen J (2016) Ecological functions of *Trichoderma* spp. and their secondary metabolites in the rhizosphere: interactions with plants. FEMS Microbiol Ecol 92:fiw036. https://doi.org/10.1093/femsec/fiw036

Contreras-Cornejo HA, del Val E, Macías-Rodríguez L, Alarcón A, González-Esquivel CE, Larsen J (2018a) *Trichoderma atroviride*, a maize root associated fungus, increases the parasitism rate of the fall armyworm *Spodoptera frugiperda* by its natural enemy *Campoletis sonorensis.* Soil Biol Biochem 122:196–202. https://doi.org/10.1016/j.soilbio.2018.04.013

Contreras-Cornejo HA, Macías-Rodríguez L, del Val E, Larsen J (2018b) The root endophytic fungus *Trichoderma atroviride* induces foliar herbivory resistance in maize plants. Appl Soil Ecol 124:45–53. https://doi.org/10.1016/j.apsoil.2017.10.004

Contreras-Cornejo HA, Viveros-Bremauntz F, del Val E, Macías-Rodríguez L, López-Carmona DA, Alarcón A, González-Esquivel CE, Larsen J (2020a, 2020) Alterations of foliar arthropod communities in a maize agroecosystem induced by the root-associated fungus *Trichoderma harzianum.* J Pest Sci. https://doi.org/10.1007/s10340-020-01261-3

Contreras-Cornejo HA, Macías-Rodríguez L, del Val E, Larsen J (2020b) Interactions of *Trichoderma* with plants, insects, and plant pathogen microorganisms: chemical and molecular bases. In: Mérillon JM, Ramawat KM (eds) Co-evolution of secondary metabolites. Springer Nature, Switzerland AG. https://doi.org/10.1007/978-3-319-76887-8_23-1

Coppola M, Cascone P, Di Lelio I, Woo SL, Lorito M, Rao R, Pennacchio F, Guerrieri E, Digilio MC (2019a) *Trichoderma atroviride* P1 colonization of tomato plants enhances both direct and indirect defense barriers against insects. Front Physiol 10:813. https://doi.org/10.3389/fphys.2019.00813

Coppola M, Diretto G, Digilio MC, Woo SL, Giuliano G, Molisso D, Pennacchio F, Lorito M, Rao R (2019b) Transcriptome and metabolome reprogramming in tomato plants by *Trichoderma harzianum* primes and enhances defense responses against aphids. Front Physiol 10:745

Coutinho BG, Mitter B, Talbi C, Sessitsch A, Bedmar EJ, Halliday N, James EK, Cámara M, Venturi V (2013) Regulon studies and *in planta* role of the BraI/R quorum-sensing system in the plant-beneficial *Burkholderia* cluster. Appl Environ Microbiol 79:4421–4432. https://doi.org/10.1128/AEM.00635-13

Cristescu SM, De Martinis D, Hekkert STL, Parker DH, Harren FJ (2002) Ethylene production by *Botrytis cinerea* in vitro and in tomatoes. Appl Environ Microbiol 68:5342–5350. https://doi.org/10.1128/AEM.68.11.5342-5350.2002

De Billy F, Grosjean C, May S, Bennett M, Cullimore JV (2001) Expression studies on *AUX1*-like genes in *Medicago truncatula* suggest that auxin is required at two steps in early nodule development. Mol Plant-Microbe Interact 14:267–277. https://doi.org/10.1094/MPMI.2001.14.3.267

De Weert S, Vermeiren H, Mulders IHM, Kuiper I, Hendrickx N, Bloemberg GV, Vanderleyden J, De Mot R, Lugtenberg BJJ (2002) Flagella-driven chemotaxis towards exudate components is an important trait for tomato colonization by *Pseudomonas fluorescens.* Mol Plant-Microbe Interact 15:1173–1180. https://doi.org/10.1094/MPMI.2002.15.11.1173

Derksen H, Rampitsch C, Daayf F (2013) Signaling cross-talk in plant disease resistance. Plant Sci 207:79–87. https://doi.org/10.1016/j.plantsci.2013.03.004

Devos S, Laukens K, Deckers P, Van Der Straeten D, Beeckman T, Inzé D, Van Onckelen HV, Witters E, Prinsen E (2006) A hormone and proteome approach to picturing the initial metabolic events during *Plasmodiophora brassicae* infection on Arabidopsis. Mol Plant-Microbe Interact 19:1431–1443. https://doi.org/10.1094/MPMI-19-1431

Dharmasiri N, Dharmasiri S, Estelle M (2005) The F-box protein TIR1 is an auxin receptor. Nature 435:441–445. https://doi.org/10.1038/nature03543

Dicke M, van Loon JJA, Soler R (2009) Chemical complexity of volatiles from plants induced by multiple attack. Nat Chem Biol 5:317–324. https://doi.org/10.1038/nchembio.169

Ditengou FA, Lapeyrie F (2000) Hypaphorine from the ectomycorrhizal fungus *Pisolithus tinctorius* counteracts activities of indole-3-acetic acid and ethylene but not synthetic auxins in eucalypt seedlings. Mol Plant-Microbe Interact 13:151–158. https://doi.org/10.1094/MPMI.2000.13.2.151

Ditengou FA, Müller A, Rosenkranz M, Felten J, Lasok H, van Doorn MM, Legué V, Palme K, Schnitzler JP, Polle A (2015) Volatile signalling by sesquiterpenes from ectomycorrhizal fungi reprogrammes root architecture. Nat Commun 6:6279. https://doi.org/10.1038/ncomms7279

Domingo C, Andrés F, Tharreau D, Iglesias DJ, Talón M (2009) Constitutive expression of *OsGH3.1* reduces auxin content and enhances defense response and resistance to a fungal pathogen in rice. Mol Plant-Microbe Interact 22:201–210. https://doi.org/10.1094/MPMI-22-2-0201

Dotson BR, Verschut V, Flärdh K, Becher PG, Rasmusson AG (2020) The *Streptomyces* volatile 3-octanone alters auxin/cytokinin and growth in *Arabidopsis thaliana* via the gene family KISS ME DEADLY. bioRxiv. 2020.02.15.949685. https://doi.org/10.1101/2020.02.15.949685

Estrada-Rivera M, Rebolledo-Prudencio OG, Pérez-Robles DA, Rocha-Medina MDC, González-López MDC, Casas-Flores S (2019) Trichoderma histone deacetylase HAD-2 modulates multiple responses in Arabidopsis. Plant Physiol 179:1343–1361. https://doi.org/10.1104/pp.18.01092

Farag MA, Zhang H, Ryu CM (2013) Dynamic chemical communication between plants and bacteria through airborne signals: induced resistance by bacterial volatiles. J Chem Ecol 39: 1007–1018. https://doi.org/10.1007/s10886-013-0317-9

Felten J, Kohler A, Morin E, Bhalerao RP, Palme K, Martin F, Dintengou FA, Legué V (2009) The Ectomycorrhizal fungus *Laccaria bicolor* stimulates lateral root formation in Poplar and Arabidopsis through auxin transport and signaling. Plant Physiol 151:1991–2005. https://doi.org/10.1104/pp.109.147231

Fincheira P, Quiroz A (2018) Microbial volatiles as plant growth inducers. Microbiol Res 208:63–75. https://doi.org/10.1016/j.micres.2018.01.002

Finkelstein R, Lynch T, Reeves W, Petitfils M, Mostachetti M (2011) Accumulation of the transcription factor ABA-insensitive (ABI)4 is tightly regulated post-transcriptionally. J Exp Bot 62:3971–3979. https://doi.org/10.1093/jxb/err093

Flores-Cortez I, Winkler R, Ramírez-Ordorica A, Elizarraraz-Anaya MIC, Carrillo-Rayas MT, Valencia-Cantero E, Macías-Rodríguez L (2019) A mass spectrometry-based study shows that volatiles emitted by *Arthrobacter agilis* UMCV2 increase the content of brassinosteroids in *Medicago truncatula* in response to iron deficiency stress. Molecules 24:3011. https://doi.org/10.3390/molecules24163011

Foo E, Ross JJ, Jones WT, Reid JB (2013) Plant hormones in arbuscular mycorrhizal symbioses: an emerging role for gibberellins. Ann Bot 111:769–779. https://doi.org/10.1093/aob/mct041

Frugier F, Kosuta S, Murray JD, Crespi M, Szczyglowski K (2008) Cytokinin: secret agent of symbiosis. Trends Plant Sci 13:115–120. https://doi.org/10.1016/j.tplants.2008.01.003

García-Gómez P, Almagro G, Sánchez-López AM, Bahaji A, Ameztoy K, Ricarte-Bermejo A, Baslam M, Antolín MC, Urdiain A, López-Belchi MD, López-Gómez MD, Morán JF, Garrido J, Muñoz FJ, Baroja-Fernández E, Pozueta-Romero J (2019) Volatile compounds other than CO_2 emitted by different microorganisms promote distinct posttranscriptionally regulated responses in plants. Plant Cell Environ 42:1729–1746. https://doi.org/10.1111/pce.13490

Garnica-Vergara A, Barrera-Ortiz S, Muñoz-Parra E, Raya-Gonzalez J, Méndez-Bravo A, Macías-Rodríguez L, Ruiz-Herrera LF, López-Bucio J (2016) The volatile 6-pentyl-2H-pyran-2-one from *Trichoderma atroviride* regulates *Arabidopsis thaliana* root morphogenesis via auxin signaling and *ETHYLENE INSENSITIVE 2* functioning. New Phytol 209:1496–1512. https://doi.org/10.1111/nph.13725

Giron D, Frago E, Glevarec G, Pieterse CMJ, Dicke M (2013) Cytokinins as key regulators in plant-microbiome-insect interactions: connecting plant growth and defense. Funct Ecol 27:599–609. https://doi.org/10.1111/1365-2435.12042

Gonzalez-Rizzo S, Crespi M, Frugier F (2006) The *Medicago truncatula* CRE1 cytokinin receptor regulates lateral root development and early symbiotic interaction with *Sinorhizobium meliloti*. Plant Cell 18:2680–2693. https://doi.org/10.1105/tpc.106.043778

Götz C, Fekete A, Gebefuegi I, Forczek ST, Fuksová K, Li X, Englmann M, Gryndler M, Hartmann A, Matucha M, Schmitt-Kopplin P, Schröder P (2007) Uptake, degradation and chiral discrimination of *N*-acyl-D/L-homoserine lactones by barley (*Hordeum vulgare*) and yam bean (*Pachyrhizus erosus*) plants. Anal Bioanal Chem 389:1447–1457. https://doi.org/10.1007/s00216-007-1579-2

Gouda S, Kerry RG, Das G, Paramithiotis S, Shin HS, Patra JK (2018) Revitalization of plant growth promoting rhizobacteria for sustainable development in agriculture. Microbiol Res 206:131–140. https://doi.org/10.1016/j.micres.2017.08.016

Groenhagen U, Baumgartner R, Bailly A, Gardiner A, Eberl L, Schulz S, Weisskopf L (2013) Production of bioactive volatiles by different *Burkholderia ambifaria* strains. J Chem Ecol 39:892–906. https://doi.org/10.1007/s10886-013-0315-y

Großkinsky DK, van der Graaff E, Roitsch T (2014) Abscisic acid–cytokinin antagonism modulates resistance against *Pseudomonas syringae* in tobacco. Phytopathology 104:1283–1288. https://doi.org/10.1094/PHYTO-03-14-0076-R

Guo Y, Jud W, Ghirardo A, Antritter F, Benz JP, Schnitzler JP, Rosenkranz M (2020) Sniffing fungi-phenotyping of volatile chemical diversity in *Trichoderma* species. New Phytol 227:244–259. https://doi.org/10.1111/nph.16530

Gutiérrez-Luna FM, López-Bucio J, Altamirano-Hernández J, Valencia-Cantero E, Reyes de la Cruz H, Macías-Rodríguez L (2010) Plant growth-promoting rhizobacteria modulate root-system architecture in *Arabidopsis thaliana* through volatile organic compound emission. Symbiosis 51:75–83. https://doi.org/10.1007/s13199-010-0066-2

Hamayun M, Khan SA, Khan MA, Khan AL, Kang SM, Kim SK, Joo GJ, Lee IJ (2009) Gibberellin production by pure cultures of a new strain of *Aspergillus fumigatus*. World J Microbiol Biotechnol 25:1785–1792. https://doi.org/10.1007/s11274-009-0078-3

Haney CH, Samuel BS, Bush J, Ausubel FM (2015) Associations with rhizosphere bacteria can confer an adaptive advantage to plants. Nat Plants 1:15051. https://doi.org/10.1038/nplants.2015.51

Hao HT, Zhao X, Shang QH, Wang Y, Guo ZH, Zhang YB, Xie ZK, Wang RY (2016) Comparative Digital gene expression analysis of the *Arabidopsis* response to volatiles emitted by *Bacillus amyloliquefaciens*. PLoS One 11:e0158621. https://doi.org/10.1371/journal.pone.0158621

Hardoim PR, van Overbeek LS, Elsas JD (2008) Properties of bacterial endophytes and their proposed role in plant growth. Trends Microbiol 16:463–471. https://doi.org/10.1016/j.tim.2008.07.008

Henrissat B, Driguez H, Viet C, Schülein M (1985) Synergism of cellulases from *Trichoderma reesei* in the degradation of cellulose. Nat Biotechnol 3:722–726. https://doi.org/10.1038/nbt0885-722

Herrera-Paredes S, Lebeis SL (2016) Giving back to the community: microbial mechanisms of plant-soil interactions. Funct Ecol 30:1043–1052. https://doi.org/10.1111/1365-2435.12684

Hershey DM, Lu X, Zi J, Peters RJ (2014) Functional conservation of the capacity for *ent*-kaurene biosynthesis and an associated operon in certain rhizobia. J Bacteriol 196:100–106. https://doi.org/10.1128/JB.01031-13

Hinsinger P, Plassard C, Tang C, Jaillard B (2003) Origins of root-mediated pH changes in the rhizosphere and their responses to environmental constraints: a review. Plant Soil 248:43–59. https://doi.org/10.1023/A:1022371130939

Hirano K, Tanaka MU, Matsuoka M (2008) GID1 mediated gibberellin signaling in plants. Trends Plant Sci 13:192–199. https://doi.org/10.1016/j.tplants.2008.02.005

Huang Y, Li H, Hutchison CE, Laskey J, Kieber JJ (2003) Biochemical and functional analysis of CTR1, a protein kinase that negatively regulates ethylene signaling in *Arabidopsis*. Plant J 33: 221–233. https://doi.org/10.1046/j.1365-313x.2003.01620.x

Hwang HH, Yang FJ, Cheng TF, Chen YC, Lee YL, Tsai YL, Lai EM (2013) The Tzs protein and exogenous cytokinin affect virulence gene expression and bacterial growth of *Agrobacterium tumefaciens*. Phytopathology 103:888–899. https://doi.org/10.1094/PHYTO-01-13-0020-R

Jaroszuk-Ściseł J, Tyśkiewicz R, Nowak A, Ozimek E, Majewska M, Hanaka A, Tyśkiewicz K, Pawlik A, Janusz G (2019) Phytohormones (auxin, gibberellin) and ACC deaminase in vitro synthesized by the mycoparasitic *Trichoderma* DEMTkZ3A0 strain and changes in the level of auxin and plant resistance markers in wheat seedlings inoculated with this strain conidia. Int J Mol Sci 20:4923. https://doi.org/10.3390/ijms20194923

Jia YJ, Kakuta Y, Sugawara M, Igarashi T, Oki N, Kisaki M, Shoji T, Kanetuna Y, Horita T, Matsui H, Honma M (1999) Synthesis and degradation of 1 aminocyclopropane 1 carboxylic acid by *Penicillium citrinum*. Biosci Biotechnol Biochem 63:542–549. https://doi.org/10.1271/bbb.63.542

Joshi MV, Loria R (2007) *Streptomyces turgidiscabies* possesses a functional cytokinin biosynthetic pathway and produces leafy galls. Mol Plant-Microbe Interact 20:751–758. https://doi.org/10.1094/MPMI-20-7-0751

Juan-Ovejero R, Briones MJI, Öpik M (2020) Fungal diversity in peatlands and its contribution to carbon cycling. Appl Soil Ecol 146:103393. https://doi.org/10.1016/j.apsoil.2019.103393

Kazan K (2015) Diverse roles of jasmonates and ethylene in abiotic stress tolerance. Trends Plant Sci 20:219–229. https://doi.org/10.1016/j.tplants.2015.02.001

Kepinski S, Leyser O (2005) The *Arabidopsis* F-box protein TIR1 is an auxin receptor. Nature 435: 446–451. https://doi.org/10.1038/nature03542

Khatabi B, Molitor A, Lindermayr C, Pfiffi S, Durner J, von Wettstein D, Kogel KH, Schäfer P (2012) Ethylene supports colonization of plant roots by the mutualistic fungus *Piriformospora indica*. PLoS One 7:e35502. https://doi.org/10.1371/journal.pone.0035502

Kieber JJ, Rothenberg M, Roman G, Feldmann KA, Ecker JR (1993) CTR1, a negative regulator of the ethylene response pathway in *Arabidopsis*, encodes a member of the raf family of protein kinases. Cell 72:427–441. https://doi.org/10.1016/0092-8674(93)90119-b

Kim H, Lee YH (2020) The rice microbiome: a model platform for crop holobiome. Phytobiomes J 4:5–18. https://doi.org/10.1094/PBIOMES-07-19-0035-RVW

Ko D, Kang J, Kiba T, Park J, Kojima M, Do J, Kim KY, Kwon M, Endler A, Song WY, Martinoia E, Sakakibara H, Lee Y (2014) Arabidopsis ABCG14 is essential for the root-to-shoot translocation of cytokinin. Proc Natl Acad Sci U S A 111:7150–7155. https://doi.org/10.1073/pnas.1321519111

Kundu A, DasGupta M (2018) Silencing of putative cytokinin receptor histidine kinase1 inhibits both inception and differentiation of root nodules in *Arachis hypogaea*. Mol Plant-Microbe Interact 31:187–199. https://doi.org/10.1094/MPMI-06-17-0144-R

Lee B, Farag MA, Park HB, Kloepper JW, Lee SH, Ryu CM (2012) Induced resistance by a long-chain bacterial volatile: elicitation of plant systemic defense by a C13 volatile produced by *Paenibacillus polymyxa*. PLoS One 7:e48744. https://doi.org/10.1371/journal.pone.0048744

Lee J, Han CT, Hur Y (2013) Molecular characterization of the *Brassica rapa* auxin-repressed, superfamily genes, *BrARP1* and *BrDRM1*. Mol Biol Rep 40:197–209. https://doi.org/10.1007/s11033-012-2050-9

Lemfack MC, Gohlke BO, Toguem SMT, Preissner S, Piechulla B, Preissner R (2018) mVOC 2.0: a database of microbial volatiles. Nucleic Acids Res 46:D1261–D1265. https://doi.org/10.1093/nar/gkx1016

Li N, Wang W, Bitas V, Subbarao K, Liu X, Kang S (2018) Volatile compounds emitted by diverse *Verticillium* species enhance plant growth by manipulating auxing signaling. Mol Plant-Microbe Interact 31:1021–1031. https://doi.org/10.1094/MPMI-11-17-0263-R

Liu Y, Wang P, Pan G, Crowley D, Li L, Zheng J, Zhang X, Zheng J (2016) Functional and structural responses of bacterial and fungal communities from paddy fields following long-term rice cultivation. J Soil Sediment 16:1460–1471. https://doi.org/10.1007/s11368-015-1343-8

Liu K, Newman M, McInroy JA, Hu CH, Kloepper JW (2017) Selection and assessment of plant growth-promoting rhizobacteria for biological control of multiple plant diseases. Phytopathology 107:928–936. https://doi.org/10.1094/PHYTO-02-17-0051-R

Lohar DP, Schaff JE, Laskey JG, Kieber JJ, Bilyeu KD, Bird DM (2004) Cytokinins play opposite roles in lateral root formation, and nematode and Rhizobial symbioses. Plant J 38:203–214. https://doi.org/10.1111/j.1365-313X.2004.02038.x

Lombardi N, Vitale S, Turrà D, Reverberi M, Fanelli C, Vinale F, Marra R, Ruocco M, Pascale A, d'Errico G, Woo SL, Lorito M (2018) Root exudates of stressed plants stimulate and attract *Trichoderma* soil fungi. Mol Plant-Microbe Interact 31:982–994. https://doi.org/10.1094/MPMI-12-17-0310-R

López-Bucio J, Campos-Cuevas JC, Hernández-Calderón E, Velásquez-Becerra C, Farías-Rodríguez R, Valencia-Cantero E (2007) *Bacillus megaterium* rhizobacteria promote growth and alter root-system architecture through an auxin and ethylene independent signaling mechanism in *Arabidopsis thaliana*. Mol Plant-Microbe Interact 20:207–217. https://doi.org/10.1094/MPMI-20-2-0207

Lugtenberg B, Kamilova F (2009) Plant-growth-promoting rhizobacteria. Annu Rev Microbiol 63: 541–546. https://doi.org/10.1146/annurev.micro.62.081307.162918

Lundberg DS, Lebeis SL, Herrera Paredes S, Yourstone S, Gehring J, Malfatti S, Tremblay J, Engelbrektson A, Kunin V, del Rio TG, Edgar RC, Eickhorst T, Ley RE, Hugenholtz P, Green S, Dangl JL (2012) Defining the core *Arabidopsis thaliana* root microbiome. Nature 488:86–90. https://doi.org/10.1038/nature11237

Macías-Rodríguez L, Guzmán-Gómez A, García-Juárez P, Contreras-Cornejo HA (2018) *Trichoderma atroviride* promotes tomato development and alters the root exudation of carbohydrates, which stimulates fungal growth and the biocontrol of the phytopathogen *Phytophthora cinnamomi* in a tripartite interaction system. FEMS Microbiol Ecol 94(9):fiy137. https://doi.org/10.1093/femsec/fiy137

MacMillan J (2002) Occurrence of gibberellins in vascular plants, fungi, and bacteria. J Plant Growth Regul 20:387–442. https://doi.org/10.1007/s003440010038

Manzoni S, Schimel JP, Porporato A (2012) Responses of soil microbial communities to water stress: results from a meta-analysis. Ecology 93:930–938. https://doi.org/10.1890/11-0026.1

Martínez-Medina A, Pescador L, Fernández I, Rodríguez-Serrano M, García JM, Romero-Puertas MC, Pozo MJ (2019) Nitric oxide and phytoglobin PHYTOGB1 are regulatory elements in the *Solanum lycopersicum-Rhizophagus irregularis* mycorrhizal symbiosis. New Phytol 223:1560–1574. https://doi.org/10.1111/nph.15898

Meldau DG, Meldau S, Hoang LH, Underberg S, Wünsche H, Baldwin IT (2013) Dimethyl disulfide produced by the naturally associated bacterium *Bacillus* sp B55 promotes *Nicotiana attenuata* growth by enhancing sulfur nutrition. Plant Cell 25:2731–2747. https://doi.org/10.1105/tpc.113.114744

Meng X, Miao Y, Liu Q, Ma L, Guo K, Liu D, Ran W, Shen Q (2019) *Tg*SWO from *Trichoderma guizhouense* NJAU4742 promotes growth in cucumber plants by modifying the root morphology and the cell wall architecture. Microb Cell Factories 18:148. https://doi.org/10.1186/s12934-019-1196-8

Minami R, Uchiyama K, Murakami T, Kawai J, Mikami K, Yamada T, Yokoi D, Ito H, Matsui H, Honma M (1998) Properties, sequence, and synthesis in *Escherichia coli* of

1 aminocyclopropane 1 carboxylate deaminase from *Hansenula saturnus*. J Biochem 123: 1112–1118. https://doi.org/10.1093/oxfordjournals.jbchem.a022050

Minerdi D, Bossi S, Maffei M, Gullino M, Garibaldi A (2011) *Fusarium oxysporum* and its bacterial consortium promote lettuce growth and expansin A5 gene expression through microbial volatile organic compounds (MVOC) emission. FEMS Microbiol Ecol 76:342–351

Mitchell EAD, Gilbert D, Buttler A, Amblard C, Grosvernier P, Gobat JM (2003) Structure of microbial communities in *Sphagnum* peatlands and effect of atmospheric carbon dioxide enrichment. Microb Ecol 46:187–199. https://doi.org/10.1007/BF03036882

Montejano-Ramírez V, Martínez-Cámara R, García-Pineda E, Valencia-Cantero E (2018) Rhizobacterium *Arthrobacter agilis* UMCV2 increases organ-specific expression of *FRO* genes in conjunction with genes associated with the systemic resistance pathways of *Medicago truncatula*. Acta Physiol Plant 40:138. https://doi.org/10.1007/s11738-018-2712-x

Morrone D, Chambers J, Lowry L, Kim G, Anterola A, Bender K, Peters RJ (2009) Gibberellin biosynthesis in bacteria: separate *ent*-copalyl diphosphate and *ent*-kaurene synthases in *Bradyrhizobium japonicum*. FEBS Lett 583:475–480. https://doi.org/10.1016/j.febslet.2008.12.052

Murphy JF, Zehnder GW, Schuster DJ, Sikora EJ, Polston JE, Kloepper JW (2000) Plant growth-promoting rhizobacteria mediated protection in tomato against *Tomato mottle virus*. Plant Dis 84:779–784. https://doi.org/10.1094/PDIS.2000.84.7.779

Nagahama K, Yoshino K, Matsuoka M, Sato M, Tanase S, Ogawa T, Fukuda H (1994) Ethylene production by strains of the plant pathogenic bacterium *Pseudomonas syringae* depends upon the presence of indigenous plasmids carrying homologous genes for the ethylene forming enzyme. Microbiology 140:2309–2313. https://doi.org/10.1099/13500872-140-9-2309

Nagahashi G, Douds DD (2000) Partial separation of root exudate compounds and their effects upon the growth of germinated spores of AM fungi. Mycol Res 104:1453–1464. https://doi.org/10.1017/S0953756200002860

Nett RS, Contreras T, Peters RJ (2017) Characterization of CYP115 as a gibberellin 3-oxidase indicates that certain rhizobia can produce bioactive gibberellin A4. ACS Chem Biol 12:912–917. https://doi.org/10.1021/acschembio.6b01038

Newton AC, Fitt BDL, Atkins SD, Walters DR, Daniel TJ (2010) Pathogenesis, parasitism and mutualism in the trophic space of microbe-plant interactions. Trends Microbiol 18:365–373. https://doi.org/10.1016/j.tim.2010.06.002

Niu DD, Liu HX, Jiang CH, Wang YP, Wang QY, Jin HL, Guo JH (2011) The plant growth–promoting rhizobacterium *Bacillus cereus* AR156 induces systemic resistance in *Arabidopsis thaliana* by simultaneously activating salicylate- and jasmonate/ethylene-dependent signaling pathways. Mol Plant-Microbe Interact 24:533–542. https://doi.org/10.1094/MPMI-09-10-0213

Okon Y, Labandera-Gonzalez CA (1994) Agronomic applications of *Azospirillum*: an evaluation of 20 years worldwide field inoculation. Soil Biol Biochem 26:1591–1601. https://doi.org/10.1016/0038-0717(94)90311-5

Oldroyd GED, Murray JD, Poole PS, Downie JA (2011) The rules of engagement in the legume-rhizobial symbiosis. Annu Rev Genet 45:119–144. https://doi.org/10.1146/annurev-genet-110410-132549

Orozco-Mosqueda MC, Velázquez-Becerra C, Macías-Rodríguez LI, Santoyo G, Flores-Cortez I, Alfaro-Cuevas R, Valencia-Cantero E (2013) Arthrobacter agilis UMCV2 induces iron acquisition in Medicago truncatula (strategy I plant) in vitro via dimethylhexadecylamine emisión. Plant Soil 362:51–66. https://doi.org/10.1007/s11104-012-1263-y

Ortíz-Castro R, Martínez-Trujillo M, López-Bucio J (2008) N-acyl-L-homoserine lactones: a class of bacterial quorum-sensing signals alter post-embryonic root development in *Arabidopsis thaliana*. Plant Cell Environ 31:1497–1509. https://doi.org/10.1111/j.1365-3040.2008.01863.x

Ortíz-Castro R, Contreras-Cornejo HA, Macías-Rodríguez L, López-Bucio J (2009) The role of microbial signals in plant growth and development. Plant Signal Behav 4:701–712. http://www.landesbioscience.com/journals/psb/article/9047

Park YS, Dutta S, Ann M, Raaijmakers JM, Park K (2015) Promotion of plant growth by *Pseudomonas fluorescens* strain SS101 via novel volatile organic compounds. Biochem Biophys Res Commun 461:361–365. https://doi.org/10.1016/j.bbrc.2015.04.039

Parry G, Estelle M (2006) Auxin receptors: a new role for F-box proteins. Curr Opin Cell Biol 18: 152–156. https://doi.org/10.1016/j.ceb.2006.02.001

Payne CM, Knott BC, Mayes HB, Hansson H, Himmel ME, Sandgren M, Ståhlberg J, Beckham GT (2015) Fungal cellulases. Chem Rev 115:1308–1448. https://doi.org/10.1021/cr500351c

Pérez-Flores P, Valencia-Cantero E, Altamirano-Hernández J, Pelagio-Flores R, López-Bucio J, García-Juárez P, Macías-Rodríguez L (2017) *Bacillus methylotrophicus* M4-96 isolated from maize (*Zea mays*) rhizoplane increases growth and auxin content in *Arabidopsis thaliana* via emission of volatiles. Protoplasma 254:2201–2213. https://doi.org/10.1007/s00709-017-1109-9

Podlešáková K, Fardoux J, Patrel D, Bonaldi K, Novák O, Strnad M, Giraud E, Spíchal L, Nouwen N (2013) Rhizobial synthesized cytokinins contribute to but are not essential for the symbiotic interaction between photosynthetic bradyrhizobia and *Aeschynomene* legumes. Mol Plant-Microbe Interact 26:1232–1238. https://doi.org/10.1094/MPMI-03-13-0076-R

Ponzio C, Gols R, Pieterse CMJ, Dicke M (2013) Ecological and phytohormonal aspects of plant volatile emission in response to single and dual infestations with herbivores and phytopathogens. Funct Ecol 27:587–598. https://doi.org/10.1111/1365-2435.12035

Qin X, Liu JH, Zhao SS, Chen XJ, Guo ZJ, Peng YL (2013) Gibberellin 20-oxidase gene *OsGA20ox3* regulates plant stature and disease development in rice. Mol Plant-Microbe Interact 26:227–239. https://doi.org/10.1094/MPMI-05-12-0138-R

Raghavendra AS, Gonugunta VK, Christmann A, Grill E (2010) ABA perception and signalling. Trends Plant Sci 15:395–401. https://doi.org/10.1016/j.tplants.2010.04.006

Ramakrishna W, Yadav R, Li K (2019) Plant growth promoting bacteria in agriculture: two sides of a coin. Appl Soil Ecol 138:10–18. https://doi.org/10.1016/j.apsoil.2019.02.019

Ramírez-Ordorica A, Valencia-Cantero E, Flores-Cortes I, Carrillo-Rayas MT, Elizarraraz-Anaya MIC, Montero-Vargas J, Winkler R, Macías-Rodríguez L (2020) Metabolomic effects of the colonization of *Medicago truncatula* by the facultative edophyte *Arthrobacter agilis* UMCV2 in a foliar inoculation system. Sci Rep 10:8426. https://doi.org/10.1038/s41598-020-65314-4

Rasmann S, Turlings TCJ (2016) Root signals that mediate mutualistic interactions in the rhizosphere. Curr Opin Plant Biol 32:62–68. https://doi.org/10.1016/j.pbi.2016.06.017

Ray P, Chi MH, Guo Y, Chen C, Adam C, Kuo A, LaButti K, Lipzen A, Barry KW, Grigoriev IV, Tang Y, Craven KD (2018) Genome sequence of the plant growth promoting fungus *Serendipita vermifera* subsp. *bescii*: the first native strain from North America. Phytobiomes 2:62–63. https://doi.org/10.1094/PBIOMES-04-17-0017-A

Real-Santillán RO, del Val E, Cruz-Ortega R, Contreras-Cornejo HA, González-Esquivel CE, Larsen J (2019) Increased maize growth and P uptake promoted by arbuscular mycorrhizal fungi coincide with higher foliar herbivory and larval biomass of the fall armyworm *Spodoptera frugiperda*. Mycorrhiza 29:615–622. https://doi.org/10.1007/s00572-019-00920-3

Reboutier D, Bianchi M, Brault M, Roux C, Dauphin A, Rona JP, Legué V, Lapeyrie F, Bouteau F (2002) The indolic compound hypaphorine produced by ectomycorrhizal fungus interferes with auxin action and evokes early responses in nonhost *Arabidopsis thaliana*. Mol Plant-Microbe Interact 15:932–938. https://doi.org/10.1094/MPMI.2002.15.9.932

Redford AJ, Bowers RM, Knight R, Linhart Y, Fierer N (2010) The ecology of the phyllosphere: geographic and phylogenetic variability in the distribution of bacteria on tree leaves. Environ Microbiol 12:2885–2893. https://doi.org/10.1111/j.1462-2920.2010.02258.x

Romero FM, Marina M, Pieckenstain FL (2014) The communities of tomato (*Solanum lycopersicum* L.) leaf endophytic bacteria, analyzed by 16S-ribosomal RNA gene pyrosequencing. FEMS Microbiol Lett 351:187–194. https://doi.org/10.1111/1574-6968.12377

Ryu CM, Farag MA, Hu CH, Reddy MS, Wei HX, Paré PW, Kloepper JW (2003) Bacterial volatiles promote growth in *Arabidopsis*. Proc Natl Acad Sci U S A 100:4927–4932. https://doi.org/10.1073/pnas.0730845100

Ryu CM, Farag MA, Hu CH, Reddy MS, Kloepper JW, Paré PW (2004) Bacterial volatiles induce systemic resistance in *Arabidopsis*. Plant Physiol 134:1017–1026. https://doi.org/10.1104/pp.103.026583

Sánchez-Rodríguez C, Rubio-Somoza I, Sibout R, Persson S (2010) Phytohormones and the cell wall in *Arabidopsis* during seedling growth. Trends Plant Sci 15:291–301. https://doi.org/10.1016/j.tplants.2010.03.002

Santoro MV, Zygadlo J, Giordano W, Banchio E (2011) Volatile organic compounds from rhizobacteria increase biosynthesis of essential oils and growth parameters in peppermint (*Mentha piperita*). Plant Physiol Biochem 49:1177–1182. https://doi.org/10.1016/j.plaphy.2011.07.016

Santoyo G, Moreno-Hagelsieb G, Orozco-Mosqueda MC, Glick BR (2016) Plant growth-promoting bacterial endophytes. Microbiol Res 183:92–99. https://doi.org/10.1016/j.micres.2015.11.008

Sasse J, Martinoia E, Northen T (2018) Feed your friends: do plant exudates shape the root microbiome? Trends Plant Sci 23:25–41. https://doi.org/10.1016/j.tplants.2017.09.003

Schäfer P, Pfiffi S, Voll LM, Zajic D, Chandler PM, Waller F, Scholz U, Kühnemann JP, Sonnewald S, Sonnewald U, Kogel KH (2009) Phytohormones in plant root *Piriformospora indica* mutualism. Plant Signal Behav 4:669–671. https://doi.org/10.4161/psb.4.7.9038

Schausberger P, Peneder S, Jürschik S, Hoffmann D (2012) Mycorrhiza changes plant volatiles to attract spider mite enemies. Funct Ecol 26:441–449. https://doi.org/10.1111/j.1365-2435.2011.01947.x

Sessitsch A, Coenye T, Sturz AV, Vandamme P, Barka EA, Salles JF, Van Elsas JD, Faure D, Reiter B, Glick BR, Wang-Pruski G, Nowak J (2005) *Burkholderia phytofirmins* sp. nov., a novel plant-associated bacterium with plant beneficial properties. Int J Syst Evol Microbiol 55:1187–1192. https://doi.org/10.1099/ijs.0.63149-0

Sevilla M, Burris RH, Gunapala N, Kennedy C (2001) Comparison of benefit to sugarcane plant growth and 15N2 incorporation following inoculation of sterile plants with *Acetobacter diazotrophicus* wild-type and Nif- mutant strains. Mol Plant-Microbe Interact 14:358–366. https://doi.org/10.1094/MPMI.2001.14.3.358

Shi TQ, Peng H, Zeng SY, Ji RY, Shi K, He H, Ji XJ (2017) Microbial production of plant hormones: opportunities and challenges. Bioengineered 8:124–128. https://doi.org/10.1080/21655979.2016.1212138

Siemens J, Keller I, Sarx J, Kunz S, Schuller A, Nagel W, Schmülling T, Parniske M, Ludwig-Müller J (2006) Transcriptome analysis of *Arabidopsis* clubroots indicate a key role for cytokinins in disease development. Mol Plant-Microbe Interact 19:480–494. https://doi.org/10.1094/MPMI-19-0480

Spaepen S, Vanderleyden J, Remans R (2007) Indole-3-acetic acid in microbial and microorganism-plant signaling. FEMS Microbiol Rev 31:425–448. https://doi.org/10.1111/j.1574-6976.2007.00072.x

Splivallo R, Fischer U, Göbel C, Fewsner I, Petr K (2009) Truffles regulate root morphogenesis via the production of auxin and ethylene. Plant Physiol 150:2018–2019. https://doi.org/10.1104/pp.109.141325

Steindler L, Bertani I, De Sordi L, Schwager S, Eberl L, Venturi V (2009) LasI/R and RhlI/R quorum sensing in a strain of *Pseudomonas aeruginosa* beneficial to plants. Appl Environ Microbiol 75:5131–5140. https://doi.org/10.1128/AEM.02914-08

Stringlis IA, Ke Y, Feussner K, de Jonge R, Van Bentum S, Van Verk MC, Berendsen RL, Bakker PAHM, Feussner I, Pieterse CMJ (2018) MYB72-dependent coumarin exudation shapes root microbiome assembly to promote plant health. Proc Natl Acad Sci U S A 115:E5213–E5222. https://doi.org/10.1073/pnas.1722335115

Swarup R, Friml J, Marchant A, Ljung K, Sandberg G, Palme K, Bennett M (2001) Localization of the auxin permease AUX1 suggests two functionally distinct hormone transport pathways operate in the *Arabidopsis* root apex. Genes Dev 15:2648–2653. https://doi.org/10.1101/gad.210501

Tahir HAS, Gu Q, Wu H, Raza W, Hanif A, Wu L, Colman MV, Gao X (2017) Plant growth promotion by volatile organic compounds produced by *Bacillus subtilis* SYST2. Front Microbiol 8:171. https://doi.org/10.3389/fmicb.2017.00171

Tatsukami Y, Ueda M (2016) Rhizobial gibberellin negatively regulates host nodule number. Sci Rep 6:27998. https://doi.org/10.1038/srep27998

Thormann MN, Rice AV, Beilman DW (2007) Yeasts in peatlands: a review of richness and roles in peat decomposition. Wetlands 27:761–772. https://doi.org/10.1672/0277-5212(2007)27[761:YIPARO]2.0.CO;2

Timm CM, Campbell AG, Utturkar SM, Jun SR, Parales RE, Tan WA, Robeson MS, Lu TYS, Jawdy S, Brown SD, Ussery DW, Schadt CW, Tuskan GA, Doktycz MJ, Weston DJ, Pelletier DA (2015) Metabolic functions of *Pseudomonas fluorescens* strains from *Populus deltoids* depend on rhizosphere or endosphere isolation compartment. Front Microbiol 6:1118. https://doi.org/10.3389/fmicb.2015.01118

Tsuchisaka A, Theologis A (2004) Unique and overlapping expression patterns among the Arabidopsis 1 amino cyclopropane 1 carboxylate synthase gene family members. Plant Physiol 136:2982–3000

Tyc O, Song C, Dickschat JS, Vos M, Garbeva P (2017) The ecological role of volatile and soluble secondary metabolites produced by soil bacteria. Trends Microbiol 25:280–292. https://doi.org/10.1016/j.tim.2016.12.002

Vadassery J, Ritter C, Venus Y, Camehl I, Varma A, Shahollari B, Novák O, Strand M, Ludwig Müller J, Oelmüller R (2008) The role of auxins and cytokinins in the mutualistic interaction between Arabidopsis and *Piriformospora indica*. Mol Plant-Microbe Interact 21:1371–1383. https://doi.org/10.1094/MPMI-21-10-1371

Vaishnav A, Kumari S, Jain S, Varma A, Choudhary DK (2015) Putative bacterial volatile-mediated growth in soybean (*Glycine max* L. Merrill) and expression of induced proteins under salt stress. J Appl Microbiol 119:539–551. https://doi.org/10.1111/jam.12866

Van Dam NM, Bouwmeester HJ (2016) Metabolomics in the rhizosphere: tapping into below-ground chemical communication. Trends Plant Sci 21:256–265. https://doi.org/10.1016/j.tplants.2016.01.008

Varbanova M, Yamaguchi S, Yang Y, McKelvey K, Hanada A, Borochov R, Yu F, Jikumaru Y, Ross J, Cortes D, Ma CJ, Noel JP, Mander L, Shulaev V, Kamiya Y, Rodermel S, Weiss D, Pichersky E (2007) Methylation of gibberellins by *Arabidopsis* GAMT1 and GAMT2. Plant Cell 19:32–45. https://doi.org/10.1105/tpc.106.044602

Velázquez-Becerra C, Macías-Rodríguez LI, López-Bucio J, Altamirano-Hernández J, Flores-Cortez I, Valencia-Cantero E (2011) A volatile organic compound analysis from *Arthrobacter agilis* identifies dimethylhexadecylamine, an amino-containing lipid modulating bacterial growth and *Medicago sativa* morphogenesis in vitro. Plant Soil 339:329–340. https://doi.org/10.1007/s11104-010-0583-z

Vicente-Hernández A, Salgado-Garciglia R, Valencia-Cantero E, Ramírez-Ordorica A, Hernández-García A, García-Juárez P, Macías-Rodríguez L (2019) *Bacillus methylotrophicus* M4-96 stimulates the growth of strawberry (*Fragaria* × *ananassa* 'Aromas') plants in vitro and slows *Botrytis cinerea* infection by two different methods of interaction. J Plant Growth Regul 38:765–777. https://doi.org/10.1007/s00344-018-9888-6

Villalobos-Escobedo JM, Esparza-Reynoso S, Pelagio-Flores R, López-Ramírez F, Ruiz-Herrera LF, López-Bucio J, Herrera-Estrella A (2020) The fungal NADPH oxidase is an essential element for the molecular dialog between *Trichoderma* and *Arabidopsis*. Plant J 103:2178–2192. https://doi.org/10.1111/tpj.14891

Viterbo A, Landau U, Kim S, Chernin L, Chet I (2010) Characterization of ACC deaminase from the biocontrol and plant growth promoting agent *Trichoderma asperellum* T203. FEMS Microbiol Lett 305:42–48. https://doi.org/10.1111/j.1574-6968.2010.01910.x

Vlot AC, Dempsey DA, Klessig DF (2009) Salicylic acid, a multifaceted hormone to combat disease. Annu Rev Phytopathol 47:177–206. https://doi.org/10.1146/annurev.phyto.050908.135202

Von Rad U, Klein I, Dobrev PI, Kottova J, Zazimalova E, Fekete A, Hartmann A, Schmitt-Kopplin P, Durner J (2008) Response of *Arabidopsis thaliana* to *N*-hexanoyl-DL-homoserine-lactone, a bacterial quorum sensing molecule produced in the rhizosphere. Planta 229:73–85. https://doi.org/10.1007/s00425-008-0811-4

Waller F, Mukherjee K, Deshmukh SD, Achatz B, Sharma M, Schäfer P, Kogel KH (2008) Systemic and local modulation of plant responses by *Piriformospora indica* and related Sebacinales species. Plant Physiol 165:60–70. https://doi.org/10.1016/j.jplph.2007.05.017

Wang Y, Ohara Y, Nakayashiki H, Tosa Y, Mayama S (2005) Microarray analysis of the gene expression profile induced by the endophytic plant growth-promoting rhizobacteria, *Pseudomonas fluorescens* FPT9601-T5 in *Arabidopsis*. Mol Plant-Microbe Interact 18:385–396. https://doi.org/10.1094/MPMI-18-0385

Wang S, Wang S, Sun Q, Yang L, Zhu Y, Yuan Y, Hua J (2017) A role of cytokinin transporter in *Arabidopsis* immunity. Mol Plant-Microbe Interact 30:325–333. https://doi.org/10.1094/MPMI-01-17-0011-R

Wang KD, Borrego EJ, Kenerley CM, Kolomiets MV (2020) Oxylipins other than jasmonic acid are xylem-resident signals regulating systemic resistance induced by *Trichoderma virens* in maize. Plant Cell 32:166–185. https://doi.org/10.1105/tpc.19.00487

Weingart H, Völksch B, Ullrich MS (1999) Comparison of ethylene production by *Pseudomonas syringae* and *Ralstonia solanacearum*. Phytopathology 89:360–365. https://doi.org/10.1094/PHYTO.1999.89.5.360

Xie X, Zhang H, Paré PW (2009) Sustained growth promotion in Arabidopsis with long-term exposure to the beneficial soil bacterium Bacillus subtilis (GB03). Plant Signal Behav 4:948–953. https://doi.org/10.4161/psb.4.10.9709

Xie SS, Wu HJ, Zang HY, Wu LM, Zhu QQ, Gao XW (2014) Plant growth promotion by spermidine-producing *Bacillus subtilis* OKB105. Mol Plant-Microbe Interact 27:655–663. https://doi.org/10.1094/MPMI-01-14-0010-R

Yamaguchi S (2008) Gibberellin metabolism and its regulation. Annu Rev Plant Biol 59:225–251. https://doi.org/10.1146/annurev.arplant.59.032607.092804

Yuan CL, Zhang LM, Wang JT, Hu HW, Shen JP, Cao P, He JZ (2019) Distributions and environmental drivers of archaea and bacteria in paddy soils. J Soils Sediments 19:23–37. https://doi.org/10.1007/s11368-018-1997-0

Zhang H, Sun Y, Xie X, Kim MS, Dowd SE, Paré PW (2009) A soil bacterium regulates plant acquisition of iron via deficiency-inducible mechanisms. Plant J 58:568–577. https://doi.org/10.1111/j.1365-313X.2009.03803.x

Zhao Y (2008) The role of local biosynthesis of auxin and cytokinin in plant development. Curr Opin Plant Biol 11:16–22. https://doi.org/10.1016/j.pbi.2007.10.008

Zhao Y, Li C, Ge J, Xu M, Zhu Q, Wu T, Guo A, Xie J, Dong H (2014) Recessive mutation identifies auxin-repressed protein ARP1, which regulates growth and disease resistance in tobacco. Mol Plant-Microbe Interact 7:638–654. https://doi.org/10.1094/MPMI-08-13-0250-R

Zhao Q, Li M, Jia Z, Liu F, Ma H, Huang Y, Song S (2016) AtMYB44 positively regulates the enhanced elongation of primary roots induced by *N*-3-oxo-hexanoyl-homoserine lactone in *Arabidopsis thaliana*. Mol Plant-Microbe Interact 29:774–785. https://doi.org/10.1094/MPMI-03-16-0063-R

Zhu Y, Nomura T, Xu Y, Zhang Y, Peng Y, Mao B, Hanada A, Zhou H, Wang R, Li P, Zhu X, Mander LN, Kamiya Y, Yamaguchi S, He Z (2006) ELONGATED UPPERMOST INTERNODE encodes a cytochrome P450 monooxygenase that epoxidizes gibberellins in a novel deactivation reaction in rice. Plant Cell 18:442–456. https://doi.org/10.1105/tpc.105.038455

Zipfel C, Oldroyd GED (2017) Plant signalling in symbiosis and immunity. Nature 543:328–336. https://doi.org/10.1038/nature22009

Zou C, Li Z, Yu D (2010) *Bacillus megaterium* strain XTBG34 promotes plant growth by producing 2-pentylfuran. J Microbiol 48:460–466. https://doi.org/10.1007/s12275-010-0068-z

Zúñiga A, Poupin MJ, Donoso R, Ledger T, Guiliani N, Gutiérrez RA, González B (2013) Quorum sensing and indole-3-acetic acid degradation play a role in colonization and plant growth

promotion of *Arabidopsis thaliana* by *Burkholderia phytofirmans* PsJN. Mol Plant-Microbe Interact 26:546–553. https://doi.org/10.1094/MPMI-10-12-0241-R
Zúniga A, Donoso RA, Ruiz D, Ruz GA, González B (2017) Quorum-sensing systems in the plant growth-promoting bacterium *Paraburkholderia phytofirmans* PsJN exhibit cross-regulation and are involved in biofilm formation. Mol Plant-Microbe Interact 30:557–565. https://doi.org/10.1094/MPMI-01-17-0008-R

Chapter 5
Quorum Sensing in the Rhizosphere

Monica Cortez, Dave Handy, Alyssa Headlee, Cynthia Montanez, Samantha Pryor, Kirstin Cutshaw, Katherine Vanselow, Alejandro Perez, Joshua Weissman, Eric Ziegler, Brooke Wheeler, and Andrew Palmer

Abstract The rhizosphere is a biologically and chemically rich and diverse environment that arises through the interplay between host plant and the microorganisms that inhabit this space. The diverse microorganisms which call this region home are crucial players in determining the success and failure of the scaffold (i.e., host plant) they inhabit. Rhizosphere interactions can impact agricultural yields, disease resistance, nutrient utilization, nutrient uptake, ecological robustness, and secondary

M. Cortez
Graduate Division of Biomedical and Biological Sciences Program in Biochemistry, Cell, and Developmental Biology, Emory University, Atlanta, GA, USA

D. Handy · K. Cutshaw
Department of Ocean Engineering and Marine Sciences, Florida Institute of Technology, Melbourne, FL, USA

A. Headlee
Department of Mathematical Sciences, Florida Institute of Technology, Melbourne, FL, USA

C. Montanez · S. Pryor · J. Weissman
Department of Aerospace, Physics and Space Sciences, Florida Institute of Technology, Melbourne, FL, USA

K. Vanselow · A. Perez · E. Ziegler
Department of Biomedical and Chemical Engineering and Sciences, Florida Institute of Technology, Melbourne, FL, USA

B. Wheeler
College of Aeronautics, Florida Institute of Technology, Melbourne, FL, USA

Aldrin Space Institute, Florida Institute of Technology, Melbourne, FL, USA

A. Palmer (✉)
Department of Ocean Engineering and Marine Sciences, Florida Institute of Technology, Melbourne, FL, USA

Department of Biomedical and Chemical Engineering and Sciences, Florida Institute of Technology, Melbourne, FL, USA

Aldrin Space Institute, Florida Institute of Technology, Melbourne, FL, USA

Institute of Global Ecology, Florida Institute of Technology, Melbourne, FL, USA
e-mail: apalmer@fit.edu

B. A. Horwitz, P. K. Mukherjee (eds.), *Microbial Cross-talk in the Rhizosphere*, Rhizosphere Biology, https://doi.org/10.1007/978-981-16-9507-0_5

metabolite production. How this mixture of cues and signals directs the actions of potentially deleterious or beneficial microbial associations with host plants is crucial to improving agricultural yields, food safety, and our general understanding of terrestrial ecology. Frequently, symbiotic microorganisms (both pathogenic and mutualistic) within the rhizosphere tightly regulate phenotypic switching to behaviors that are relevant to their host plants based on population density, a phenomenon known as quorum sensing (QS). QS has emerged as a crucial regulatory strategy for rhizosphere behaviors such as nitrogen fixation, as well as biofilm and virulence factor production. Here we review a variety of known QS mechanisms, how biotic and abiotic factors influence QS size and persistence, the effectiveness of QS at the root surface, and provide relevant examples of QS microorganisms within the rhizosphere. We also explore how host plants have evolved to detect and respond to QS signals, as well as the potential significance of this discovery. Finally, we consider how to integrate QS processes into existing models for biotic and abiotic cycles also present at the rhizosphere.

Keywords Rhizosphere · Quorum sensing · *N*-Acyl-L-homoserine lactones · AHLs

5.1 Introduction

The narrow region of soil proximal to the surface of roots is an extremely rich and diverse ecosystem relative to its surroundings. This region, the rhizosphere, can contain $\approx 10^{11}$ microbial cells/gram of root tissue with as many as 10^4 different species inhabiting it (Egamberdieva et al. 2008; Mendes et al. 2011). The localization and composition of the microorganisms that populate the rhizosphere can also vary both spatially and temporally over the life of the host plant (Baudoin et al. 2003; Chaparro et al. 2014; Donn et al. 2015; Emmett et al. 2017; Nuccio et al. 2020; Steindler et al. 2009). Such changes presumably reflect the evolving relationship between host and microbiome over the lifetime of this holobiont, i.e., the combined assemblage of the host and its microbial cohort.

Within this system, both the host plants and their microbial co-habitants are intimately connected in a number of ways—ecologically, physiologically, and biochemically—and have co-evolved since the origin of terrestrial plants (Berendsen et al. 2012; Lambers et al. 2009). The diverse microorganisms which call this narrow region home are crucial players in determining the success and failure of their host plants. Indeed, the composition and behavior of this microbial community can greatly impact agricultural yields, disease resistance, nutrient utilization, nutrient uptake, ecological robustness, and secondary metabolite production (Barea et al. 2002; Donn et al. 2015; Dutta and Podile 2010; Huang et al. 2014; Lynch 2007; Mendes et al. 2013; Pingali 2012).

The interactions between species, as well as their host plant in this diverse microbiome, depend on crucial metabolic pathways/cycles, which reside across multiple kingdoms. For example, photosynthates derived from the Calvin cycle in the host plant provide materials to microorganisms in otherwise carbon-limited soil

(Badri and Vivanco 2009; Guyonnet et al. 2018; Sasse et al. 2018; Walker et al. 2003). Meanwhile, microorganisms can provide vital nitrogen or other materials to the host, facilitating plant growth (Alami et al. 2000; Fang et al. 2011; Herman et al. 2006; Smith and Smith 2011). Growth of the root system expands the rhizosphere "neighborhood," while growth of the aerial portions of the plant provides additional photosynthates to the residents. The result is a chemically heterogeneous, yet complementary, environment composed of a mixture of secondary metabolites, including signaling molecules.

How this mixture of cues and signals directs the actions of potentially deleterious (pathogenic) or beneficial (mutualistic) microbial associations with host plants is crucial to improving agricultural yields, food safety, and our general understanding of terrestrial ecology. In many cases, symbiotic microorganisms (both pathogenic and mutualistic) display a robust phenotypic switch between their host and non-host related behaviors. Such phenotypic switching is typically coupled to cell density, a phenomenon known as quorum sensing (QS). QS refers to a collection of strategies by which a number of unicellular microorganisms employ the synthesis, diffusion, and perception of low-molecular-weight signals to serve as proxies for cell density. The exchange of these signals allows QS bacteria to limit specific behaviors to cell densities at which their expression will be most beneficial. For example, exopolysaccharides (EPS) produced by a single bacterium are unlikely to form a biofilm of any real utility, while a biofilm formed by 10^5 cells/ml is another matter (Boedicker et al. 2009).

The term "quorum sensing" has been widely applied to a variety of systems, often leaving some question as to the validity of its usage. It is therefore useful to define some parameters for the phenomena we will refer to as QS, utilizing six "rules" defined by Winters et al. (2019). First, QS signals must be transmitted into the surrounding environment (extracellularly). Second, QS responses are integrated through one, or more, *specific* response pathways. Third, QS signals must be active at physiologically relevant concentrations. Fourth, the act of QS should improve the fitness for the community involved, i.e., their expression provides a selective advantage to the organism. Fifth, a critical threshold of the QS signal is necessary for activation of the QS response. Finally, and perhaps most importantly, the accumulation of QS signals, as well as the activation of the response pathways as a result, must be positively correlated with cell density.

QS either directly or indirectly impacts a variety of phenotypes important to the rhizosphere, such as nodulation, Ti plasmid conjugation, nitrogen cycling, antibiotic resistance, biofilm formation, virulence factor production, and more (for specific references see sections below). Identifying QS strategies and the phenotypes they regulate is crucial to our understanding of the rhizosphere microbiome. In this chapter, we will discuss (1) rhizosphere relevant QS strategies and phenotypes, (2) temporal and spatial dynamics of QS within the rhizosphere, and (3) how plants can detect, respond to, and manipulate these quorums.

5.2 QS Strategies Within the Rhizosphere

QS has been observed across all three Domains of life (Bacteria, Archaea, and Eukaryota) as a mechanism for coordinating the behavior of individuals within a unicellular community. While the exact molecular players and chemical signals used vary significantly across the different QS strategies, the general mechanism is largely conserved (Fig. 5.1) (Deng et al. 2010; Montgomery et al. 2013; Zhang et al. 2012). In all cases, QS is driven by the synthesis, diffusion, and perception of one or more chemical signals, generically referred to as autoinducers (AIs) (Ahlgren et al. 2011; Lindemann et al. 2011). Chemically speaking, AIs show broad structural diversity (Fig. 5.2) with little similarity between the various classes of compounds, and with an equally broad range of effective concentrations (10^{-9} to 10^{-5} M). The conservation of strategy and mechanism, despite the variation in sensitivity and signal type, underscores the convergent evolution of QS as a solution for phenotypic switching among unicellular organisms regardless of clade.

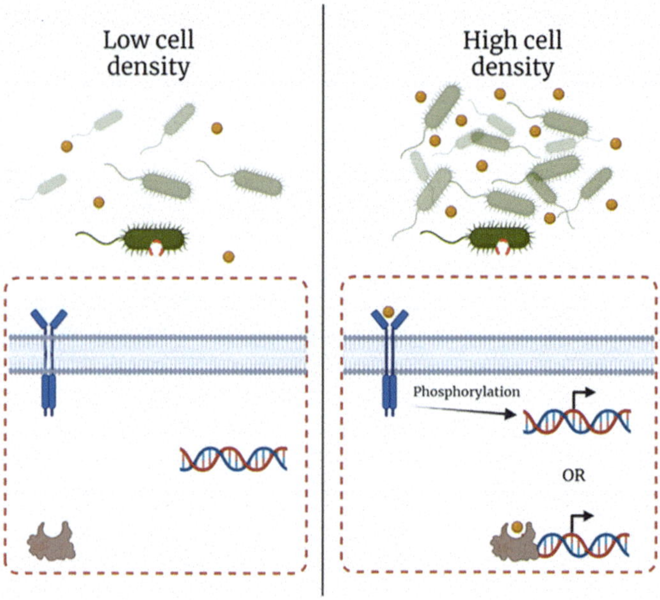

Fig. 5.1 Generic Strategy for QS. QS signals generically classified as autoinducers (AIs, gold spheres) are synthesized at a constitutive level at low cell densities and either diffuse or are actively transported out of the cell. As cell density increases, the corresponding AI concentration also increases. As AI concentration increases they are ultimately perceived by receptors located at the cell surface or intracellularly. Intracellular receptors typically function as transcriptional activators, directly binding to DNA and initiating transcription of the genes associated with the high-density phenotype. Receptors at the cell surface typically initiate a phosphorylation cascade ultimately resulting in similar changes in gene expression. In both routes, successful detection of these signals triggers increased biosynthesis of the AI in question, amplifying the effect and ensuring a cooperative switch between phenotypes. (Created with BioRender.com)

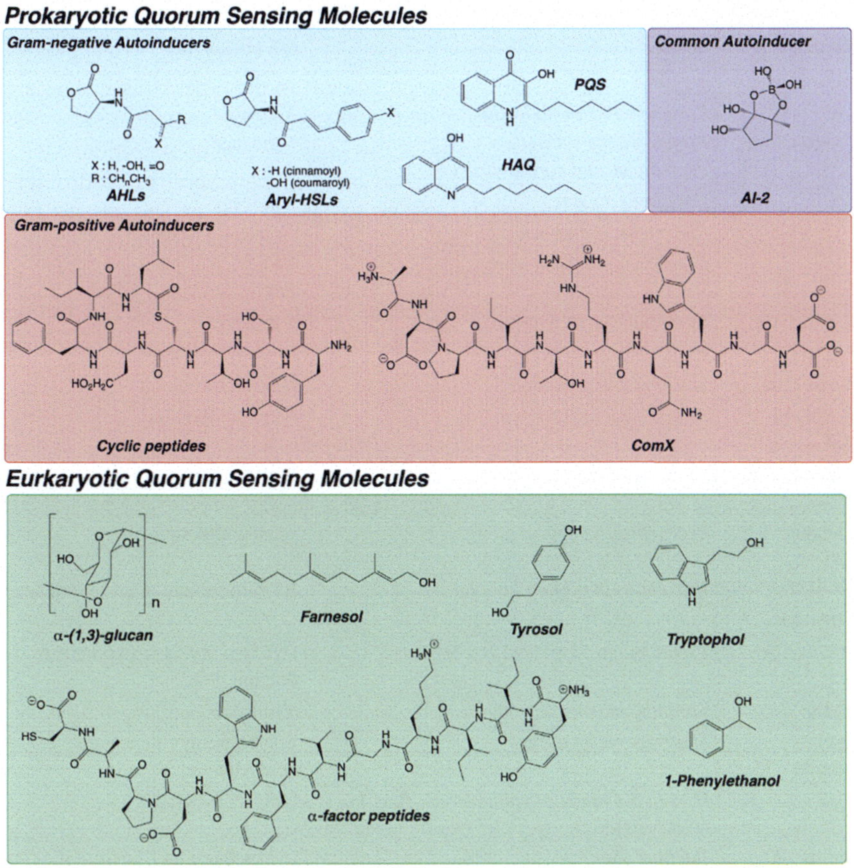

Fig. 5.2 Common prokaryotic and eukaryotic autoinducers. The chemical scaffolds capable of inducing QS (autoinducers) can vary significantly in structure and have evolved multiple times. Other than the proposed "universal autoinducer," AI-2, QS between Gram-negative and Gram-positive bacteria appears to rely on significantly different types of chemical scaffolds. The former (Gram-negative) utilize low-molecular-weight monomers, while the latter (Gram-positive) rely on small peptide sequences. As QS in eukaryotes, other than fungi, has only recently been identified, the specificity of these QS signals remains unclear. It may well be the case that *Chlamydomonas reinhardtii* and other QS algae utilize similar signals

AIs are synthesized at a constitutive rate when the population is below the QS threshold, and effectively serve as a proxy for cell density. These molecules either diffuse or are actively transported across the cell membrane, and then move through the environment, primarily through diffusion (Moore et al. 2014; Pearson et al. 1999). Perception of these signals by receptors either at the cell surface, or intracellularly, results in the following: (1) expression of the genes associated with the QS phenotype and (2) increased synthesis of the associated QS signal, hence the term "autoinducer." The latter leads to the accumulation of additional signal and therefore

Table 5.1 Sample AIs, species which utilize them, and the QS Phenotypes they regulate

Autoinducer	Species	QS phenotypes
AHLs		
C4	*Pseudomonas aeruginosa*	Virulence
3-Oxo-C6	*Pseudomonas syringae* *Pectobacterium carotovorum*	Virulence
C8	*Burkholderia* spp.	Virulence
3-Oxo-C8	*Rhizobium radiobacter*	Conjugal plasmid transfer
C10	*Pseudomonas fluorescens*	Virulence or plant-growth promoting activity (PGPA)
3-Oxo-C10	*Burkholderia vietnamiensis*	Virulence
3-Oxo-C12	*Pseudomonas aeruginosa*	Biofilm formation
3-oxo-C16	*Sinorhizobium meliloti*	Biofilm formation
Isovaleryl-HSL	*Bradyrhizobia japonica*	Nitrogen fixation
p-Coumaryl-HSL	*Rhosopseudomonas palustris*	PGPA
Cinnamoyl-HSL	*Bradyrhizobia* spp.	Stem nodulation
PA-HSL	*Prosthecomicrobium hirschii*	Biofilm formation and virulence
Other AIs		
PQS/HAQ	*Pseudomonas* spp., and *B. cepacia*	Cytotoxicity, siderophore production, PGPA
DSFs	*Xanthomonas* spp.	Virulence
Cyclic-peptide	*Staphylococcus aureus*	Sporulation/virulence and competence
AI-2	*E. coli*, *Salmonella typhimurium*, *Bacillus velezensis*	Universal signal
ComX	*Bacillus* spp.	Competence and surfactant production
Farnesol Tyrosol Tryptophol 1-Phenylethanol	*Candida albicans*, *C. parapsilosis*, *C. tropicalis*, *Saccharomyces cerevisiae*, *Aspergillus nidulans*, and *A. fumigatus*	Germ tube formation, virulence, biofilm regulation, iron transport, PGPA
α-(1,3)-Glucan	*Histoplasma capsulatum*	Virulence
Unknown	*Chlamydomonas reinhardtii*	Motility

increased activity in the QS circuit, ultimately increasing the changes in the expression of gens associated with the QS-phenotype. The result being a rapid and population-wide switch in behavior in the QS population. This typically manifests in a sigmoidal response within the population once QS is initiated (Fig. 5.1). Here we will briefly review the chemical signals and mechanisms that regulate QS systems common within the rhizosphere. Table 5.1 summarizes a variety of rhizosphere-associated microorganisms that engage in QS behaviors, the AIs they utilize, and the phenotypes regulated by the phenomenon.

5.2.1 Gram-Negative QS

5.2.1.1 Acyl and Aryl-L-Homoserine Lactones (AHLs)

QS was first discovered in *Vibrio fischeri*, a Gram-negative bacteria, which colonizes the bobtail squid *Euprymna scolopes* (Lupp and Ruby 2005). In this microorganism, QS controls bioluminescence, a phenomenon which provides counter-illumination to the host squid, as a co-evolved strategy to limit predation (Jones and Nishiguchi 2004). This mutualistic symbiosis is regulated by members of the *N*-acyl or *N*-aryl-L-homoserine lactone (AHLs) family of AIs, which appears to be the most prevalent QS-strategy overall, having been extensively observed in both marine and terrestrial environments. Indeed, various accounts suggest that AHL-mediated QS is likely to occur in thousands of different bacterial species (Ahlgren et al. 2011).

Structurally speaking, the majority of AHLs are of the *N*-acyl, rather than *N*-aryl type, consisting of a conserved L-homoserine lactone head group, variable oxidation states at the 3 position (-H, -OH, =O), and a 4–18 carbon acyl tail, with the potential for one or more units of unsaturation (Fig. 5.2). Variations in the 3-C oxidation state as well as the acyl tail length and degree of unsaturation are responsible for dictating species specificity. One, as of yet unique, variation on this structure has been observed in *Bradyrhizobia japonica*, which utilize branched *N*-acyl-L-homoserine lactones like isovaleryl-homoserine lactone to regulate QS (Lindemann et al. 2011). For the purposes of nomenclature, we will define the substitution at the 3 position first, followed by the chain length. For example, an AHL with a 12-carbon acyl tail and a =O at the third carbon would be designated as 3-oxo-C12. Similarly, an AHL that is fully reduced with an -H at the 3 position and a 12-carbon acyl tail is simply referred to as C12. Similar nomenclature is found in a number of existing references (e.g., Yates et al. 2002; Huang et al. 2007; Amara et al. 2009).

In addition, to their *N*-acyl counterparts, it is now clear that *N*-aryl-L-homoserine lactones, in which an aromatic functionality replaces the standard carbon tail, occur with some regularity within the rhizosphere (Ahlgren et al. 2011). Examples include the *p*-coumaryl-HSL of *Rhodopseudomonas palustris* and cinnamoyl-HSL from some species of *Bradyrhizobia* (Ahlgren et al. 2011; Schaefer et al. 2008). Initial hypotheses regarding the source of these aromatic scaffolds suggested the tantalizing possibility that they were derived from host cell wall phenolic compounds (Palmer and Blackwell 2008). This was based, at least in part, on the role these compounds play in plant-plant signaling (Fuller et al. 2017; Keyes et al. 2007; Liang et al. 2016; Palmer et al. 2004). However, it has since become clear that these molecules are typically derived through biosynthetic pathways endogenous to the microbe, although examples in which an exogenous source is required have been identified (Dong et al. 2017).

We note the structural similarities between these *N*-aryl-L-homoserine lactones to many of the synthetic AHL analogues (SAHLAs) synthesized by Blackwell et al. and others (Geske et al. 2008; Welsh and Blackwell 2016) to agonize or antagonize specific *N*-acyl-L-homoserine lactones. Similarly, branched *N*-acyl-L-homoserine lactones, like the isovaleryl-homoserine lactone in *B. japonica* (Lindemann et al. 2011), can be utilized as QS antagonists (Mattmann et al. 2008) in *N*-acyl-L-homoserine QS circuits as well. This raises the very real possibility for cross talk and differential regulation of QS circuits in the mixed microbial populations that inhabit the rhizosphere. Moreover, recent studies on the effects of SAHLAs on plants themselves suggest that a subset of these aryl ligands may function as phytohormones in plants like *Arabidopsis thaliana*, raising the possibility that these *N*-aryl-L-homoserine lactones, like their *N*-acyl counterparts, may also possess some interkingdom activity (Palmer et al. 2018). Proposed mechanisms for, as well as the effects of, AHL perception by plants will be discussed below (see Sect. 5.5).

AHL QS "circuits" typically consist of genes for the synthase, receptor, and those products required for the observed phenotype (biofilm, virulence factor production, etc.). Both the acyl and aryl HSLs are the product of a general family of enzymes known as AHL-synthases (or LuxI-type proteins) and are perceived by intracellular receptors (LuxR-type proteins). For example, QS-mediated bioluminescence in *V. fischeri* is regulated by the LuxIR QS circuit, containing the AHL synthase (LuxI) as well as the AHL receptor (LuxR) (Ahlgren et al. 2011). The product of the AHL-synthase LuxI (3-oxo-C6) is freely diffusible across the cell membrane, while more hydrophobic AHLs, i.e., those with longer acyl tails like 3-oxo-C12, may depend on influx or efflux pumps for transport (Pearson et al. 1999). AHL binding to their cognate LuxR-type receptor protein induces a conformational change and homodimerization. The newly formed complex then functions as a transcriptional activator, recruiting the appropriate machinery to the promoter region of the appropriate QS operon.

5.2.1.2 Coordinating Multiple AHL Circuits

Pseudomonas aeruginosa is a ubiquitous, opportunistic pathogen widely utilized as a model for more complex QS-regulation of bacterial behaviors, as it depends on multiple AHL as well as non-AHL-based QS circuits, for the regulation of behavior. Two of these circuits are AHL based, las (LasI/LasR) and rhl (RhlI/RhlR), and utilize the 3-oxo-C12 and C4 AHLs, respectively. Activation of the las circuit, via binding of 3-oxo-C12 to LasR, activates the downstream rhl circuit. Together these circuits coordinate the production of biofilm, which facilitates persistence, as well as a variety of virulence factors. Among the compounds generated by *P. aeruginosa* as a function of QS are rhamnolipids, amphiphilic glycolipids, which at low concentrations facilitate bacterial movement across surfaces, and at high concentrations assist in biofilm formation, both of which may assist in root colonization or persistence (Abbasi et al. 2012; Abdel-Mawgoud et al. 2010). These molecules also display antimicrobial activity towards Gram-positive microorganisms, potentially reducing competition during infection (Vatsa et al. 2010). Tight regulation over rhamnolipid production is not surprising given their potential role in initiating host defense responses (Sanchez et al. 2012).

5.2.1.3 Orphan luxR-Type Receptors

P. aeruginosa also maintains an "orphan" AHL-receptor, one that lacks a cognate synthase, known as QscR. This orphan receptor has lower ligand-binding specificity compared to LasR and RhlR, but shows greater substrate promiscuity, as it is able to detect nanomolar concentrations of C8, C10, 3-oxo-C10, C12, 3-oxo-C12, and C14 HSLs (Papenfort and Bassler 2016). Such orphan receptors may allow these microorganisms to monitor AHL production by other species within a community. Indeed, in the presence of *Pseudomonas fluorescens*, a plant growth promoting organism, and *Burkholderia vietnamiensis*, an opportunistic human pathogen frequently associated with the rhizosphere, QscR preferentially binds to the AHLs they produce, C10 and 3-oxo-C10, respectively. Productive AHL-binding to QscR represses the LasR circuit, and indirectly RhlR, raising the possibility that this process allows different species of microorganisms to coordinate the transition to virulence within the rhizosphere (Chugani et al. 2001; Fuqua 2006; Ha et al. 2012). Selective activation of the QscR QS circuit may even be exploited as a mechanism for reducing virulence factor production (Mattmann et al. 2008). Such orphan receptors are not unique to *P. aeruginosa*, having been observed in *E. coli* (SdiA), and thought to be widespread among the Proteobacteria (Venturi et al. 2018). Ultimately, orphan receptors may provide a strategy to integrate population density information in parallel to existing QS circuits and provide additional information to the "decision making" process of populations within the rhizosphere.

5.2.1.4 *Pseudomonas* Quinolone Signals (PQS)

In addition to the AHL-based QS circuits, other structurally distinct AIs have been identified that control phenotypic switching. For example, a variety of alkylquinolones, generically classified as *Pseudomonas* quinolone signals (PQS) (Fig. 5.2), serve as regulators for an assortment of important density-dependent phenotypes in a variety of *Pseudomonas* spp. such as iron acquisition (siderophores), cytotoxicity, and modulation of host immune responses, all of which can be important in regulating host-microbial associations, as well as the composition of the microbiome (Rutherford and Bassler 2012). PQS signals may also play a role in regulating the behavior of plant-growth promoting bacteria, such as *Pseudomonas putida* (Fernández-Piñar et al. 2011). PQS signals, like the 4-hydroxy-2-alkylquinolines (HAQs) were first discovered in *P. aeruginosa* where they are detected by the LysR-type receptor, PqsR (Sifri 2008). Related receptors and signals have since been identified in a variety of other *Pseudomonas* spp. as well as members of the genus *Burkholderia* (Coulon et al. 2019). Indeed, approximately 1/3 of the known members of the *Burkholderia cepacia* complex, a common rhizosphere inhabitant associated with plant-growth promoting capabilities, have the potential to produce and/or detect PQSs (Zhang and Xie 2007).

Like the R-type proteins of AHLs, binding of PQS to PqsR, or other LysR-type receptors, results in DNA binding to the promoter region of the pqsABCDE-operon and transcription of the related genes (Rutherford and Bassler 2012). Furthermore, disruption of the PqsR circuit reduces 3-oxo-C12 and elastase production, suggesting significant cross talk exists between these structurally distinct QS circuits. Manipulation of this circuit can impact biofilm formation in co-cultured species of *Bacillus* and *Candida.* The latter underscores the potential for interkingdom communication within the rhizosphere via PQS, though such interactions remain less clear at this time (Reen et al. 2011).

5.2.1.5 Diffusible Signal Factor (DSF)

A relatively new class of Gram-negative QS signals known as Diffusible Signal Factors (DSFs) (Fig. 5.2) have been identified in a number of soil microbes. These fatty acids are synthesized by homologues of the enzyme RpfF (Zhou et al. 2015). In the plant pathogen *Xanthomonas campestris*, this synthase produces four structurally distinct DSFs that participate in virulence regulation (Barber et al. 1997). Binding to homologues of the receptor kinase RpfG results in signal transduction and the expression of the QS-associated genes (Tang et al. 1991). Homologues of RpfG are widely distributed among the Xanthomonads, an extremely well-established group of plant pathogens, suggesting that DSF-mediated virulence is likely common in the rhizosphere. Furthermore, DSFs are able to modulate QS activity in human pathogens like *Burkholderia cenocepacia* as well as potentially disrupt QS in fungi, such as *Candida albicans* (Boon et al. 2008; Ryan et al. 2009).

5.2.2 *Gram-Positive Bacterial Quorum Sensing*

5.2.2.1 Cyclic Peptides

In contrast to the scaffolds utilized by Gram-negative bacteria, most Gram-positive prokaryotes utilize a variety of oligopeptides as AIs for QS. The first class of peptide-based AIs were cyclic peptides (CPs) (Fig. 5.2); however, other peptide systems, such as ComX (Fig. 5.2), have also been discovered (see below). Peptides are generally not diffusible across the cell membrane, so these bacteria utilize ATP-binding cassette (ABC) transporters to secrete peptides outside of the cell (Novick and Geisinger 2008). These signals are perceived by receptor kinases at the cell surface, which trigger a phosphorylation cascade and, ultimately, the activation of gene expression. Many Gram-positive bacteria utilize multiple peptides to communicate and express various genes. QS in Gram-positive bacteria is usually used to control competence, sporulation, and/or virulence. The best characterized CP-based QS systems to date are those in the genus *Staphylococcus*, in particular *S. aureus*. While commonly found in the rhizosphere, they are not typically considered major players in rhizosphere dynamics or plant-microbial interactions. However, soils contaminated with this human pathogen remain a potential reservoir for disease (Mendes et al. 2013).

5.2.2.2 ComX

In addition to cyclic peptides, a number of Gram-positive species within the rhizosphere are likely to deploy the ComX peptide sequence to regulate phenotypes such as biofilm production and competence (DNA uptake). Production and perception of this peptide depends on the ComQXPA operon, a 4 gene operon that codes for and perceives the ComX peptide sequence (Kalamara et al. 2018; Oslizlo et al. 2015; Tortosa et al. 2001). Like cyclic peptides, ComX perception is dependent on extracellular binding to ComP, a receptor kinase, which results in a phosphorylation cascade and the necessary changes in gene expression. First discovered in *Bacillus* spp., bioinformatics support the presence of this signaling system in at least 20 other species within the phylum Firmicutes, common constituents of the rhizosphere (Dogsa et al. 2014). While it may be premature to state with certainty, it appears that non-cyclic peptides, like ComX, predominate over cyclic peptides as the QS signals for choice among Gram-positive bacteria within the rhizosphere.

5.2.3 AI-2: A Universal QS Signal for Proteobacteria

The QS strategies described above generally appear divided between Gram-positive species, which deploy a variety of peptide-based AIs, and Gram-negative species that depend on a variety of non-peptide low-molecular weight signals. The one clear exception to this at present is the furanosyl borate diester, known as autoinducer-2 (AI-2) (Xavier and Bassler 2003) (Fig. 5.2). Unlike other AIs in which species specificity is dictated by slight changes in AI structure (e.g., chain length, oxidation state), the AI-2 structure is not variable and appears conserved between the species which utilize it. This has prompted the suggestion that AI-2 may work as interspecies regulator of density-dependent behaviors in which mixed species may be desirable, such as mixed species biofilm production. It remains unclear exactly how, or if, this satisfies the rule that QS must provide improved fitness to the population in question, which has prompted some doubts to the label of this as a "universal" QS system.

First discovered in the marine bacterium *Vibrio harveyi* (Cao and Meighen 1989), AI-2 has since been found in a number of soil microorganisms including *E. coli* and *Salmonella typhimurium*, potential human pathogens which show increased survival within the rhizosphere relative to bulk soil (Ongeng et al. 2011; Taga et al. 2008). More recently, AI-2 activity has been directly correlated with successful root colonization by the plant growth promoting *Bacillus velezensis* SQR9, underscoring the potential importance of this signal in plant-microbial associations within the rhizosphere (Xiong et al. 2020). AI-2 is produced at a constitutive level by the highly conserved AI-2 synthase LuxS (Schauder et al. 2001; Xavier and Bassler 2003). However, in a sharp contrast to most QS circuits in which the synthase and receptor are often coupled in a circuit, there appear to be a variety of mechanisms by which species may detect AI-2, suggesting that different species may respond to this "universal signal" in different ways (Federle 2009). This also challenges the rule that QS requires specific perception and response mechanisms. We note that like AHL-based QS systems, orphan receptors for AI-2 have also been observed, underscoring its potential role as universal monitoring system for QS in mixed microbial systems (Pereira et al. 2008).

5.2.4 Eukaryotic QS and Control in the Rhizosphere

While QS appears to predominately occur in prokaryotes, a role for this phenomenon in regulating the behaviors of certain eukaryotes such as fungi, and possibly algae, has been observed. While current examples of fungal QS within the rhizosphere are limited, it is worth considering how this phenomenon may impact fungal processes within this environment, given their potential significance to both plants and rhizosphere-dwelling microorganisms. For example, communities of fungi and microbes may extend the functional range of the rhizosphere well beyond the host root surface further increasing the chemical and biological complexity of this environment. QS may ultimately prove critical to managing processes in this environment. Here we briefly explore the potential of eukaryotic QS within the rhizosphere and its potential implications to prospective host plants.

Fungal assemblages play vital roles in decomposition as well as both mutualistic and pathogenic associations with a wide assortment of plants (Barea et al. 2002; Briones 2018; Klamer et al. 2002; Miransari 2011; Moll et al. 2015; van der Wal et al. 2013). Like fungi in other environments, these eukaryotes scavenge for available nutrients through filamentous growth and sporulate to confer longevity, behaviors frequently regulated by QS. One of the first described and most commonly observed AIs among fungi is the sesquiterpene alcohol, Farnesol, which has been observed in *Candida albicans*, *C. parapsilosis*, *C. tropicalis*, *Saccharomyces cerevisiae*, *Aspergillus nidulans*, and *A. fumigatus* (Cao et al. 2005; Hornby et al. 2001; Joo and Jetten 2010; Mosel et al. 2005; Weber et al. 2010) (Fig. 5.2). In *C. albicans* biofilms, Farnesol inhibits germ tube formation, while encouraging the dispersal of yeast-phase cells from biofilms. It has also been associated with increased drug resistance, iron transport, and the production of heat shock proteins. A role for Farnesol in virulence among some fungi is also clear, suggesting AI activity may well vary among subspecies, just as these signals often do in prokaryotic QS (Shea 2006).

Other AIs such as Tyrosol, tryptophol, and 1-phenylethanol have also been characterized as potential AIs in *C. albicans*, as well as other fungi, with roles in driving hyphae growth, inhibiting biofilm formation, and other phenotypes (Alem et al. 2006; Chen 2006; Chen et al. 2004; Hornby et al. 2001; Kruppa 2009; Weber et al. 2010) (Fig. 5.2). Many of these molecules, as well as one or more aspects of these QS behaviors, are conserved among *Aspergillus* spp., which have been associated with plant growth promoting activity (Islam et al. 2014). Similarly, α-(1,3)-glucan has been identified as an AI, regulating the switch to virulence in *Histoplasma capsulatum* (Kugler et al. 2000), a soil pathogenic fungi of animals. Their presence in soil further underscores the potential for other eukaryotic QS systems within the soil (Parniske 2008). As with AHL-based QS, communication between fungal colonies appears possible and dictated by specific signals (Alem et al. 2006; Chen et al. 2004; Hornby et al. 2001; Kruppa 2009).

The mutualistic symbiosis between arbuscular mycorrhizal fungi (AMF) and the roots of the majority of terrestrial plants is an ancient and crucial relationship, and here too QS may play an important role. Indeed, a number of plant mutualistic microbes with QS circuits, such as members of the Sinorhizobia family, have been associated with AMF spores (Palla et al. 2018). Branching in these fungi allows them to increase their surface area both for nutrient uptake as well as interaction with prospective hosts and with bacteria. AM-branching can be stimulated by the presence of strigolactones, secondary metabolites exuded by the roots of many plants (Dun et al. 2009). Strigolactones have been proposed to function in a manner analogous to AIs in some species of mosses where they also regulate branching in a population density-dependent manner (Proust et al. 2011). The potential for QS to occur in multicellular plants is not restricted to strigolactones, but a different class of proposed AIs, *p*-benzoquinones, have been proposed to play an important role in the development of the root system architecture between multiple plants (Fuller et al. 2017). While it remains unclear if either strigolactones or *p*-benzoquinones represent "true QS" phenomena, i.e., meeting all six of the rules stated at the beginning of this review, the potential impact of their existence on our understanding of root system architecture, as well as plant-plant, plant-fungal, and plant-microbial interactions warrants consideration. Intriguingly, both of the prospective AIs associated specifically with plants were originally derived from research into the same obligate parasitic plant *Striga asiatica* (witchweed) underscoring the utility of these model plant systems for the discovery of important signaling pathways.

Finally, many species of photosynthetic algae contribute a variety of exopolysaccharides to soil which can help shape the exometabolome and impact plant growth (Lewin 1956). Members of the genus Chlamydomonas are eukaryotic photoautotrophs commonly found in soil and freshwater, and members of this group are model organisms for cellular motility, photosynthesis, toxicology, and more (Harris 2001; Taylor et al. 2016). Emerging evidence confirms that members of this genus may release low-molecular weight exudates capable of modulating bacterial QS, potentially allowing them to modulate the density-dependent phenotypes of other microorganisms in the rhizosphere, though whether these accumulate at physiologically relevant concentrations remains unclear (Teplitski et al. 2004). Furthermore, we have recently observed the existence of QS-like control over motility in at least two members of this genus, potentially revealing an additional density-dependent process at work within the rhizosphere (Folcik et al. 2020a, b).

Even the algal communities within soil may yet prove to be active participants in QS processes.

5.3 Frequency and Dynamics of QS Within the Rhizosphere

While there are a number of critical phenotypes regulated by QS within the heterogenous environment of the rhizosphere, the density of bacteria required and the frequency with which these events occur remain ill-defined. The number of cells required for QS to occur within a particular environment is highly variable and critically dependent on both biotic and abiotic factors (Praneenararat et al. 2012; Sifri 2008). Biological considerations include: (1) the rates of AI production and release into the environment, (2) the distance between participating cells, and (3) the concentration at which the AI associates with its cognate receptor. Among the abiotic considerations, the chemical composition of the environment is of particular importance because this can impact AI stability and diffusion, potentially limiting or increasing signal availability (Yates et al. 2002; Ziegler et al. 2019).

As each cell produces AIs at a constitutive level, it essentially serves as the vote for that cell within the population. However, if the AI quickly degrades within the environment, then these votes can be "silenced," impacting the number of cells required to establish a threshold. For example, QS-active cells within a viscous biofilm can maintain concentrations of AHLs at least an order of magnitude higher than in aqueous conditions (Alberghini et al. 2009). The result is typically an increase in quorum sensing, and the production of phenotypes that are often going to further increase growth of the colony. In some species, QS can limit the production of the EPS allowing these organisms to then disperse into new niches (Nadell et al. 2008).

5.3.1 From Phyllosphere to Rhizosphere: QS Lessons from Across Plant Surfaces

The surface of leaves, the phyllosphere, is obviously a different environment than the rhizosphere, but can serve as a useful framework for understanding QS in its biological context. Work by Dulla, Lindow, and Quiñones established that aggregates of *Pseudomonas syringae*, which range from 0.5 to 0.8 μm by 1.5 to 3.0 μm, can induce QS over distances ranging from 4 to 78 μm (Dulla and Lindow 2008; Gantner et al. 2006; Quiñones et al. 2005). According to the authors, these distances are functionally analogous to "two players communicating to one another across a soccer field" (Dulla and Lindow 2008). These findings confirm that direct contact between the participants is not required and that signals may be freely diffusible over moderate distances on the microscopic level.

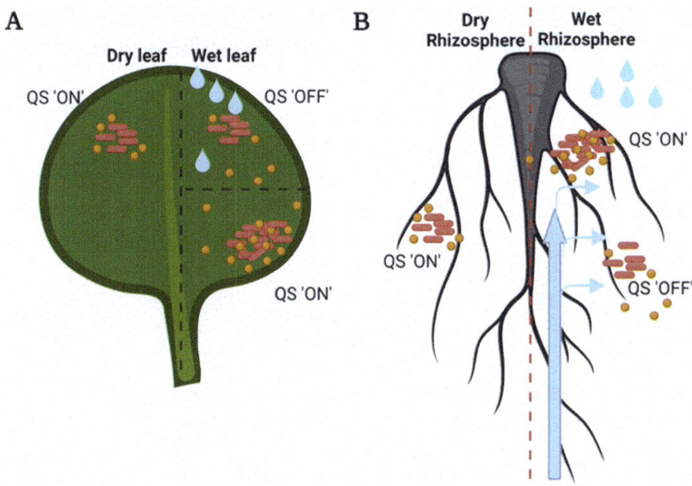

Fig. 5.3 Leaf and root environment dictate the amount of cells required for QS. (**a**) On dry leaves, microbial cells (purple) and autoinducers (orange) remain close together allowing QS to occur more rapidly and at lower concentrations. Moisture from a variety of sources such as rain, guttation, etc. can drive AI diffusion away from the colony, increasing the concentration required for QS to occur. (**b**) Theoretically, moisture levels in the rhizosphere would similarly impact QS processes. For example, in dry soil, AIs remain proximal to the microbial population allowing QS to be reached at a lower threshold. The introduction of water from hydraulic lift, rain, or other sources can redistribute AIs and microbes across the root surface, potentially raising the concentration required for QS to occur. Phenomena like guttation and hydraulic lift are coupled to stomatal opening and closing/circadian rhythms. This suggests redistribution of quorums across the surfaces of plants may happen on a daily basis. (Created with BioRender.com)

Utilizing a QS reporter strain of the pathogen *Pseudomonas syringae*, Dulla and Lindow confirmed that QS is significantly affected by water availability on leaves. In this study, *P. syringae* was engineered to express Red Fluorescent Protein (RFP) in response to the accumulation of 3-oxo-C6. At 24 h post inoculation, cells showed little growth, and $<1\%$ of cells displayed QS on either wet or dry leaves. However, 48 h post-inoculation, approximately 70% of the cells on dry leaves showed QS-mediated increases in fluorescence. On wet leaves, QS initiation was delayed by 24 h relative to dry leaf controls, with 86% of the population showing activity after 72 h. The simplest explanation for this difference is that moisture along the surface of wet leaves causes AHLs to diffuse further across the surface, reducing their effective concentration. As a result, a higher concentration of a specific AHL is required for QS to occur (Dulla and Lindow 2008) (Fig. 5.3).

Moisture conditions impact not only the rate at which a quorum can develop but also the number of cells required. There appears to be at least a twofold difference between the concentrations of cells and time required for QS on wet versus dry leaves, with dry leaves supporting QS in $\approx$20 cells in only 24 h. Conversely, wet leaves required $>$40 cells for activity and over 48 h of incubation. In an extreme example, a 2000 cell aggregate on the surface of a wet leaf still required more than

24 h for quorum sensing to occur. The simplest explanation is that the introduction of water results in increased diffusion of the autoinducer (reducing effective concentration) while potentially accelerating AHL degradation (see Sect. 5.3.2). Routine moisture deposition (dew and/or guttation) may well establish a cycle to QS events along these surfaces, which follow a day-night cycle, not unlike the control of QS-mediated symbiosis between *V. fischeri* and host (Hirsch and Mcfall-Ngai 2000). This underscores the importance of trying to observe QS under native conditions and with native phenotypes as much as possible to provide a better understanding of the dynamics of this communication.

Considerably less is known about the specifics of quorum sizes within the rhizosphere as this environment is substantially more difficult to visualize than the surfaces associated with the phyllosphere in a non-disruptive fashion. However, based on our understanding of soil cycles and other environmental factors, we can make several predictions about QS cycles which might exist in the rhizosphere. For example, just as rain wets the surface of a leaf, the roots of plants routinely redistribute water across the rhizosphere via hydraulic lift and related phenomena (Dawson 1993; Dirksen and Raats 1985). Hypothetically, changes in water concentration and distribution within the rhizosphere play as important of a role in this environment as it does across leaves (Fig. 5.3). Such changes in rhizosphere water distribution would be heavily linked to stomatal opening and closing which is coupled to day-night cycles. Indeed, we propose that quorum sensing across all surfaces of a plants are significantly impacted by the circadian clock (Hubbard et al. 2018). The potential for substantial changes for water redistribution across the root surface may explain why some QS bacteria couple phenotypic switching to spatial isolation, as in the case of nodulation, which is isolated by root hair curling, or some pathogens which require evidence of a damaged root surface, such as *Agrobacterium tumefaciens*, permitting them to infiltrate their prospective host (Fig. 5.4).

5.3.2 AI Stability

In addition to the impact of the environment, the stability of specific AIs plays a significant role in how and when QS occurs on any given surface. AI persistence is an important consideration as their continual accumulation could potentially activate QS prematurely, or have other unintended consequences. Of the known AIs among prokaryotes and eukaryotes, the stability of AHLs has received the greatest attention to date, and significantly less is known about the fates of other AIs. This remains an important and under-investigated area of research that is crucial to understanding how quorums are maintained under native conditions.

AHL stability has been well characterized under a variety of conditions and is generally positively correlated with hydrophobicity, i.e., acyl chain length. The primary mechanism of pH-dependent AHL degradation is through hydrolysis of the lactone ring (Yates et al. 2002) (Fig. 5.5). For example, at room temperature, under low acid conditions ($pH \leq 5$) the C6 AHL has a half-life of several days, while

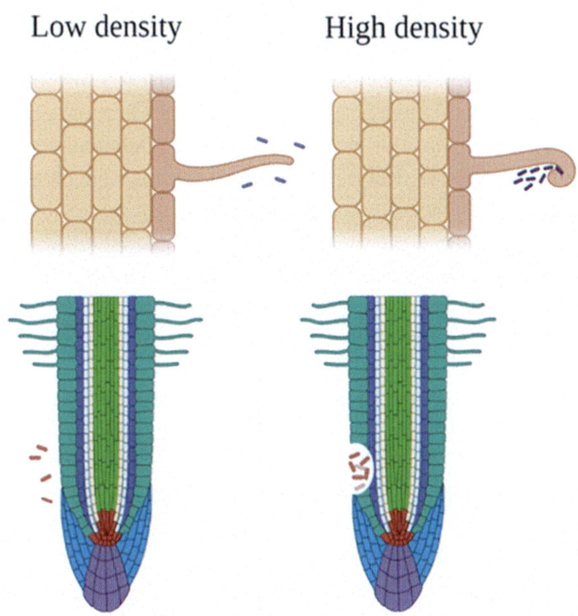

Fig. 5.4 Host plant root creates opportunities for QS. In both mutualistic (top) as well as pathogenic (bottom) associations, many organisms fail to reach densities sufficient for productive QS to occur (left column). However, changes in the root surface can cause localized increases in microbial density, allowing QS to occur (right column). In the case of nodulation (top), bacterial-derived Nod factors induces root hair curling, rapidly increasing microbial density in that area. Similarly, many plant pathogens are sensitive to structural fragments associated with host root wounding to provide a site away from the immediate rhizosphere, increasing cell density and allowing QS to occur here as well. (Created with BioRender.com)

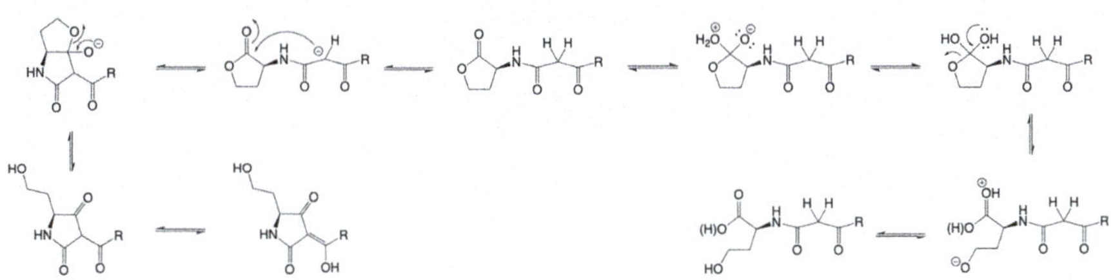

Fig. 5.5 AHL hydrolysis and tetramic acid formation mechanisms—Two distinct routes for spontaneous AHL degradation (center) have been observed under biologically relevant conditions. (Right) AHL hydrolysis is the better studied of these two routes and occurs readily in the presence of water. The rate of this hydrolysis is both chain length and pH-dependent. (Left) The spontaneous rearrangement of AHLs to form tetramic acids has yet to be observed in nature, but has been documented under biologically relevant conditions in a laboratory setting

under alkaline conditions (pH $\geq$ 8) this reaction can be measured in minutes (Ziegler et al. 2019). Conversely, longer chain AHLs may have a half-life an order of magnitude or higher than those for short chain AHLs, and the presence of the

3-keto functionality also increases stability (Crowe et al. 2017; Sánchez-Sanz et al. 2018).

A number of studies have established the potential for some plant exudates as well as other potential rival microorganisms (such as Gram-positive microorganisms) to disrupt quorums through lactone hydrolysis (Delalande et al. 2005; Uroz 2003). In addition to lactone hydrolysis, laboratory studies have identified that AHLs may also degrade through the formation of tetramic acids, an essentially irreversible process; however, the formation of these under native conditions has yet to be observed (Kaufmann et al. 2005) (Fig. 5.5). However, the selective antimicrobial properties of tetramic acids could allow degraded AHLs to serve a new role in policing the composition of the rhizosphere population (Lowery et al. 2009).

Unlike AHLs, which can easily degrade spontaneously within the environment, the mechanisms that underscore peptide AI degradation are considerably less clear. In the case of the ComX AI, the accumulation of the peptide in *B. subtilis* cultures stimulates the production of exoproteases capable of degrading this AI, thereby limiting signal accumulation (Spacapan et al. 2018). However, the regulation of cyclic peptide AIs does not appear to have evolved a specific strategy. In the case of AI-2, the concentration of this signal appears to be regulated by the activity of ABC-transporter proteins and/or metabolizing of these signals thereby controlling the transition between phenotypes (Taga et al. 2003, 2008).

5.4 Examples of QS in the Rhizosphere

Having identified some specific QS molecules, modes of action, and their complex interplay within the diverse environment of the rhizosphere, we now provide some additional and important examples of QS which directly impact host-microbial associations, rhizosphere dynamics, and related phenomena.

5.4.1 Nitrogen Fixation

QS also plays an important role in regulating distinct components of the nitrogen cycle. For example, DeAngelis et al. screened 533 bacterial isolates from the rhizosphere of an *Avena fatua* (wild oat) field for the production of exoenzymes involved in the depolymerization of nitrogen, i.e., the release of nitrogen from biological polymers such as chitin and protein. Approximately 40% of these isolates displayed chitinase and/or protease activities while 30 unique isolates were observed to produce AHLs through the use of two biosensor strains: *Agrobacterium tumefaciens* pAHL-Ice and *Chromobacterium violaceum* CVOblu (Chu et al. 2011; Deangelis et al. 2008). The addition of the AHL lactonase *aiiD* to the isolates inhibited chitinase and/or protease activities, strongly supporting a role for QS in the regulation of their production (Deangelis et al. 2008).

In contrast to the more generalized nitrogen fixation process described above, the legume-rhizobia symbiosis (nodulation) is a well-established, highly organized, and tightly regulated interkingdom, mutualistic symbiosis. Inside the nodule, members of the paraphyletic group collectively known as *Rhizobium* spp. produce the enzyme nitrogenase, which makes atmospheric nitrogen available to their legume hosts (Kuzma et al. 1993; Marketon et al. 2002). Nodule initiation depends on the exchange of a variety of chemical signals between the host legume and compatible rhizobia species, including flavonoids and Nod factors (Oldroyd and Downie 2008). Nodule initiation occurs when Nod factors, generated by rhizobia, reach the surface of a prospective host plant. The presence of Nod factors induces root hair curling, capturing the bacteria within this new microenvironment, ultimately leading to nodule formation (Esseling et al. 2003). In this instance, QS does not appear to regulate the production of bacterial flavonoids, neither is it directly in control of nitrogenase production. Rather, root hair curling around rhizobia effectively increases their cell density, and initiates QS-mediated production of the biofilm necessary to form viable nodules.

The best characterized system for the study of QS in nodulation is between the legume *Medicago truncatula* and *Sinorhizobium meliloti*. As with most rhizobia, *S. meliloti* utilizes unusually long-chain AHLs, often 14–18 carbons in length, which are the products of the AHL-synthase SinI and perceived by the receptor SinR. The significance of this initial QS event and the production of the biofilm it regulates are crucial for nodulation, as disruption of either SinI or SinR results in significantly reduced nodule number as well as a significant number of "immature" nodules. The latter is characterized by the nodule remaining white in color, rather than turning pink from the accumulation of leghemoglobin, suggesting incomplete nodulation formation (González and Marketon 2003). Conversely, the addition of exogenous AHLs has been observed to increase the total number of nodules (Veliz-Vallejos et al. 2014). Work in our lab confirmed that the exogenous addition of AHLs compatible with SinR (C14, C16 for example) showed a significant increase (10–20%) in total nodule number relative to controls (Palmer et al. 2016). Our study confirmed that the SinR circuit can be activated by the addition of SAHLAs, suggesting nodulation as a potential target for improving QS (Palmer et al. 2016).

5.4.2 Rhizobium radiobacter *(A.k.a.* Agrobacterium tumefaciens*)*

In contrast to the mutualistic symbiosis of nodulation described above, other members of the *Rhizobiaceae* family utilize QS in the service of virulence. Specifically, in *Rhizobium radiobacter* (formerly *Agrobacterium tumefaciens*) and related pathogens, QS is crucial for virulence. In these pathogens, virulence is manifested in the host through the formation of crown gall or hairy root tumors, structural growths in which dysregulated host cell division occurs. Formation of these structures depends on interkingdom gene transfer from the pathogen to host plant cells. Viable cells for transfer and infection are identified by the presence of H^+, sugars, and phenols, all of which serve as viable molecular markers for a wound site where invasion can occur (He et al. 2009; Hu et al. 2013; Lin et al. 2008; Palmer et al. 2004). The genes encoding for the transfer machinery, as well as the genes to be transferred themselves, are housed on Ti or Ri-megaplasmids (<250 kbp) which are generally unfavorable to maintain during the large periods of time when the bacteria are not in the presence of a viable host (Lang and Faure 2014; White and Winans 2005).

At any given time, it is unclear how many members of a particular colony carry these megaplasmids, a potential obstacle to successful infection (Platt et al. 2014). It is under these conditions that QS is employed by *R. radiobacter* to facilitate infection. Like mutualistic *Rhizobia* species, QS occurs along a microenvironment in the rhizosphere, in this case a wound site. This site limits AHL diffusion and accumulates H^+, effectively stabilizing AHL, raising the available concentration of ligand. The primary AHL in this system is 3-oxo-C8, the product of TraI and perceived by TraR. However, alternative AHLs, such as 3-OH-C8, have been observed in some strains, such as *R. radiobacter* P4 (Mhedbi-Hajri et al. 2016). Successful detection of 3-oxo-C8 by TraR results in the upregulation of genes associated with transfer of the Ti plasmid between members of the colony. The result is the distribution of the Ti plasmid to the *R. radiobacter* within the wound site.

5.4.3 Burkholderia

Burkholderia spp. are Gram-negative bacteria which have been associated with both mutualistic as well as pathogenic phenotypes. In pathogenic species, like *B. glumae*, pathogenesis is dependent on flagellar motility, as well as rhamnolipid production, both of which are regulated, in part, by the C8 AHL, the product of the synthase TofI and perceived by the receptor TofR (Goo et al. 2010; Jang et al. 2014; Nickzad et al. 2015; Vatsa et al. 2010). Evaluation of rhamnolipid production by Nickzad et al. confirmed these molecules are crucial for regulating swarming motility and potentially reducing competition within the environment. TofI− mutant motility could be restored by the addition of exogenous AI supporting this characterization (Nickzad et al. 2015). Conservation of this circuit and motility regulation among other pathogenic species of *Burkholderia* is highly likely. For example, a similar luxIR-type circuit and the genes for rhamnolipid production have already been observed in *B. thailandensis* along with some antimicrobial activity (Costa et al. 2011; Dubeau et al. 2009).

5.4.4 Pseudomonas syringae

A causative agent of wild-fire disease in soybean and tobacco, *P. syringae* is a well-established plant pathogen in which virulence has been shown to be, at least in part, dependent on QS (Quiñones et al. 2005). In this pathogen, synthesis and detection of 3-oxo-C6, as well as similar short chain AHLs, can regulate motility, as well as EPS and cell wall degrading enzyme production. SAHLAs have previously been employed to inhibit maceration of *Phaseolus vulgaris* (green beans) during wild-type infections with *P. syringae* underscoring the potential utility of these compounds to inhibit pathogenic virulence. We note that these studies confirmed that a single application of SAHLA was unable to sufficiently inhibit virulence but rather required multiple dosages (Palmer et al. 2011a).

5.4.5 Pseudomonas fluorescens

Strains of *P. fluorescens* have been identified in which QS mediates both pathogenic as well as mutualistic associations with prospective host plants. AHLs in these organisms can vary in length, oxidation state, and unsaturations, highlighting the diversity of AHL QS within the Pseudomonads (Cui et al. 2005). In the case of *P. fluorescens* 5064, as well as some related isolates, QS mediates the release of surfactins which are critical for colonization of broccoli prior to infection (Cui et al. 2005), akin to what is observed in *P. aeruginosa*. Alternatively, plant growth promoting strains of this organism have been associated with the QS-mediated production of Mupirocin, a polyketide antibiotic with greater activity towards Gram-positive organisms (El-Sayed et al. 2001) or the antifungal lipopeptide sclerosin (Berry et al. 2014). In these cases, mutualistic strains of *P. fluorescens* may "police" the rhizosphere for potentially unwanted microorganisms.

5.4.6 Pectobacterium carotovorum

The soft rot pathogen *Pectobacterium carotovorum*, formerly *Erwinia carotorum*, employs AHL-mediated QS to regulate the production of a variety of cell wall degrading enzymes to facilitate host invasion and resource acquisition. However, in an intriguing departure from the traditional AHL-QS circuit, the luxR-type receptors of *P. carotovorum* (ExpR1 and ExpR2) form homodimers and bind to DNA in the absence of their cognate AHLs (Cui et al. 2006) (Fig. 5.5). Binding of the dimerized apoproteins triggers the expression of a global inhibitor of virulence factor production RsmA (Mukherjee et al. 1996). AHL binding to either ExpR1 or ExpR2 increases dissociation of the receptor from the promoter, the loss of *rsmA* transcription, and results in the upregulation of virulence factor production

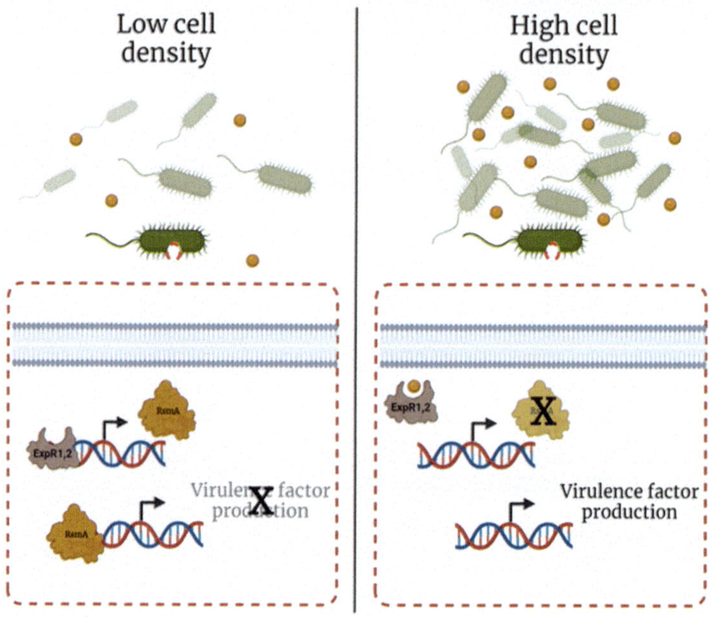

Fig. 5.6 *P. carotovorum* and the dissociative mechanism of AHL-mediated QS. In *P. carotovorum*, the AHL-free LuxR-homologues ExpR1 and ExpR2 bind to DNA and initiate the transcription of RsmA, a transcriptional repressor that blocks the expression of a variety of virulence factors. AHL-binding to the apoproteins of ExpR1 and ExpR2 causes them to dissociate from DNA, preventing further RsmA production. As RsmA levels in the cell decline, virulence factor expression is initiated. (Created with BioRender.com)

(Fig. 5.6). ExpR2, the major regulator of virulence of these two QS circuits, shows increased substrate promiscuity, binding several short chain AHLs (3-oxoC6, C6, 3-oxo-C8, and C8) with equal efficacy (Palmer et al. 2011b). This is one of the few systems in which inhibition of QS through the use of SAHLAs has been successfully evaluated in wild-type bacteria during host infection assays (potatoes) (Palmer et al. 2011a).

5.4.7 Bacillus subtilis

The Gram-positive soil dwelling microbe, *Bacillus subtilis*, is a well-known rhizosphere inhabitant capable of inhibiting Gram-negative QS through the production of the AHL-lactonase aiiA (Rajamani et al. 2011). Like *Pseudomonas aeruginosa* and other species, *B. subtilis* maintains several QS circuits that regulate overlapping or even sometimes apparently contradictory responses depending on environmental conditions (Kalamara et al. 2018). QS regulated phenotypes can include swarming as well as biofilm and exoprotease production (Dahl et al. 1992; Kearns and Losick 2004). Biofilm production as well as competence (DNA uptake) appears to rely on the behavior of a ComX QS system (Kalamara et al. 2018; Oslizlo et al. 2015; Tortosa et al. 2001). This system competes with the Rap-Phr QS circuit, also peptide based, which at high cell densities drives the cells towards sporulation (Omer Bendori et al. 2015). This underscores the potential for QS to coordinate responses BEFORE nutrient deficiency settles in. Recently, a study of the competitive pathogenesis between *B. subtilis* and the fungal pathogen *Setophoma terrestris*, both of which can infect the onion rhizosphere, suggested this interaction specifically selected for strains of the former with specific mutations to the ComX QS circuit. While the resulting mutants were QS-deficient they showed increased production of antifungal compounds. Taken together, these results suggest that QS is actively subject to evolutionary pressure within the rhizosphere (Albarracín Orio 2020).

5.4.8 Xanthomonas *Spp.*

DSF-mediated QS plays an important part in the virulence of many members of the *Xanthomonas* phytopathogen group, including *Xanthomonas campestris* pv. *campestris* (*Xcc*) (Barber et al. 1997), *Xanthomonas oryzae* pv. *oryzae* (*Xoo*) (Chatterjee and Sonti 2002), *Xanthomonas citri* subsp. *citri* (*Xac*) (Siciliano et al. 2006), and *Xanthomonas axonopodis* pv. *glycines* (Xag) (Thowthampitak et al. 2008). These are pathogens of cruciferous plants, rice, citrus, and soybean, respectively (Kakkar et al. 2015). Among the common virulence factors in this group is xanthan, which suppresses host plant defense responses while DSF itself elicits innate immunity in plants (Kakkar et al. 2015).

5.4.9 Partners in QS

Studies of the soil photoheterotroph *Rhodopseudomonas palustris* determined that this organism produces *p-coumaroyl*-HSL to regulate QS, but is unable to produce the precursor *p*-coumaric acid on its own (Ahlgren et al. 2011; Schaefer et al. 2008). In an interesting, and likely not unique case, QS in at least one species of the genus *Bradyrhizobia* utilizes the cinnamoyl-HSL to which the *R. palustris* QS circuit responds. While the extent of QS phenotypes under the regulation of these systems remains to be defined, QS in *R. palustris* may be associated with improved host plant immunity to potential pathogens such as tobacco mosaic virus (TMV) (Du et al. 2020). A similar circumstance has been observed in the opportunistic pathogen *Prosthecomicrobium hirschii*, which utilizes the phenylacetyl-homoserine lactone (PA-HSL) to regulate aggregation, biofilm formation, and pigment production (Liao et al. 2018). As with *R. palustris*, the source of the aryl precursor to this AHL must be derived from an exogenous source. It may yet be the case that some aryl-AHL QS molecules are derived from prospective host cell wall phenolic compounds (Palmer and Blackwell 2008).

5.5 AI Sensitivity and QS Manipulation in Plants

5.5.1 *Plant Sensitivity to AIs*

Despite the number of different AI-types found in soil-associated bacteria, little is known about potential host-plant sensitivity to any of these signals other than AHLs. Specifically, the effects of *N*-acyl-L-homoserine lactones have been well characterized to date, while little information regarding the effects of naturally occurring *N*-aryl-L-homoserine lactones are available. Responses of the former have been characterized in a number of model plant systems such as *A. thaliana*, *Medicago truncatula*, and *S. tuberosum* (tomato) where they have been shown to modulate growth, engage auxin and ethylene-associated responses, and prime defense responses.

Foundational work in this area was conducted by Mathesius et al. who exposed seedlings of *M. truncatula* to either 3-oxo-C12 or 3-oxo-C16:1 derived from the pathogen *P. aeruginosa* and the symbiont *Sinorhizobium meliloti* respectively. This study identified a common set of approximately 100 proteins that were similarly regulated by both AHLs, while an additional 30–35 proteins appeared differentially regulated based on the AHL selected. Furthermore, the extent of these responses could be modulated by AHL concentration. Together, these results suggested that plants are able to discriminate not only between specific AHLs but the relative concentration of the microbes that produce them (Mathesius et al. 2003).

Subsequent studies, from our lab and others, have confirmed that these plants typically display a biphasic response to AHLs, with low concentrations (<1 μM) generally improving transpiration and growth, while higher concentrations (>50 μM) inhibit growth through increased ethylene production (Ortíz-Castro et al. 2008; Palmer et al. 2014; von Rad et al. 2008). Unlike most Gram-negative bacteria, which generally manifest clear preferences for specific AHL signals, the extent of plant responses to these signals is correlated only with the length of the acyl chain not the oxidation state at the 3-position. Indeed, plants are even sensitive to the hydrolyzed lactone (ring-open) degradation products of AHLs, which are largely inactive in bacteria. Revisiting the work of Mathesius et al. in this context underscores a role for plant detection of AHLs in the process of nodulation, specifically, the observation that hydrolyzed (ring open) C14 and C16 treatments increased the rate at which nodules form in seedlings of *M. truncatula* (Palmer et al. 2016).

Sensitivity to hydrolyzed AHLs suggests plants may have the ability to detect the potential for bacterial quorums; i.e., BEFORE AHLs reach their threshold concentration and initiate phenotypic switching. How this impacts rhizosphere colonization and extract production remains largely unclear, and is an area ripe for future exploration. One possibility is that regions of higher hydrolyzed-AHL concentrations within the rhizosphere may allow growing roots to detect relative densities of bacteria ahead. In this way, plants may select for the bacterial density, if not the type of microorganisms, into which they grow. Such "information" could be integrated with other secondary metabolites to influence root system architecture as well as the starting number of bacteria from which to establish a new rhizosphere.

The discovery of plant-sensitivity to ring-open AHLs directly led to the discovery that the hydrolytic products of AHLs, the free L-homoserine, duplicated the growth effects described above suggesting cleavage of the amide bond might be required for the response to AHLs. This cleavage is catalyzed by one or more members of the fatty acid amide hydrolase (FAAH) family (Palmer et al. 2014). In that study, we initially proposed that increased L-homoserine concentrations were directly metabolized into the amplified ethylene levels observed at higher AHL exposures in *A. thaliana*. However, the introduction of deuterium at non-exchangeable positions on the lactone ring of 3-oxo-C12 exogenously supplied to plants accounted for $<5\%$ of the total increase in ethylene production. A specific mechanism by which the accumulation of this amino acid drives these changes remains unclear. Furthermore, as the relative tissue levels of FAAH may vary as a function of time, it is possible that this may help temporal and spatial organization of QS microbes within the rhizosphere as well as phyllosphere. Finally, the recent discovery of a second family of FAAH enzymes, which may potentially cleave aryl-L-homoserine lactones, has the potential to expand the QS signals on which plants can eavesdrop (Aziz and Chapman 2020).

5.5.2 *Manipulation of QS by Plants*

In addition to their ability to detect and respond to AHLs, plants themselves have been shown to both agonize and antagonize QS circuits. Work by Walker et al. confirmed that the presence of fungal elicitors stimulated the release of rosmarinic acid (RA) into the exudate of *A. thaliana* as well as sweet basil (*Ocimum basilicum*) at concentrations capable of inhibiting QS-mediated biofilm formation of *P. aeruginosa* (Walker et al. 2004). Similarly, early work by Mathesius et al. confirmed a variety of root exudate fractions are able to agonize or antagonize a suite of LuxR-type receptors in test strains *V. harveyi* BB170 and *E. coli* pSB1075. Moreover, these studies confirmed significant changes in extract composition in response to the presence of AHLs (Mathesius et al. 2003). Such changes were not generic in response to the presence of AHLs, but rather, varied based on the structure of the AHL added. Specifically, this study investigated the effects of C6 and 3-oxo-C12. Such extracts are also capable of modulating the activity of the AI-2 QS circuit as well, suggesting the potential for broad regulation of QS. More recently, a variety of specific flavonoids have been shown to agonize or antagonize specific quorum circuits, indicating these common components of root exudates may play a critical role in regulating QS within the rhizosphere (Szoboszlay et al. 2016).

Clearly, the root itself plays an important role in the establishment and maintenance of quorums within the rhizosphere. At present, our understanding of this process is strongly limited by our ability to spatially resolve the production of these molecules in real time. As a result it is difficult to resolve where and when along the root surface exudates accumulate at concentrations able to agonize or antagonize QS. However, given the potential impact of QS on host plants it is highly likely that plants have evolved mechanisms to manipulate phenotypic switching in bacteria. Finally, our understanding of this is almost exclusively limited to AHL-mediated QS, but it seems equally unlikely that plants would not attempt similar efforts to control Gram-positive or even fungal QS.

5.6 Final Thoughts

The full extent of the QS events which occur within the rhizosphere as well as their impacts on the host plants with which they coexist are far from clear. However, it is apparent that mutualistic as well as pathogenic associations with plants are critically dependent on this phenomenon. By leveraging advances in sequencing and screening, the identification and characterization of both prokaryotic and eukaryotic QS circuits within the rhizosphere is likely to continue for years to come. Based on the lines of evidence presented here, major questions persist that will likely guide researchers in the coming years if our models of QS are to become increasingly more accurate. First, how do environmental factors, both biotic and abiotic, influence the persistence and distribution of QS signals? Which of the daily cycles of prospective host plants that impact distribution and composition of the molecules within the rhizosphere, can impact QS dynamics within the soil? How do these diverse signaling networks interact with one another to impact the broader chemical environment within the rhizosphere? Do AIs and other constituents within the rhizosphere have potential synergistic or inhibitory effects? Lastly, how do plants perceive the various QS dialogues occurring within the rhizosphere? While obvious responses to the presence of AHLs have been observed in plants, similar responses to the presence of other classes of AIs have not been equally well characterized. Yet, given the potential significance of successful QS on host plants, it seems unlikely that AIs other than AHLs are generally invisible to them. Ultimately the complex interplay between different AIs, the metabolic pathways they regulate, and how other organisms detect and modify these signals are important for understanding the rhizosphere holobiont in which plants, algae, fungi, and bacteria functionally coexist.

References

Abbasi H, Hamedi MM, Lotfabad TB, Zahiri HS, Sharafi H, Masoomi F et al (2012) Biosurfactant-producing bacterium, Pseudomonas aeruginosa MA01 isolated from spoiled apples: physicochemical and structural characteristics of isolated biosurfactant. J Biosci Bioeng 113:211–219. https://doi.org/10.1016/j.jbiosc.2011.10.002

Abdel-Mawgoud AM, Lépine F, Déziel E (2010) Rhamnolipids: diversity of structures, microbial origins and roles. Appl Microbiol Biotechnol 86:1323–1336. https://doi.org/10.1007/s00253-010-2498-2

Ahlgren NA, Harwood CS, Schaefer AL, Giraud E, Greenberg EP (2011) Aryl-homoserine lactone quorum sensing in stem-nodulating photosynthetic bradyrhizobia. Proc Natl Acad Sci U S A 108:7183–7188. https://doi.org/10.1073/pnas.1103821108

Alami Y, Achouak W, Marol C, Heulin T (2000) Rhizosphere soil aggregation and plant growth promotion of sunflowers by an exopolysaccharide-producing Rhizobium sp. strain isolated from sunflower roots. Appl Environ Microbiol 66:3393–3398. https://doi.org/10.1128/AEM.66.8.3393-3398.2000

Albarracín Orio AG, Petras D, Tobares RA, Aksenov AA, Wang M, Juncosa F, Sayago P, Moyano AJ, Dorrestein PC, Smania AM (2020) Fungal–bacterial interaction selects for quorum sensing mutants with increased production of natural antifungal compounds. Commun Biol 3(1):1–9. https://doi.org/10.1038/s42003-020-01342-0

Alberghini S, Polone E, Corich V, Carlot M, Seno F, Trovato A et al (2009) Consequences of relative cellular positioning on quorum sensing and bacterial cell-to-cell communication. FEMS Microbiol Lett 292:149–161. https://doi.org/10.1111/j.1574-6968.2008.01478.x

Alem MAS, Oteef MDY, Flowers TH, Douglas LJ (2006) Production of tyrosol by Candida albicans biofilms and its role in quorum sensing and biofilm development. Eukaryot Cell 5: 1770–1779. https://doi.org/10.1128/EC.00219-06

Amara N, Mashiach R, Amar D, Krief P, Spieser SAH, Bottomley MJ et al (2009) Covalent inhibition of bacterial quorum sensing. J Am Chem Soc 131:10610–10619. https://doi.org/10.1021/ja903292v

Aziz M, Chapman KD (2020) Fatty acid amide hydrolases: an expanded capacity for chemical communication? Trends Plant Sci 25:236–249. https://doi.org/10.1016/j.tplants.2019.11.002

Badri DV, Vivanco JM (2009) Regulation and function of root exudates. Plant Cell Environ 32: 666–681. https://doi.org/10.1111/j.1365-3040.2009.01926.x

Barber CE, Tang JL, Feng JX, Pan MQ, Wilson TJG, Slater H et al (1997) A novel regulatory system required for pathogenicity of Xanthomonas campestris is mediated by a small diffusible signal molecule. Mol Microbiol 24:555–566. https://doi.org/10.1046/j.1365-2958.1997.3721736.x

Barea JM, Azcón R, Azcón-Aguilar C (2002) Mycorrhizosphere interactions to improve plant fitness and soil quality. Antonie van Leeuwenhoek 81:343–351. https://doi.org/10.1023/A:1020588701325

Baudoin E, Benizri E, Guckert A (2003) Impact of artificial root exudates on the bacterial community structure in bulk soil and maize rhizosphere. Soil Biol Biochem 35:1183–1192. https://doi.org/10.1016/S0038-0717(03)00179-2

Berendsen RL, Pieterse CMJ, Bakker PAHM (2012) The rhizosphere microbiome and plant health. Trends Plant Sci 17:478–486. https://doi.org/10.1016/j.tplants.2012.04.001

Berry CL, Nandi M, Manuel J, Brassinga AKC, Fernando WGD, Loewen PC et al (2014) Characterization of the Pseudomonas sp. DF41 quorum sensing locus and its role in fungal antagonism. Biol Control 69:82–89. https://doi.org/10.1016/j.biocontrol.2013.11.005

Boedicker JQ, Vincent ME, Ismagilov RF (2009) Microfluidic confinement of single cells of bacteria in small volumes initiates high-density behavior of quorum sensing and growth and reveals its variability. Angew Chem Int Ed 48:5908–5911. https://doi.org/10.1002/anie.200901550

Boon C, Deng Y, Wang L-H, He Y, Xu J-L, Fan Y et al (2008) A novel DSF-like signal from Burkholderia cenocepacia interferes with Candida albicans morphological transition. ISME J 2: 27–36. https://doi.org/10.1038/ismej.2007.76

Briones MJI (2018) The serendipitous value of soil fauna in ecosystem functioning: the unexplained explained. Front Environ Sci 6. https://doi.org/10.3389/fenvs.2018.00149

Cao JG, Meighen EA (1989) Purification and structural identification of an autoinducer for the luminescence system of Vibrio harveyi. J Biol Chem 264:21670–21676. https://doi.org/10.1016/S0021-9258(20)88238-6

Cao Y-Y, Cao Y-B, Xu Z, Ying K, Li Y, Xie Y et al (2005) cDNA microarray analysis of differential gene expression in Candida albicans biofilm exposed to farnesol. Antimicrob Agents Chemother 49:584–589. https://doi.org/10.1128/AAC.49.2.584-589.2005

Chaparro JM, Badri DV, Vivanco JM (2014) Rhizosphere microbiome assemblage is affected by plant development. ISME J 8:790–803. https://doi.org/10.1038/ismej.2013.196

Chatterjee S, Sonti RV (2002) rpfF Mutants of Xanthomonas oryzae pv. oryzae are deficient for virulence and growth under low iron conditions. Mol Plant-Microbe Interact 15:463–471. https://doi.org/10.1094/MPMI.2002.15.5.463

Chen H (2006) Feedback control of morphogenesis in fungi by aromatic alcohols. Genes Dev 20: 1150–1161. https://doi.org/10.1101/gad.1411806

Chen H, Fujita M, Feng Q, Clardy J, Fink GR (2004) Tyrosol is a quorum-sensing molecule in Candida albicans. Proc Natl Acad Sci U S A 101:5048–5052. https://doi.org/10.1073/pnas.0401416101

Chu W, Vattem DA, Maitin V, Barnes MB, McLean RJC (2011) Bioassays of quorum sensing compounds using Agrobacterium tumefaciens and Chromobacterium violaceum. Methods Mol Biol 692:3–19. https://doi.org/10.1007/978-1-60761-971-0_1

Chugani SA, Whiteley M, Lee KM, D'Argenio D, Manoil C, Greenberg EP (2001) QscR, a modulator of quorum-sensing signal synthesis and virulence in Pseudomonas aeruginosa. Proc Natl Acad Sci U S A 98:2752–2757. https://doi.org/10.1073/pnas.051624298

Costa SGVAO, Déziel E, Lépine F (2011) Characterization of rhamnolipid production by Burkholderia glumae. Lett Appl Microbiol 53:620–627. https://doi.org/10.1111/j.1472-765X.2011.03154.x

Coulon PML, Groleau M-C, Déziel E (2019) Potential of the Burkholderia cepacia complex to produce 4-hydroxy-3-methyl-2-alkyquinolines. Front Cell Infect Microbiol 9. https://doi.org/10.3389/fcimb.2019.00033

Crowe D, Nicholson A, Fleming A, Carey E, Sánchez-Sanz G, Kelleher F (2017) Conformational studies of Gram-negative bacterial quorum sensing 3-oxo N-acyl homoserine lactone molecules. Bioorg Med Chem 25:4285–4296. https://doi.org/10.1016/j.bmc.2017.06.006

Cui X, Harling R, Mutch P, Darling D (2005) Identification of N-3-hydroxyoctanoyl-homoserine lactone production in Pseudomonas fluorescens 5064, pathogenic to broccoli, and controlling biosurfactant production by quorum sensing. Eur J Plant Pathol 111:297–308. https://doi.org/10.1007/s10658-004-4171-z

Cui Y, Chatterjee A, Hasegawa H, Chatterjee AK (2006) Erwinia carotovora subspecies produce duplicate variants of ExpR, LuxR homologs that activate rsmA transcription but differ in their interactions with N-acylhomoserine lactone signals. J Bacteriol 188:4715–4726. https://doi.org/10.1128/JB.00351-06

Dahl MK, Msadek T, Kunst F, Rapoport G (1992) The phosphorylation state of the DegU response regulator acts as a molecular switch allowing either degradative enzyme synthesis or expression of genetic competence in Bacillus subtilis. J Biol Chem 267:14509–14514

Dawson TE (1993) Hydraulic lift and water use by plants: implications for water balance, performance and plant-plant interactions. Oecologia 95:565–574. https://doi.org/10.1007/BF00317442

Deangelis KM, Lindow SE, Firestone MK (2008) Bacterial quorum sensing and nitrogen cycling in rhizosphere soil. FEMS Microbiol Ecol 66:197–207. https://doi.org/10.1111/j.1574-6941.2008.00550.x

Delalande L, Faure D, Raffoux A, Uroz S, D'Angelo-Picard C, Elasri M et al (2005) N-hexanoyl-L-homoserine lactone, a mediator of bacterial quorum-sensing regulation, exhibits plant-dependent stability and may be inactivated by germinating Lotus corniculatus seedlings. FEMS Microbiol Ecol 52:13–20. https://doi.org/10.1016/j.femsec.2004.10.005

Deng Y, Wu J, Eberl L, Zhang L-H (2010) Structural and functional characterization of diffusible signal factor family quorum-sensing signals produced by members of the Burkholderia cepacia complex. Appl Environ Microbiol 76:4675–4683. https://doi.org/10.1128/AEM.00480-10

Dirksen C, Raats PAC (1985) Water uptake and release by alfalfa roots. Agron J 77:621–626. https://doi.org/10.2134/agronj1985.00021962007700040025x

Dogsa I, Choudhary KS, Marsetic Z, Hudaiberdiev S, Vera R, Pongor S et al (2014) ComQXPA quorum sensing systems may not be unique to Bacillus subtilis: a census in prokaryotic genomes. PLoS One 9:e96122. https://doi.org/10.1371/journal.pone.0096122

Dong SH, Frane ND, Christensen QH, Greenberg EP, Nagarajan R, Nair SK (2017) Molecular basis for the substrate specificity of quorum signal synthases. Proc Natl Acad Sci U S A 114:9092–9097. https://doi.org/10.1073/pnas.1705400114

Donn S, Kirkegaard JA, Perera G, Richardson AE, Watt M (2015) Evolution of bacterial communities in the wheat crop rhizosphere. Environ Microbiol 17:610–621. https://doi.org/10.1111/1462-2920.12452

Du X, Huang R, Zhang Z, Zhang D, Cheng J, Tian P et al (2020) Rhodopseudomonas palustris quorum sensing molecule pC-HSL induces systemic resistance to TMV infection via up-regulation of NbSIPK/NbWIPK expressions in Nicotiana benthamiana. Phytopathology. https://doi.org/10.1094/phyto-05-20-0177-r

Dubeau D, Déziel E, Woods DE, Lépine F (2009) Burkholderia thailandensis harbors two identical rhl gene clusters responsible for the biosynthesis of rhamnolipids. BMC Microbiol 9. https://doi.org/10.1186/1471-2180-9-263

Dulla G, Lindow SE (2008) Quorum size of Pseudomonas syringae is small and dictated by water availability on the leaf surface. Proc Natl Acad Sci U S A 105:3082–3087. https://doi.org/10.1073/pnas.0711723105

Dun EA, Brewer PB, Beveridge CA (2009) Strigolactones: discovery of the elusive shoot branching hormone. Trends Plant Sci 14:364–372. https://doi.org/10.1016/j.tplants.2009.04.003

Dutta S, Podile AR (2010) Plant growth promoting Rhizobacteria (PGPR): the bugs to debug the root zone. Crit Rev Microbiol 36:232–244. https://doi.org/10.3109/10408411003766806

Egamberdieva D, Kamilova F, Validov S, Gafurova L, Kucharova Z, Lugtenberg B (2008) High incidence of plant growth-stimulating bacteria associated with the rhizosphere of wheat grown on salinated soil in Uzbekistan. Environ Microbiol 10:1–9. https://doi.org/10.1111/j.1462-2920.2007.01424.x

El-Sayed AK, Hothersall J, Thomas CM (2001) Quorum-sensing-dependent regulation of biosynthesis of the polyketide antibiotic mupirocin in Pseudomonas fluorescens NCIMB 10586 The GenBank accession numbers for the sequences determined in this work are AF318063 (mupA), AF318064 (mupR) and AF318065 (mupl). Microbiology 147:2127–2139. https://doi.org/10.1099/00221287-147-8-2127

Emmett BD, Youngblut ND, Buckley DH, Drinkwater LE (2017) Plant phylogeny and life history shape rhizosphere bacterial microbiome of summer annuals in an agricultural field. Front Microbiol 8. https://doi.org/10.3389/fmicb.2017.02414

Esseling JJ, Lhuissier FGP, Emons AMC (2003) Nod factor-induced root hair curling: continuous polar growth towards the point of nod factor application. Plant Physiol 132:1982–1988. https://doi.org/10.1104/pp.103.021634

Fang S, Gao X, Deng Y, Chen X, Liao H (2011) Crop root behavior coordinates phosphorus status and neighbors: from field studies to three-dimensional in situ reconstruction of root system architecture. Plant Physiol 155:1277–1285. https://doi.org/10.1104/pp.110.167304

Federle MJ (2009) Autoinducer-2-based chemical communication in bacteria: complexities of interspecies signaling. In: Bacterial sensing and signaling. Karger, Basel, pp 18–32. https://doi.org/10.1159/000219371

Fernández-Piñar R, Cámara M, Dubern J-F, Ramos JL, Espinosa-Urgel M (2011) The Pseudomonas aeruginosa quinolone quorum sensing signal alters the multicellular behaviour of Pseudomonas putida KT2440. Res Microbiol 162:773–781. https://doi.org/10.1016/j.resmic.2011.06.013

Folcik AM, Cutshaw K, Haire T, Goode J, Shah P, Zaidi F et al (2020a) Quorum sensing behavior in the model unicellular eukaryote Chlamydomonas reinhardtii. iScience 23:101714. https://doi.org/10.1016/j.isci.2020.101714

Folcik AM, Haire T, Cutshaw K, Riddle M, Shola C, Nassani S et al (2020b) Computer-assisted tracking of Chlamydomonas species. Front Plant Sci 10:1616. https://doi.org/10.3389/FPLS.2019.01616

Fuller A, Young P, Pierce D, Kitson-Finuff J, Jain P, Schneider K et al (2017) Redox-mediated quorum sensing in plants. PLoS One 12(9):e0182655

Fuqua C (2006) The QscR quorum-sensing regulon of Pseudomonas aeruginosa: an orphan claims its identity. J Bacteriol. https://doi.org/10.1128/JB.188.9.3169-3171.2006

Gantner S, Schmid M, Dürr C, Schuhegger R, Steidle A, Hutzler P et al (2006) In situ quantitation of the spatial scale of calling distances and population density-independent N-acylhomoserine lactone-mediated communication by rhizobacteria colonized on plant roots. FEMS Microbiol Ecol 56:188–194. https://doi.org/10.1111/j.1574-6941.2005.00037.x

Geske GD, O'Neill JC, Blackwell HE (2008) Expanding dialogues: from natural autoinducers to non-natural analogues that modulate quorum sensing in Gram-negative bacteria. Chem Soc Rev 37:1432–1447. https://doi.org/10.1039/b703021p

González JE, Marketon MM (2003) Quorum sensing in nitrogen-fixing rhizobia. Microbiol Mol Biol Rev 67:574–592. https://doi.org/10.1128/MMBR.67.4.574

Goo E, Kang Y, Kim H, Hwang I (2010) Proteomic analysis of quorum sensing-dependent proteins in Burkholderia glumae. J Proteome Res 9:3184–3199. https://doi.org/10.1021/pr100045n

Guyonnet JP, Cantarel AAM, Simon L, Haichar FZ (2018) Root exudation rate as functional trait involved in plant nutrient-use strategy classification. Int J Bus Innov Res 17:8573–8581. https://doi.org/10.1002/ece3.4383

Ha C, Park SJ, Im S-J, Park S-J, Lee J-H (2012) Interspecies signaling through QscR, a quorum receptor of Pseudomonas aeruginosa. Mol Cells 33:53–59. https://doi.org/10.1007/s10059-012-2208-2

Harris EH (2001) Chlamydomonas as a model organism. Annu Rev Plant Physiol Plant Mol Biol 52:363–406. https://doi.org/10.1146/annurev.arplant.52.1.363

He F, Nair GR, Soto CS, Chang Y, Hsu L, Ronzone E et al (2009) Molecular basis of ChvE function in sugar binding, sugar utilization, and virulence in Agrobacterium tumefaciens. J Bacteriol 191:5802–5813. https://doi.org/10.1128/JB.00451-09

Herman DJ, Johnson KK, Jaeger CH, Schwartz E, Firestone MK (2006) Root influence on nitrogen mineralization and nitrification in Avena barbata rhizosphere soil. Soil Sci Soc Am J 70:1504–1511. https://doi.org/10.2136/sssaj2005.0113

Hirsch AM, Mcfall-Ngai MJ (2000) Fundamental concepts in symbiotic interactions: light and dark, day and night, squid and legume. J Plant Growth Regul 19:113–130. https://doi.org/10.1007/s003440000025

Hornby JM, Jensen EC, Lisec AD, Tasto JJ, Jahnke B, Shoemaker R et al (2001) Quorum sensing in the dimorphic fungus Candida albicans is mediated by farnesol. Appl Environ Microbiol 67:2982–2992. https://doi.org/10.1128/AEM.67.7.2982-2992.2001

Hu X, Zhao J, DeGrado WF, Binns AN (2013) Agrobacterium tumefaciens recognizes its host environment using ChvE to bind diverse plant sugars as virulence signals. Proc Natl Acad Sci U S A 110:678–683. https://doi.org/10.1073/pnas.1215033110

Huang Y-L, Dobretsov S, Ki J-S, Yang L-H, Qian P-Y (2007) Presence of acyl-homoserine lactone in subtidal biofilm and the implication in larval behavioral response in the polychaete Hydroides elegans. Microb Ecol 54:384–392. https://doi.org/10.1007/s00248-007-9210-9

Huang X-F, Chaparro JM, Reardon KF, Zhang R, Shen Q, Vivanco JM (2014) Rhizosphere interactions: root exudates, microbes, and microbial communities. Botany 92:267–275. https://doi.org/10.1139/cjb-2013-0225

Hubbard CJ, Brock MT, van Diepen LT, Maignien L, Ewers BE, Weinig C (2018) The plant circadian clock influences rhizosphere community structure and function. ISME J 12:400–410. https://doi.org/10.1038/ismej.2017.172

Islam S, Akanda AM, Sultana F, Hossain MM (2014) Chilli rhizosphere fungus Aspergillus spp. PPA1 promotes vegetative growth of cucumber (Cucumis sativus) plants upon root colonisation. Arch Phytopathol Plant Protect 47:1231–1238. https://doi.org/10.1080/03235408.2013.837633

Jang MS, Goo E, An JH, Kim J, Hwang I (2014) Quorum sensing controls flagellar morphogenesis in Burkholderia glumae. PLoS One 9:e84831. https://doi.org/10.1371/journal.pone.0084831

John M, Maurizio S, Poeta D (2006) Lipid signaling in pathogenic fungi. Curr Opin Microbiol 9(4):352–358. https://doi.org/10.1016/j.mib.2006.06.003

Jones BW, Nishiguchi MK (2004) Counterillumination in the Hawaiian bobtail squid, Euprymna scolopes Berry (Mollusca: Cephalopoda). Mar Biol 144:1151–1155. https://doi.org/10.1007/s00227-003-1285-3

Joo JH, Jetten AM (2010) Molecular mechanisms involved in farnesol-induced apoptosis. Cancer Lett 287:123–135. https://doi.org/10.1016/j.canlet.2009.05.015

Kakkar A, Nizampatnam NR, Kondreddy A, Pradhan BB, Chatterjee S (2015) Xanthomonas campestris cell–cell signalling molecule DSF (diffusible signal factor) elicits innate immunity in plants and is suppressed by the exopolysaccharide xanthan. J Exp Bot 66:6697–6714. https://doi.org/10.1093/jxb/erv377

Kalamara M, Spacapan M, Mandic-Mulec I, Stanley-Wall NR (2018) Social behaviours by Bacillus subtilis: quorum sensing, kin discrimination and beyond. Mol Microbiol 110:863–878. https://doi.org/10.1111/mmi.14127

Kaufmann GF, Sartorio R, Lee S-H, Rogers CJ, Meijler MM, Moss JA et al (2005) Revisiting quorum sensing: discovery of additional chemical and biological functions for 3-oxo-N-acylhomoserine lactones. Proc Natl Acad Sci U S A 102:309–314. https://doi.org/10.1073/pnas.0408639102

Kearns DB, Losick R (2004) Swarming motility in undomesticated Bacillus subtilis. Mol Microbiol 49:581–590. https://doi.org/10.1046/j.1365-2958.2003.03584.x

Keyes WJ, Palmer AG, Erbil WK, Taylor JV, Apkarian RP, Weeks ER et al (2007) Semagenesis and the parasitic angiosperm Striga asiatica. Plant J 51:707–716

Klamer M, Roberts MS, Levine LH, Drake BG, Garland JL (2002) Influence of elevated CO_2 on the fungal community in a coastal scrub oak forest soil investigated with terminal-restriction fragment length polymorphism analysis. Appl Environ Microbiol 68:4370–4376. https://doi.org/10.1128/AEM.68.9.4370-4376.2002

Kruppa M (2009) Quorum sensing and Candida albicans. Mycoses 52:1–10. https://doi.org/10.1111/j.1439-0507.2008.01626.x

Kugler S, Sebghati TS, Eissenberg LG, Goldman WE (2000) Phenotypic variation and intracellular parasitism by Histoplasma capsulatum. Proc Natl Acad Sci U S A 97:8794–8798. https://doi.org/10.1073/pnas.97.16.8794

Kuzma MM, Hunt S, Layzell DB (1993) Role of oxygen in the limitation and inhibition of nitrogenase activity and respiration rate in individual soybean nodules. Plant Physiol 101:161–169. https://doi.org/10.1104/pp.101.1.161

Lambers H, Mougel C, Jaillard B, Hinsinger P (2009) Plant-microbe-soil interactions in the rhizosphere: an evolutionary perspective. Plant Soil 321:83–115. https://doi.org/10.1007/s11104-009-0042-x

Lang J, Faure D (2014) Functions and regulation of quorum-sensing in Agrobacterium tumefaciens. Front Plant Sci 5:14. https://doi.org/10.3389/fpls.2014.00014

Lewin RA (1956) Extracellular polysaccharides of green algae. Can J Microbiol 2:665–672. https://doi.org/10.1139/m56-079

Liang L, Liu Y, Jariwala J, Lynn DG, Palmer AG (2016) Detection and adaptation in parasitic angiosperm host selection. Am J Plant Sci 7:1275–1290. https://doi.org/10.4236/ajps.2016.78123

Liao L, Schaefer AL, Coutinho BG, Brown PJB, Greenberg EP (2018) An aryl-homoserine lactone quorum-sensing signal produced by a dimorphic prosthecate bacterium. Proc Natl Acad Sci U S A 115:7587–7592. https://doi.org/10.1073/pnas.1808351115

Lin Y-H, Binns AN, Lynn DG (2008) The initial steps in Agrobacterium tumefaciens pathogenesis: chemical biology of host recognition. In: Tzfira T, Citovsky V (eds) Agrobacterium: from biology to biotechnology. Springer, New York, pp 221–241

Lindemann A, Pessi G, Schaefer AL, Mattmann ME, Christensen QH, Kessler A et al (2011) Isovaleryl-homoserine lactone, an unusual branched-chain quorum-sensing signal from the soybean symbiont Bradyrhizobium japonicum. Proc Natl Acad Sci U S A 108:16765–16770. https://doi.org/10.1073/pnas.1114125108

Lowery CA, Park J, Gloeckner C, Meijler MM, Mueller RS, Boshoff HI et al (2009) Defining the mode of action of tetramic acid antibacterials derived from Pseudomonas aeruginosa quorum sensing signals. J Am Chem Soc 131:14473–14479. https://doi.org/10.1021/ja9056079

Lupp C, Ruby EG (2005) Vibrio fischeri uses two quorum-sensing systems for the regulation of early and late colonization factors. J Bacteriol 187:3620–3629. https://doi.org/10.1128/JB.187.11.3620-3629.2005

Lynch JP (2007) Turner review no. 14. Roots of the second green revolution. Aust J Bot 55:493–512. https://doi.org/10.1071/BT06118

Marketon MM, Gronquist MR, Eberhard A, Gonza JE (2002) Production of novel N-acyl homoserine lactones. J Bacteriol 184:5686–5695. https://doi.org/10.1128/JB.184.20.5686

Mathesius U, Mulders S, Gao M, Teplitski M, Caetano-anolle G, Rolfe BG et al (2003) Extensive and specific responses of a eukaryote to bacterial quorum-sensing signals. Proc Natl Acad Sci U S A 100:1444–1449. https://doi.org/10.1073/pnas.262672599

Mattmann ME, Geske GD, Worzalla GA, Chandler JR, Sappington KJ, Greenberg EP et al (2008) Synthetic ligands that activate and inhibit a quorum-sensing regulator in Pseudomonas aeruginosa. Bioorg Med Chem Lett 18:3072–3075. https://doi.org/10.1016/j.bmcl.2007.11.095

Mendes R, Kruijt M, de Bruijn I, Dekkers E, van der Voort M, Schneider JHM et al (2011) Deciphering the rhizosphere microbiome for disease-suppressive bacteria. Science 332:1097–1100. https://doi.org/10.1126/science.1203980

Mendes R, Garbeva P, Raaijmakers JM (2013) The rhizosphere microbiome: significance of plant beneficial, plant pathogenic, and human pathogenic microorganisms. FEMS Microbiol Rev 37: 634–663. https://doi.org/10.1111/1574-6976.12028

Mhedbi-Hajri N, Yahiaoui N, Mondy S, Hue N, Pélissier F, Faure D et al (2016) Transcriptome analysis revealed that a quorum sensing system regulates the transfer of the pAt megaplasmid in Agrobacterium tumefaciens. BMC Genomics 17:1–14. https://doi.org/10.1186/s12864-016-3007-5

Miransari M (2011) Interactions between arbuscular mycorrhizal fungi and soil bacteria. Appl Microbiol Biotechnol 89:917–930. https://doi.org/10.1007/s00253-010-3004-6

Moll J, Goldmann K, Kramer S, Hempel S, Kandeler E, Marhan S et al (2015) Resource type and availability regulate fungal communities along arable soil profiles. Microb Ecol 70:390–399. https://doi.org/10.1007/s00248-015-0569-8

Montgomery K, Charlesworth JC, LeBard R, Visscher PT, Burns BP (2013) Quorum sensing in extreme environments. Life (Basel, Switzerland) 3:131–148. https://doi.org/10.3390/life3010131

Moore JD, Gerdt JP, Eibergen NR, Blackwell HE (2014) Active efflux influences the potency of quorum sensing inhibitors in Pseudomonas aeruginosa. Chembiochem 15:435–442. https://doi.org/10.1002/cbic.201300701

Mosel DD, Dumitru R, Hornby JM, Atkin AL, Nickerson KW (2005) Farnesol concentrations required to block germ tube formation in Candida albicans in the presence and absence of serum. Appl Environ Microbiol 71:4938–4940. https://doi.org/10.1128/AEM.71.8.4938-4940.2005

Mukherjee A, Cui Y, Liu Y, Dumenyo CK, Chatterjee AK (1996) Global regulation in Erwinia species by Erwnia carotovora rSmA, a homologue of Escherichia coli csrA: repression of secondary metabolites, pathogenicity and hypersensitive reaction. www.microbiologyresearch.org. Accessed 10 Dec 2020

Nadell CD, Xavier JB, Levin SA, Foster KR (2008) The evolution of quorum sensing in bacterial biofilms. PLoS Biol 6:0171–0179. https://doi.org/10.1371/journal.pbio.0060014

Nickzad A, Lépine F, Déziel E (2015) Quorum sensing controls swarming motility of Burkholderia glumae through regulation of rhamnolipids. PLoS One 10:e0128509. https://doi.org/10.1371/journal.pone.0128509

Novick RP, Geisinger E (2008) Quorum sensing in staphylococci. Annu Rev Genet 42:541–564. https://doi.org/10.1146/annurev.genet.42.110807.091640

Nuccio EE, Starr E, Karaoz U, Brodie EL, Zhou J, Tringe SG et al (2020) Niche differentiation is spatially and temporally regulated in the rhizosphere. ISME J 14:999–1014. https://doi.org/10.1038/s41396-019-0582-x

Oldroyd GED, Downie JA (2008) Coordinating nodule morphogenesis with rhizobial infection in legumes. Annu Rev Plant Biol 59:519–546. https://doi.org/10.1146/annurev.arplant.59.032607.092839

Omer Bendori S, Pollak S, Hizi D, Eldar A (2015) The RapP-PhrP quorum-sensing system of Bacillus subtilis strain NCIB3610 affects biofilm formation through multiple targets, due to an atypical signal-insensitive allele of RapP. J Bacteriol 197:592–602. https://doi.org/10.1128/JB.02382-14

Ongeng D, Muyanja C, Ryckeboer J, Geeraerd AH, Springael D (2011) Rhizosphere effect on survival of Escherichia coli O157:H7 and Salmonella enterica serovar Typhimurium in manure-amended soil during cabbage (Brassica oleracea) cultivation under tropical field conditions in Sub-Saharan Africa. Int J Food Microbiol 149:133–142. https://doi.org/10.1016/j.ijfoodmicro.2011.06.009

Ortíz-Castro R, Martínez-Trujillo M, López-Bucio J (2008) N-acyl-L-homoserine lactones: a class of bacterial quorum-sensing signals alter post-embryonic root development in Arabidopsis thaliana. Plant Cell Environ 31:1497–1509. https://doi.org/10.1111/j.1365-3040.2008.01863.x

Oslizlo A, Stefanic P, Vatovec S, Beigot Glaser S, Rupnik M, Mandic-Mulec I (2015) Exploring ComQXPA quorum-sensing diversity and biocontrol potential of Bacillus spp. isolates from tomato rhizoplane. Microb Biotechnol 8:527–540. https://doi.org/10.1111/1751-7915.12258

Palla M, Battini F, Cristani C, Giovannetti M, Squartini A, Agnolucci M (2018) Quorum sensing in rhizobia isolated from the spores of the mycorrhizal symbiont Rhizophagus intraradices. Mycorrhiza 28:773–778. https://doi.org/10.1007/s00572-018-0847-7

Palmer AG, Blackwell HE (2008) Deciphering a protolanguage for bacteria-host communication. Nat Chem Biol 4:452–454. https://doi.org/10.1038/nchembio0808-452

Palmer AG, Gao R, Maresh J, Erbil WK, Lynn DG (2004) Chemical biology of multi-host/pathogen interactions: chemical perception and metabolic complementation. Annu Rev Phytopathol 42:439–464. https://doi.org/10.1146/annurev.phyto.41.052002.095701

Palmer AG, Streng E, Blackwell HE (2011a) Attenuation of virulence in pathogenic bacteria using synthetic quorum-sensing modulators under native conditions on plant hosts. ACS Chem Biol 6:1348–1356. https://doi.org/10.1021/cb200298g

Palmer AG, Streng E, Jewell KA, Blackwell HE (2011b) Quorum sensing in bacterial species that use degenerate autoinducers can be tuned by using structurally identical non-native ligands. Chembiochem 12:138–147. https://doi.org/10.1002/cbic.201000551

Palmer AG, Senechal AC, Mukherjee A, Ané J-M, Blackwell HE (2014) Plant responses to bacterial N-acyl L-homoserine lactones are dependent on enzymatic degradation to L-homoserine. ACS Chem Biol 9:1834–1845. https://doi.org/10.1021/cb500191a

Palmer AG, Mukherjee A, Stacy DM, Lazar S, Ané JM, Blackwell HE (2016) Interkingdom responses to bacterial quorum sensing signals regulate frequency and rate of nodulation in legume–rhizobia symbiosis. Chembiochem 17:2199–2205. https://doi.org/10.1002/cbic.201600373

Palmer AG, Senechal AC, Haire TC, Mehta NP, Valiquette SD, Blackwell HE (2018) Selection of appropriate autoinducer analogues for the modulation of quorum sensing at the host-bacterium interface. ACS Chem Biol 13:3115–3122. https://doi.org/10.1021/acschembio.8b00676

Papenfort K, Bassler BL (2016) Quorum sensing signal–response systems in Gram-negative bacteria. Nat Rev Microbiol 14:576–588. https://doi.org/10.1038/nrmicro.2016.89

Parniske M (2008) Arbuscular mycorrhiza: the mother of plant root endosymbioses. Nat Rev Microbiol 6:763–775. https://doi.org/10.1038/nrmicro1987

Pearson JP, Van Delden C, Iglewski BH (1999) Active efflux and diffusion are involved in transport of Pseudomonas aeruginosa cell-to-cell signals. J Bacteriol 181:1203–1210

Pereira CS, McAuley JR, Taga ME, Xavier KB, Miller ST (2008) Sinorhizobium meliloti, a bacterium lacking the autoinducer-2 (AI-2) synthase, responds to AI-2 supplied by other bacteria. Mol Microbiol 70:1223–1235. https://doi.org/10.1111/j.1365-2958.2008.06477.x

Pingali PL (2012) Green revolution: impacts, limits, and the path ahead. Proc Natl Acad Sci U S A 109:12302–12308. https://doi.org/10.1073/pnas.0912953109

Platt TG, Morton ER, Barton IS, Bever JD, Fuqua C (2014) Ecological dynamics and complex interactions of Agrobacterium megaplasmids. Front Plant Sci 5:1–15. https://doi.org/10.3389/fpls.2014.00635

Praneenararat T, Palmer AG, Blackwell HE (2012) Chemical methods to interrogate bacterial quorum sensing pathways. Org Biomol Chem 10:8189–8199. https://doi.org/10.1039/c2ob26353j

Proust H, Hoffmann B, Xie X, Yoneyama K, Schaefer DG, Yoneyama K et al (2011) Strigolactones regulate protonema branching and act as a quorum sensing-like signal in the moss Physcomitrella patens. Development 138:1531–1539. https://doi.org/10.1242/dev.058495

Quiñones B, Dulla G, Lindow SE (2005) Quorum sensing regulates exopolysaccharide production, motility, and virulence in Pseudomonas syringae. Mol Plant-Microbe Interact 18:682–693. https://doi.org/10.1094/MPMI-18-0682

Rajamani S, Teplitski M, Kumar A, Krediet CJ, Sayre RT, Bauer WD (2011) N-Acyl homoserine lactone lactonase, aiia, inactivation of quorum-sensing agonists produced by Chlamydomonas reinhardtii (Chlorophyta) and characterization of aiia transgenic algAE. J Phycol 47:1219–1227. https://doi.org/10.1111/j.1529-8817.2011.01049.x

Reen FJ, Mooij MJ, Holcombe LJ, McSweeney CM, McGlacken GP, Morrissey JP et al (2011) The Pseudomonas quinolone signal (PQS), and its precursor HHQ, modulate interspecies and interkingdom behaviour. FEMS Microbiol Ecol 77:413–428. https://doi.org/10.1111/j.1574-6941.2011.01121.x

Rutherford ST, Bassler BL (2012) Bacterial quorum sensing: its role in virulence and possibilities for its control. Cold Spring Harb Perspect Med 2:a012427–a012427. https://doi.org/10.1101/cshperspect.a012427

Ryan RP, McCarthy Y, Watt SA, Niehaus K, Dow JM (2009) Intraspecies signaling involving the diffusible signal factor BDSF (cis-2-dodecenoic acid) influences virulence in Burkholderia cenocepacia. J Bacteriol 191:5013–5019. https://doi.org/10.1128/JB.00473-09

Sanchez L, Courteaux B, Hubert J, Kauffmann S, Renault J-H, Clément C et al (2012) Rhamnolipids elicit defense responses and induce disease resistance against biotrophic, hemibiotrophic, and necrotrophic pathogens that require different signaling pathways in Arabidopsis and highlight a central role for salicylic acid. Plant Physiol 160:1630–1641. https://doi.org/10.1104/pp.112.201913

Sánchez-Sanz G, Crowe D, Nicholson A, Fleming A, Carey E, Kelleher F (2018) Conformational studies of Gram-negative bacterial quorum sensing acyl homoserine lactone (AHL) molecules: the importance of the n $\rightarrow \pi^*$ interaction. Biophys Chem 238:16–21. https://doi.org/10.1016/j.bpc.2018.04.002

Sasse J, Martinoia E, Northen T (2018) Feed your friends: do plant exudates shape the root microbiome? Trends Plant Sci 23:25–41. https://doi.org/10.1016/j.tplants.2017.09.003

Schaefer AL, Greenberg EP, Oliver CM, Oda Y, Huang JJ, Bittan-Banin G et al (2008) A new class of homoserine lactone quorum-sensing signals. Nature 454:595–599. https://doi.org/10.1038/nature07088

Schauder S, Shokat K, Surette MG, Bassler BL (2001) The LuxS family of bacterial autoinducers: biosynthesis of a novel quorum-sensing signal molecule. Mol Microbiol 41:463–476. https://doi.org/10.1046/j.1365-2958.2001.02532.x

Siciliano F, Torres P, Sendin L, Bermejo C, Filippone P, Vellice G et al (2006) Analysis of the molecular basis of Xanthomonas axonopodis pv. citri pathogenesis in Citrus limon. Electron J Biotechnol 9. https://doi.org/10.2225/vol9-issue3-fulltext-20

Sifri CD (2008) Quorum sensing: bacteria talk sense. Clin Infect Dis. https://doi.org/10.1086/592072

Smith SE, Smith FA (2011) Roles of arbuscular mycorrhizas in plant nutrition and growth: new paradigms from cellular to ecosystem scales. Annu Rev Plant Biol 62:227–250. https://doi.org/10.1146/annurev-arplant-042110-103846

Spacapan M, Danevčič T, Mandic-Mulec I (2018) ComX-induced exoproteases degrade ComX in Bacillus subtilis PS-216. Front Microbiol 9. https://doi.org/10.3389/fmicb.2018.00105

Steindler L, Bertani I, De Sordi L, Schwager S, Eberl L, Venturi V (2009) LasI/R and RhlI/R quorum sensing in a strain of Pseudomonas aeruginosa beneficial to plants. Appl Environ Microbiol 75:5131–5140. https://doi.org/10.1128/AEM.02914-08

Szoboszlay M, White-Monsant A, Moe LA (2016) The effect of root exudate 7,4′-dihydroxyflavone and naringenin on soil bacterial community structure. PLoS One 11:e0146555. https://doi.org/10.1371/journal.pone.0146555

Taga ME, Miller ST, Bassler BL (2003) Lsr-mediated transport and processing of AI-2 in Salmonella typhimurium. Mol Microbiol 50:1411–1427. https://doi.org/10.1046/j.1365-2958.2003.03781.x

Taga ME, Semmelhack JL, Bassler BL (2008) The LuxS-dependent autoinducer AI-2 controls the expression of an ABC transporter that functions in AI-2 uptake in Salmonella typhimurium. Mol Microbiol 42:777–793. https://doi.org/10.1046/j.1365-2958.2001.02669.x

Tang J-L, Liu Y-N, Barber CE, Dow JM, Wootton JC, Daniels MJ (1991) Genetic and molecular analysis of a cluster of rpf genes involved in positive regulation of synthesis of extracellular enzymes and polysaccharide in Xanthomonas campestris pathovar campestris. Mol Gen Genet MGG 226:409–417. https://doi.org/10.1007/BF00260653

Taylor NS, Merrifield R, Williams TD, Chipman JK, Lead JR, Viant MR (2016) Molecular toxicity of cerium oxide nanoparticles to the freshwater alga Chlamydomonas reinhardtii is associated with supra-environmental exposure concentrations. Nanotoxicology 10:1–10. https://doi.org/10.3109/17435390.2014.1002868

Teplitski M, Chen H, Rajamani S, Gao M, Merighi M, Sayre RT et al (2004) Chlamydomonas reinhardtii secretes compounds that mimic bacterial signals and interfere with quorum sensing regulation in bacteria. Plant Physiol 134:137–146. https://doi.org/10.1104/pp.103.029918

Thowthampitak J, Shaffer BT, Prathuangwong S, Loper JE (2008) Role of rpfF in virulence and exoenzyme production of Xanthomonas axonopodis pv. glycines, the causal agent of bacterial pustule of soybean. Phytopathology 98:1252–1260. https://doi.org/10.1094/PHYTO-98-12-1252

Tortosa P, Logsdon L, Kraigher B, Itoh Y, Mandic-Mulec I, Dubnau D (2001) Specificity and genetic polymorphism of the Bacillus competence quorum-sensing system. J Bacteriol 183:451–460. https://doi.org/10.1128/JB.183.2.451-460.2001

Uroz S (2003) Novel bacteria degrading N-acylhomoserine lactones and their use as quenchers of quorum-sensing-regulated functions of plant-pathogenic bacteria. Microbiology 149:1981–1989. https://doi.org/10.1099/mic.0.26375-0

van der Wal A, Geydan TD, Kuyper TW, de Boer W (2013) A thready affair: linking fungal diversity and community dynamics to terrestrial decomposition processes. FEMS Microbiol Rev 37:477–494. https://doi.org/10.1111/1574-6976.12001

Vatsa P, Sanchez L, Clement C, Baillieul F, Dorey S (2010) Rhamnolipid biosurfactants as new players in animal and plant defense against microbes. Int J Mol Sci 11:5095–5108. https://doi.org/10.3390/ijms11125095

Veliz-Vallejos DF, van Noorden GE, Yuan M, Mathesius U (2014) A Sinorhizobium meliloti-specific N-acyl homoserine lactone quorum-sensing signal increases nodule numbers in Medicago truncatula independent of autoregulation. Front Plant Sci 5:551. https://doi.org/10.3389/fpls.2014.00551

Venturi V, Subramoni S, Sabag-Daigle A, Ahmer BMM (2018) Methods to study solo/orphan quorum-sensing receptors. Methods Mol Biol 1673:145–159. https://doi.org/10.1007/978-1-4939-7309-5_12

von Rad U, Klein I, Dobrev PI, Kottova J, Zazimalova E, Fekete A et al (2008) Response of Arabidopsis thaliana to N-hexanoyl-dl-homoserine-lactone, a bacterial quorum sensing molecule produced in the rhizosphere. Planta 229:73–85. https://doi.org/10.1007/s00425-008-0811-4

Walker TS, Bais HP, Grotewold E, Vivanco JM (2003) Root exudation and rhizosphere biology. Plant Physiol 132:44–51. https://doi.org/10.1104/pp.102.019661

Walker TS, Bais HP, De E, Schweizer HP, Rahme LG, Fall R et al (2004) Pseudomonas aeruginosa-plant root interactions. Pathogenicity, biofilm formation, and root exudation. Plant Physiol 134:320–331. https://doi.org/10.1104/pp.103.027888.such

Weber K, Schulz B, Ruhnke M (2010) The quorum-sensing molecule E,E-farnesol-its variable secretion and its impact on the growth and metabolism of Candida species. Yeast 27:727–739. https://doi.org/10.1002/yea.1769

Welsh MA, Blackwell HE (2016) Chemical probes of quorum sensing: from compound development to biological discovery. FEMS Microbiol Rev 40:774–794. https://doi.org/10.1093/femsre/fuw009

White CE, Winans SC (2005) Identification of amino acid residues of the Agrobacterium tumefaciens quorum-sensing regulator TraR that are critical for positive control of transcription. Mol Microbiol 55:1473–1486. https://doi.org/10.1111/j.1365-2958.2004.04482.x

Winters M, Arneborg N, Appels R, Howell K (2019) Can community-based signalling behaviour in Saccharomyces cerevisiae be called quorum sensing? A critical review of the literature. FEMS Yeast Res 19. https://doi.org/10.1093/femsyr/foz046

Xavier KB, Bassler BL (2003) LuxS quorum sensing: more than just a numbers game. Curr Opin Microbiol 6:191–197. https://doi.org/10.1016/S1369-5274(03)00028-6

Xiong Q, Liu D, Zhang H, Dong X, Zhang G, Liu Y et al (2020) Quorum sensing signal autoinducer-2 promotes root colonization of Bacillus velezensis SQR9 by affecting biofilm formation and motility. Appl Microbiol Biotechnol 104:7177–7185. https://doi.org/10.1007/s00253-020-10713-w

Yates EA, Philipp B, Buckley C, Atkinson S, Chhabra SR, Sockett RE et al (2002) N-acylhomoserine lactones undergo lactonolysis in a pH-, temperature-, and acyl chain length-dependent manner during growth of Yersinia pseudotuberculosis and Pseudomonas aeruginosa. Infect Immun 70:5635–5646. https://doi.org/10.1128/IAI.70.10.5635-5646.2002

Zhang L, Xie G (2007) Diversity and distribution of Burkholderia cepacia complex in the rhizosphere of rice and maize. FEMS Microbiol Lett 266:231–235. https://doi.org/10.1111/j.1574-6968.2006.00530.x

Zhang G, Zhang F, Ding G, Li J, Guo X, Zhu J et al (2012) Acyl homoserine lactone-based quorum sensing in a methanogenic archaeon. ISME J 6:1336–1344. https://doi.org/10.1038/ismej.2011.203

Zhou L, Yu Y, Chen X, Diab AA, Ruan L, He J et al (2015) The multiple DSF-family QS signals are synthesized from carbohydrate and branched-chain amino acids via the FAS elongation cycle. Sci Rep 5:13294. https://doi.org/10.1038/srep13294

Ziegler EW, Brown AB, Nesnas N, Palmer AG (2019) Abiotic hydrolysis kinetics of N-acyl-L-homoserine lactones: natural silencing of bacterial quorum sensing signals. Eur J Org Chem. https://doi.org/10.1002/ejoc.201900322

Chapter 6
Metabolomics Approaches for Studying the *Trichoderma*-Plant Interactions

David Barbosa Medeiros, Alisdair R. Fernie, and Yariv Brotman

Abstract The aim of metabolomics is to simultaneously detect, quantify, and annotate a large number of metabolites in a given biological sample. The application of metabolomics—mostly based on mass spectrometry coupled with either gas or liquid chromatography—provides access to an array of diverse compounds. Many of these play crucial roles in interactions between plants and beneficial microorganisms, as well as influence the plant's response to abiotic or biotic stress. An increasing number of studies use metabolomics for understanding key features that underlie these interactions. Key studies have provided insight into the priming phenomenon landscape, pinpointed key signal molecules, showed how the interactions are formed and maintained, and unraveled the metabolic changes that plants undergo and that result in increased growth and yield—the hallmark of many beneficial interactions. In this chapter we review some of the key technical issues to be addressed when conducting such metabolomics studies. We give examples of how such studies promote our understanding of these complex and fascinating biological interactions—many times when integrated with other OMICS approaches, of how they allow the identification of biomolecules that can be used for agricultural improvement, and of small molecules that will promote such interactions and could be used as biostimulants.

Keywords Metabolomics · Mass-spectrometry · Gas and liquid chromatography · Metabolic pathways

D. B. Medeiros · A. R. Fernie
Max-Planck-Institute of Molecular Plant Physiology, Potsdam, Germany

Y. Brotman (✉)
Department of Life Sciences, Ben-Gurion University of the Negev, Beersheva, Israel
e-mail: brotmany@post.bgu.ac.il

B. A. Horwitz, P. K. Mukherjee (eds.), *Microbial Cross-talk in the Rhizosphere*, Rhizosphere Biology, https://doi.org/10.1007/978-981-16-9507-0_6

6.1 Introduction

The use of living organisms (antagonists) to reduce the pathogenic activity, known as biological control, is an important approach in the management of plant diseases. Advantages of using biological control among others are the reduction in the use of chemical pesticides, it does not generate pathogen resistance, reduce environment chemical contamination, and it is suitable for use in organic farming. *Trichoderma* genus is considered the greatest biological control against plant fungal diseases within a range of 25 fungal antagonists (Thambugala et al. 2020). It comprises filamentous fungi, saprophytic, avirulent, and opportunistic plant symbionts that are mostly rhizospheric. The ability of *Trichoderma* spp. to parasitize other fungi, mainly soil pathogens, such as Fusarium and Rhizoctonia or the oomycete Pythium, is behind its commercial success as a biopesticide, being the basis of more than 60% of all registered biopesticides available (Pozo et al. 2021).

Trichoderma spp. can also improve plant growth and tolerance against abiotic stresses and induce plant systemic resistance against pathogens and pests (Martínez-Medina et al. 2017). At the cellular level *Trichoderma* alters plant specific gene expression (De Palma et al. 2019), which program protein levels and ultimately the metabolite contents. It is often noted that one must study protein levels, since they determine the activity of the cell, and not only transcript levels, which may be more easily accessible to measurement. This argument needs to be extended to metabolite levels. The metabolome is defined as the entire set of low molecular weight compounds of an organism and its composition might be considered as the ultimate form of the phenotypic signature of a living organism, downstream to the genetic variance, transcriptomic and proteomic components. Therefore, studying the metabolome of a given biological system is essential if one would like to reach a complete understanding of biological events. The ambitious aim of metabolomics is to follow all detectable metabolites of a certain biological system. In recent years, studies of the plant-*Trichoderma* interactions have implemented these approaches by applying the recent advances in plant metabolomics methods. However, it is evident that there is still a lot of room for improvements to reach a detailed and comprehensive and view of all plant metabolites responsive to this beneficial interaction.

In this chapter, we focus on how metabolomics approaches can guide our understanding of the molecular processes underlying the plant-*Trichoderma* interaction. A recurring theme is the question of how *Trichoderma* remodels the plant responses, without causing disease. Some, but not all, of the answers will come from the transcriptome and proteome levels. As a logical extension studying the changes of the plant metabolome will facilitate unrevealing the molecular mechanisms involved in the plant-*Trichoderma* interactions.

Over the past two decades novel methods used for metabolite detection that offer robust, accurate, and sensitive analysis of several hundreds to thousands of compounds have been established (Kopka et al. 2004; Lisec et al. 2006). While methods for the measurement of individual metabolites by spectrophotometric assays or simple separation methods using chromatography have been used for a long time, the analysis of several compounds only started to become feasible with the hyphenation of separation methods to various detection systems (Fernie et al. 2004). The separation methods which are commonly applied for metabolite profiling include gas chromatography (GC), liquid chromatography (LC), and capillary electrophoresis (CE). Chromatography requires relatively lengthy extraction (and, in the case of GC, derivatization to render the metabolites volatile) in return for highly detailed results. Two major detection systems dominate metabolomics studies: MS, which is more sensitive, and NMR that can better quantify metabolites and detect conformational isomers (Obata and Fernie 2012; Alseekh and Fernie 2018).

Metabolites detected and quantified by mass spectrometry-based methods are usually divided into three main groups based on their physicochemical properties and molecular mass as follows: (1) Polar, low molecular mass metabolites (mostly primary metabolites) that are detected by gas chromatography-mass spectrometry (GC-MS); (2) Polar, high molecular mass metabolites (mostly secondary metabolites) that are detected by liquid chromatography-mass spectrometry (LC-MS); (3) Non-polar (lipids), metabolites that are detected by either LC-MS or GC-MS.

Classification of the metabolite pool into primary, secondary, and lipid metabolites is based not only on their physicochemical properties, but also on their biological functions. The primary metabolite group contains those that are in the core functions of each cell, and depleting metabolites from this group will cause immediate damage to the normal function of the organism, including death. Examples for some of the major groups of primary metabolites are amino acids, sugars and sugars alcohol derivatives, the citric cycle intermediates, polyamines. The secondary metabolites group is defined as metabolites that are not directly involved in the normal growth and development of a given organism. Unlike primary metabolites, absence of secondary metabolites does not result in immediate death. In plants a repertoire of secondary metabolites can be synthesized, for example, in response to abiotic and biotic stress (examples of plant secondary metabolites are carotenoids, flavonoids and atropine). Lipids are the plethora of chemically different molecules formed by combining fatty acids with different backbones, together commonly referred to as the lipidome (Wenk 2005). Lipids are often defined by their hydrophobicity, their inability to dissolve in water. This class of metabolites covers a broad spectrum of diverse substances ranging from slightly polar, e.g., glycosylated sphingolipids (Merrill Jr. et al. 2009), to highly non-polar lipids like, e.g., triacylglycerol (Kuksis 2007).

Measuring the changes in the level of metabolites that belong to each of the abovementioned classes (primary, secondary, and lipid metabolites) certainly contributes to our understanding of the processes taking place during the interaction of beneficial microorganisms with plants in different ways. For instance, monitoring the changes in primary metabolites will contribute to the understanding of the carbohydrate trafficking from plants to the beneficial microorganism. This could be achieved by stable isotope labeling of the plants before the onset of the interaction. Moreover, using stable isotope labeling allows one to follow the metabolic fluxes and resolve the dynamics of metabolic pathways occurring during plant-microbial interactions. Many secondary metabolites are involved with plant defenses against biotic threats and abiotic stress tolerance. Thus, analysis of the pool of the secondary metabolites contributes to our understanding of: (1) metabolic pathways responsive to pathogens and abiotic stress or as result of plant priming by beneficial microorganism; (2) metabolites that are involved with establishment and maintenance of the beneficial interactions. Lastly, lipids are a *major* form of *energy storage* in *plants* and are also the building blocks of the plant mechanical defense layers (cutin and waxes). Therefore, analyzing the changes in the lipid composition may also contribute to our understanding of the growth promotion effect of many plant-microbial interactions as well as plant responses to pathogens. Finally, in all the mentioned classes of metabolites there are precursor molecules for synthesis of phytohormones. Thus, their study will also highlight hormonal signaling process taking place during the plant-microbe interaction.

Despite the important advances in the metabolomics field, a comprehensive analysis of the metabolism on one single analytical platform is far from possible. The high chemical complexity and the wide range of in vivo concentrations, from sub-nanomolar to millimolar (Miggiels et al. 2019), make a full comprehensive analysis currently impossible. The plant kingdom is currently estimated to have ca. 1 million different metabolites, but predictions using combinatorial approaches imply that the set of putative compounds with molecular weight of <2000 Da could be as large as 600 million structures (Kind and Fiehn 2007). While the current estimate of the number of plant compounds seems to be underestimated, microbial metabolomes, based on genome-wide metabolic models, seem to cover more realistically the number of compounds in those organisms (Perez de Souza et al. 2021).

The combination of multiple techniques has proven to be the best way to extend metabolite coverage and published studies reported the detection of from 76 up to more than 2000 metabolites or metabolic mass traces with several hundreds of samples analyzed per experiment (Keurentjes et al. 2006; Schauer and Fernie 2006; Meyer et al. 2007), demonstrating recent technological advances.

A variety of software tools aiding in the unbiased or reference-based evaluation of metabolomics experiments have been developed to keep pace with the technological progress (Fiehn et al. 2005; Lisec et al. 2006; De Vos et al. 2007; Styczynski et al. 2007; Luedemann et al. 2008). Thus, the recent advances in *high-throughput metabolomics techniques allow the Trichoderma research community to obtain vast information on the small molecules that are involved in the beneficial interaction between Trichoderma and plants.*

6.2 Metabolic Approaches in the Study of the Interaction of Beneficial Microorganisms with Plants

Over the last few years several studies demonstrated the power of metabolomics in the analysis of different plant-pathogen interactions (Hagemeier et al. 2001; Desbrosses et al. 2005; Berger et al. 2007; Ward et al. 2010). This approach provided a first view of system-wide changes in plant metabolism upon pathogen challenge, and identification of specific metabolites that play a key function in plant innate immune response. For example, by using LC-MS analysis glucosinolate metabolites were shown to mediate broad-spectrum antifungal and anti-bacterial defense (Bednarek et al. 2009; Clay et al. 2009).

To date, the metabolomics changes of only a few plant species during their interactions with different plant-beneficial microorganisms (for example, plant-growth-promoting bacteria, and arbuscular mycorrhizal fungi) have been reported. Those studies only show the tip of the iceberg in terms of the potential of the use of metabolomic approaches. Thus, we would like to give few examples from those studies, in order to show what could be achieved in the study of plant-*Trichoderma* interaction using a metabolomics approach. The study by Barsch et al. (2006a, b) employed GC-MS analysis to follow the metabolite profiles of nodulated alfalfa plants, indicating that distinct stages of nodule organogenesis are accompanied by global physiological adaptations.

GC-MS analysis was also used to study the changes in the primary metabolites of in vitro-grown poplars upon interaction with the plant-growth-promoting bacteria (PGPR) of the genus *Paenibacillus* (Scherling et al. 2009). Infection of poplar strongly affects the composition of 11 primary metabolites. Among them are increased asparagine and urea levels, as well as depleted sugars and organic acids of the tricarboxylic acid cycle. These observations coincide with the fact that the *Paenibacillus* strain P22 is able to grow without nitrogen in the medium, indicating nitrogen fixation from the air, which is also known from other *Paenibacillus* spp. The abovementioned studies used GC-MS analysis, a method that covers just a small part of the metabolome (mostly primary metabolites). Nevertheless, in these studies researchers could obtain substantial informative data on the nature of the interaction between the PGPR and the plant. GC-MS analyses have been applied for investigation of plant-bacteria interactions in nodules (Desbrosses et al. 2005; Barsch et al. 2006a, b).

A good example of how a metabolic approach is used to dissect the interaction of beneficial microorganisms with plants comes from the work of Van de Mortel (van de Mortel et al. 2012) on *Pseudomonas fluorescens* SS101. In this study, which combined genome-wide transcriptomic and untargeted metabolomic analyses, the authors showed that in roots and leaves of *Arabidopsis* plants treated with the beneficial rhizobacterium *Pf.*SS101, approximately 1910 genes and 50 metabolites were differentially regulated relative to untreated plants. Integration of both sets of "omics" data pointed to a prominent role of camalexin and glucosinolates in the *Pf.* SS101-induced resistance response. Subsequent bioassays with seven Arabidopsis mutants (*myb51*, *cyp79B2*, *cyp79B3*, *cyp81F2*, *pen2*, *cyp71A12*, *cyp71A13*, and *myb28*, *myb29*) disrupted in the biosynthesis pathways for these plant secondary metabolites showed that camalexin and glucosinolates are indeed required for the induction of resistance to the pathogen *Pseudomonas syringae* by *Pf.*SS101. Similar changes in gene expression patterns and metabolite levels were recorded in *Arabidopsis* roots colonized by either *Pf.*SS101 or *Trichoderma* (Brotman et al. 2013). *Indole glucosinolate metabolites showed elevated levels in Arabidopsis roots* *colonized by either Trichoderma or Pf.*SS101. *Moreover, some gene expression patterns leading to same metabolic pathways showed similar behavior upon microorganism challenge,* thus illustrating the powerful possibilities of using "omics" tools in order to dissect the interaction and gain novel understanding on conserved patches across different genera of plant-beneficial microorganisms.

By screening of *Arabidopsis* mutants that do not show induced growth promotion mediated by *Piriformospora indica*, the OXI1 Kinase Pathway was identified as an essential component of the beneficial response (Camehl et al. 2011). OXI1 Kinase activity has been previously shown to be necessary for oxidative burst-mediated signaling in Arabidopsis (Rentel et al. 2004). Moreover, *P. indica* stimulates the lipid phosphatidic acid (PA) synthesis, but not H_2O_2 production in Arabidopsis plants. *P. indica* regulates plant growth via PA-stimulated PDK1 activation that subsequently triggers activation of the OXI1 pathway (Camehl et al. 2011). These results provides a link between the *P. indica*-induced positive growth phenotype and the primary metabolism.

6.3 Metabolomics to Inform *Trichoderma* spp. Biology

Metabolomics has been applied as a tool to better understand *Trichoderma* spp. metabolism. For instance, GC-MS analysis of a *T. asperellum* strain, named GDFS1009, identified nine primary metabolites believed to be fungicides or precursors/intermediates of fungicides, insecticides, or herbicides such as acetamide, ethanolamine, ethylamine, diethylamine, ethylene glycol, glycine, *O*-toluic acid, citric acid, and malic acid as well as a variety of antimicrobial secondary metabolites, including polyketides and alkanes (Wu et al. 2017). In addition, a functional analysis of the *T. arundinaceum TRI6* homolog (involved in the biosynthesis of trichothecene) indicated that TRI6 affects the expression of not only *TRI* genes, but also other secondary biosynthetic genes, including mevalonate-related genes (Lindo et al. 2018). Metabolomics analysis confirmed that *TRI6* is required for trichothecene production in *T. arundinaceum* and *TRI6* deletion increases the antimicrobial metabolite aspinolide (Lindo et al. 2018). The disruption of another *T. arundinaceum* gene TRI gene (*TRI4*) which encodes a P450 mono-oxygenase reduced the biocontrol activity of this *tri4* mutant. A deeper examination of the secondary metabolism of this mutant confirmed a terpene:polyketide cross-pathway in *T. arundinaceum*, in which the increase of aspinolides might compensate for the loss of harzianum A in the mutant (Izquierdo-Bueno et al. 2018). Loss of harzianum A was previously shown to result in a drastic reduction in the biocontrol activity against the phytopathogenic fungi *Botrytis cinerea* and *Rhizoctonia solani* (Malmierca et al. 2013). More recently, it was found that TRI10 is required for wild-type expression of *tri* genes and trichothecene synthesis during the first 12 h of growth of *T. arundinaceum*. Comparison of the effect of *TRI10* deletion in *T. arundinaceum* and *Fusarium* revealed similarities in the genetic regulation of trichothecene biosynthesis in these two fungi with different lifestyles (Lindo et al. 2019). Unlike trichothecenes levels

TRI10 deletion increased ergosterol and aspinolides contents, which is likely to be caused by an increase of intracellular pool of farnesyl diphosphate, a precursor of the trichothecene and other terpenoids (e.g., ergosterol) due to the lack of trichothecene production (Lindo et al. 2019).

One of the main secondary metabolites produced by *T. atroviride* is the 6-pentyl-α-pyrone, which is an organic compound with antifungal and plant-growth-promoting activities. The 6-pentyl-α-pyrone biosynthesis was previously proposed to involve a lipoxygenase, but targeted gene deletion of a lipoxygenase-encoding gene LOX1 revealed that this gene is not essential for pentyl-α-pyrone biosynthesis, nor for the ability of *T. atroviride* to parasitize and antagonize host fungi. Instead, LOX1 seems to be directly involved in the production of several metabolites, including oxylipins and volatile organic compounds, as well as in the induction of systemic resistance against the plant pathogenic fungus *B. cinerea* in *Arabidopsis thaliana* plants (Speckbacher et al. 2020).

The mechanisms of *T. asperellum* tolerance to the organophosphorus pesticide dichlorvos was investigated too. It is important to understand this relationship since combining bio- and chemical control could be a strategy to reduce the use of chemical pesticides. *T. asperellum* TJ01 has limited capacity to absorb dichlorvos, which might explain certain resistance to the pesticide, but does not avoid major impacts on the tricarboxylic acid cycle with reduced citrate and isocitrate levels whereas succinate, fumarate, and malate were unaltered. Changes in amino acids, lipids, carbohydrates, and secondary metabolites, e.g., flavonoids and alkaloids, were also observed (Wu et al. 2018).

6.4 Effect of *Trichoderma* spp. on the Pathogen Metabolome

Trichoderma spp. can produce a wide range of enzymes such as amylase, glucanases, and chitinases, as well as mainly terpenes, pyrones, gliotoxin, gliovirin, and peptaibols, which may express antifungal activity (Vinale et al. 2014). Those compounds are involved in the communication and interaction with other organisms. The antagonistic mechanisms against pathogens of *Trichoderma* spp. involve the production of lytic enzymes, mycoparasitism, and competition for nutrients and space. For instance, the contact with *T. viride* induced pigmentation and cell wall hydrolysis in *Schizophyllum commune* with concomitant increase in phenoloxidase activity. The metabolite profiling also showed increased levels of oxidative stress indicators, phenolic compounds, antioxidant γ-amino butyric acid, pyridoxine, and osmoprotective sugar alcohols (Ujor et al. 2012). Dual assays using *Fusarium oxysporum* f. sp. *conglutinans* with three different biocontrol agents, *T. harzianum*, *Bacillus amyloliquefaciens*, and *Pseudomonas aeruginosa*, showed that the biocontrol activity of *T. harzianum* and *B. amyloliquefaciens* is due to their ability to suppress *F. oxysporum* mycotoxin beauvericin production, suggesting that this mechanism is not exclusive to *Trichoderma* species (Palyzová et al. 2019). This negative impact of *T. harzianum* on mycotoxin production was also observed in *Fusarium culmorum*. In addition, the protein profile was affected in *F. culmorum* treated with *T. harzianum* extract. In particular, proteins required for carbohydrate metabolism, such as glyceraldehyde-3-phosphate dehydrogenase, fructose-bisphosphate aldolase and enolase, which play a major role in the rapid growth and high pathogenicity *F. culmorum* are reduced (Mironenka et al. 2021).

6.5 Impacts of *Trichoderma*-Plant Interactions Through Metabolomics Lens

6.5.1 On Roots and Rhizosphere

Some rhizosphere-competent strains of *Trichoderma* can colonize entire root surfaces with morphological features reminiscent of those seen during mycoparasitism, and can be defined as opportunistic avirulent plant symbionts. *Trichoderma* spp. have developed strategies to build a mutually beneficial relationship with plants by using sucrose or other nutrients from the plants and boosting plant immunity against invading pathogens and improve photosynthetic abilities in return. This inter kingdom communication is achieved through biochemical signaling, in which fungi produce compounds able to alter plant transcriptome, proteome, and metabolome (Alfiky and Weisskopf 2021).

At the colonization step *Trichoderma* spp. produce hormonal signals, e.g., auxins, which promote plant growth, including the formation of lateral roots, facilitating more colonization due to increased surface area (Contreras-Cornejo et al. 2009). *Trichoderma* spp. secrete expansin for enhancing its penetration into the first few cell layers of epidermis and root outer cortex through cellulose-binding modules along with endopolygalacturonase (Morán-Diez et al. 2009). Interestingly, it was shown that *T. koningii* reduces the biosynthesis of the isoflavonoid phytoalexin vesticol during colonization of *Lotus japonicus* roots, which may explain the role of plant resistance in determining host selectivity (Masunaka et al. 2011). Moreover, the inoculation of the strain *T. harzianum* CCTCC-RW0024 in maize rhizosphere could increase plant-growth-promoting acidobacteria and, in turn, plant growth. This strain could also increase the plant-beneficial microbiome in maize rhizosphere (Saravanakumar et al. 2017).

A comprehensive analysis using two complementary metabolomics approaches LC-MS and GC-MS was performed to investigate the modulation of root exudation imposed by *T. atroviride* AT10 (application in the substrate or seeds). It showed a distinctive and specific root exudation pattern as a function of the fungal interaction and the means of application (Lucini et al. 2019). An increase in ergosterol and cholesterol lipids, as well as impairment of other membrane lipids, was demonstrated under fungal association. In addition, phytohormone and phenylpropanoid pathway were also shown to have an important role in the response to fungal inoculation. For instance, coumaric acid was reduced in all of the inoculated plants, suggesting a modification of the carbon flux in the phenylpropanoid pathway, which might have led to the increase in phenolic compounds downstream (Lucini et al. 2019). The metabolic fingerprinting of maize roots, mycelia, and fungal culture supernatants performed using LC coupled to diode array detection and quadrupole time-of-flight tandem mass spectrometry showed that metabolic composition of *T. virens*-colonized roots differed profoundly from that of non-colonized roots. Several specialized metabolites derived from the shikimate pathway, including an aromatic amino acid, and several flavonoids and benzoxazinoids were modulated by the interaction (Schweiger et al. 2021).

6.5.2 On the Systemic Defense Responses

Yedidia et al. (2003) provide evidence for the induction of a systemic response against angular leaf spot of cucumber (*Pseudomonas syringae* pv. *lachrymans*) following application of *T. asperellum* T203 to the root system. Disease symptoms were reduced in plants pretreated with *T. asperellum*. This was further supported by the accumulation of secondary metabolites of a phenolic nature. The phenolic compounds showed an increased capacity to inhibit bacterial growth in vitro. The bulk of the antimicrobial activity was found in the acid-hydrolyzed extract containing the phenolics in their aglycone form. High-performance liquid chromatography (HPLC) analysis of phenolic compounds showed a marked change in their profile in the challenged, pre-elicited plants relative to that in challenged controls. These findings were further supported by gene expression assays, showing the induced expression of two genes that are directly involved in secondary metabolite synthesis: the phenylpropanoid pathway gene encoding phenylalanine ammonia lyase (*PAL*) and the lipoxygenase pathway gene encoding hydroxyperoxide lyase (*HPL*).

Early studies employed metabolomics approaches to measure the changes in metabolite levels in leaves of *Arabidopsis* plants colonized by *T. asperelloides* T203 and the onset of ISR by *Pseudomonas syringae* (Brotman et al. 2012), as well as plants colonized by *Trichoderma hamatum* T382 and the onset of ISR by *B. cinerea* (Mathys et al. 2012). Using one of the most intensely studied systems in plant pathogen interactions, the Arabidopsis-*Pseudomonas* model pathosystem (Brotman et al. 2012), showed that induction of a systemic response by application of *T. asperellum* T203 to the root system of *Arabidopsis* increased resistance against *Pseudomonas syringae* pv. *tomato* (DC3000). Using GC-MS analysis, they could track metabolic changes in cross-comparison of four distinct treatments: (1) control untreated plants, (2) plants treated with *T. asperellum*, (3) plants inoculated with *Pseudomonas*, and (4) plants treated with *T. asperellum* following inoculation with *Pseudomonas*. Thus, by measuring *61 known metabolites they could determine the* changes in the metabolic profile unique to each of the conditions, and therefore could assess the metabolic signature caused by the priming effect. Root colonization by this beneficial fungus substantially alters the plant metabolic profile including significant changes in amino acids, sugars, and TCA cycle intermediates. Correlation analysis of the 61 metabolites (Brotman et al. 2012) clearly separated the data according to four groups corresponding to: control, *T. asperelloides*, *P. syringae*, and combined Trichoderma and Pseudomonas treatments. Highly correlated metabolites are metabolites that might act together in a coordinate fashion in the plant response to stimuli and thus give information on the specific pathways that are changing in the course of plant-*Trichoderma* interaction. The activation of plant defenses and growth promotion require*s* increased energy supply that ultimately must come from photosynthesis, and probably needs to be accompanied by greater respiratory rates (Shoresh et al. 2010). In a proteomic study of the T22-maize system, the most commonly affected proteins differentially expressed in *Trichoderma* inoculated plants were those involved in carbohydrate metabolism, especially those in the glycolytic, tricarboxylic acid (TCA), or respiratory pathways (Shoresh and Harman 2008).

An interesting link between the inoculation of *cucumber* and *Arabidopsis plants by foliar bacterial pathogens after priming* with *Trichoderma* (as discussed above) is the increased level of aromatic amino acids, including phenylalanine in *Arabidopsis*, and increased expression of phenylalanine ammonia lyase (PAL) gene, in cucumber. This may indicate an activation of the phenylpropanoid pathway that leads to the synthesis of secondary metabolites with antimicrobial properties.

Gene expressions were monitored in Arabidopsis roots during the early stages (9–48 h from the onset of the interaction) of colonization by *T. asperelloides* or *T. harzianum* (Moran-Diez et al. 2012; Brotman et al. 2013). *Both studies show alteration in genes that correspond to metabolic pathways. Among them are genes in the metabolic biosynthesis pathways of indole glucosinolates (IGS) and* camalexin. It is noteworthy *that the gene cyp71A13 showed increased expression in root colonization with T. asperelloides but showed decreased expression upon colonization with T. harzianum, 24 h after the onset of the interaction.* This provides evidence for a model where *Trichoderma* spp. fine-tune the expression of *CYP71A13,* that *in turn promotes* camalexin biosynthesis, to allow colonization. *Moreover the cyp79B2/cyp79B3 genes that possess redundant enzymatic activity* (the conversion of tryptophan to indole-3-acetaldoxime, a precursor of IGS and the antimicrobial molecule camalexin) *show different expression patterns. cyp79B2 shows increased expression upon T. harzianum colonization and cyp79B3 shows increased expression upon T. asperelloides colonization.*

Expression of the ethylene dependent transcription factor *MYB51*, involved in the activation of genes during plant defense responses (Clay et al. 2009), increased in response to colonization 24 h after the application of *T. asperelloides* to the roots. This is consistent with *the activation of* IGS biosynthetic pathway in the *Arabidopsis* root. The influence of *Trichoderma* on the level of IGS in *Arabidopsis* roots was further tested by targeted LC-IT/ESIMS analysis for quantification of the content of three key IGS metabolites in *Trichoderma* treated and control *Arabidopsis* roots 24 h after colonization. A significant increase in the level of 4-methoxy-I3G and methoxy-3-indolyl-methylglucosinolate and decrease in the level of their precursor indolyl-methyl glucosinolate was observed (Brotman et al. 2013). Interestingly, besides *cyp79B3* two other CYP genes, *cyp71B15* and *cyp71A13*, which function in camalexin biosynthesis (Nafisi et al. 2007) are significantly affected by *T. asperelloides* root colonization.

During the interaction, microbe-associated molecular pattern (MAMPs) secreted from *Trichoderma* trigger plant immunity and activate host basal defense responses. Several *Trichoderma* MAMPs have been identified (Hermosa et al. 2012) and induce two types of changes in the host. The first changes occur at the site of colonization: the roots. The second, long distance changes, occur in the upper part of the plants by an induced systemic resistance (ISR) response. These responses require a fine regulation of hormone signaling pathways. For instance, root colonization of tomato plants with *T. harzianum* rendered the leaves more resistant against *B. cinerea*. However this increased resistance was impaired in tomato plants defective in producing defense-related hormones such as jasmonic acid (JA), ethylene (ET), salicylic acid (SA), and abscisic acid (ABA) as well as the peptide prosystemin (Martinez-Medina et al. 2013). Interestingly, the multi hormonal response seems to be a plastic and adaptive mechanism depending on the parasitism stage. It was shown that *T. harzianum* induced resistance to the root knot nematode *Meloidogyne incognita* both locally and systemically at multiple stages of the parasitism, including invasion, galling, and reproduction. At first, SA-related defenses were increased, limiting the nematode invasion, then enhanced JA-related defenses repressed galling and reproduction by alleviating the deregulation of JA-dependent immunity elicited by the nematodes (Martínez-Medina et al. 2017). Changes in phytohormones observed in Arabidopsis leaves, especially the decreased SA and increased ABA levels, caused by the interaction with *Trichoderma gamsii* also negatively affected the feeding behavior of *Trichoplusia ni* (Zhou et al. 2018).

Targeted metabolite profiling using HPLC and spectrophotometric analyses was used to demonstrate that *T. longibrachiatum* isolated from desert soil can confer beneficial agronomic traits to onion (*Allium cepa* L.) and induce defense against *F. oxysporum* f. sp. *cepa* (Abdelrahman et al. 2016). The metabolome of *T. longibrachiatum*-primed onion and primed onion challenged with *F. oxysporum* f. sp. *cepa* displayed significant accumulation of 25 abiotic and biotic stress-responsive metabolites. Specifically, increased levels of sugars (e.g., fructose), flavonols (e.g., quercetin and kaempferol), amino acids, ascorbic acid, and cysteine sulfoxides. These results indicate that *T. longibrachiatum* ISR in onion involves an upregulation of a number of primary and secondary metabolite pathways, including phenylpropanoids, carbohydrate metabolism, and sulfur assimilation (Abdelrahman et al. 2016).

Among the effector metabolites that beneficial microbes produce during the interaction with plant and other microbes, secondary metabolites have received some attention too (Kumar and Khurana 2021). Secondary metabolites comprise different classes of natural compounds with low molecular weight and a wide range of biological functions. Metabolomics analyses of the interactions between plants, phytopathogens, and beneficial fungi have helped in the identification of several fungal secondary metabolites that positively affect plant metabolism.

Trichoderma spp. secondary metabolism seems to play a role in the ISR. During the biosynthesis of secondary metabolites, specific enzymes participate in the formation of hydroxyl and epoxy groups, belonging to the p450 mono-oxygenases family. A study using a strain of *T. virens* overexpressing *TvCyt2* (encodes a P450 mono-oxygenase) showed that the mutant strains increased terpene-like molecules while the amount in the wild-type strain and null mutant strains were very low or absent (Ramírez-Valdespino et al. 2017). Sesquiterpenes were previously shown to be essential for the establishment of the plant-microbe interaction (Ditengou et al. 2015), reinforcing the importance of secondary metabolites in biological interactions. Moreover, increased activation of the JA-mediated defense by plants suggests induction of the plant defense response (Ramírez-Valdespino et al. 2017).

Changes in the metabolome of tomato plants treated with *T. harzianum* or its purified secondary metabolite harzianic acid in the presence or the absence of the soil-borne pathogen *Rhizoctonia solani* demonstrated the ability of *T. harzianum* to activate defense responses in infected tomato plants (Manganiello et al. 2018). This included an upregulation of genes related to detoxification of reactive oxygen species and ethylene, jasmonic acid, and salicylic acid signaling pathways. Untargeted metabolomics also showed that 25 annotated compounds increased when plants were either infected with *T. harzianum* or treated with harzianic acid. Several of the putatively identified compounds belong to the class of steroidal glycoalkaloids (Manganiello et al. 2018), previously reported to be involved in the response to fungal attack (Friedman 2002). Targeted and untargeted semi-polar metabolites detected using LC-MS approaches revealed a wide alteration of the tomato leaf metabolome of plants treated with *T. harzianum* and subsequently infested by the aphid *Macrosiphum euphorbiae*. A transcriptome reprogramming with impacts on metabolic processes, regulating gene expression and defense responses was observed (Coppola et al. 2019). An increased expression of glycolytic enzymes was suggested to redirect the higher sugar flux in treated plants to increase the carbon supply to biosynthetic pathways involved in the production of plant resistance-secondary metabolites. Indeed, the metabolomics analysis revealed the induction of biochemical defenses elicited by *T. harzianum* when coupled with insect feeding. Defense-related secondary metabolites were accumulated in this tripartite interaction compared to samples with only the aphid infestation. The defense barrier array involved alkaloids (α-/β-tomatine) that could be responsible for the reduction in aphid survival together with late defense gene products (polyphenol oxidase, leucine aminopeptidase, miraculin, and many others), phenolic acids, and flavonoids (Coppola et al. 2019).

Volatile organic compounds (VOCs) produced by soil-borne microorganisms also play crucial roles in fungal interactions with plants and phytopathogens. VOCs have been characterized in *Trichoderma* spp. using a head space-solid phase microextraction gas chromatography–mass spectrometry approach (Speckbacher et al. 2021), revealing that the mechanisms against phytopathogens seem to be strain- and pathosystem-dependent (Lazazzara et al. 2021). Reduced severity of grapevine (*Vitis vinifera*) downy mildew disease, which is caused by *Plasmopara viticola*, was found to be VOC-mediated when analyzing grapevine leaf disks treated with *T. asperellum*, *T. harzianum*, and *T. atroviride*. Of 31 detected compounds, 6-pentyl-2*H*-pyran-2-one (6PP) and 2-pentylfuran were the main ones inducing the accumulation of callose and expression of defense-related genes after *P. viticola* inoculation, suggesting that these VOCs are elicitors of grapevine defense (Lazazzara et al. 2021). Intriguingly, GC-MS analysis revealed that the release of 19 VOCs from olive trees was significantly increased upon treatment with different *Trichoderma* strains or their secondary metabolites harzianic acid and 6-pentyl-α-pyrone. The fingerprint of each plant-fungus/metabolite interaction was found reflecting emission of specific VOC by the plant (Dini et al. 2021). For instance, when treated with *T. harzianum* strain M10, 6PP, or harzianic acid there was an enhanced synthesis of monoterpenes by controlling the MEP pathway. *T. asperellum*, *T. virens*, or harzianic acid increased formation of the hydrocarbon aldehyde (nonanal). Last but not least, treatment with *T. harzianum*, *T. virens*, 6PP, or harzianic acid enhanced aromatic amino acids synthesis (Dini et al. 2021). Altogether, these results suggest a great impact of Trichoderma strains and their metabolites on important biosynthetic pathways such as methylerythritol 1-phosphate, lipid-signaling, and shikimate pathways, thereby triggering a systemic response in olive trees.

6.6 Integrating Transcriptomic, Proteomics, and Metabolomic Information onto Diagrams of Metabolic Pathways

When handling the large amounts of data obtained from the application of high-throughput profiling technologies (such as global gene expression analysis, proteomic and metabolic profiling) there is a need to address the issue of data visualization. Furthermore the data need to be mapped to genes and metabolic pathways in order to obtain comprehensive understanding in the context of the biological processes. This then permits interpretation of the results and placing them in the context of the biological system. In the following section we will briefly highlight a few current approaches to develop models of genetic and molecular networks for the systems of plant-*Trichoderma* interactions. For this purpose in recent years, an increasing number of tools have been developed in order to map and visualize plant genomics and metabolomics data. (For review of bioinformatics tools developed for plants, see Baginsky et al. 2010; Pitzschke and Hirt 2010; Mochida and Shinozaki 2011).

The main drawback of those tools is that they were mostly developed for work with model organisms, although there are expanding data base resources for crop plants. Therefore when using those tools to study the interaction of *Trichoderma* with non-model plants they required some modification. An example of such a study is the work by Palmieri et al. (2012) who performed proteomic analysis in order to identify the proteins whose abundance depends on the systemic resistance induced in grapevine treated with *Trichoderma harzianum* T39, upon infection by the oomycete

Plasmopara viticola, agent of downy mildew. In order to map the proteomic data onto diagrams of metabolic pathways and display it, the authors used the MapMan tool (Thimm et al. 2004). Thus, it was possible to visualize the metabolic processes affected during the complex interaction and map them onto different pathways, such as amino acid biosynthesis, secondary metabolism, and photosynthetic processes. Moreover the authors developed an in-house pathway of biotic and abiotic stress responses using the MapMan *Arabidopsis* biotic stress pathway as a template and manually integrating it with other correlated MapMan pathways. This illustrates the power of integrative "omics" tools that contain data based on several model plants in mapping data obtained from the combined interactions of a crop plant, *Trichoderma*, and a pathogen.

In order to map the genes that show altered expression into signaling cascade and metabolic pathways, Brotman et al. (2013) used MapMan software to highlight the biological processes affected during *Trichoderma* colonization. This was combined with gene ontology analysis using agriGO: a GO analysis toolkit that has been developed for the agricultural research community (Du et al. 2010), and KEGG metabolic pathways (Kanehisa et al. 2012) for insight on specific metabolic pathways. Mapping metabolites into specific pathways is necessary in order to elucidate the biological meaning of the metabolic changes. To illustrate that, the results obtained from the study of Brotman et al. (2012) (the results of this study are discussed above in Sect. 6.2) were mapped using the KEGG metabolic pathways for alanine, aspartate, and glutamate metabolism. Thus, the changes that accrue in specific metabolites in *Arabidopsis* leaves during root colonization by *Trichoderma*, *Pseudomonas* infection, and *Pseudomonas* infection after *Trichoderma* priming are presented in the wider context of the pathway.

6.7 Future Perspectives and Concluding Remarks

The latest advances in metabolomics technologies allow rapid and simultaneous detection and quantification of large number of metabolites. Coupling metabolite profiling with other resources (such as, transcriptomic, proteomic and genomic) offers new possibilities in the study of plant-*Trichoderma* interactions. Here we explore some of the topics for future research using metabolic techniques.

Trichoderma spp. are able to colonize diverse plant species (Harman et al. 2004). Different plant species, however, have different repertoires of secondary metabolites (for example, glucosinolates are unique to some plants of the order Brassicales, and the genus *Drypetes*). Thus, it will be interesting to make comparisons across different plant species to identify metabolites that change upon colonization. Moreover, it will be very interesting to compare the metabolic profile of different accessions from a given plant species, because diverse accessions can interact in a different manner with *Trichoderma*. It has been shown that the beneficial effect of *Trichoderma* spp. on plant is modulated by the plant genotype, and treatment with the biocontrol agent can sometimes even be detrimental (Tucci et al. 2011). Hence, metabolic profiling might help to reveal the complex genetic recognition and activation of downstream signaling in the plant-*Trichoderma* interaction. Due to the large number of metabolites that are detected in untargeted metabolic profiling, the metabolic signature can give an indication about even small differences between very similar biological systems. This might allow characterization of the interaction of plants with different *Trichoderma* spp.

Furthermore, secondary metabolites originating from the *Trichoderma* spp. associated with plants are a rich source of compounds, some potentially novel (Mukherjee et al. 2012). Trichothecenes, secreted from *Trichoderma*, are involved in biocontrol activity and in the induction of plant defense-related genes (Malmierca et al. 2012). Several of the proteins involved in the biosynthetic pathway of trichothecenes in *Trichoderma* show significant differences in functionality, compared to their *Fusarium* orthologues (Malmierca et al. 2012). As different fungi have different arsenals of metabolites adapted to diverse lifestyles and growth strategies (Ohm et al. 2012), metabolite profiling will allow identification of metabolites, synthesized by *Trichoderma*, during the 3-way interaction with plants and pathogens. These are likely to be involved in induction of resistance and/or may have *mycotoxin* properties. Furthermore, studying the metabolic profiling of *Trichoderma* spp. interacting with plants and plant pathogens can significantly contribute to our understanding of the mechanism underlying the beneficial interaction. This information will be vital for efforts to apply *Trichoderma* as a biocontrol agent, or *Trichoderma* based products for commercial use.

Acknowledgments I am grateful to Dr. Ada Viterbo for comments on the manuscript.

References

Abdelrahman M, Abdel-Motaal F, El-Sayed M, Jogaiah S, Shigyo M, Ito S-i, Tran L-SP (2016) Dissection of Trichoderma longibrachiatum-induced defense in onion (Allium cepa L.) against Fusarium oxysporum f. sp. cepa by target metabolite profiling. Plant Sci 246:128–138

Alfiky A, Weisskopf L (2021) Deciphering Trichoderma–plant–pathogen interactions for better development of biocontrol applications. J Fungi 7(1):61

Alseekh S, Fernie AR (2018) Metabolomics 20 years on: what have we learned and what hurdles remain? Plant J. 94:933–942. https://doi.org/10.1111/TPJ.13950

Baginsky S, Hennig L, Zimmermann P, Gruissem W (2010) Gene expression analysis, proteomics, and network discovery. Plant Physiol 152:402–410

Barsch A, Carvalho HG, Cullimore JV, Niehaus K (2006a) GC-MS based metabolite profiling implies three interdependent ways of ammonium assimilation in Medicago truncatula root nodules. J Biotechnol 127:79–83

Barsch A, Tellstrom V, Patschkowski T, Kuster H, Niehaus K (2006b) Metabolite profiles of nodulated alfalfa plants indicate that distinct stages of nodule organogenesis are accompanied by global physiological adaptations. Mol Plant-Microbe Interact 19:998–1013

Bednarek P, Pislewska-Bednarek M, Svatos A, Schneider B, Doubsky J, Mansurova M, Humphry M, Consonni C, Panstruga R, Sanchez-Vallet A, Molina A, Schulze-Lefert P (2009) A glucosinolate metabolism pathway in living plant cells mediates broad-spectrum antifungal defense. Science 323:101–106

Berger S, Sinha AK, Roitsch T (2007) Plant physiology meets phytopathology: plant primary metabolism and plant-pathogen interactions. J Exp Bot 58:4019–4026
Brotman Y, Lisec J, Meret M, Chet I, Willmitzer L, Viterbo A (2012) Transcript and metabolite analysis of the Trichoderma-induced systemic resistance response to Pseudomonas syringae in Arabidopsis thaliana. Microbiology 158:139–146
Brotman Y, Landau U, Cuadros-Inostroza A, Takayuki T, Fernie A, Chet I, Willmitzer L (2013) Trichoderma-plant root colonization: escaping early plant defense responses and activation of the antioxidant machinery for saline stress tolerance. PLoS Pathog 9(3):e1003221
Camehl I, Drzewiecki C, Vadassery J, Shahollari B, Sherameti I, Forzani C, Munnik T, Hirt H, Oelmuller R (2011) The OXI1 kinase pathway mediates Piriformospora indica-induced growth promotion in Arabidopsis. PLoS Pathog 7:e1002051
Clay NK, Adio AM, Denoux C, Jander G, Ausubel FM (2009) Glucosinolate metabolites required for an Arabidopsis innate immune response. Science 323:95–101
Contreras-Cornejo HA, Macías-Rodríguez L, Cortés-Penagos C, López-Bucio J (2009) Trichoderma virens, a plant beneficial fungus, enhances biomass production and promotes lateral root growth through an auxin-dependent mechanism in Arabidopsis. Plant Physiol 149:1579–1592
Coppola M, Diretto G, Digilio MC, Woo SL, Giuliano G, Molisso D, Pennacchio F, Lorito M, Rao R (2019) Transcriptome and metabolome reprogramming in tomato plants by Trichoderma harzianum strain T22 primes and enhances defense responses against aphids. Front Physiol 10:745
De Palma M, Salzano M, Villano C, Aversano R, Lorito M, Ruocco M, Docimo T, Piccinelli AL, D'Agostino N, Tucci M (2019) Transcriptome reprogramming, epigenetic modifications and alternative splicing orchestrate the tomato root response to the beneficial fungus Trichoderma harzianum. Hortic Res 6:5
De Vos RC, Moco S, Lommen A, Keurentjes JJ, Bino RJ, Hall RD (2007) Untargeted large-scale plant metabolomics using liquid chromatography coupled to mass spectrometry. Nat Protoc 2:778–791
Desbrosses GG, Kopka J, Udvardi MK (2005) Lotus japonicus metabolic profiling. Development of gas chromatography-mass spectrometry resources for the study of plant-microbe interactions. Plant Physiol 137:1302–1318
Dini I, Marra R, Cavallo P, Pironti A, Sepe I, Troisi J, Scala G, Lombari P, Vinale F (2021) Trichoderma strains and metabolites selectively increase the production of volatile organic compounds (VOCs) in olive trees. Metabolites 11(4):213
Ditengou FA, Müller A, Rosenkranz M, Felten J, Lasok H, van Doorn MM, Legué V, Palme K, Schnitzler J-P, Polle A (2015) Volatile signalling by sesquiterpenes from ectomycorrhizal fungi reprogrammes root architecture. Nat Commun 6:6279
Du Z, Zhou X, Ling Y, Zhang Z, Su Z (2010) agriGO: a GO analysis toolkit for the agricultural community. Nucleic Acids Res 38:W64–W70
Fernie AR, Trethewey RN, Krotzky AJ, Willmitzer L (2004) Metabolite profiling: from diagnostics to systems biology. Nat Rev Mol Cell Biol 5:763–769
Fiehn O, Wohlgemuth G, Scholz M (2005) Setup and annotation of metabolomic experiments by integrating biological and mass spectrometric metadata. Proc Data Integr Life Sci 3615:224–239
Friedman M (2002) Tomato glycoalkaloids: role in the plant and in the diet. J Agric Food Chem 50:5751–5780
Hagemeier J, Schneider B, Oldham NJ, Hahlbrock K (2001) Accumulation of soluble and wall-bound indolic metabolites in Arabidopsis thaliana leaves infected with virulent or avirulent Pseudomonas syringae pathovar tomato strains. Proc Natl Acad Sci U S A 98:753–758
Harman GE, Howell CR, Viterbo A, Chet I, Lorito M (2004) Trichoderma species—opportunistic, avirulent plant symbionts. Nat Rev Microbiol 2:43–56
Hermosa R, Viterbo A, Chet I, Monte E (2012) Plant-beneficial effects of Trichoderma and of its genes. Microbiology 158:17–25

Izquierdo-Bueno I, Moraga J, Cardoza RE, Lindo L, Hanson JR, Gutiérrez S, Collado IG (2018) Relevance of the deletion of the Tatri4 gene in the secondary metabolome of Trichoderma arundinaceum. Org Biomol Chem 16:2955–2965

Kanehisa M, Goto S, Sato Y, Furumichi M, Tanabe M (2012) KEGG for integration and interpretation of large-scale molecular data sets. Nucleic Acids Res 40:D109–D114

Keurentjes JJ, Fu J, de Vos CH, Lommen A, Hall RD, Bino RJ, van der Plas LH, Jansen RC, Vreugdenhil D, Koornneef M (2006) The genetics of plant metabolism. Nat Genet 38:842–849

Kind T, Fiehn O (2007) Seven Golden Rules for heuristic filtering of molecular formulas obtained by accurate mass spectrometry. BMC Bioinformatics 8:105

Kopka J, Fernie A, Weckwerth W, Gibon Y, Stitt M (2004) Metabolite profiling in plant biology: platforms and destinations. Genome Biol 5:109

Kuksis A (2007) Lipidomics in triacylglycerol and cholesteryl ester oxidation. Front Biosci 12:3203–3246

Kumar N, Khurana SMP (2021) 3—Trichoderma-plant-pathogen interactions for benefit of agriculture and environment. In: Jogaiah S (ed) Biocontrol agents and secondary metabolites. Woodhead Publishing, Sawston, UK, pp 41–63

Lazazzara V, Vicelli B, Bueschl C, Parich A, Pertot I, Schuhmacher R, Perazzolli M (2021) Trichoderma spp. volatile organic compounds protect grapevine plants by activating defense-related processes against downy mildew. Physiol Plant 172(4):1950–1965

Lindo L, McCormick SP, Cardoza RE, Brown DW, Kim H-S, Alexander NJ, Proctor RH, Gutiérrez S (2018) Effect of deletion of a trichothecene toxin regulatory gene on the secondary metabolism transcriptome of the saprotrophic fungus Trichoderma arundinaceum. Fungal Genet Biol 119:29–46

Lindo L, McCormick SP, Cardoza RE, Kim H-S, Brown DW, Alexander NJ, Proctor RH, Gutiérrez S (2019) Role of Trichoderma arundinaceum tri10 in regulation of terpene biosynthetic genes and in control of metabolic flux. Fungal Genet Biol 122:31–46

Lisec J, Schauer N, Kopka J, Willmitzer L, Fernie AR (2006) Gas chromatography mass spectrometry-based metabolite profiling in plants. Nat Protoc 1:387–396

Lucini L, Colla G, Miras Moreno MB, Bernardo L, Cardarelli M, Terzi V, Bonini P, Rouphael Y (2019) Inoculation of Rhizoglomus irregulare or Trichoderma atroviride differentially modulates metabolite profiling of wheat root exudates. Phytochemistry 157:158–167

Luedemann A, Strassburg K, Erban A, Kopka J (2008) TagFinder for the quantitative analysis of gas chromatography—mass spectrometry (GC-MS)-based metabolite profiling experiments. Bioinformatics 24:732–737

Malmierca MG, Cardoza RE, Alexander NJ, McCormick SP, Hermosa R, Monte E, Gutierrez S (2012) Involvement of Trichoderma trichothecenes in the biocontrol activity and induction of plant defense-related genes. Appl Environ Microbiol 78:4856–4868

Malmierca MG, Cardoza RE, Alexander NJ, McCormick SP, Collado IG, Hermosa R, Monte E, Gutiérrez S (2013) Relevance of trichothecenes in fungal physiology: disruption of tri5 in Trichoderma arundinaceum. Fungal Genet Biol 53:22–33

Manganiello G, Sacco A, Ercolano MR, Vinale F, Lanzuise S, Pascale A, Napolitano M, Lombardi N, Lorito M, Woo SL (2018) Modulation of tomato response to Rhizoctonia solani by Trichoderma harzianum and its secondary metabolite harzianic acid. Front Microbiol 9:1966

Martinez-Medina A, Fernandez I, Sánchez-Guzmán M, Jung S, Pascual J, Pozo M (2013) Deciphering the hormonal signalling network behind the systemic resistance induced by Trichoderma harzianum in tomato. Front Plant Sci 4:206

Martínez-Medina A, Fernandez I, Lok GB, Pozo MJ, Pieterse CMJ, Van Wees SCM (2017) Shifting from priming of salicylic acid- to jasmonic acid-regulated defences by Trichoderma protects tomato against the root knot nematode Meloidogyne incognita. New Phytol 213:1363–1377

Masunaka A, Hyakumachi M, Takenaka S (2011) Plant growth-promoting fungus, Trichoderma koningi suppresses isoflavonoid phytoalexin vestitol production for colonization on/in the roots of Lotus japonicus. Microbes Environ 26:128–134

Mathys J, De Cremer K, Timmermans P, Van Kerckhove S, Lievens B, Vanhaecke M, Cammue BP, De Coninck B (2012) Genome-wide characterization of ISR induced in Arabidopsis thaliana by Trichoderma hamatum T382 against Botrytis cinerea infection. Front Plant Sci 3:108

Merrill AH Jr, Stokes TH, Momin A, Park H, Portz BJ, Kelly S, Wang E, Sullards MC, Wang MD (2009) Sphingolipidomics: a valuable tool for understanding the roles of sphingolipids in biology and disease. J Lipid Res 50(Suppl):S97–S102

Meyer RC, Steinfath M, Lisec J, Becher M, Witucka-Wall H, Torjek O, Fiehn O, Eckardt A, Willmitzer L, Selbig J, Altmann T (2007) The metabolic signature related to high plant growth rate in Arabidopsis thaliana. Proc Natl Acad Sci U S A 104:4759–4764

Miggiels P, Wouters B, van Westen GJP, Dubbelman A-C, Hankemeier T (2019) Novel technologies for metabolomics: more for less. TrAC Trends Anal Chem 120:115323

Mironenka J, Różalska S, Soboń A, Bernat P (2021) Trichoderma harzianum metabolites disturb Fusarium culmorum metabolism: metabolomic and proteomic studies. Microbiol Res 249:126770

Mochida K, Shinozaki K (2011) Advances in omics and bioinformatics tools for systems analyses of plant functions. Plant Cell Physiol 52:2017–2038

Morán-Diez E, Hermosa R, Ambrosino P, Cardoza RE, Gutiérrez S, Lorito M, Monte E (2009) The ThPG1 endopolygalacturonase is required for the Trichoderma harzianum–plant beneficial interaction. Mol Plant-Microbe Interact 22:1021–1031

Moran-Diez E, Rubio B, Dominguez S, Hermosa R, Monte E, Nicolas C (2012) Transcriptomic response of Arabidopsis thaliana after 24 h incubation with the biocontrol fungus Trichoderma harzianum. J Plant Physiol 169:614–620

Mukherjee PK, Horwitz BA, Kenerley CM (2012) Secondary metabolism in Trichoderma—a genomic perspective. Microbiology 158:35–45

Nafisi M, Goregaoker S, Botanga CJ, Glawischnig E, Olsen CE, Halkier BA, Glazebrook J (2007) Arabidopsis cytochrome P450 monooxygenase 71A13 catalyzes the conversion of indole-3-acetaldoxime in camalexin synthesis. Plant Cell 19:2039–2052

Obata T, Fernie AR (2012) The use of metabolomics to dissect plant responses to abiotic stresses. Cell Mol Life Sci 69:3225–3243. https://doi.org/10.1007/S00018-012-1091-5

Ohm RA, Feau N, Henrissat B, Schoch CL, Horwitz BA, Barry KW, Condon BJ, Copeland AC, Dhillon B, Glaser F, Hesse CN, Kosti I, Labutti K, Lindquist EA, Lucas S, Salamov AA, Bradshaw RE, Ciuffetti L, Hamelin RC, Kema GH, Lawrence C, Scott JA, Spatafora JW, Turgeon BG, de Wit PJ, Zhong S, Goodwin SB, Grigoriev IV (2012) Diverse lifestyles and strategies of plant pathogenesis encoded in the genomes of eighteen dothideomycetes fungi. PLoS Pathog 8:e1003037

Palmieri MC, Perazzolli M, Matafora V, Moretto M, Bachi A, Pertot I (2012) Proteomic analysis of grapevine resistance induced by Trichoderma harzianum T39 reveals specific defence pathways activated against downy mildew. J Exp Bot 63:6237–6251

Palyzová A, Svobodová K, Sokolová L, Novák J, Novotný Č (2019) Metabolic profiling of Fusarium oxysporum f. sp. conglutinans race 2 in dual cultures with biocontrol agents Bacillus amyloliquefaciens, Pseudomonas aeruginosa, and Trichoderma harzianum. Folia Microbiol 64:779–787

Perez de Souza L, Alseekh S, Scossa F, Fernie AR (2021) Ultra-high-performance liquid chromatography high-resolution mass spectrometry variants for metabolomics research. Nat Methods. https://doi.org/10.1038/s41592-021-01116-4

Pitzschke A, Hirt H (2010) Bioinformatic and systems biology tools to generate testable models of signaling pathways and their targets. Plant Physiol 152:460–469

Pozo MJ, Zabalgogeazcoa I, Vazquez de Aldana BR, Martinez-Medina A (2021) Untapping the potential of plant mycobiomes for applications in agriculture. Curr Opin Plant Biol 60:102034

Ramírez-Valdespino CA, Porras-Troncoso MD, Corrales-Escobosa AR, Wrobel K, Martínez-Hernández P, Olmedo-Monfil V (2017) Functional characterization of TvCyt2, a member of the p450 monooxygenases from Trichoderma virens relevant during the association with plants and mycoparasitism. Mol Plant-Microbe Interact 31:289–298

Rentel MC, Lecourieux D, Ouaked F, Usher SL, Petersen L, Okamoto H, Knight H, Peck SC, Grierson CS, Hirt H, Knight MR (2004) OXI1 kinase is necessary for oxidative burst-mediated signalling in Arabidopsis. Nature 427:858–861

Saravanakumar K, Li Y, Yu C, Wang Q-q, Wang M, Sun J, Gao J-x, Chen J (2017) Effect of Trichoderma harzianum on maize rhizosphere microbiome and biocontrol of Fusarium Stalk rot. Sci Rep 7:1771

Schauer N, Fernie AR (2006) Plant metabolomics: towards biological function and mechanism. Trends Plant Sci 11:508–516

Scherling C, Ulrich K, Ewald D, Weckwerth W (2009) A metabolic signature of the beneficial interaction of the endophyte Paenibacillus sp isolate and in vitro-grown poplar plants revealed by metabolomics. Mol Plant-Microbe Interact 22:1032–1037

Schweiger R, Padilla-Arizmendi F, Nogueira-López G, Rostás M, Lawry R, Brown C, Hampton J, Steyaert JM, Müller C, Mendoza-Mendoza A (2021) Insights into metabolic changes caused by the Trichoderma virens–maize root interaction. Mol Plant-Microbe Interact 34(5):524–537

Shoresh M, Harman GE (2008) The molecular basis of shoot responses of maize seedlings to Trichoderma harzianum T22 inoculation of the root: a proteomic approach. Plant Physiol 147:2147–2163

Shoresh M, Harman GE, Mastouri F (2010) Induced systemic resistance and plant responses to fungal biocontrol agents. Annu Rev Phytopathol 48:21–43. https://doi.org/10.1146/annurev-phyto-073009-114450

Speckbacher V, Ruzsanyi V, Martinez-Medina A, Hinterdobler W, Doppler M, Schreiner U, Böhmdorfer S, Beccaccioli M, Schuhmacher R, Reverberi M, Schmoll M, Zeilinger S (2020) The lipoxygenase Lox1 is involved in light- and injury-response, conidiation, and volatile organic compound biosynthesis in the mycoparasitic fungus Trichoderma atroviride. Front Microbiol 11:2004

Speckbacher V, Zeilinger S, Zimmermann S, Mayhew CA, Wiesenhofer H, Ruzsanyi V (2021) Monitoring the volatile language of fungi using gas chromatography-ion mobility spectrometry. Anal Bioanal Chem 413:3055–3067

Styczynski MP, Moxley JF, Tong LV, Walther JL, Jensen KL, Stephanopoulos GN (2007) Systematic identification of conserved metabolites in GC/MS data for metabolomics and biomarker discovery. Anal Chem 79:966–973

Thambugala KM, Daranagama DA, Phillips AJL, Kannangara SD, Promputtha I (2020) Fungi vs. fungi in biocontrol: an overview of fungal antagonists applied against fungal plant pathogens. Front Cell Infect Microbiol 10:604923

Thimm O, Blasing O, Gibon Y, Nagel A, Meyer S, Kruger P, Selbig J, Muller LA, Rhee SY, Stitt M (2004) MAPMAN: a user-driven tool to display genomics data sets onto diagrams of metabolic pathways and other biological processes. Plant J 37:914–939

Tucci M, Ruocco M, De Masi L, De Palma M, Lorito M (2011) The beneficial effect of Trichoderma spp. on tomato is modulated by the plant genotype. Mol Plant Pathol 12:341–354

Ujor VC, Monti M, Peiris DG, Clements MO, Hedger JN (2012) The mycelial response of the white-rot fungus, Schizophyllum commune to the biocontrol agent, Trichoderma viride. Fungal Biol 116:332–341

van de Mortel JED, de Vos RCD, Dekkers E, Pineda AD, Guillod L, Bouwmeester KD, van Loon JJD, Dicke MPD, Raaijmakers JMD (2012) Metabolic and transcriptomic changes induced in Arabidopsis by the rhizobacterium Pseudomonas fluorescens SS101. Plant Physiol 160(4):2173–2188

Vinale F, Manganiello G, Nigro M, Mazzei P, Piccolo A, Pascale A et al (2014) A novel fungal metabolite with beneficial properties for agricultural applications. Molecules 19:9760–9772. https://doi.org/10.3390/MOLECULES19079760

Ward JL, Forcat S, Beckmann M, Bennett M, Miller SJ, Baker JM, Hawkins ND, Vermeer CP, Lu C, Lin W, Truman WM, Beale MH, Draper J, Mansfield JW, Grant M (2010) The metabolic transition during disease following infection of Arabidopsis thaliana by Pseudomonas syringae pv. tomato. Plant J 63(3):443–457

Wenk MR (2005) The emerging field of lipidomics. Nat Rev Drug Discov 4:594–610

Wu Q, Sun R, Ni M, Yu J, Li Y, Yu C, Dou K, Ren J, Chen J (2017) Identification of a novel fungus, Trichoderma asperellum GDFS1009, and comprehensive evaluation of its biocontrol efficacy. PLoS One 12:e0179957

Wu Q, Ni M, Wang G, Liu Q, Yu M, Tang J (2018) Omics for understanding the tolerant mechanism of Trichoderma asperellum TJ01 to organophosphorus pesticide dichlorvos. BMC Genomics 19:596

Yedidia I, Shoresh M, Kerem Z, Benhamou N, Kapulnik Y, Chet I (2003) Concomitant induction of systemic resistance to Pseudomonas syringae pv. lachrymans in cucumber by Trichoderma asperellum (T-203) and accumulation of phytoalexins. Appl Environ Microbiol 69:7343–7353

Zhou D, Huang X-F, Guo J, dos Santos ML, Vivanco JM (2018) Trichoderma gamsii affected herbivore feeding behaviour on Arabidopsis thaliana by modifying the leaf metabolome and phytohormones. Microb Biotechnol 11:1195–1206

Chapter 7
Crosstalk Between Wilt-Causing Fungi, Plants and Their Microbiome

Davide Spadaro and Maria Lodovica Gullino

Abstract Healthy soils are fundamental to sustainable crop production and to sustainable disease management of soilborne pathogens. Among the most common soilborne diseases, there are Fusarium wilts, caused by *Fusarium oxysporum*. *Fusarium* spp. is a trans-kingdom pathogen that includes plant pathogens, human pathogens, and saprotrophic isolates. *F. oxysporum*, with its wide array of *formae speciales*, is a wonderful model organism to explore the crosstalks between plants and their microbiota. *Fusarium* genomes are compartmentalized into regions responsible for essential functions (core genome) and for host specialization and pathogen virulence (lineage specific chromosomes). In *F. oxysporum*, different avirulence genes, coding for effectors, were identified, matching with resistance genes in the corresponding hosts. The information gained by genome sequencing could be used to design advanced diagnostic tools. Due to long persistence in the soil of *F. oxysporum*, the main strategy to manage Fusarium wilt is to reduce its inoculum. In addition, several control approaches are used to manage the disease. Suppressive soils are soils where disease development is minimal. Often saprophytic *F. oxysporum* have been isolated from suppressive soils, which have been widely exploited for their activity, based on rhizosphere competence, against soilborne pathogens. In *Fusarium*, horizontal chromosome transfer, and also the interaction with ectosymbiotic bacteria, could turn non-pathogenic strain into a pathogen and vice versa. *F. oxysporum* can also have additive or synergistic activity with plant-parasitic nematodes in the soil. Advances in the understanding of the host-pathogen-microbiota interactions contribute to the improvement of crop protection strategies to manage Fusarium wilts.

D. Spadaro
Centre of Competence for the Innovation in the Agro-Environmental Sector—AGROINNOVA, University of Turin, Grugliasco, TO, Italy

Department of Agricultural, Forestry and Food Sciences (DiSAFA), University of Torino, Grugliasco, TO, Italy

M. L. Gullino (✉)
Centre of Competence for the Innovation in the Agro-Environmental Sector—AGROINNOVA, University of Turin, Grugliasco, TO, Italy
e-mail: marialodovica.gullino@unito.it

B. A. Horwitz, P. K. Mukherjee (eds.), *Microbial Cross-talk in the Rhizosphere*, Rhizosphere Biology, https://doi.org/10.1007/978-981-16-9507-0_7

Keywords Bacteria · Biocontrol agents · Diagnostics · *Forma specialis* · *Fusarium oxysporum* · Microbiota · Nematodes · Pathogenicity · Soil health · Soil suppressiveness

7.1 Plant and Soil Microbiota

Plants own a specific microflora, essential to guarantee their health status (Ash and Mueller 2016). In the plant microflora, beneficial bacteria and fungi coexist with endophytic, saprotrophic and pathogenic microorganisms. The strict relationship between microbiota and the plant favoured the development of the meta-organisms or holobiont concept, where host-microbe systems constitute complete biological entities.

Beside the well-known effects of environmental factors such as temperature, relative humidity, and light, a complex and less explored system at the soil level, including biotic and abiotic factors, such as soil structure, microbiota (including pathogens and symbionts), and plant genotype, could affect plant productivity. This chapter will focus on the interactions occurring at the soil level.

Healthy soils are fundamental to sustainable crop production and, consequently, to sustainable disease management, as they affect the soilborne pathogen density, the structure of beneficial microbiota, and the availability of nutrients for the plants. The soils harbour several microorganisms, including saprotrophic and pathogenic ones. Among the most common soilborne microorganisms, there are isolates of *Fusarium oxysporum*. *F. oxysporum*, due to the economic importance of the diseases incited on hundreds of relevant hosts, as well as because of the number of studies carried out (Gullino et al. 2012; Dean et al. 2012), is a wonderful model organism to explore the crosstalks between plants and their microbiota.

7.2 *Fusarium oxysporum*

Fusarium oxysporum von Schlechtendal is generally considered a "species complex"—a collection of clonal lines belonging to the *Fusarium* genus—and it is considered a trans-kingdom pathogen, including both plant pathogens and human pathogens, together with saprotrophic isolates. *F. oxysporum* is characterized by high biological and genetic diversity, demonstrated by a broad range of host plants (O'Donnell et al. 2009; Edel-Hermann and Lecomte 2019).

F. oxysporum does not have a perfect stage, but it produces three types of asexual spores: macroconidia, microconidia and chlamydospores (Fig. 7.1). Hyaline pluriseptate macroconidia are produced on branched conidiophores, clustered in sporodochia. Hyaline mono- and bicellular microconidia are formed on short conidiophores on the aerial mycelium. Spherical chlamydospores, characterized by a thick and brown wall, are produced in hyphae and permit the fungus to survive for years under difficult conditions such as dry soil. Chlamydospores germinate in response to root exudates.

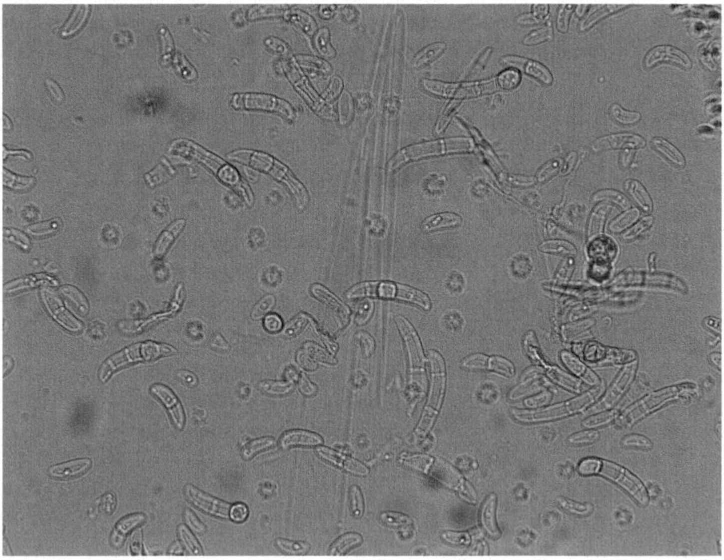

Fig. 7.1 Macroconidia and microconidia of *Fusarium oxysporum*

Interactions between plant roots and *F. oxysporum* can be neutral, beneficial, or detrimental for the host. Besides the original saprophytic forms, several pathogenic forms differentiated, which are specialized on a single host plant or on a limited number of host species.

7.3 Plant Diseases

F. oxysporum is able to cause vascular wilts or root and crown rots on many plant species, including agronomically important crops. *F. oxysporum* agents of wilting penetrate the roots and reach the xylem vessels, which are colonized, resulting in plant yellowing and wilting (Bishop and Cooper 1983; Di Pietro et al. 2003). Other *F. oxysporum* strains cause tissue discolouration evolving into black necrotic spots that end in root and hypocotyl rotting. Rot diseases can be named basal rots, stem rots, or crown and root rots.

7.4 *Formae Speciales*

The host range of pathogenic isolates permitted the development of the concept of *formae speciales* (ff. spp.), each *forma specialis* (f. sp.) including isolates characterized by the same host profile. The concept of *forma specialis* is particularly useful as it permits to identify a set of isolates with the same host range, but without constituting a genetically homogeneous group. Isolates inside a *forma specialis* can have both a monophyletic and polyphyletic origin. The identification of the origin has direct implications on the disease control strategies, if genetic resistance is considered (Leslie 2012).

In the study of the evolution of pathogenic forms of *F. oxysporum*, a fundamental role is played by the frequency of origination and by the temporal distance from the origin. If the origin of the *forma specialis* is in a remote past, the current isolates have co-evolved with their hosts. In such context, it is expected that isolates that are pathogenic towards a common host have a common ancestor and *formae speciales* form monophyletic groups. Alternatively, *formae speciales* could also be associated to distinct evolution lines deriving from different ancestors. By analysing isolates of *F. oxysporum* f. sp. *cubense*, agent of Fusarium wilt of banana, two genetically distinct lines were evident (Fourie et al. 2009). The *forma specialis* concept was established to distinguish strains with similar morphology but different host range (Gordon 1965). Edel-Hermann and Lecomte (2019) reported 106 well documented ff. spp., 37 insufficiently documented ff. spp. and 58 host plants without a characterized f. sp.

F. oxysporum has potential gene-for-gene relationships with several hosts. While the monophyletic origin of *formae speciales* is explained by the co-evolution of pathogens with their hosts, the horizontal gene transfer of a whole chromosome could explain the polyphyletic origin. It has been demonstrated that the transfer of a pathogenicity chromosome from *F. oxysporum* f. sp. *lycopersici* to a non-pathogenic isolate of *F. oxysporum* can confer pathogenicity on tomato (Ma et al. 2010). The *forma specialis lycopersici* of *F. oxysporum* (Fig. 7.2), abbreviated *Fol* here and in the rest of the chapter, is considered a model to study biology, epidemiology and genetics of the strains of *F. oxysporum* pathogenic on plants.

F. oxysporum formae speciales were initially restricted to one plant species. Later, for several *formae speciales*, the host range was shown to be wider. Most of the described *formae speciales* are pathogenic to one host plant only (e.g. *F. oxysporum* f. sp. *basilici*; Chiocchetti et al. 1999), whereas the rest are characterized by a broader host range. Several *formae speciales* are pathogenic on several species within a genus, or several genera of a botanical family, or plants belonging to different families (e.g. *F. oxysporum* f. sp. *raphani*; Srinivasan et al. 2010).

To differentiate root rot strains from vascular wilt strains, the term *radicis* was introduced. Some plants can be attacked by two *formae speciales* causing different diseases, i.e. tomato is susceptible to the *formae speciales lycopersici* causing wilt and *radicis-lycopersici* causing rot (Jarvis 1988). Similarly, cucumber is susceptible to the *formae speciales cucumerinum* and *radicis-cucumerinum* (Lievens et al. 2007) and pepper is susceptible to the *formae speciales capsici* and *radicis-capsici* (Lomas-Cano et al. 2016).

Identification of the *forma specialis* is typically achieved through pathogenicity testing on different plant species (O'Donnell et al. 2009; Lievens et al. 2012). Furthermore, the determination of races is based on pathogenicity tests on different

Fig. 7.2 Fusarium wilt of tomato, caused by *Fusarium oxysporum* f. sp. *lycopersici*

cultivars of a single plant species (Lievens et al. 2012). Although these bioassays are efficient, they require a lot of time and labour. Moreover, due to the high number of described *formae speciales*, several potential hosts and cultivars should be tested to achieve a conclusive identification (Fravel et al. 2003). In addition, results of pathogenicity assays are often not clear as a wide range of factors such as the environment, the host genetics, and the disease index adopted can affect them (Hopkins et al. 1992; Bertoldo et al. 2015).

7.5 Vegetative Compatibility Groups

Each *forma specialis* of *F. oxysporum* could consist of one or more clonal lines corresponding to vegetative compatibility groups (VCGs) (Puhalla 1985; Gordon and Martyn 1997; Katan 1999; Katan and Di Primo 1999). Vegetative compatibility grouping is based on the ability to form heterokaryons between different fungal individuals. Strains belonging to the same VCG have identical alleles at their compatibility VIC loci. Anastomosis enables the exchange of nuclear material (Glass et al. 2000); therefore members of the same VCG generally belong to the same clonal line. When several VCGs belong to the same f. sp. (Katan 1999; Katan and Di Primo 1999), this is generally not monophyletic, suggesting that pathogenicity emerged independently several times during evolution: multiple independent lines evolved polyphyletically through a convergent approach (Baayen et al. 2000; O'Donnell et al. 1998; Michielse and Rep 2009).

7.6 Physiological Races

Certain *formae speciales* are further divided into physiological races, characterized by host specialization at cultivar level. A physiological race is a biotype, which can be distinguished from other biotypes by physiological features, such as pathogenicity. All these close interactions between host roots and fungal pathogen are strictly regulated at the gene level (Gordon and Martyn 1997). Races can be numbered in chronological order of discovery, or defined according to the effector genes they have or the host resistance genes they overcome. When a gene-for-gene relationship is not known, races are defined according to their pathogenic profile on differential cultivars.

7.7 Genomics and Pathogenomics

The first genome of a *F. oxysporum* isolate was sequenced by the Broad Institute (Ma et al. 2010). The presence of several transposons in the genome of *F. oxysporum* favours the generation of mutations able to modify the expression of different effectors (Daboussi et al. 2002; Daboussi and Capy 2003).

Comparative analyses have revealed that *Fusarium* genomes are compartmentalized into regions responsible for essential functions (core genome) and for host specialization and pathogen virulence (adaptive/accessory genome, on lineage specific (LS) chromosomes) (Ma et al. 2010). Ma et al. (2010) sequenced a strain of *Fol* and demonstrated that this pathogen contains four unique chromosomes that represent over one-quarter of its genome. The characteristics of the genes in LS regions suggest a distinct evolutionary origin of such regions.

Horizontal gene transfer (HGT) is another mechanism involved in the genetic diversity between pathogenic strains (Ma et al. 2013). HGT between pathogenic and saprophytic *F. oxysporum* was demonstrated under controlled conditions (Ma et al. 2010), but competition among *F. oxysporum* strains in the rhizosphere may provide favourable conditions for HGT.

The acquisition of foreign genes (xenologs) could happen in two ways: transfer of an entire plasmid or chromosome, or integration into chromosomal complement of a species (Ma et al. 2013). In *Fusarium* the transfer of entire chromosomes is more common (Coleman et al. 2009; Ma et al. 2010). Usually the transferred chromosomes are small and are not required for survival under standard conditions. Supernumerary chromosomes share some recognizable features: (1) a lack of housekeeping genes involved in primary metabolism, (2) a high level of G + C content (compared to normal chromosomes), (3) a lack of synteny with related species, and (4) a large number of transposable elements (TEs) (Ma et al. 2013).

HGT of supernumerary chromosomes improved for host-specific virulence had an important part in the polyphyletic distribution of host specificity within *F. oxysporum* species complex and the quick insurgence of new pathogens (Ma 2014). Comparative genome analysis also suggests that *Fusarium* has the genetic potential to produce more secondary metabolites than previously thought (Ma et al. 2013). Multiple evolutionary processes, including vertical inheritance, HGT, gene duplication, and gene deletion, could have triggered the present distribution of secondary metabolite biosynthetic gene clusters and production in *Fusarium* (Ma et al. 2013). Furthermore, assigning definitive roles to genes involved in virulence processes could be very stimulating, due to the often minor contributions of individual genes to pathogenicity and their tendency to have pleiotropic effects. In addition, deletion of genes involved in the production of asexual and sexual spore types tends to have pleiotropic effects. Transcriptomic studies indicate that a large number of genes are differentially expressed in the sexual cycle (Ma et al. 2013).

The LS regions contain over 74% of the transposable elements and 95% of the DNA transposons present in the *Fol* genome. In the *Fol* genome, repetitive sequences are recognized as retroelements, long interspersed nuclear elements, short interspersed nuclear elements, and DNA transposons. The predicted genes in the *Fol* LS regions belong to "secreted effectors and virulence factors", "transcription factors", and "proteins involved in signal transduction" categories. Among these genes related to pathogenicity there are effector proteins and enzymes expressed during early infection stages.

7.8 The Infection Process: Penetration and Colonization

F. oxysporum strains are present in the rhizosphere of many plant species growing in field soils. The infection process of *F. oxysporum* is characterized by different phases: root recognition, root attachment and colonization, root cortex penetration and colonization, and mycelial proliferation in the xylem vessels (Di Pietro et al. 2003).

The infection process starts when infection hyphae adhere to and penetrate the host roots. This is not a specific process because the fungus can adhere to the surface of both host and non-host plants (Vidhyasekaran 1997). The fungal pathogen enters the apical region of the root where the endodermis is not completely differentiated.

During the colonization, the mycelium grows intercellularly through the root cortex, to reach the xylem vessels. Inside the xylem system, the fungus remains exclusively in a lievitoid form, using the vascular system to colonize the host. Plant wilting is the first disease symptom, due to the hyphal proliferation in the xylem, synthesis of toxins and degrading enzymes, plant production of gels, gums and tyloses to avoid fungal diffusion. Wilting is due to vessel occlusion, which causes strong water stress. Symptoms turn then to vein clearing, leaf epinasty, chlorosis, necrosis, and abscission. The most evident internal symptom is the vascular browning. Severely infected plants wilt and die, whereas plants less diseased may become stunted and unproductive (Agrios 2005).

7.9 Genes Involved in the Infection Process

In recent years, forward and reverse genetics approaches, together with genome sequencing, started to shed a light on the comprehension of the molecular mechanisms at the basis of pathogenesis. Specifically, characterization of targeted knockouts demonstrated that both mitogen-activated protein kinase (MAPK) and cyclic AMP-protein kinase A (PKA) signalling cascades are fundamental for conferring virulence, not only to *F. oxysporum* but also to other plant pathogenic fungi, due to the high level of conservation (Michielse and Rep 2009; Rispail et al. 2009). cAMP-PKA and MAPK pathways are strictly regulated processes which control a variety of critical events, such as fungal growth, development, reproduction and pathogenicity. The signal transduction involving both cAMP-PKA and MAPK initiates upon ligand/stimulus binding to membrane bound or G-protein coupled receptors (GPCR). Following this, heterotrimeric G-proteins, constituted by α and a βγ subunits, are activated: GDP bound to the Gα subunit is replaced by GTP, and dissociation of Gα and Gβγ occurs. The now independent subunits are capable of starting a sequential activation of the signalling cascades, resulting in the expression of specific genes in response to a particular signal (Neer 1995; Gilman 1987).

In *F. oxysporum*, the inactivation of *fmk1* gene, encoding a MAP kinase, produced avirulent mutants characterized by wettable hyphae unable to penetrate the tomato roots (Di Pietro et al. 2001). Mutant strains lacking the G-protein β subunit (*fgb1*) showed a strongly reduced virulence (Delgado-Jarana et al. 2005); likewise, disruption of the G-protein α subunit *fga1* resulted in modified colony morphology, reduced conidiation, and reduced pathogenicity. Furthermore, *fga2* deletion caused a complete pathogenicity loss (Jain et al. 2002, 2005). A mucin-like protein, Msb2, which is located upstream of Fmk1, showed its key role for invasive growth and virulence in *F. oxysporum* (Pérez-Nadales and Di Pietro 2011). Mucins are transmembrane proteins thought to be sensors of the surrounding environment, thus capable of initiating intracellular signalling transduction (Carraway et al. 2003). Many other mutants were obtained and characterized in *F. oxysporum* enlightening the putative role of specific genes in the pathogenesis and infection processes. In particular, genes encoding crucial elements for the integrity of the cell wall such as chitin synthases (chs2, chs7, chsV, and chsVb), a GTPase (rho1), and a β-1,3-glucanosyltransferase (gas1) proved to be necessary for full virulence, thus indicating that the cell wall plays a key role in invasive growth (Martin-Urdiroz et al. 2008; Martínez-Rocha et al. 2008).

In addition, plant pathogens produce and secrete a variety of cell wall degrading enzymes (CWDE), such as polygalacturonases, pectate lyases, xylanases and proteases that can promote hyphal invasion and infection, although their role in pathogenesis is still controversial. Most probably due to functional redundancy, inactivation of single CWDE encoding genes failed to reveal any effect on virulence (Ruiz-Roldán and Di Pietro 2012). However, CWDEs are subjected to carbon catabolite repression, relieved in yeast by the activity of the protein kinase Snf1. Deletion of Snf1 in *F. oxysporum* provoked a downregulation of CWDEs together with a reduced virulence (Ospina-Giraldo et al. 2003).

Both nitrogen and iron limitation have been suggested to modulate virulence gene expression with a lack in pathogenicity following target inactivation of nitrogen and iron regulators Fnr1 and HapX (Divon et al. 2006). Several other classes of functional genes have been confirmed to be indispensable for pathogenesis: an argininosuccinate lyase (ARG1), the pH responsive transcription factor (pacC), a transcriptional regulator (FOW2), a mitochondrial carrier protein (FOW1), a chloride channel (CLC1), an F-box protein (FRP1), and a chloride conductance regulatory protein (FPD1) (Caracuel et al. 2005; Inoue et al. 2002; Imazaki et al. 2007; Namiki et al. 2001; Jonkers et al. 2009; Canero and Roncero 2008; Kawabe et al. 2004). A special attention must be dedicated to the efflux pumps in fungi, which are classified in ATP-binding cassette (ABC) and major facilitator superfamily (MFS) transporters. By using different sources of energy (ATP hydrolysis or proton motive force) these two classes of transmembrane proteins translocate across the cell membrane secondary metabolites, virulence factors, natural toxic compounds or antifungal drugs. For these reasons, their role in pathogenicity was demonstrated in *F. oxysporum*.

Finally, metabolic pathways involved in lipid metabolism and protein translocation/degradation are also thought to confer full virulence to *F. oxysporum* as confirmed by deletion of peroxin genes PEX1, PEX10, PEX12 and PEX26, implicated in β-oxidation of fatty acids and peroxide detoxification (Michielse and Rep 2009).

7.10 Pathogenicity Genes and Their Interaction with Resistance Genes

To cause disease, fungal pathogens should grow on living host tissue. On the other hand, plants have a broad spectrum of defence responses to block the pathogen invasion (Ruiz-Roldán and Di Pietro 2012).

To favour host colonization, many plant-pathogen produce and secrete small proteins that suppress or reduce host defences. These small secreted proteins are the most common fungal avirulence factors and are also called effectors (van der Does and Rep 2007). Some effectors are recognized by the innate immune system of the plant, leading to disease resistance.

In several crops, for example, in tomato (but also in melon, cucumber, pea and bean), polymorphic Resistance (R) genes, conferring resistance against *Fol*, are present. These resistance genes include *I-1* (I stands for immunity), *I-2*, and *I-3* (Huang and Lindhout 1997). Races are named according to the resistance gene, able to block them: the *I* gene and the *I-1* gene are effective against race 1, *I-2* is effective against race 2, and *I-3* is effective against race 3 (Rep et al. 2005).

The existence of major R genes in tomato host and corresponding races in the pathogen (Flor 1971) lead to hypothesize that *Fol* should contain avirulence genes matching R genes in host. During the last years, different avirulence genes were identified in *Fol*. Their products were found in xylem sap of infected plants, and are necessary to mediate resistance by R genes. These genes are called *SIX* (*S*ecreted *I*n *X*ylem). *SIX* proteins were renamed *Avr* in correspondence with the immunity resistance gene recognizing them: *Avr1* (*Six4*) is recognized by *I* and *I-1* gene (Houterman et al. 2008); *Avr2* (*Six3*) is recognized by *I-2* (Houterman et al. 2009); and *Avr3* (*Six1*) is recognized by *I-3* (Rep et al. 2004). *Avr2*, *Avr3* and *Six6* (Houterman et al. 2009; Rep 2005) are effectors, which contribute to the virulence of *Fol* (Takken and Rep 2010). Their loss enables *Fol* races to escape disease resistance mediated by respective *I* counterpart (Houterman et al. 2008).

All these *SIX* genes, except *SIX4*, are located on chromosome 14, which is a LS chromosome. These genes are conserved in *Fol* strains, but they are absent in other *formae speciales*. LS chromosome 14 is most probably essential for pathogenicity of *Fol* on tomato.

7.11 Tools for Diagnostics and Detection

The *formae speciales* are not differentiated by morphological features, but they can be distinguished by molecular markers (Spadaro et al. 2020).The TEF1 gene is widely used to determine the species of *Fusarium* (Amatulli et al. 2012; Poli et al. 2012). The translation elongation factor 1-α (TEF1), the RNA polymerase II subunits (RPB1 andRPB2), the Large Subunit (LSU) of the nuclear ribosomal RNA, or polygalacturonases genes, and the intergenic spacer region (IGS) of the ribosomal operon are useful target regions to differentiate *formae speciales* (Bertoldo et al. 2015; Mbofung et al. 2007; O'Donnell et al. 2013, 2015; Srinivasan et al. 2010, 2012). Starting from the sequence of these genic regions, PCR- and qPCR-based assays have been developed to identify and quantify *formae speciales* (Mbofung and Pryor 2010; Haegi et al. 2013).

The abundance of transposable elements on the *F. Oxysporum* genome has been exploited to design specific assays: Inter-retrotransposon amplified polymorphism (IRAP) is a technique based on the amplification of genomic regions between long

Fig. 7.3 Section of lettuce plant showing vascular browning caused by *Fusarium oxysporum* f. sp. *lactucae*

terminal repeats (Fernandez et al. 1998). It was applied to develop specific PCR primers for *F. oxysporum* f. sp. *lactucae* (Fig. 7.3; Pasquali et al. 2007).

The previously described regions were used to develop LAMP assays that combined with crude extraction methods show great potential for on-site detection. In particular, LAMP assays have been developed for the identification of species of *Fusarium* (Franco Ortega et al. 2018a), *formae speciales* within *F. oxysporum* species complex (Franco Ortega et al. 2018b), or races within a *forma specialis* (Ayukawa et al. 2016).

Sequence resources are available in the FUSARIUM-ID database (http://isolate.fusariumdb.org/blast.php; Park et al. 2010) and the Fusarium MLST database (https://fusarium.mycobank.org/; O'Donnell et al. 2010).

When *formae speciales* are polyphyletic, it is hard to identify specific molecular markers (Baayen et al. 2000), but pathogenicity-related genes, such as SIX genes, could be exploited for their specificity. One of the first diagnostic tools based on a host-specific virulence gene was developed for *Fol* (Lievens et al. 2009). Several SIX genes are currently known in different *formae speciales*.

The available genomes could be used to obtain sequences specific for a species, a *forma specialis*, or a race. Comparative genomics is useful to identify host specificity in *F. oxysporum:* seven *formae speciales* were differentiated based on their effector pattern extracted from comparative genomics analysis (van Dam et al. 2017).

7.12 Disease Management

Due to long persistence in the soil of *F. oxysporum* in chlamydospore form, the main strategy to manage Fusarium wilt is to reduce the inoculum with a holistic management, paying attention to all the components: plant residues, alternative hosts, contamination of water, wind, insects, and machinery, and use of healthy propagation material (Katan 2006). In addition, chemical, physical, biological, cultural, and genetic approaches are used to manage Fusarium wilts. The main strategies include chemical or physical soil disinfestation, soil solarization, use of composts, induction of host resistance, use of fungicides or biofungicides, sanitation, soilless culture, and use of resistant plant cultivars if available. Unfortunately, breeding for resistance is difficult when no dominant resistance genes (R) are identified or with dioecious host species. In addition, new races may develop that are able to bypass host resistance (Lievens et al. 2008).

During the last years, plant pathologists recognized the necessity of an eco-friendly approach to disease control, leading to an integrated disease management system, combining different strategies and methods together (Katan et al. 2012). The term disease management has gradually replaced disease control, as it more appropriately denotes a continuous process.

7.13 Suppressive Soils

Suppressive ability of some soils has been recognized since the twentieth century all over the world and the mechanism by which disease suppression is brought about has been studied in the last 50 years. Suppressive soils to soilborne pathogens are soils where disease development is minimal even in the presence of the three key elements favouring a disease: a virulent pathogen, a susceptible host, and conducive environmental conditions. Soil suppressiveness is a complex phenomenon involving soil structure, nutrient and water availability, microbiota, and plant genotype. Suppressive soils are well described in the case of *formae speciales* of *F. oxysporum* (Yuen et al. 1985; Garibaldi and Gullino 1987; Janvier et al. 2007). Natural soils own a general disease suppression capacity compared to pasteurized soils, due to their living microbiota (Schlatter et al. 2017). The involvement of biological factors or their metabolites, including saprophytic *F. oxysporum*, in the mechanism of soil suppressiveness was demonstrated. It is important to notice that a soil suppressive to a certain pathogen may not necessarily be suppressive to another one, demonstrating the specific interactions of the soil-plant-microbiota. Generally, all soils have a level of suppressive activity towards some pathogens, but the cultural practices can significantly modify this level (Garibaldi et al. 1990).

Suppressive soils to Fusarium wilts are known since long time, but their complex mechanisms have been demonstrated with omics technologies only recently (Ou et al. 2019). Enrichment and activity-specific microorganisms can favour soil suppressiveness (Kinkel et al. 2011). Suppressive and conducive soils to diseases harbour specific microbiota (Bowen et al. 2019).

Soil bacteria, such as *Pseudomonas* spp. and *Alcaligenes* spp. in the USA (Kloepper et al. 1980), and soil fungi, like *Fusarium* spp. in France and Italy (Janvier et al. 2007; Garibaldi and Gullino 1987), were demonstrated to play a role in the suppression of *F. oxysporum*. Antagonistic *Fusarium* spp., isolated from the plant rhizosphere from suppressive soils, showed high rhizosphere competence and were able to control *F. oxysporum* on different crops (Gullino and Garibaldi 2007). An efficient disease control potential is present in the soil microbiota; therefore a deep comprehension of the composition of the microbiome of suppressive soils is desirable.

7.14 Non-pathogenic *Fusarium oxysporum* and Rhizosphere Competition

Many isolates found in soil are not pathogenic (Gordon and Martyn 1997; Recorbet et al. 2003), and it is relevant to understand if such ubiquitous non-pathogenic *Fusarium* spp. can evolve to become pathogenic.

Saprophytic *F. oxysporum*, often isolated from suppressive soils, have been used to control soilborne pathogens, due to their rhizosphere competence. Induced systemic resistance has been demonstrated, particularly for endophytic *F. oxysporum*. Several tools have been developed to track and monitor the occurrence of antagonistic strains of *F. oxysporum* into the soil (Gullino et al. 1995).

In *Fusarium*, horizontal chromosome transfer could turn a saprophytic strain into a pathogen (van der Does et al. 2008). Moreover, other factors also could explain the pathogenic and non-pathogenic feature of some strains, such as the interaction with ectosymbiotic bacteria living on hyphal surface which are capable of silencing the expression of genes involved in fungal pathogenesis, and modifying the hyphal morphology.

Non-pathogenic *F. oxysporum* strains can penetrate the roots, without invading the vascular system or causing disease. Saprophytic strains, isolated from disease-suppressive soils, can protect plants against pathogenic *F. oxysporum* isolates and can be useful in biocontrol (Gullino et al. 2012).

7.15 Bacteria–*Fusarium oxysporum* Interactions

Non-pathogenic *F. oxysporum*, isolated from Italian soils suppressive towards *F. oxysporum* (Aloi et al. 1994), reduced the severity of Fusarium wilts on different crops (Gilardi et al. 2005). Observations of fungal cultures evidenced that a bacterial layer was associated to external layers of fungal hyphae. Physical interactions between rhizobacteria and soil fungi is a well-known phenomenon involving the interactions of plant, pathogen, and antagonist (Bertaux et al. 2003; Partida-Martinez and Hertweck 2005; Artursson et al. 2006) and should be considered when studying biological control. A large amount of rod-shaped bacteria can be found on the hyphae of non-pathogenic *F. oxysporum*. The bacterial consortium was able to silence the virulence of a strain, which behaved as an antagonist (Minerdi et al.

2008). Ectosymbiotic bacteria were able to silence the expression of three fungal genes (*fmk*1, *chs*V and *pl*1) involved in pathogenesis, changing the hyphal morphology and the capacity to penetrate lettuce roots (Woesten et al. 1994). Moreover, the antagonistic *F. oxysporum* strain was able to release volatile organic compounds (VOCs), mainly sesquiterpenes, which seemed to be responsible for the acquired antagonistic activity of this strain. Physical association between fungi and bacteria is common in nature. A sort of dialogue can be established between the organisms, creating a physiologically interdependent network with characteristics different from those typically observed in the single components (Frey-Klett et al. 2011; Tarkka et al. 2009).

7.16 Potential Biocontrol Agents

Management of Fusarium wilts relied mainly on chemical soil disinfestation and selection of resistant cultivars (Katan et al. 2012). However, nowadays many fumigants have been banned or considered as environmentally dangerous. As for the use of resistant cultivars, new more virulent races of the pathogen have arisen to bypass the host resistance. Therefore, the interest in biological control of Fusarium wilt by applying selected strains of antagonistic bacteria or strains of *F. oxysporum* has been rapidly increasing (Fravel et al. 2003). Strains of *F. oxysporum* which are not capable of invading the host vascular system or causing disease have been isolated from suppressive soils.

Mechanisms responsible for antagonistic activity can be direct: antibiosis via inhibition of the pathogen by diffusible or volatile antibiotics (such as cyclic lipopeptides; Kim et al. 2015) or biosurfactants (Vitullo et al. 2012), competition for nutrients and/or infection and colonization sites, and mycoparasitism, which involves the production of extracellular cell wall-degrading enzymes such as chitinases, β-1,3-glucanases, or proteases (Bloemberg and Lugtenberg 2001; Whipps 2001; Spadaro and Gullino 2005).

An indirect mechanism involves the induction of host defence (Walters 2012). Antagonistic microorganisms can induce pathogenesis-related (PR) genes encoding chitinase and β-1,3-glucanases (Edreva 2005; Walters and Daniell 2007), which show antifungal activity (Bargabus et al. 2002; Rep et al. 2002). For this reason, nursery treatment with biocontrol agents is gaining importance as an effective strategy to control soilborne diseases on plantlets, before transplant. In a recent study, the preventative nursery application of biocontrol agents, such as *Bacillus subtilis* and *Trichoderma* sp., induced a significant reduction of *F. oxysporum* f. sp. *lycopersici* on tomato, without impacting the non-target microbial communities and by favouring the expression of PR proteins in the host (Cucu et al. 2020).

The management of antagonistic microorganisms in the plant rhizosphere has become a major challenge for both plant protection and plant growth promotion. Antagonists are naturally occurring organisms with genetic traits, which enable them to interfere with the pathogen growth, survival, infection, or ability to colonize plants

(Chernin and Chet 2002). Biocontrol of soilborne diseases is particularly complex because it works in the dynamic environment present at the interface between root and rhizosphere. Biological control represents a valuable control option for *Fusarium* wilts (Nel et al. 2006; Shishido et al. 2005; Panina et al. 2007).

7.17 *Fusarium oxysporum* and Nematodes

The interaction between *F. oxysporum* and plant-parasitic nematodes was reported for the first time in 1892, by Atkinson who reported that *Meloidogyne incognita* was associated with increased severity of *F. oxysporum* f. sp. *vasinfectum* disease on cotton and of *F. oxysporum* f. sp. *albedinis* on date palm (Greco et al. 1980). Interactions between nematodes and fungal pathogens can be additive or synergistic (Mai and Abawi 1987): in the first ones, the host damage equals the sum of the damage caused by each organism alone, whereas in the second ones the host damage is higher than the sum caused by the single organisms. Nematodes may provide entry points for the fungus through wounds produced in the root during nematode penetration (Starr et al. 1989). Moreover, the enhanced susceptibility of crops to *F. oxysporum* due to nematode infection appears related to the changes in the host physiology—i.e. an increased susceptibility—brought about by nematode parasitism (Hillocks 1985).

Recently, scientists have explored the role of plant and microbial volatile organic compounds (VOCs) on nematode parasitism, finding that VOCs produced by *Fusarium* spp. isolated from nematode egg masses and plant roots are toxic to *M. incognita* (Estupiñan-López et al. 2018). Moreover, root microbiota may assist plants in fighting nematodes. Plant root exudates are important to determine host-microbe communication to select and enrich specific sets of antagonistic microorganisms in the rhizosphere (Topalović et al. 2020). A positive interaction between host roots and their beneficial microbiota is related to a lower nematode parasitism (Gao et al. 2008).

7.18 Conclusions

Advances in biochemistry, physiology and molecular biology permit to have a deeper understanding of the host-pathogen-microbiota interactions, with several positive effects on the improvement of crop protection strategies to manage Fusarium wilts. Omics technologies permit to obtain interesting results regarding the genetics of *F. oxysporum*, even though the genetic structure is difficult to study with traditional tools, due to the lack of the sexual reproduction. To control soilborne pathogens, including *F. oxysporum*, an integrated strategy should be adopted, including prevention practice and early diagnostics tools. In the recent past, a few chemical strategies, based on soil disinfestation, could easily control a wide range of soilborne diseases. In the framework of a sustainable crop protection strategy, soil health should be guaranteed by including the health of the soil microbiota. A correct disease management strategy needs to consider the impact on the soil and crop health, but also on the agroecological environment, the natural resources, and the human health.

Acknowledgements This work was partially funded by the European Union's Horizon 2020 research and innovation programme under grant agreement No 817946 "Exploiting the multifunctional potential of belowground biodiversity in horticultural farming—EXCALIBUR" and by the Italian Ministry for Education, University and Research for funding the project "A gnotobiotic-based approach to unravel the role of the plant microbiome and develop synthetic communities increasing plant growth and stress tolerance—NATURE" (PRIN 2017 Programme). The revision of Prof. Angelo Garibaldi is kindly recognized.

References

Agrios GN (2005) Plant pathology, 5th edn. Academic, San Diego, p 952

Aloi C, Conti A, Garibaldi A (1994) Fusarium antagonisti contro la tracheofusariosi del garofano. Colture Protette 4:79–82

Amatulli MT, Spadaro D, Gullino ML, Garibaldi A (2012) Conventional and real-time PCR for the identification of *Fusarium fujikuroi* and *Fusarium proliferatum* from diseased rice tissues and seeds. Eur J Plant Pathol 134:401–408. https://doi.org/10.1007/s10658-012-9998-0

Artursson V, Finlay RD, Jansson JK (2006) Interactions between arbuscular mycorrhizal fungi and bacteria and their potential for stimulating plant growth. Environ Microbiol 8:1–10. https://doi.org/10.1111/j.1462-2920.2005.00942.x

Ash C, Mueller K (2016) Manipulating the microbiota. Science 352:530–531. https://doi.org/10.1126/science.352.6285.530

Atkinson GF (1892) Some diseases of cotton. 3. Frenching. Bull Ala Agric Exp Stn 41:19–29

Ayukawa Y, Komatsu K, Kashiwa T, Akai K, Yamada M, Teraoka T, Arie T (2016) Detection and differentiation of Fusarium oxysporum f. sp. lycopersici race 1 using loop-mediated isothermal amplification with three primer sets. Lett Appl Microbiol 63:202–209. https://doi.org/10.1111/lam.12597

Baayen RP, O'Donnell K, Bonants PJM, Cigelnik E, Kroon LPNM, Roebroeck EJA, Waalwijk C (2000) Gene genealogies and AFLP analyses in the *Fusarium oxysporum* complex identify monophyletic and nonmonophyletic *formaespeciales* causing wilt and rot disease. Phytopathology 90:891–900. https://doi.org/10.1094/PHYTO.2000.90.8.891

Bargabus RL, Zidack NK, Sherwood JW, Jacobsen BJ (2002) Characterization of systemic resistance in sugar beet elicited by a non-pathogenic, phyllosphere-colonizing *Bacillus mycoides*, biological control agent. Physiol Mol Plant Pathol 61:289–298. https://doi.org/10.1006/pmpp.2003.0443

Bertaux J, Schmid M, Prevost-Boure NC, Churin JL, Hartmann A, Garbaye J, Frey-Mett P (2003) *In situ* identification of intracellular bacteria related to *Paenibacillus* spp. in the mycelium of the ectomycorrhizal fungus *Laccaria bicolor* S238N. Appl Environ Microbiol 69:4243–4248. https://doi.org/10.1128/AEM.69.7.4243-4248.2003

Bertoldo C, Gilardi G, Spadaro D, Gullino ML, Garibaldi A (2015) Genetic diversity and virulence of Italian strains of *Fusarium oxysporum* isolated from *Eustoma grandiflorum*. Eur J Plant Pathol 141:83–97. https://doi.org/10.1007/s10658-014-0526-2

Bishop CD, Cooper RM (1983) An ultrastructural study of root invasion in three vascular wilt diseases. Physiol Mol Plant Pathol 22:15–27. https://doi.org/10.1016/S0048-4059(83)81034-0

Bloemberg GV, Lugtenberg BJJ (2001) Molecular basis of plant growth promotion and biocontrol by rhizobacteria. Curr Opin Plant Biol 4:343–350. https://doi.org/10.1016/s1369-5266(00)00183-7

Bowen A, Orr R, McBeath AV, Pattison A, Nelson PN (2019) Suppressiveness or conduciveness to Fusarium wilt of bananas differs between key Australian soils. Soil Res 57:158–165. https://doi.org/10.1071/SR18159

Canero DC, Roncero MIG (2008) Influence of the chloride channel of *Fusarium oxysporum* on extracellular laccase activity and virulence on tomato plants. Microbiology 154:1474–1481. https://doi.org/10.1099/mic.0.2007/015388-0

Caracuel Z, Martínez-Rocha AL, Di Pietro A, Madrid MP, Roncero MI (2005) *Fusarium oxysporum* gas1 encodes a putative beta-1,3-glucanosyltransferase required for virulence on tomato plants. Mol Plant-Microbe Interact 18:1140–1147. https://doi.org/10.1094/MPMI-18-1140

Carraway KL, Ramsauer VP, Haq B, Carothers Carraway CA (2003) Cell signaling through membrane mucins. BioEssays 25:66–71. https://doi.org/10.1002/bies.10201

Chernin L, Chet I (2002) Microbial enzymes in biocontrol of plant pathogens and pests. In: Burns R, Dick R (eds) Enzymes in the environment: activity, ecology, and applications. Marcel Dekker, New York, pp 171–225

Chiocchetti A, Ghignone S, Minuto A, Gullino ML, Garibaldi A, Migheli Q (1999) Identification of *Fusarium oxysporum* f. sp. *basilici* isolated from soil, basil seed, and plants by RAPD analysis. Plant Dis 83:576–581. https://doi.org/10.1094/PDIS.1999.83.6.576

Coleman JJ, Rounsley SD, Rodriguez-Carres M, Kuo A, Wasmann CC, Grimwood J, Schmutz J, Taga M, White GJ, Zhou S, Schwartz DC, Freitag M, Ma LJ, Danchin EG, Henrissat B, Coutinho PM, Nelson DR, Straney D, Napoli CA, Barker BM, Gribskov M, Rep M, Kroken S, Molnár I, Rensing C, Kennell JC, Zamora J, Farman ML, Selker EU, Salamov A, Shapiro H, Pangilinan J, Lindquist E, Lamers C, Grigoriev IV, Geiser DM, Covert SF, Temporini E, Vanetten HD (2009) The genome of *Nectria haematococca*: contribution of supernumerary chromosomes to gene expansion. PLoS Genet 5:e1000618. https://doi.org/10.1371/journal.pgen.1000618

Cucu MA, Gilardi G, Pugliese M, Gullino ML, Garibaldi A (2020) An assessment of the modulation of the population dynamics of pathogenic Fusarium oxysporum f. sp. lycopersici in the tomato rhizosphere by means of the application of Bacillus subtilis QST 713, Trichoderma sp. TW2 and two composts. Biol Control 142:104158. https://doi.org/10.1016/j.biocontrol.2019.104158

Daboussi MJ, Capy P (2003) Transposable elements in filamentous fungi. Annu Rev Microbiol 57:275–299. https://doi.org/10.1146/annurev.micro.57.030502.091029

Daboussi MJ, Davière JM, Graziani S, Langin T (2002) Evolution of the Fot1 transposons in the genus *Fusarium*: discontinuous distribution and epigenetic inactivation. Mol Biol Evol 19:510–520. https://doi.org/10.1093/oxfordjournals.molbev.a004106

Dean R, van Kan JAL, Pretorius ZA, Hammond-Kosack KE, Di Pietro A, Spanu PD, Rudd JJ, Dickman M, Kahmann R, Ellis J, Foster GD (2012) The top 10 fungal pathogens in molecular plant pathology. Mol Plant Pathol 13:414–430. https://doi.org/10.1111/j.1364-3703.2011.00783.x

Delgado-Jarana JS, Martinez-Rocha AL, Roldan-Rodriguez R, Roncero MIG, Di Pietro A (2005) *Fusarium oxysporum* G-protein beta subunit Fgb1 regulates hyphal growth, development, and virulence through multiple signalling pathways. Fungal Genet Biol 42:61–72. https://doi.org/10.1016/j.fgb.2004.10.001

Di Pietro A, Garcia-Maceira FI, Meglecz E, Roncero MIG (2001) A MAP kinase of the vascular wilt fungus *Fusarium oxysporum* is essential for root penetration and pathogenesis. Mol Microbiol 39:1140–1152. https://doi.org/10.1111/j.1365-2958.2001.02307.x

Di Pietro A, Madrid MP, Caracuel Z, Delgado-Jarana J, Roncero MIG (2003) *Fusarium oxysporum*: exploring the molecular arsenal of a vascular wilt fungus. Mol Plant Pathol 4:315–325. https://doi.org/10.1046/j.1364-3703.2003.00180.x

Divon HH, Ziv C, Davydov O, Yarden O, Fluhr R (2006) The global nitrogen regulator, FNR1, regulates fungal nutrition-genes and fitness during *Fusarium oxysporum* pathogenesis. Mol Plant Pathol 7:485–497. https://doi.org/10.1111/j.1364-3703.2006.00354.x

Edel-Hermann V, Lecomte C (2019) Current status of *Fusarium oxysporum formae speciales* and races. Phytopathology 109:512–530. https://doi.org/10.1094/PHYTO-08-18-0320-RVW

Edreva A (2005) Pathogenesis-related proteins: research progress in the last 15 years. Gen Appl Plant Physiol 31:105–124

Estupiñan-López L, Campos VP, da Silva Júnior JC, Pedroso MP, Terra WC, da Silva JCP (2018) Volatile compounds produced by *Fusarium* spp. isolated from *Meloidogyne paranaensis* egg masses and corticous root tissues from coffee crops are toxic to *Meloidogyne incognita*. Trop Plant Pathol 43:183–193. https://doi.org/10.1007/s40858-017-0202-0

Fernandez D, Ouinten M, Tantaoui A, Geiger JP, Daboussi MJ, Langin T (1998) Fot 1 insertions in the *Fusarium oxysporum* f. sp. *albedinis* genome provide diagnostic PCR targets for detection of the date palm pathogen. Appl Environ Microbiol 64:633–636. https://doi.org/10.1128/AEM.64.2.633-636.1998

Flor HH (1971) Current status of the gene-for-gene concept. Annu Rev Phytopathol 9:275–296. https://doi.org/10.1146/annurev.py.09.090171.001423

Fourie G, Steenkamp ET, Gordon TR, Viljoen A (2009) Evolutionary relationships among the *Fusarium oxysporum* f. sp. *cubense* vegetative compatibility groups. Appl Environ Microbiol 75:4770–4781. https://doi.org/10.1128/AEM.00370-09

Franco Ortega S, Tomlinson J, Hodgetts J, Spadaro D, Gullino ML, Boonham N (2018a) Development of loop-mediated isothermal amplification assays for the detection of seedborne fungal pathogens, *Fusarium fujikuroi* and *Magnaporthe oryzae*, in rice seeds. Plant Dis 102:1549–1558. https://doi.org/10.1094/PDIS-08-17-1307-RE

Franco Ortega S, Tomlinson J, Gilardi G, Spadaro D, Gullino ML, Garibaldi A, Boonham N (2018b) Rapid detection of *Fusarium oxysporum* f. sp. *lactucae* on soil, lettuce seeds and plants using loop-mediated isothermal amplification. Plant Pathol 67:1462–1473. https://doi.org/10.1111/ppa.12855

Fravel D, Olivain C, Alabouvette C (2003) *Fusarium oxysporum* and its biocontrol. New Phytol 157:493–502. https://doi.org/10.1046/j.1469-8137.2003.00700.x

Frey-Klett P, Burlinson P, Deveau A, Barret M, Tarkka M, Sarniguet A (2011) Bacterial-fungal interactions: hyphens between agricultural, clinical, environmental, and food microbiologists. Microbiol Mol Biol Rev 75:583–609. https://doi.org/10.1128/MMBR.00020-11

Gao X, Yin B, Borneman J, Becker JO (2008) Assessment of parasitic activity of *Fusarium* strains obtained from a *Heterodera schachtii*-suppressive soil. J Nematol 40:1–6

Garibaldi A, Gullino ML (1987) Fusarium wilt of carnation: present situation, problems and perspectives. Acta Hortic 216:45–54. https://doi.org/10.17660/ActaHortic.1987.216.6

Garibaldi A, Guglielmone L, Gullino ML (1990) Rhizosphere competence of antagonistic Fusaria isolated from suppressive soils. Symbiosis 9:401–404

Gilardi G, Tinivella F, Gullino ML, Garibaldi A (2005) Seed dressing to control *Fusarium oxysporum* f. sp. *lactucae*. J Plant Dis Protect 112:240–246. https://doi.org/10.1007/BF02980861

Gilman AG (1987) G-proteins - transducers of receptor-generated signals. Annu Rev Biochem 56:615–649. https://doi.org/10.1146/annurev.bi.56.070187.003151

Glass NL, Jacobson DJ, Shiu PKT (2000) The genetics of hyphal fusion and vegetative incompatibility in filamentous ascomycetes. Annu Rev Genet 34:165–186. https://doi.org/10.1146/annurev.genet.34.1.165

Gordon WL (1965) Pathogenic strains of *Fusarium oxysporum*. Can J Bot 43:1309–1318. https://doi.org/10.1139/b65-138

Gordon TR, Martyn RD (1997) The evolutionary biology of *Fusarium oxysporum*. Annu Rev Phytopathol 35:111–128. https://doi.org/10.1146/annurev.phyto.35.1.111

Greco N, Surico G, Lamberti F (1980) Indagini preliminari su una possibile interazione tra *Fusarium oxysporum* f. sp. *albedinis* e *Meloidogyne incognita* su palma da dattero. Nemat Medit 8:113–119

Gullino ML, Garibaldi A (2007) Critical aspects in management of fungal diseases of ornamental plants and directions in research. Phytopathol Mediterr 46:135–149. https://doi.org/10.14601/Phytopathol_Mediterr-2150

Gullino ML, Migheli Q, Mezzalama M (1995) Risk analysis for biological control agents: antagonistic *Fusarium* spp. as a case study. Plant Dis 12:1193–1201. https://doi.org/10.1094/PD-79-1193

Gullino ML, Katan J, Garibaldi A (2012) Fusarium wilts of greenhouse vegetable and ornamental crops. The American Phytopathological Society, St. Paul, p 243

Haegi A, Catalano V, Luongo L, Vitale S, Scotton M, Ficcadenti N, Belisario A (2013) A newly developed real-time PCR assay for detection and quantification of *Fusarium oxysporum* and its use in compatible and incompatible interactions with grafted melon genotypes. Phytopathology 103:802–810. https://doi.org/10.1094/PHYTO-11-12-0293-R

Hillocks RJ (1985) The effect of root-knot nematode on resistance to Fusarium wilt in the stems of cotton plants. Ann Appl Biol 107:213–218. https://doi.org/10.1111/j.1744-7348.1985.tb01564.x

Hopkins DL, Lobinske RJ, Larkin RP (1992) Selection for *Fusarium oxysporum* f. sp. *niveum* race 2 in monocultures of watermelon cultivars resistant to Fusarium wilt. Phytopathology 82:290–293. https://doi.org/10.1094/Phyto-82-290

Houterman PM, Cornelissen BJC, Rep M (2008) Suppression of plant resistance gene-based immunity by a fungal effector. PLoS Pathog 4:e1000061. https://doi.org/10.1371/journal.ppat.1000061

Houterman PM, Ma L, van Ooijen G, de Vroomen MJ, Cornelissen BJC, Takken FLW, Rep M (2009) The effector protein Avr2 of the xylem-colonizing fungus *Fusarium oxysporum* activates the tomato resistance protein I-2 intracellularly. Plant J 58:970–978. https://doi.org/10.1111/j.1365-313X.2009.03838.x

Huang CC, Lindhout P (1997) Screening for resistance in wild *Lycopersicon* species to Fusarium oxysporum f. sp. lycopersici race 1 and race 2. Euphytica 93:145–153. https://doi.org/10.1023/A:1002943805229

Imazaki I, Kurahashi M, Iida Y, Tsuge T (2007) Fow2, a Zn(II)2Cys6-type transcription regulator, controls plant infection of the vascular wilt fungus *Fusarium oxysporum*. Mol Microbiol 63: 737–753. https://doi.org/10.1111/j.1365-2958.2006.05554.x

Inoue I, Namiki F, Tsuge T (2002) Plant colonization by the vascular wilt fungus *Fusarium oxysporum* requires FOW1, a gene encoding a mitochondrial protein. Plant Cell 14:1869–1883. https://doi.org/10.1105/tpc.002576

Jain S, Akiyama K, Mae K, Ohguchi T, Takata R (2002) Targeted disruption of a G protein alpha subunit gene results in reduced pathogenicity in *Fusarium oxysporum*. Curr Genet 41:407–413. https://doi.org/10.1007/s00294-002-0322-y

Jain S, Akiyama K, Takata R, Ohguchi T (2005) Signaling via the G protein alpha subunit FGA2 is necessary for pathogenesis in *Fusarium oxysporum*. FEMS Microbiol Lett 243:165–172. https://doi.org/10.1016/j.femsle.2004.12.009

Janvier C, Villeneuve F, Alabouvette A, Edel-Hermann V, Mateille T, Steinberg C (2007) Soil health through soil disease suppression: which strategy from descriptors to indicators? Soil Biol Biochem 39:1–23. https://doi.org/10.1016/j.soilbio.2006.07.001

Jarvis WR (1988) Fusarium crown and root rot of tomatoes. Phytoprotection 69:49–64

Jonkers W, Andrade Rodrigues CD, Rep M (2009) Impaired colonization and infection of tomato roots by the Δfrp1 mutant of *Fusarium oxysporum* correlates with reduced CWDE gene expression. Mol Plant-Microbe Interact 22:507–518. https://doi.org/10.1094/MPMI-22-5-0507

Katan T (1999) Current status of vegetative compatibility groups in *Fusarium oxysporum*. Phytoparasitica 27:51–64. https://doi.org/10.1007/BF02980727

Katan J (2006) Strategies for the management of soil-borne pathogens: phytopathological and environmental aspects. Proc Congr Mediterr Phytopathol Union 232–235

Katan T, Di Primo P (1999) Current status of vegetative compatibility groups in *Fusarium oxysporum*. Phytoparasitica 27:273–277. https://doi.org/10.1007/BF02981483

Katan J, Gullino ML, Garibaldi A (2012) Strategies and tactics for Fusarium wilt management. In: Gullino ML, Katan J, Garibaldi A (eds) Fusarium wilts of greenhouse vegetable and ornamental crops. The American Phytopathological Society, St. Paul, pp 61–66

Kawabe M, Mizutani K, Yoshida T, Teraoka T, Yoneyama K, Yamaguchi I, Arie T (2004) Cloning of the pathogenicity-related gene FPD1 in *Fusarium oxysporum* f. sp. *lycopersici*. J Gen Plant Pathol 70:16–20. https://doi.org/10.1007/s10327-003-0089-0

Kim YG, Kang HK, Kwon KD, Seo CH, Lee HB, Park Y (2015) Antagonistic activities of novel peptides from *Bacillus amyloliquefaciens* PT14 against *Fusarium solani* and *Fusarium oxysporum*. J Agric Food Chem 63:10380–10387. https://doi.org/10.1021/acs.jafc.5b04068

Kinkel LL, Bakker MG, Schlatter DC (2011) A coevolutionary framework for managing disease-suppressive soils. Annu Rev Phytopathol 49:47–67. https://doi.org/10.1146/annurev-phyto-072910-095232

Kloepper JW, Leong J, Teintze M, Schroth MN (1980) Pseudomonas siderophores: a mechanism explaining disease-suppressive soils. Curr Microbiol 4:317–320. https://doi.org/10.1007/BF02602840

Leslie JF (2012) Genetics and *Fusarium oxysporum*. In: Gullino ML, Katan J, Garibaldi A (eds) Fusarium wilts of greenhouse vegetable and ornamental crops. The American Phytopathological Society, St. Paul, pp 39–46

Lievens B, Claes L, Vakalounakis DJ, Vanachter AC, Thomma BP (2007) A robust identification and detection assay to discriminate the cucumber pathogens *Fusarium oxysporum* f. sp. *cucumerinum* and f. sp. *radicis-cucumerinum*. Environ Microbiol 9:2145–2161. https://doi.org/10.1111/j.1462-2920.2007.01329.x

Lievens B, Rep M, Thomma BPH (2008) Recent developments in the molecular discrimination of *formae speciales* of *Fusarium oxysporum*. Pest Manag Sci 64:781–788. https://doi.org/10.1002/ps.1564

Lievens B, van Baarlen P, Verreth C, van Kerckhove S, Rep M, Thomma BP (2009) Evolutionary relationships between *Fusarium oxysporum* f. sp. *lycopersici* and *F. oxysporum* f. sp. *radicis-lycopersici* isolates inferred from mating type, elongation factor-1alpha and exopolygalacturonase sequences. Mycol Res 113:1181–1191. https://doi.org/10.1016/j.mycres.2009.07.019

Lievens B, Hanssen IM, Rep M (2012) Recent developments in the detection and identification of *formae speciales* and races of *Fusarium oxysporum*: from pathogenicity testing to molecular diagnostics. In: Gullino ML, Katan J, Garibaldi A (eds) Fusarium wilts of greenhouse vegetable and ornamental crops. The American Phytopathological Society, St. Paul, pp 47–55

Lomas-Cano T, Boix-Ruiz A, de Cara-García M, Marín-Guirao JI, Palmero-Llamas D, Camacho-Ferre F, Tello-Marquina JC (2016) Etiological and epidemiological concerns about Pepper root and lower stem rot caused by *Fusarium oxysporum* f. sp. *radicis-capsici* f. sp. *nova*. Phytoparasitica 44:283–293. https://doi.org/10.1007/s12600-016-0522-5

Ma LJ (2014) Horizontal chromosome transfer and rational strategies to manage Fusarium vascular wilt diseases. Mol Plant Pathol 15:763–766. https://doi.org/10.1111/mpp.12171

Ma LJ, van der Does HC, Borkovich KA, Coleman JJ, Daboussi MJ, Di Pietro A, Dufresne M, Freitag M, Grabherr M, Henrissat B, Houterman PM, Kang S, Shim WB, Woloshuk C, Xie X, Xu JR, Antoniw J, Baker SE, Bluhm BH, Breakspear A, Brown DW, Butchko RA, Chapman S, Coulson R, Coutinho PM, Danchin EG, Diener A, Gale LR, Gardiner DM, Goff S, Hammond-Kosack KE, Hilburn K, Hua-Van A, Jonkers W, Kazan K, Kodira CD, Koehrsen M, Kumar L, Lee YH, Li L, Manners JM, Miranda-Saavedra D, Mukherjee M, Park G, Park J, Park SY, Proctor RH, Regev A, Ruiz-Roldan MC, Sain D, Sakthikumar S, Sykes S, Schwartz DC, Turgeon BG, Wapinski I, Yoder O, Young S, Zeng Q, Zhou S, Galagan J, Cuomo CA, Kistler

HC, Rep M (2010) Comparative genomics reveals mobile pathogenicity chromosomes in *Fusarium*. Nature 464:367–373. https://doi.org/10.1038/nature08850

Ma LJ, Geiser DM, Proctor RH, Rooney AP, O'Donnell K, Trail F, Gardiner DM, Manners JM, Kazan K (2013) Fusarium pathogenomics. Annu Rev Microbiol 67:399–416. https://doi.org/10.1146/annurev-micro-092412-155650

Mai WF, Abawi GS (1987) Interactions among rootknot nematodes and Fusarium wilt fungi on host plants. Annu Rev Phytopathol 25:317–338. https://doi.org/10.1146/annurev.py.25.090187.001533

Martínez-Rocha AL, Roncero MI, López-Ramirez A, Mariné M, Guarro J, Martínez-Cadena G, Di Pietro A (2008) Rho1 has distinct functions in morphogenesis, cell wall biosynthesis and virulence of *Fusarium oxysporum*. Cell Microbiol 10:1339–1351. https://doi.org/10.1111/j.1462-5822.2008.01130.x

Martin-Urdiroz M, Roncero MIG, Antonio Gonzalez-Reyes J, Ruiz-Roldan C (2008) ChsVb, a class VII chitin synthase involved in septation, is critical for pathogenicity in *Fusarium oxysporum*. Eukaryot Cell 7:112–121. https://doi.org/10.1128/EC.00347-07

Mbofung GCY, Pryor BM (2010) A PCR-based assay for detection of *Fusarium oxysporum* f. sp. *lactucae* in lettuce seed. Plant Dis 94:860–866. https://doi.org/10.1094/PDIS-94-7-0860

Mbofung GY, Hong SG, Pryor BM (2007) Phylogeny of *Fusarium oxysporum* f. sp. *lactucae* inferred from mitochondrial small subunit, elongation factor 1-alpha, and nuclear ribosomal intergenic spacer sequence data. Phytopathology 97:87–98. https://doi.org/10.1094/PHYTO-97-0087

Michielse CB, Rep M (2009) Pathogen profile update: *Fusarium oxysporum*. Mol Plant Pathol 10:311–324. https://doi.org/10.1111/j.1364-3703.2009.00538.x

Minerdi D, Moretti M, Gilardi G, Barberio C, Gullino ML, Garibaldi A (2008) Bacterial ectosymbionts and virulence silencing in a *Fusarium oxysporum* strain. Environ Microbiol 10:1725–1741. https://doi.org/10.1111/j.1462-2920.2008.01594.x

Namiki F, Matsunaga M, Okuda M, Inoue I, Nishi K et al (2001) Mutation of an arginine biosynthesis gene causes reduced pathogenicity in *Fusarium oxysporum* f. sp *melonis*. Mol Plant-Microbe Interact 14:580–584. https://doi.org/10.1094/MPMI.2001.14.4.580

Neer EJ (1995) Heterotrimeric g-proteins - organizers of transmembrane signals. Cell 80:249–257. https://doi.org/10.1016/0092-8674(95)90407-7

Nel B, Steinberg C, Labuschagne N, Viljoen A (2006) The potential of non-pathogenic *Fusarium oxysporum* and other biological control organisms for suppressing Fusarium wilt of banana. Plant Pathol 55:217–223. https://doi.org/10.1111/j.1365-3059.2006.01344.x

O'Donnell K, Kistler HC, Cigelnik E, Ploetz RC (1998) Multiple evolutionary origins of the fungus causing Panama disease of banana: concordant evidence from nuclear and mitochondrial gene genealogies. Proc Natl Acad Sci U S A 95:2044–2049. https://doi.org/10.1073/pnas.95.5.2044

O'Donnell K, Gueidan C, Sink S, Johnston PR, Crous PW, Glenn A, Riley R, Zitomer NC, Colyer P, Waalwijk C, Tv L, Moretti A, Kang S, Kim HS, Geiser DM, Juba JH, Baayen RP, Cromey MG, Bithell S, Sutton DA, Skovgaard K, Ploetz R, Corby Kistler H, Elliott M, Davis M, Sarver BA (2009) A two-locus DNA sequence database for typing plant and human pathogens within the *Fusarium oxysporum* species complex. Fungal Genet Biol 46:936–948. https://doi.org/10.1016/j.fgb.2009.08.006

O'Donnell K, Sutton DA, Rinaldi MG, Sarver BA, Balajee SA, Schroers HJ, Summerbell RC, Robert VA, Crous PW, Zhang N, Aoki T, Jung K, Park J, Lee YH, Kang S, Park B, Geiser DM (2010) An Internet-accessible DNA sequence database for identifying fusaria from human and animal infections. J Clin Microbiol 48:3708–3718. https://doi.org/10.1128/JCM.00989-10

O'Donnell K, Rooney AP, Proctor RH, Brown DW, McCormick SP, Ward TJ, Frandsen RJ, Lysøe E, Rehner SA, Aoki T, Robert VA, Crous PW, Groenewald JZ, Kang S, Geiser DM (2013) Phylogenetic analyses of *RPB1* and *RPB2* support a middle cretaceous origin for a clade comprising all agriculturally and medically important fusaria. Fungal Genet Biol 52:20–31. https://doi.org/10.1016/j.fgb.2012.12.004

O'Donnell K, Ward TJ, Robert VARG, Crous PW, Geiser DM, Kang S (2015) DNA sequence-based identification of *Fusarium*: current status and future directions. Phytoparasitica 43:583–595. https://doi.org/10.1007/s12600-015-0484-z

Ospina-Giraldo MD, Mullins E, Kang S (2003) Loss of function of the *Fusarium oxysporum* SNF1 gene reduces virulence on cabbage and Arabidopsis. Curr Genet 44:49–57. https://doi.org/10.1007/s00294-003-0419-y

Ou Y, Penton CR, Geisen S, Shen Z, Sun Y, Lv N, Wang B, Ruan Y, Xiong W, Li R, Shen Q (2019) Deciphering underlying drivers of disease suppressiveness against pathogenic *Fusarium oxysporum*. Front Microbiol 10:2535. https://doi.org/10.3389/fmicb.2019.02535

Panina Y, Fravel DR, Baker CJ, Scherbakova LA (2007) Biocontrol and plant pathogenic *Fusarium oxysporum*-induced changes in phenolic compounds in tomato leaves and roots. J Phytopathol 55:475–481. https://doi.org/10.1111/j.1439-0434.2007.01260.x

Pasquali M, Dematheis F, Gullino ML, Garibaldi A (2007) Identification of race 1 of Fusarium oxysporum f. sp. lactucae on Lettuce by inter-retrotransposon sequence-characterized amplified region technique. Phytopathology 97:987–996. https://doi.org/10.1094/PHYTO-97-8-0987

Park B, Park J, Cheong KC, Choi J, Jung K, Kim D et al (2010) Cyber infrastructure for Fusarium: three integrated platforms supporting strain identification, phylogenetics, comparative genomics, and knowledge sharing. Nucleic Acids Res 39:D640–D646. https://doi.org/10.1093/nar/gkq1166

Partida-Martinez LP, Hertweck C (2005) Pathogenic fungus harbours endosymbiotic bacteria for toxin production. Nature 437:884–888. https://doi.org/10.1038/nature03997

Pérez-Nadales E, Di Pietro A (2011) The membrane mucin Msb2 regulates invasive growth and plant infection in *Fusarium oxysporum*. Plant Cell 23:1171–1185. https://doi.org/10.1105/tpc.110.075093

Poli A, Gilardi G, Spadaro D, Gullino ML, Garibaldi A (2012) Molecular characterization of *Fusarium oxysporum* f. sp. *cichorii* pathogenic on chicory (*Cichorium intybus*). Phytoparasitica 40:383–391. https://doi.org/10.1007/s12600-012-0239-z

Puhalla JE (1985) Classification of strains of *Fusarium oxysporum* on the basis of vegetative compatibility. Can J Bot 63:179–183. https://doi.org/10.1139/b85-020

Recorbet G, Steinberg C, Olivain C, Edel V, Trouvelot S, Dumas-Gaudot E, Gianinazzi S, Alabouvette C (2003) Wanted: pathogenesis-related marker molecules for Fusarium oxysporum. New Phytol 159:73–92. https://doi.org/10.1046/j.1469-8137.2003.00795.x

Rep M (2005) Small proteins of plant-pathogenic fungi secreted during host colonization. FEMS Microbiol Lett 253:19–27. https://doi.org/10.1016/j.femsle.2005.09.014

Rep M, Dekker HL, Vossen JH, de Boer AD, Houterman PM, Speijer D, Back JW, de Koster CG, Cornelissen BJC (2002) Mass spectrometric identification of isoforms of PR proteins in xylem sap of fungus-infected tomato. Plant Physiol 130:904–917. https://doi.org/10.1104/pp.007427

Rep M, van der Does HC, Meijer M, van Wijk R, Houterman PM et al (2004) A small, cysteine-rich protein secreted by *Fusarium oxysporum* during colonization of xylem vessels is required for I-3-mediated resistance in tomato. Mol Microbiol 53:1373–1383. https://doi.org/10.1111/j.1365-2958.2004.04177.x

Rep M, Meijer M, Houterman PM, van der Does HC, Cornelissen BJC (2005) *Fusarium oxysporum* evades I-3-mediated resistance without altering the matching avirulence gene. Mol Plant-Microbe Interact 18:15–23. https://doi.org/10.1094/MPMI-18-0015

Rispail N, Soanes DM, Ant C, Czajkowski R, Grünler A, Huguet R, Perez-Nadales E, Poli A, Sartorel E, Valiante V, Yang M, Beffa R, Brakhage AA, Gow NA, Kahmann R, Lebrun MH, Lenasi H, Perez-Martin J, Talbot NJ, Wendland J, Di Pietro A (2009) Comparative genomics of MAP kinase and calcium-calcineurin signalling components in plant and human pathogenic fungi. Fungal Genet Biol 46:287–298. https://doi.org/10.1016/j.fgb.2009.01.002

Ruiz-Roldán MC, Di Pietro A (2012) Host/pathogen interactions. In: Gullino ML, Katan J, Garibaldi A (eds) Fusarium wilts of greenhouse vegetable and ornamental crops. American Phytopathological Society, St. Paul, pp 21–38

Schlatter D, Kinkel L, Thomashow L, Weller D, Paulitz T (2017) Disease suppressive soils: new insights from the soil microbiome. Phytopathology 107:1284–1297. https://doi.org/10.1094/PHYTO-03-17-0111-RVW

Shishido M, Miwa C, Usami T, Amemiya Y, Jahnson KB (2005) Biological control efficiency of Fusarium wilt of tomato by non-pathogenic *Fusarium oxysporum* Fo-B2 in different environments. Phytopathology 95:1072–1080. https://doi.org/10.1094/PHYTO-95-1072

Spadaro D, Gullino ML (2005) Improving the efficacy of biocontrol agents against soilborne pathogens. Crop Prot 24:601–613. https://doi.org/10.1016/j.cropro.2004.11.003

Spadaro D, Agustí N, Franco Ortega S, Ma HR (2020) Diagnostics and identification of diseases, insects and mites. In: Gullino ML, Albajes R, Nicot P (eds) Integrated pest and disease management in greenhouse crops. Springer, Dordrecht, pp 231–258

Srinivasan K, Gilardi G, Spadaro D, Garibaldi A, Gullino ML (2010) Molecular characterization through IGS sequencing of *formae speciales* of *Fusarium oxysporum* pathogenic on lamb's lettuce. Phytopathol Mediterr 49:309–320. https://doi.org/10.14601/Phytopathol_Mediterr-5460

Srinivasan K, Spadaro D, Poli A, Gilardi G, Gullino ML, Garibaldi A (2012) Genetic diversity and pathogenicity of *Fusarium oxysporum* isolated from wilted rocket plants in Italy. Phytoparasitica 40:157–170. https://doi.org/10.1007/s12600-011-0207-z

Starr JL, Jeger MJ, Martyn RD, Schilling K (1989) Effects of *Meloidogyne incognita* and *Fusarium oxysporum* f. sp. *vasinfectum* on plant mortality and yield of cotton. Phytopathology 79:640–646. https://doi.org/10.1094/Phyto-79-640

Takken F, Rep M (2010) The arms race between tomato and *Fusarium oxysporum*. Mol Plant Pathol 11:309–314. https://doi.org/10.1111/j.1364-3703.2009.00605.x

Tarkka MT, Sarniguet A, Frey-Klett P (2009) Inter-kingdom encounters: recent advances in molecular bacterium–fungus interactions. Curr Genet 55:233–243. https://doi.org/10.1007/s00294-009-0241-2

Topalović O, Hussain M, Heuer H (2020) Plants and associated soil microbiota cooperatively suppress plant-parasitic nematodes. Front Microbiol 11:313. https://doi.org/10.3389/fmicb.2020.00313

van Dam P, de Sain M, Ter Horst A, van der Gragt M, Rep M (2017) Use of comparative genomics-based markers for discrimination of host specificity in *Fusarium oxysporum*. Appl Environ Microbiol 84:e01868–e01817. https://doi.org/10.1128/AEM.01868-17

van der Does HC, Rep M (2007) Virulence genes and the evolution of host specificity in plant-pathogenic fungi. Mol Plant-Microbe Interact 20:1175–1182. https://doi.org/10.1094/MPMI-20-10-1175

Van Der Does HC, Lievens B, Claes L, Houterman PM, Cornelissen BJC, Rep M (2008) The presence of a virulence locus discriminates *Fusarium oxysporum* isolates causing tomato wilt from other isolates. Environ Microbiol 10:1475–1485. https://doi.org/10.1111/j.1462-2920.2007.01561.x

Vidhyasekaran P (1997) Fungal pathogenesis in plants and crops: molecular biology and host defence mechanisms. Marcel Dekker Inc., New York, p 536

Vitullo D, Di Pietro A, Romano A, Lanzotti V, Lima G (2012) Role of new bacterial surfactins in the antifungal interaction between Bacillus amyloliquefaciens and *Fusarium oxysporum*. Plant Pathol 61:689–699. https://doi.org/10.1111/j.1365-3059.2011.02561.x

Walters DR (2012) Induced resistance as a management tool. In: Gullino ML, Katan J, Garibaldi A (eds) Fusarium wilts of greenhouse vegetable and ornamental crops. American Phytopathological Society, St Paul, pp 101–107

Walters D, Daniell T (2007) Microbial induction of resistance to pathogens. In: Walters D, Newton A, Lyon G (eds) Induced resistance for plant defence. Blackwell Publishing, Oxford, pp 143–156

Whipps J (2001) Microbial interactions and biocontrol in the rhizosphere. J Exp Bot 52:487–511. https://doi.org/10.1093/jxb/52.suppl_1.487

Woesten HAB, Schuren FHJ, Wessels JGH (1994) Interfacial self-assembly of a hydrophobin into an amphipathic protein membrane mediates fungal attachment to hydrophobic surfaces. EMBO J 13:5848–5854. https://doi.org/10.1002/j.1460-2075.1994.tb06929.x

Yuen GY, Schroth MN, McCain AH (1985) Reduction of Fusarium wilt of carnation with suppressive soils and antagonistic bacteria. Plant Dis 69:1071–1075. https://doi.org/10.1094/PD-69-1071

Chapter 8
Biocontrol from the Rhizosphere: Probiotic Pseudomonads

Anne J. Anderson

Abstract Plants evolved in association with microbes and some are selectively nurtured with the payback that they are probiotics in protecting the plant from potential microbial pathogens. Root colonization by the biocontrol pseudomonads permits the establishment of overlapping spheres of protection, from induced systemic resistance in the plant tissues to a physical layer of biofilm cells over the root surface. Production of water-soluble antimicrobial agents from the biofilm offers protection within the rhizosphere and diffusion of antimicrobial volatiles extends the zones of impact. The processes of biofilm formation on root surfaces can be initiated by the chemotaxis of soil-borne pseudomonads to the root surface by recognition of key metabolites. Adhesion of the pseudomonad cells follows through specialized surface surface adaptions. Specific metabolites in the rhizosphere solutions promote the transition from a mobile to a sessile lifestyle. Other metabolites signal release of motile cells from a mature biofilm to initiate new colonization sites, thus sustaining the beneficial root-microbe interactions. Plant and pseudomonad products in the rhizosphere manipulate the defense mechanisms boosting plant resilience to pathogen challenge. Examination of these processes reveals that many rhizosphere metabolites are more than just an intermediate in a pathway, rather they have key roles in conditioning plant health.

Keywords Biofilm · Metabolites · Plant resistance · Root exudates · Stress

8.1 Introduction

Biocontrol pseudomonads pack many punches from the rhizosphere to benefit plant performance. Diverse plants grown globally under a range of growth conditions are colonized by biocontrol-active pseudomonads. The title of this chapter "Biocontrol from the Rhizo*sphere*" is appropriate because the pseudomonads master processes that fan out multidirectionally from the rhizosphere for whole plant protection.

A. J. Anderson (✉)
Department of Biological Engineering, Utah State University, Logan, UT, USA

B. A. Horwitz, P. K. Mukherjee (eds.), *Microbial Cross-talk in the Rhizosphere*, Rhizosphere Biology, https://doi.org/10.1007/978-981-16-9507-0_8

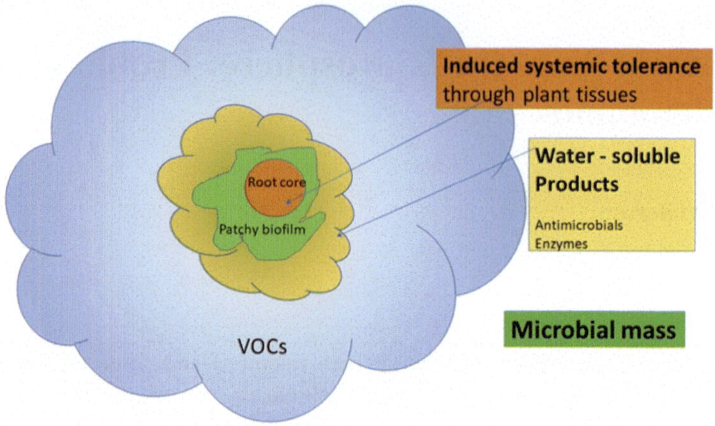

Fig. 8.1 Colonization of the root with certain pseudomonads establishes multidirectional and overlapping zones of defense for the plant. Induced systemic tolerance within the plant tissues is induced by root colonization by the protectant pseudomonads. The formation of biofilms of the pseudomonads at the root surface offers physical protection, root hydration, and concentration of microbial product. The composition of the metabolites in the rhizosphere solutions is altered by colonization and would alter nutrient availability for soil pathogens. Water-soluble microbial products such as antimicrobials and enzymes could move further in the water channels of the rhizosphere to contact soil organisms. Volatile emissions from the colonizing microbes would spread in through soil pores and the air to influence aerial plant structures and adjacent roots as well as other organisms

Figure 8.1 is a cartoon depicting the different but overlapping protective zones around the plant root as influenced by microbially related rhizosphere events. Central is the core of protected plant tissues where systemic plant defense mechanisms, induced by the root-colonizing biological control pseudomonads, are manifest as improved tolerance to pathogen and abiotic challenges (Berendsen et al. 2012; Mauch-Mani et al. 2017; Timmusk et al. 2017; Ciancio et al. 2019; Saijo and Loo 2020). Outside of the root surface there are patches of biofilms that are hydrated and act as extra films of physical protection over the rhizoplane. The biofilms also act as a repository of secreted microbial products. The zone outside of the biofilm would accumulate water-soluble products, including the antibiotics or enzymes, secreted by the microbial biofilm cells, which have the potential to directly inhibit the growth of other soil microbes and fauna. These microbial antagonistic compounds would be dispersed further in the water channels of the rhizosphere matrix. The array of antimicrobials produced by biocontrol pseudomonads is discussed in many publications with attention to structure, regulation of synthesis and pathogen sensitivities (e.g., Yang and Cao 2012; Raaijmakers and Mazzola 2012; Yu et al. 2017; Biessy and Filion 2018; Biessy et al. 2019; Shah et al. 2020). Likewise, the production and multiple roles of microbial volatiles, including direct antagonism of pathogens, plant growth promotion, and activation of plant defense, is extensively reviewed (Farag et al. 2013; Chung et al. 2016; Martins et al. 2019; Kim and Anderson 2020). These volatiles constitute a third sphere of protection operative at greater distances from the rhizoplane; the volatiles would travel through air spaces to influence both roots and aerial plant structures, even to adjacent plants, as well as rhizosphere organisms. Consequently, the sphere of potential protection for the plant is extensive, reaching well outside of the root. The interplay of the array of different mechanisms promotes functional resilience in the environment.

Publications in the late 1960 on root-associated microbes with biocontrol potential sparked interest in the beneficial plant-microbe interactions. A 1965 publication by Garrett uses the term "Biological Control" in a paper included in the book "Ecology of Soil-Borne Pathogens" (Garrett 1965). A following review, by Baker in 1968, "Mechanisms for Biological Control of Soil-borne Pathogens," presents a spirited discussion of findings: antagonism of the pathogen, competition for nutrients (C, N and vitamins), lysis of pathogenic cells, and survival of the biocontrol agent (Baker 1968), all topics that are still pertinent. Biological control research, having been born by the observations of plant pathologists, now is wholistic, to include studies by the ecologist, microbiologist as well as soil and plant scientists, and those with skill in the still developing "omics" arts and large data management (e.g., Carvalhais et al. 2012; Girard et al. 2016; Levy et al. 2017; Giovannini et al. 2020). The desire to understand the bases for successful biocontrol activity to supply products for commercial agriculture in the field and greenhouses drives the research. The realization that sustainable and regenerative agriculture is required to provide quality food for the world's future population is boosting interest in the role of beneficial microbes in commercial agriculture (Ciancio et al. 2019). A recent e-book, "Harnessing Useful Rhizosphere Microorganisms for Pathogen and Pest Biocontrol" (2019), where papers are arranged by target pathogen or pest, includes several chapters on the varied roles of pseudomonads as biocontrol agents.

The different research thrusts in the studies of biocontrol by pseudomonads have evolved from the identification and characterization of potential antimicrobials, i.e., first the water-soluble antimicrobial compounds (e.g., Imperiali et al. 2017; Thomashow et al. 2019) to more recent analysis of the plethora of volatile metabolites (Weisskopf et al. 2016; Fincheira and Quiroz 2018). Other research focuses on activation of plant defense pathways that accompanies rhizosphere colonization (van der Ent et al. 2018) and the interactions of the beneficial strains in the total rhizosphere microbiome (Sasse et al. 2018; Dorosky et al. 2018). The focus for this chapter is the current understanding of events important in the colonization of the rhizosphere by probiotic pseudomonads and how it is coupled with plant defenses. Figure 8.2 summarizes the overall discussion in this paper.

Initiating some answers to the question "What is in the rhizosphere and why?" are discussions of the relationships between the plant metabolites in root exudates and regulation of biofilm building by the root-colonizing microbes. Included in this section is a discussion on the interactions between the pseudomonads and rhizosphere fungi, indicating that processes important in root colonization are also significant to plant health. To conclude, the integration of these colonization processes with the outcome of enhanced plant defenses is discussed. Findings reveal that certain simple metabolites are more than just an intermediate in a biochemical pathway, rather they also may play key roles in the biocontrol process to promote

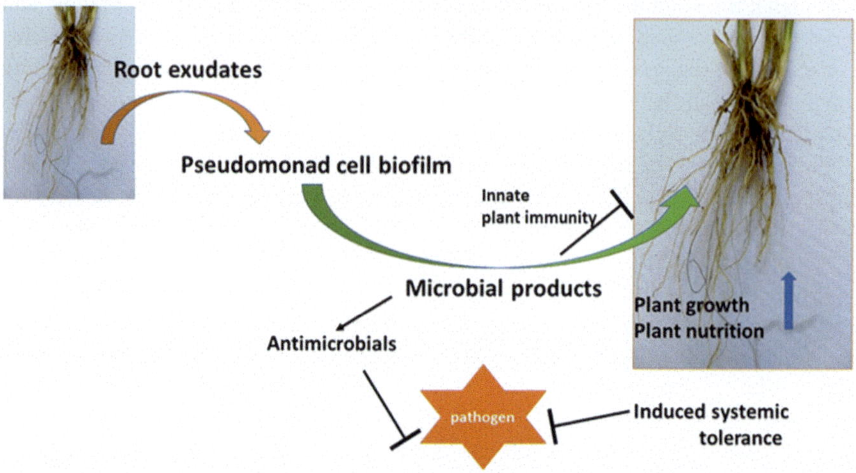

Fig. 8.2 Plant root metabolites are repurposed by root-colonizing pseudomonads to processes that protect the plant. The root metabolites nurture the formation of protective pseudomonad biofilms on root surfaces. Products from these pseudomonad cells initially repress innate plant immunity but prime the plant for induced systemic tolerance against biotic and abiotic challenges. Microbial products also have the potential for direct pathogen antagonism. Other microbial traits enhance plant growth and nutrition in part through improving access to the essential elements in soil minerals and by altering plant growth processes

plant health. In this respect, this review highlights the potential importance of two rhizosphere metabolites, arginine and gluconate, as examples of metabolites with influence over plant health.

8.2 What Is in the Rhizosphere and Why?

An ability to colonize the rhizosphere is the basic trait for many biocontrol-active pseudomonads. A paper addressing root colonization of rice grown in field soils illustrates that the plant initially "selects" microbes that are present in the soil microbiome (Edwards et al. 2015). Activation of microbial growth in the rhizosphere, due to plant root exudates, is a gated process leading to colonization of the rhizoplane that in turn allows only selected microbes to colonize the plant root internally, becoming the endophytes (Edwards et al. 2015). Findings from field-grown plants indicate that each plant root likely has a "core" microbiome selected from the soil (Lundberg et al. 2012). Wagner et al. (2016) show that for maize, plant genetics, age, and weather conditions all influence the root microbiome dependent on the local soil. Population studies led Dangl's group to the concept of a common core of positively interacting microbes associated with the root. From this information a synthetic mix of representative microbes, termed a SynCom, can be developed to investigate the effects of the rhizosphere microbiome (Herrera Paredes et al. 2018). Indeed, a mixed species SynCom for *Arabidopsis* plays a key role in the induction of plant defense mechanisms under low soil phosphate (Castrillo et al. 2017). Pseudomonads in the Arabidopsis SynCom differentially affect root elongation, suberization, and ion import into the root (Salas-González et al. 2021). Another rhizosphere SynCom for sugar cane comprises the most robust microbial colonizers; all these isolates feature strong abilities for C and N catabolism, whereas production of secondary components is a less conserved trait (Armanhi et al. 2018; de Souza et al. 2019).

8.3 The Importance of Biofilms in Root Colonization for Beneficial Plant Effects

This widespread occurrence of pseudomonads as rhizosphere colonists is in part due to their robust defenses against varied environmental challenges coupled with versatility in utilization of a range of C and N sources (Spiers et al. 2000). The formation of a mass of sessile cells within biofilms on plants, and other surfaces, is one of their survival tactics. Biofilm formation has long been associated with protection against antibiotics; cells within the biofilm are resilient compared with the sensitivity of planktonic cells (Costerton et al. 1999). Pseudomonad survival in the competitive environment of the rhizosphere, where many other microbes may secrete antibiotics to secure a niche, clearly points to biofilm tolerance as one mechanism to maintain their populations. Valentini and Filloux (2016) discuss the sequences in biofilm formation for *Pseudomonas aeruginosa*, isolates of which have biocontrol potential on plants (Yasmin et al. 2017; Ma et al. 2018), although most of the findings originate from isolates that impact human health. Figure 8.3 illustrates some of the key events for the colonization of the rhizoplane surface for the biocontrol pseudomonads. How the pseudomonad cells function in the different stages of biofilm formation is discussed in the following section.

The pseudomonads need to attain a threshold cell mass and density in the rhizosphere for biocontrol. Their nutrition comes from catabolism of the metabolites in the root exudates, supplemented with essential elements gleaned by the microbes from the soil minerals (Uroz et al. 2009). Pseudomonads excel in the plasticity of their catabolic potential and their ability to liberate metal ions and phosphates from soil minerals (Spiers et al. 2000; Dakora and Phillips 2002). It is this strong saprophytic growth potential that is a significant factor in the global colonization by the pseudomonads of the diverse plant genera. Such diversity is displayed in the natural habitats of isolates of one species, *Pseudomonas chlororaphis,* where isolates with biocontrol potential are found in many cereal and vegetable crop rhizospheres (Anderson and Kim 2018).

Root exudates contain a common core of metabolites, i.e., simple sugars, mainly glucose, sucrose, and fructose, an array of amino acids, and a mix of carboxylic

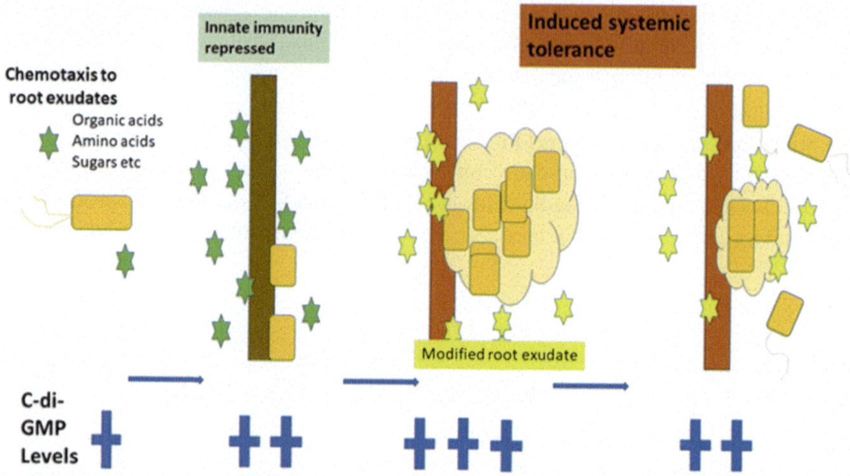

Fig. 8.3 Colonization of the plant root by biocontrol pseudomonads proceeds in stages. First, planktonic microbial cells in the rhizosphere solution move towards the root by chemotaxis to metabolites in the root exudates (green stars). Attachment of cells to the root surface using an array of bacterial surface features occurs with increases in the bacterial intracellular regulator, c-di-GMP (level is illustrated by the number of blue plus signs) in response to different external signals. Microbial cell replication is nurtured by regulated catabolism of the plant root metabolites. The concentration and composition of the root exudates are changed (shown by change from green to yellow stars) due to the metabolism of the biofilm cells. Innate immunity is suppressed but systemic tolerance to abiotic and biotic stress is induced in the plant. Changes to the c-di-GMP levels in the bacterial cells promote biofilm dissolution with release of planktonic cells to allow colonization at new plant root locations. Several metabolites in the rhizosphere solutions regulate c-di-GMP levels within the bacterial cells

organic acids including citrate, malate, fumarate, and succinate (Dennis et al. 2010; Baetz and Martinoia 2014; Tian et al. 2020). Each of these classes of compounds are readily consumed by pseudomonads supporting cell growth and replication. Genes for carbon utilization and amino acid transport are required for rhizosphere competency by *P. protogens* when colonizing *Arabidopsis* seedlings (Li et al. 2020). Transcriptomic assays for several *P. fluorescens* isolates reveal genes for the catabolism of simple carbohydrates, amino acids, organic acids, and phenolics are activated upon growth on root exudates from the model grass, *Brachypodium* (Mavrodi et al. 2021). Studies with wheat colonized by *P. chlororaphis* O6 under gnotobiotic conditions confirm extensive reduction in levels of the amino acids, and organic acids in the rhizosphere solutions of colonized plants compared with noncolonized controls (Hortin et al. unpublished). These studies support the discussion raised by Baker (1968) that one of the processes important in biocontrol is the altered C and N sources in the colonized rhizosphere; reduced nutrient supply and removal of readily catabolized metabolites would be to the detriment of pathogen growth.

Preferred use of components by the pseudomonads may play a role in the variability observed in colonization between different isolates. Catabolite repression, a process that regulates the use of one substrate over another, exists in the pseudomonads; the ability for rapid change between a diversity of sources for catabolism in the pseudomonads is possibly a strategy that enables them to outcompete other microbes, or promote compatible colonization of mixed microbial isolates (Liu et al. 2017; Park et al. 2020). Zboralski et al. (2020) find that the stronger colonization of roots of both *Arabidopsis thaliana* and potato by *P. chlororaphis* isolates versus isolates of *P. fluorescens* correlates with their better utilization of amino acids and amines in the root exudates. High variability in growth on simple metabolites is displayed by eight biocontrol pseudomonads (Mavrodi et al. 2021). The biocontrol isolate *P. chlororaphis* O6 also utilizes organic acids and the amino acids in wheat root exudates (Wright et al. 2016; McManus et al. 2018; Hortin unpublished). A mutant of isolate O6 deficient in *dctA*, which prevents uptake of the C4-carboxylic acids succinate and fumarate, reduces the extent of colonization and induction of systemic tolerance compared with the wild-type strain (Nam et al. 2003, 2006). These findings support the view that there is selective use of multiple metabolites in the root zone by rhizosphere pseudomonads.

Although the metabolites released in the root exudates represent a substantial portion of the C fixed by photosynthesis, this energy is not wasted, rather the process plays an active role in recruiting beneficial microbes (Dennis et al. 2010; Stringlis et al. 2018a, b, 2019; Tovi et al. 2019; Vives-Peris et al. 2018, 2020). Change in the composition and strength of root exudates, factors that alter with age and stress, may govern which microbes are selected for plant colonization. In citrus, exudates from plants with salt or water stress support greater growth of a *P. putida* strain than exudates from nonstressed plants (Vives-Peris et al. 2018). Higher levels of proline and a plant growth regulator, salicylate, in the citrus exudates correlate with these effects. Pathogen challenge of aerial tissues also leads to altered root exudate composition. Yuan et al. (2018) observed an altered composition of root exudates after growing generations of *Arabidopsis* sequentially in soil with challenge by the foliar pathogen, *Pseudomonas syringae*. The root exudates show increases in amino acids, long chain fatty acids, and nucleotides but decreases in sugars, alcohols, and shorter chain organic acids with pathogen challenge. Coincidently, the rhizosphere microbiome changes and the plants became more tolerant to *P. syringae*, correlating with increased levels of jasmonic acid associated with induced tolerance (Yuan et al. 2018). Similarly, infection of tomato by the pathogenic bacterium causing wilt, a *Ralstonia* sp., influences its root microbiome; an increase in caffeic acid in the root exudates is viewed as being part of a protective response (Gu et al. 2016). Secretion of an iron-chelator, the coumarin scopoletin, from *Arabidopsis thaliana* roots when colonized by the biocontrol bacterium *Pseudomonas simiae* WCS417, affects the root microbiome; microbes, such as *P. simiae* WCS417, that are tolerant of scopoletin dominate over those that are sensitive, including the fungal pathogens *Fusarium* and *Verticillium* sp. (Stringlis et al. 2018a, b; Stassen et al. 2021). Here withholding of Fe from these pathogenic fungi, by its chelation with scopoletin, becomes part of the protective response. Competition for Fe in the rhizosphere is one of the mechanisms first recognized to be responsible for pathogen suppression, as is observed in certain "suppressive soils" (Kloepper et al. 1980; Elad and Baker 1985;

Gu et al. 2020). Thus, these are examples where plant health is related to root metabolism and the recruited rhizosphere microbiome.

8.4 Biofilm Building: Attraction

Colonization of the root may begin by transport of inoculum along the developing root from the seed, or attraction of the beneficial microbe to the root surface through the active response of chemotaxis. Polar-flagellated pseudomonad cells are attracted by gradients of nutrients radiating out from the root surface because of the root exudates. Chemotaxis of the biocontrol strain *P. fluorescens* PflO-1 towards amino acids (Oku et al. 2012) and dicarboxylic acids (Oku et al. 2014) is observed. Likewise, *P. chlororaphis* PCL 1606 is chemotactic towards glucose, dicarboxylic acids, and certain amino acids, as well as the mixture of metabolites in diluted avocado root exudates. Chemotaxis is, thus, proposed to be part of the package for biocontrol success of this pseudomonad in the control of fungal rot of the avocado root (Polonio et al. 2017).

Plant secondary components also may be significant in recruiting a pseudomonad by chemotaxis. One of the main secondary compounds, nicotine, released in tobacco root exudates is a positive attractant for the biocontrol-active *P. aeruginosa* isolate NXHG29 (Ma et al. 2018). For maize root exudates, the phenolic 2,4-dihydroxy-7-methoxy-2*H*-1,4-benzoxazin-3(4*H*)-one (DIMBOA), a known allelochemical, promotes chemotaxis of a *P. putida* isolate (Neal et al. 2012). Another "feed forward" beneficial effect is chemotaxis of *P. putida* to the ethylene precursor, 1-aminocyclopropane-1-caboxylate (ACC), released in root exudates (Li et al. 2019). This finding is interesting because the efficacy of many biocontrol bacteria is correlated with their ACC deaminase activity (Glick 2004), with the concept that reducing ethylene production in the root tissues promotes root growth through limiting ethylene antagonism of indole-acetic acid (IAA) growth promotion. Plant growth stimulation and altered root morphology also is boosted by production of IAA by many of the pseudomonad root colonizers (Dimkpa et al. 2012; Duca et al. 2014); the amino acid tryptophan is the initial substrate for the microbial pathways. Thus, the roots release metabolites in their exudates that not only attract the microbe but also serve as food sources for protectant organisms. In the discussions below two rhizosphere metabolites, arginine and gluconate, are highlighted for their roles in biofilm building and plant health.

8.5 Attachment: The Role of Different Pseudomonas Cell Surface Features

Once at the root surface, attachment of the pseudomonad cells is a multifaceted process, progressing from reversible to irreversible stages to allow biofilm maturation. The flagella, essential for chemotaxis, may play a role here because initial attachments are observed to involve the polar ends of the pseudomonad where the flagella are located. The biocontrol isolate, *P. fluorescens* C7R12, has multiple polar flagella as well as aggregative fimbriae originating from the sides of the bacterial cell that function in attachment (Bergeau et al. 2019). There are preferences for colonization sites and population densities on plant roots; for three pseudomonads isolated from wheat roots, one preferred tip colonization, another locations along the root behind the tip, whereas the third had no spatial preference (Tovi et al. 2019). Preferential colonization of wheat over cucumber also is reported for these isolates (Tovi et al. 2019). The authors suggest that the variability between these isolates in motility and potential for attached growth as a static biofilm are some of the processes accounting for the differences in root colonization. However, future work could determine whether there are roles for the composition and concentration of released plant metabolites, and function of extracellular plant factors such as lectins in locational specificity for colonization.

Other pseudomonad cell surface structures implicated in "adhesion and the cohesion" of cells in a biofilm include large proteins with tandem repeats, the Lap/Map adhesins (Fuqua 2010). A given pseudomonad may produce more than one Lap-type adhesin, e.g., LapA and LapF in *P. putida* 2440 (Martínez-Gil et al. 2014). For instance, LapA is proposed to settle the cell into an adhesive state on a substrate (e.g., attachment to sand grains, Hinsa et al. 2003) whereas LapF functions to promote the cell-to-cell adhesion (Martínez-Gil et al. 2014) allowing biofilm maturation into a 3 D-structure. The LapF adhesin contributes to the hydrophobicity of the cell surface (Lahesaare et al. 2016), which could influence biofilm architecture such as the development of water channels. Substrate adhesion requires LapA to span a TolC-like pore in the outer membrane so that the C-terminus is outside the cell whereas the N terminus is within the periplasm (López-Baena et al. 2019). However, the cell is released after proteolytic cleavage by a periplasmic protease of the periplasmic N-terminus of Lap A allowing extracellular passage of the modified adhesin. Activation of the specific protease is dependent on the level of an intracellular regulatory signal, cyclic-di-GMP (c-di-GMC) that is a major factor in biofilm function.

P. fluorescens Pf0-1 has two adhesins, LapA and MapA, encoded at two different gene loci. These adhesins localize to different regions of a biofilm suggesting that they have discrete functions (Collins et al. 2020). Also, of interest is the finding that expression from the *map* gene cluster is induced under defined growth medium conditions, with arginine being an inducer. The gene encoding MapA is within a cluster that includes the genes for synthesis of an inner membrane ABC-transport structure for this adhesin. However, the release of both LapA and MapA from their outer membrane pores is by the same protease that cleaves the LapA protein, so that its extracellular release is also under c-di-GMP regulation.

Amyloid-like proteins constitute another class of microbial cell surface structures in addition to the Lap/Map proteins associated with attachment during biofilm formation. Operons encoding the formation and secretion of the Fap fibers are present in the genomes of *P. aeruginosa*, *P. fluorescens,* and *P. putida*; these amyloid proteins are detected in samples from rhizospheres colonized by *P. putida* (Rouse et al. 2018). The amyloid proteins confer greater resilience to drying and increased hydrophobicity and mechanical robustness to the biofilms (Zeng et al. 2015). Adhesion to seeds appears to involve other proteins (Molina et al. 2006), showing the plasticity yet specificity of mechanisms by which the pseudomonads adhere to plant surfaces (Yousef-Coronado et al. 2008).

Maturation of the biofilm involves entrapment of multiple cells in an extracellular matrix formed from cellular secretions. The gel-like matrix is variable in composition, but core components include microbially secreted polysaccharides, proteins including enzymes and extracellular DNA. A biofilm made by *P. brassicacearum* DF41 is associated with the killing of nematodes and this matrix involves output from a gene cluster predicted to encode biosynthesis of poly-beta-*N*-acetyl glucosamine (Nandi et al. 2016). At least three types of polysaccharides are produced by *P. aeruginosa*: alginate from an *alg* gene cluster, and other acidic polymers made by the *pel* and *psl* clusters (Colvin et al. 2011, 2012). With *P. putida* 2440, alginate and cellulose are formed; cellulose also is produced by some *P. fluorescens* strains (Spiers and Rainey 2005; Ude et al. 2006; Ardré et al. 2010). A potential advantage of using cellulose fibers in attachment is that these polymers would not be recognized as “foreign” by the plant and, thus, trigger plant defense. Other structures, such as flagellin and lipopolysaccharide, lead to activation of plant defenses and consequently these products are termed Microbe-Associated Molecular Patterns (MAMPs) (Newman et al. 2013).

The extracellular gels have features in addition to structural support for the cells in the matrix. Alginate and cellulose protect against reactive oxygen stress (ROS) (Svenningsen et al. 2018), an important trait because ROS is elevated when roots are initially colonized (Katsuwon and Anderson 1992) or the cells are stressed. Alginate has water-holding potential; mutants of *P. fluorescens* Pf0-1 lacking in alginate synthesis are less able to colonize moist soil or survive in soils under drought (Marshall et al. 2019). The genes for alginate synthesis are expressed rapidly under water limitation, and thus the presence of alginate in the biofilm matrix promotes bacterial cell survival during drought conditions (Chang et al. 2007). The negative charge on alginate serves to bind cations, and so offers protection for the embedded cells from the ions of heavy metals, such as Cu (Svenningsen et al. 2018) or Zn (Upadhyay et al. 2017). Cation trapping would also occur with the eDNA. Cations that are essential elements when bound by the biofilm matrix provide reserves for both microbe and plant cells. DNA fragments also act as damage associated molecular patterns (DAMPs) which trigger plant defense responses (Wang et al. 2016; Quintana-Rodriguez et al. 2018).

Maturation of the biofilm involves limited movement of the bacterial cells, such as swarming to enhance surface coverage. Flavonoids in root exudates enhance the flagella movements needed for swarming in beneficial *P. fluorescens* 2P24 (Yu et al. 2020). Rhamnolipids are part of the secreted surface-active agents produced by many pseudomonads that also enhance swarming, and these are under the control of the global regulatory, Gac/Rsm system (Noirot-Gros et al. 2019). Features of the Gac/Rsm system vary between pseudomonad strains (Sobrero and Valverde 2020) and connections with other cell regulators (e.g., (p)ppGpp (Wu et al. 2020) are being revealed to enhance the understanding of the network controlling cell metabolism. Noirot-Gros et al. (2019) use CRISPR-induced inactivation of defined genes in three *P. fluorescens* strains to confirm effects of c-di-GMP and Gac/Rsm regulation of swarming and biofilm production. However, an additional function is revealed in studies where a rhamnolipid decreases virulence of an *Aspergillus* pathogen. This control occurs through inhibition of a beta-glucan synthase activity so that production of fungal cell walls is impaired (Briard et al. 2017). Other directed movement of cells as the biofilm matures may involve twitching motility which employs retractable type IV pilus structures, as demonstrated for *P. aeruginosa* isolates (O'Toole and Kolter 1998). Formation of the type IV pili is under c-di-GMP control in *P. aeruginosa* (Jain et al. 2012). Twitching motility is identified as a consistent trait for biocontrol pseudomonads, the ability to maintain root tip-colonizing ability with root elongation for avocado roots (Pliego et al. 2008).

8.6 Regulation of Biofilm Formation: Tuning into the Environment

Biofilm formation is promoted by high levels of c-di-GMP through forcing a series of metabolic changes in the microbial cell that cause transition from a planktonic lifestyle to a sessile existence. Figure 8.3 illustrates the concept that changes in the intracellular level of this signal compound correlate with different stages of biofilm formation. The discussion above for Lap proteins shows how adhesion is one of the steps regulated by c-di-GMP. López-Baena et al. (2019) summarize the release of Lap proteins, as well as other types of bacterial secretion mechanisms, governed by c-di-GMP. However, as reviewed for *P. aeruginosa* (Römling et al. 2013), this secondary messenger also is involved in regulation of swarming motility through control of flagella-based movement as well as movement with type IV pili. The intracellular levels of c-di-GMP also modify the matrix composition through extracellular polysaccharide production and promote the release of cells from a mature biofilm (Römling et al. 2013; Cai et al. 2020). Additionally, c-di-GMP promotes greater tolerance to oxidative stress, as indicated by enhanced accumulation of protective enzymes, e.g., catalase in *P. putida* KT2440 (Xiao et al. 2019).

Intricate networking connects c-di-GMP levels with other extensive regulatory circuits, e.g., the stress global regulator alternative sigma factor, RpoS, and the

Gac/Rsm cell density sensing system, to ensure tight environmental regulation of biofilm formation. Indeed phosphate (P) supply is one environmental factor: biofilm formation is reduced at low P for isolate *P. fluorescens* Pf0-1. Interaction between the systems for sensing P and c-di-GMP levels results in low P modulating the release of LapA (Monds et al. 2007). Earlier Monds et al. (2001) report that another biocontrol isolate, *P. aureofaciens* PA-147-2, also forms poor biofilms in low P medium. The low P condition promotes the secretion of alkaline phosphatase, an enzyme that could liberate phosphate in surrounding soils from a complexed source such as rock phosphate or phytate. Regulation by phosphate deficiency may be important in soils where bioavailability of P is usually limited through low solubility but is enhanced by microbial activities such as organic acids, siderophores, and phosphatases (Miller et al. 2010).

Other root exudate metabolites will influence biofilm regulation. Yu et al. (2020) reveal a complex role of flavonoids in root exudates on root colonization via biofilms with *P. fluorescens* 2P24. The tested flavonoids promote EPS (cellulose) formation and swarming motility, although they repress the formation of the antimicrobial, diacetyl phloroglucinol (DAPG). The amino acid, arginine, present in root exudates (Thiele et al. 2008; Hortin et al. unpublished) also appears to be a key player as summarized in Fig. 8.4a. Arginine is superior among several amino acids in promoting c-di-GMP accumulation in the microbial cell through activation of two cyclases involved in its synthesis (Bernier et al. 2011). Rinaldo et al. (2018) confirm that for *P. aeruginosa,* arginine is a key factor in biofilm formation because of its effect on c-di-GMP.

Several of the antimicrobial products produced by biocontrol pseudomonads have an influence on biofilm formation. These metabolites include phenazines from *P. chlororaphis* 30-84 (Wang et al. 2016), DAPG for *P. brassicacearum* (Paulin et al. 2017), and 2-hexyl-5-propyl resorcinol with *P. chlororaphis* PCL 1606 (Calderón et al. 2019). Mutants of PCL 1606 lacking in resorcinol are modified in adhesion and show altered matrix structures in their biofilms (Calderón et al. 2019). Similarly, the phenazine, 2-OH phenazine-1-carboxylic acid, produced by *P. chlororaphis* 30-84, increases the eDNA content of the biofilm and modifies its biofilm structure (Wang et al. 2016). For isolate 30-84, changes in gene transcription induced by phenazine are proposed to activate cell lysis, thus enhancing enhanced eDNA levels. Thus, these products, familiar because of their antimicrobial activities, also play other subtle roles in survival of the pseudomonads.

Multiple roles of antibiotics, used in the control of mammalian bacterial diseases, are discussed in a review by Pishchany and Kolter (2020) and these activities include modifications to biofilm structure. They note the redox-active phenazines could act as backup electron acceptors, useful for cell survival in micro-anerobic zones in the biofilm. Indeed, in work with *P. aeruginosa,* Okegbe et al. (2017) find that phenazines trigger lower c-di-GMP levels which then limit matrix production in a manner connected with redox balancing. Later work (Lin et al. 2018) shows that the phenazines impact redox balancing by promoting nitrate as an alternative electron acceptor through increasing expression of a gene encoding a novel nitrate reductase. Nitrate reductase catalyzes the start of the denitrification process and many

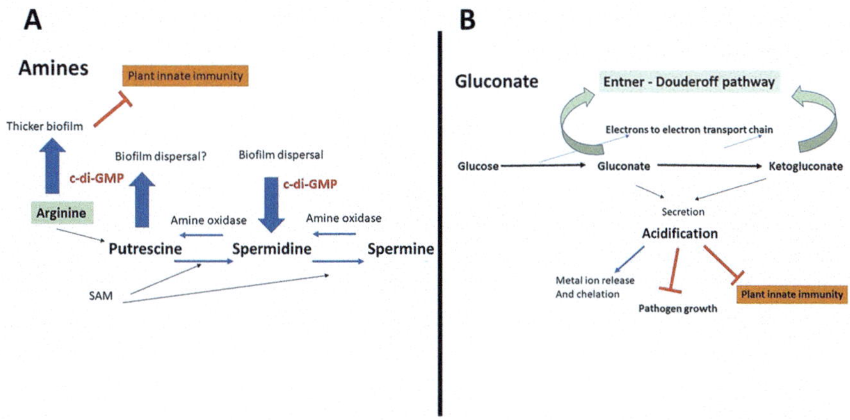

Fig. 8.4 Illustrations of the roles in rhizosphere health played by two metabolites in the rhizosphere solutions: (**a**) arginine and (**b**) gluconate. (**a**) The amino acid arginine, present in root exudates, promotes biofilm formation by increasing c-di-GMP in the pseudomonad cells. It also is a starting point for formation of polyamines, putrescine, spermidine, and spermine that are detected in rhizosphere solutions. Synthesis involves extension of the polyamine chain using C and N from *S*-adenosyl methionine (SAM). Back conversion of the polyamines involves amine oxidases that generate hydrogen peroxide in the transformation. Arginine and putrescine increase c-di-GMP in pseudomonad cells. Spermidine and spermine may promote biofilm dispersal through lowering c-di-GMP levels. (**b**) Gluconate, generated from glucose oxidation, when secreted into the rhizosphere solutions, results in localized acidification. For pseudomonad cells the oxidation of glucose to gluconate followed by its transformation to 2-keto gluconate may directly shunt electrons into the electron transport chain allowing ATP generation without loss of C. However, gluconate and 2-keto gluconate may be catabolized by the pseudomonads using the Entner-Doudoroff (ED) pathway to provide C for growth of the microbial cells colonizing the root. Localized acidification in the rhizosphere could result in inhibition of innate immunity in the plant and/or direct reduction of pathogen growth. Gluconate also could modify nutrition of the plant and microbe through promoting dissolution of minerals with release of metal ions or phosphate

pseudomonads possess this potential for converting nitrate to nitrogen gas. One intermediate in denitrification, the volatile nitric oxide (NO), is associated with lowering c-di-GMP levels leading to dissociation of the biofilm and release of flagellated cells (Cutruzzolà and Frankenberg-Dinkel 2016). NO production within plant roots may also be influential: a fluorescent sensor reveals wheat roots colonized by *P. chlororaphis* O6 accumulate NO (Adams et al. 2017). The dissociation of the biofilm, associated with changes in the matrix, stimulates release of the microbial cells bearing flagella that allows new chemotactic driven motility for colonization of new sites in the rhizosphere.

8.7 Interaction with Fungal Soil Communities: Mycelial Surface Colonization

The events involved in biofilm formation at the root surface discussed above likewise are important in colonization of other soil-borne organisms, the fungi, such that they influence plant health. These interactions, however, have received less attention for their potential to influence biocontrol than, for instance, direct antagonism of fungal growth.

Mutants of *P. putida* and *P. fluorescens* lacking flagella are unable to colonize the hyphal surface of *Phytopthora parasitica,* suggesting that essential processes of chemotaxis towards the hyphae and/or attachment are impaired. Other mutants defective in producing the potent iron-chelating siderophore, pyoverdine, fail to inhibit mycelial growth (Yang et al. 1994), again suggesting the importance of "who controls Fe availability" in the rhizosphere. An array of genes is induced in *P. putida* 06909 by contact with the mycelia of *P. parasitica* (Ahn et al. 2007). The nature of the genes activated suggests that the bacterium is nurtured by metabolites released from the mycelial surface (Ahn et al. 2007), akin to the role of root exudates. Similarly, transcriptome studies with a biocontrol strain, *Pseudomonas fluorescens* Pf29Arp, show that different genes are upregulated before and after contact with the "take-all" pathogenic fungus, *Gaeumannomyces graminis* var. *tritici* (Barret et al. 2009). These findings suggest that the bacterium can sense the fungus by mechanisms that operate at a distance, such as the detection of diffusible or volatile products. The presence of the fungus boosts growth of the biocontrol strain consistent with bacterial use of fungal exudates for nutrition (Barret et al. 2009). Microbial traits in common with those for root colonization, levels of c-di-GMP, chemotaxis and alginate production, are among those required for colonization of mycelia of a root rot fungus, *Rosalinia necatrix*, by the biocontrol pseudomonad, *P. pseudoalcaligenes* AVO110; other microbial genes triggered by growth on the mycelia are those required to use fungal exudates (amino acids, fatty acids and aromatics) for nutrition (Pliego et al. 2019). Thus, in these pathosystems, biocontrol can be linked to pseudomonad colonization of the surfaces of fungal pathogens. Location at this surface would facilitate the action of antimicrobials or cell wall-degrading enzymes produced by the biocontrol agent, to make an impact on fungal growth. However, pathogens are also aware of the weaponry in biocontrol. The production of fusaric acid by *Fusarium* species shuts down phenazine synthesis by the biocontrol pseudomonad *P. chlororaphis* PCL1391 (van Rij et al. 2005) or production of DAPG in *P. protogens* Pf5 (Quecine et al. 2015), thus aiding its virulence and thwarting biocontrol efforts.

Pseudomonad mycelial colonization is also beneficial for root interactions with plant-growth-promoting mycorrhizae. As reviewed by Frey-Klett et al. (2007), an array of different soil bacteria, including *Pseudomonas* isolates, improve mycorrhizal growth and the degree of host root colonization. Cusano et al. (2011) find that a type III secretion apparatus is required for *P. fluorescens* BBc6R8 to act successfully as a helper for ectomycorrhizal colonization by *Laccaria bicolor*. Other related work shows *Pseudomonas fluorescens* Pf29Arp, a fungal surface colonizer (Barret et al. 2009), differentially expresses type III and four distinct type VI secretion systems on healthy wheat roots and roots infected with the "take-all" fungus (Marchi et al. 2013). The type III secretion system is not widespread in other pseudomonad biocontrol agents (Loper et al. 2012). The nature of any transported products is unknown but by comparison to other systems, Marchi et al. (2013) suggest the type III transporters ferry effectors to modulate plant host defenses. Shinde et al. (2019) working in aspen with *L. bicolor* provide transcriptomic data supporting that the helper bacterium functions in the downplay of root resistance to the fungus. Giovannini et al. (2020) discuss that omics studies are needed to understand the mechanisms driving the observed beneficial outcomes in the tripartite systems of plant, vesicular mycorrhizae and pseudomonad, or other, helper bacterium.

8.8 Colonization and Induction of Plant Defense

One of the areas that has most strengthened in understanding from the suggestion in the Baker's (1968) paper is the impact of biological control microbes on plant defense mechanisms; now there are numerous examples of modulations in plant defense triggered by the biocontrol pseudomonads. In this paper, events that are key in colonization are aligned with the known changes to plant defense mechanisms with a focus on three events: the tempering of innate immunity, the priming of tissues to react faster with protective responses when stress challenged, and the initiation of systemic defenses within the plant tissue.

8.9 Tempering of Innate Immunity

One current research thrust is that the biocontrol pseudomonads must first repress the plant response of innate immunity for colonization to proceed. Innate immunity involves plant recognition of danger-associated molecular patterns (DAMPs) or of common bacterial structures, the microbial- associated molecular patterns (MAMPs) which are found in both pathogenic and commensal microbes (Newman et al. 2013; Choi and Klessig 2016; Zhou and Zhang 2020; Mermigka et al. 2020) including the biocontrol pseudomonads. Recognition of such MAMPs in the plant generates a burst of reactive oxygen involving NADPH oxidase, Ca ion influx, with MPK3/MPK6 activation culminating in the expression of host defense genes (Newman et al. 2013). Thus, because these microbial structures would induce innate immunity, suppression of the innate immunity response must be occurring to allow plant colonization by the biocontrol pseudomonads. Reviews of the players and mechanisms underlying microbial recognition from the aspect of the plant cell include Hacquard et al. (2017) and Zhou and Zhang (2020). Hacquard et al. (2017) suggest that the plant has its own layers of protection based on mechanisms that gate and stepwise restrict a microbe in its entry into the plant, with a microbe being pathogenic because all defense mechanisms are thwarted. This plant-based concept can be augmented by overlay of the spheres of protection for the plant dependent on root colonization by protectant pseudomonads (Fig. 8.1).

A diversity of mechanisms is employed by microbes to avoid being recognized by the plant receptors that activate innate immunity (Newman et al. 2013; Yu et al. 2019a, b). Li et al. (2020) in studies of *P. protegens* suggest modifications to the LPS structures are crucial in adaption of this strain to colonization of *Arabidopsis* roots. Pfeilmeier et al. (2016) working with beneficial *P. fluorescens* Pf5, as well as pathogenic pseudomonads, reveal that the initiation of innate immunity is in part alleviated by repression of flagella production by high c-di-GMP levels in the bacterial cells, i.e., conditions that would program sessile development as a biofilm during colonization. Also, the chelation of Ca^{2+}, part of the cell signaling response in innate immunity, by the acidic extracellular polysaccharides in the matrix of the biofilm is another mechanism proposed for limiting onset of innate immunity (Aslam et al. 2008).

Work with *P. simiae* WCS417 compares expression of genes in Arabidopsis roots for the first 6 h of challenge with the bacterium to responses from defined MAMPs (Stringlis et al. 2018a, b). Root colonization by WCS417 results in improved root growth and induced systemic tolerance, whereas impaired root growth is seen with MAMP challenge. Both activation and repression of plant MAMP-related genes are detected with the bacterial challenge. The evidence suggests that flagellin recognition is a major, but not the only MAMP-related stimulus. The bacterium represses more MAMP-related genes than the MAMP treatments, with genes involved in plant growth being primary in this nonrepressed category. Co-expression of genes related to auxin regulation with the defense genes in the root is key in the bacterial challenge and correlates with altered root morphology and improved growth (Stringlis et al. 2018a, b). The timing of these events, within 6 h, shows the rapidity with which the plant allows colonization by the beneficial microbe to proceed.

What are other active factors in suppression of innate immunity? Research findings raise a role for polyamines, which are detected in root exudates (Oota et al. 2020, Hortin et al. unpublished data) and are both catabolized and synthesized by pseudomonads (Park et al. 2018). Figure 8.4b is an illustration of the basic polyamine pathway: putrescine is a diamine, spermidine a triamine, and spermine a tetraamine. Arginine and/ornithine are precursors and the polyamines are interconvertible through amine oxidases, activities that also generate H_2O_2.

Polyamines influence many aspects of plant development including a complex relationship with redox balance and they are associated with activation of plant defenses (Chen et al. 2019; Liu et al. 2019; Marco et al. 2019). However, metabolic effects in the pseudomonads are pertinent. In *P. chlororaphis* O6, mutation in *gacS* reduces putrescine and spermidine levels, showing regulation by unknown mechanisms by the global Gac/Rsm regulatory system (Park et al. 2018). Further, mutations in two biosynthetic genes that eliminate polyamine production impair production of phenazines, pyrollnitrin, a pyoverdine-like siderophore, and extracellular protease. Each of these factors have potential roles in biocontrol (Park et al.

2018). Thus, polyamines in a biocontrol pseudomonad are potential players in the biocontrol mechanisms that function through antimicrobial synthesis.

Control of polyamine levels appears to be an important process for the plant. One virulence mechanism of the pathogen, *P. syringae,* is the production of a modified amine, phevamine A (a valine derivative of spermidine), that represses flagellin-stimulated innate immunity (O'Neill et al. 2018). Perhaps this modification disguises the polyamine while also reducing the levels of its precursors, putrescine and arginine. Both putrescine and arginine induce accumulation of the intracellular signal, c-di-GMP in pseudomonads (Fig. 8.4a).

Studies (Liu et al. 2018) with the rhizosphere colonist, *P. fluorescens* Pf0-1, propose that regulation of putrescine levels is required to restrict overly robust biofilms from being formed to avoid activation of innate immunity. Biofilm thickness for isolate Pf0-1 is controlled by *spuC,* a gene encoding an enzyme that catabolizes putrescine. Mutation of this gene causes hyper-formation of biofilms, consistent with heightened putrescine accumulation enhancing cellular c-di-GMP levels. Inoculation of *Arabidopsis* with the *spuC* mutant stimulates expression of the plant gene, MYB51, indicative of MAMP resistance, whereas upregulation does not occur with the wild-type strain. Consequently, Liu et al. (2018) suggest the extent of biofilm formation with the wild-type strain is modulated by polyamine levels through c-di-GMP to limit MAMP-innate immunity.

Deletion in *P. fluorescens* Pf0-1 of another gene, *morA*, encoding one of the cells phospho-diesterases that hydrolyze c-di-GMP also results in more intense biofilms (Liu et al. 2018). They postulate that MorA is responsible for balancing biofilm formation through stimulating microbial cell release from a mature biofilm, i.e., purposeful lowering of c-di-GMP functions to promote biofilm cell dispersal. A consequence of the *morA* mutation is that the cells, unlike the wild-type cells, are unable to spread from the initial sites of colonization due to induction of induced immunity. Consequently, in the parental *P. fluorescens* cell, the wild-type genes *morA* and *spuC* function in repression of MAMP-triggered resistance but through different processes.

These concepts are interesting because microscopic imaging of *P. chlororaphis* O6 colonization of wheat rhizoplanes shows the cells in patchy biofilms, not as a continuous sheet of biofilm as is observed with colonization of an artificial membrane (Anderson et al. 2018; Bonebrake et al. 2018). This finding could be consistent with the existence of mechanisms, involving cellular c-di-GMP that monitor biofilm architecture on the root with thickness being one important parameter. Recent speculations are that there is variability in c-di-GMP levels with location in the pseudomonad cell based on the activities of individual cyclases for synthesis and diesterases for breakdown of this signal (Nicastro et al. 2020). Consequently, the total cell concentration of c-di-GMP may not be as significant as a localized concentration (e.g., sensing only at a polar location, where it would influence flagella responses). Fine tuning is needed to understand which metabolites interact with the array of different esterases and cyclases within the pseudomonads. For instance, along with the potential role of certain polyamines in biofilm cell dispersal is the finding that elevated pulses of metabolites present in rhizosphere solutions could be important. Spikes of succinate, glutamate, and glucose promote the release of *P. aeruginosa* cells from their biofilm surface (Sauer et al. 2004). Follow-up studies (Basu Roy and Sauer 2014) indicate glutamate activates a mechanism resulting in reduced c-di-GMP levels to promote cell release from the biofilm. Understanding the dynamics of the players in the rhizosphere solutions that govern biofilm formation on plant roots will aid in our knowledge of the control of plant innate immunity.

Another rhizosphere metabolite potentially involved in limiting innate immunity is gluconate (Fig. 8.4b). The catabolic pathway for glucose in pseudomonads involves oxidation to transform glucose to gluconate. Pseudomonads have two potential enzymes for this process, a glucose oxidase which evolves hydrogen peroxide and a glucose dehydrogenase. Both glucose dehydrogenase and gluconate dehydrogenase, which produces 2-keto-gluconate, feed electrons directly into the electron transport chain to generate energy for the cell (Molina et al. 2019). And, phosphorylation of gluconate and ketogluconate by different kinases leads to 6-phosphogluconate which is the starting point for C flux through the Entner-Doudoroff (ED) pathway for C-assimilation into microbial mass. Flux of C through the ED pathway with final products of pyruvate and glyceraldehyde-3-phosphate correlates with higher oxidative stress tolerance in the bacterium (Chavarría et al. 2013). Gluconate is proposed as a potential inducer of the ED pathway in *P. fluorescens* (Quay et al. 1972).

Gluconate and also 2-ketogluconate are secreted as a regulated process by the pseudomonad cells. Molina et al. (2019) find that during early growth of *P. putida* KT2440 on rich lysogeny broth, glucose is converted to gluconate and ketogluconate only for energy production, because both these oxidized products are secreted (Fig. 8.4b). Only at a later stage of culture are these metabolites transported back into the cell for catabolism through the ED pathway. Secretion rates for the acids also are higher under low Fe growth conditions (Sasnow et al. 2016), where 44% of C accessed as glucose is released as gluconate. Cheng et al. (2015) report for *P. fluorescens* SBW25 that mutations in *gacS,* the sensor in the global control Gac/Rsm system, leads to enhanced secretion of gluconate. Gluconate also is secreted in clinical isolates of *P. aeruginosa* in a *rpoN* mutant, a process connected with impaired gluconate-6-phosphate dehydratase function, the first enzyme in the ED pathway (Behrends et al. 2013). Thus, it is interesting that *edd* mutants lacking gluconate-6-phosphate dehydratase in *P. chlororaphis* O6 colonize roots at lower levels and fail to induce systemic tolerance (Kim et al. 2007). Any role of high gluconate secretion by the *edd* mutant in these processes awaits resolve.

Yu et al. (2019a, b) propose that it is the acidification accompanying gluconate secretion that is involved in depression of innate immunity (Fig. 8.4b). Mutants of isolate *P. capeferrum* defective in gluconate production cause a greater activation of MAMP-indicator defense genes in the plant, such as MYB51. Gluconate shuts down flagellin recognition, this effect being mimicked by acidification of the rhizosphere to pH 3.7 versus a control of pH 5.7. Because of the link between glucose flux to both gluconate and alginate in the gel biofilm matrix (Maleki et al. 2017), acidification could be intensified by the trapping of gluconate within the biofilms on the colonized rhizoplane. The metabolites ketogluconate, pyruvate, and eDNA are other candidates for biofilm acidification (Wilton et al. 2015). The use of sensors (Fulaz et al. 2019) to access pH zones within rhizosphere biofilms may be very informative; this group found pockets of acidification down to pH 5 within in vitro biofilms of *P. fluorescens*.

The acidification of the environment by gluconate secretion also is suggested as a direct method of reducing pathogen growth. Cheng et al. find with *P. fluorescens* SBW25 that gluconate secretion increases in *gacS* mutants to inhibit pathogen growth by acidic pH (Cheng et al. 2015). They suggest that this may aid the naturally occurring Gac/Rsm mutants still to benefit plant health although they lack the array of secondary compounds produced by parental cells. Enhanced secretion of the pyoverdine siderophore in *gacS* mutants (Spencer et al. 2003) could be another process benefiting rhizosphere health in the mutants. The pyoverdines would effectively restrict the bioavailability of Fe by competing microbes. Effects of gluconate-caused acidification are noted on other soil microbes: Galet et al. (2014) find gluconate release from *P. fluorescens* BBc6R8 inhibits production of an antibiotic by *Streptomyces coelicolor,* an event that may change the rhizosphere microbiome and its biocontrol potential. It is interesting that glucose is the important precursor of the acid, thus again indicating the importance of root exudate composition (Galet et al. 2014).

It is interesting that several studies report resistance to pathogen attack in transgenic plants expressing glucose oxidase. However, Wu et al. (1995) ascribe that the hydrogen peroxide produced in this enzymatic process is the major stimulus for plant resistance rather than gluconate-caused acidification. Secretion of gluconate from microbial cells and in root exudates is also credited with solubilization of phosphate from insoluble sources in the soil for use by both plant and microbe (Miller et al. 2010; Giles et al. 2015; Oteino et al. 2015). The connection with phosphate release is intriguing in light of the studies by Dangl's group connecting the low phosphate sensor in plants with activation of plant defense genes in response to altered root exudates and root microbiome composition (Castrillo et al. 2017). Gluconate also chelates metal ions (Hortin et al. 2019) and, thus aids, albeit weakly in comparison with a siderophore, to extract Fe/Cu/Zn/K from soil minerals (Sasnow et al. 2016). Improved supplies of essential metals and phosphate would boost both plant and rhizosphere microbiome health. These processes are attributed as one mechanism important in the role of pseudomonads as plant-growth-promoting bacteria (Shen et al. 2016; Emami et al. 2019).

Activation of ROS production by activation of plant extracellular NADPH oxidases from MAMPs detection (Kimura et al. 2020) is well established as part of the plant defense mechanism. However, hydrogen peroxide also can be generated through oxidation of metabolites in the rhizosphere solutions (glycine, polyamines, aspartate, oxalate, etc.) with specific oxidases (Smirnoff and Arnaud 2019). The presence of different schemes to generate hydrogen peroxide might have another benefit in addition to its established defense role because its decomposition through the activity of catalase would liberate oxygen. Localized pockets of oxygen would boost oxidative metabolism in both plant root and pseudomonad cells, enhancing their cellular activities. Pseudomonad cells have a strong complement of catalases.

Induced catalase activity occurs as pseudomonad cells contact the root surface (Katsuwon and Anderson 1992) and when pseudomonads grow on grass root exudates (Mavrodi et al. 2021). Thus, certain rhizosphere metabolites could participate in another beneficial process, in addition to providing nutrients to support microbial growth, by promoting aerobic generation of energy for plant and pseudomonad well-being.

8.10 Priming and Systemic Defense Responses

Priming is part of the beneficial induced systemic defense responses that are triggered after probiotic pseudomonads colonize roots. The term describes a faster and greater response to initiate tolerance/resistance after a challenge, such as a pathogen attack. Early in research on priming, Beckers et al. (2009) established that MAPK3 and MAPK6, players in innate immunity, are required for priming. Increased transcription and activation of these mitogen-activated kinases are involved. Zhou and Zhang (2020) summarize the complexity and interplay of the plant cell receptors and signal transducers in plant immunity from the plants' aspect although fine details are unknown. Although innate immunity is repressed early in the contact between the pseudomonad and the plant, priming and activation of systemic resistance responses subsequently occurs with the biocontrol strains.

8.11 Biocontrol Pseudomonads Shift SA-, JA-, and Et-Plant Resistance Pathways

The systemic defense responses involve the hijacking of the hormonal pathways of the plant by the protectant microbes (Berens et al. 2017). Involvement of a salicylic acid (SA) pathway became apparent from field studies of pathogen-challenged plants by Kuc (1987) where symptoms caused systemic resistance, a type of "immune" response, to subsequently develop and even be passed to the next plant generation. The SA pathway is associated in the "immunized" plant with synthesis of microbial growth inhibitors, such as beta-glucanase and chitinase enzymes, which would cause fungal cell wall breakdown. Plant cell wall strengthening by lignification and callose deposition provides additional barriers to pathogen challenge. However, this foliar SA pathway for defense (Lebeis et al. 2015) is balanced by reduced plant growth (van Butselaar and Van den Ackerveken 2020). Activation of the SA defense pathway is reported with the rhizosphere isolate *Pseudomonas* sp. isolate 23, where foliar protection against a bacterial pathogen is induced; this pseudomonad also displays direct growth antagonism of the pathogen (Takishita et al. 2018). Communication between the plant cells to mount the defense response may involve movement of SA through the outer cuticle and wax deposits (Lim et al. 2020).

Most beneficial pseudomonads, however, initiate systemic resistance to pathogens through resistance pathways controlled by the jasmonic acid (JA)/ethylene (Et) pathways of defense (Pieterse et al. 2001; Spencer et al. 2003). Recognition of a JA-defense pathway extends back to reports of resistance in solanaceous plants against herbivores. Here the leaf damage induces a wound peptide, systemin, which signals the production of protease inhibitors to restrict insect damage (Ryan and Pearce 2003).

Microbial resistance however involves both JA and Et responses, resulting in enhanced expression of an array of defense genes that are distinct from those of the JA and SA pathways. Initially there seemed to be clear distinctions between the SA and JA/Et pathways, in the triggers and the pathogens controlled. Moreover, competition between the pathways exists with one pathway inhibiting the other, consistent with the plant being limited in how to invest its energy resources. However, the situation is more complex with cross talk between these and other plant hormones, e.g., abscisic acid and induction of novel arrays of defense genes. For instance, the quorum sensing signals, acyl homoserine lactones, which control gene expression in the Gac/Rsm regulon in some biocontrol pseudomonads (Schenk et al. 2012) trigger plant defense responses by an SA/oxylipin pathway. The activated plant defense genes are members of both pathways (Schenk et al. 2014). Activation of varied plant defense responses also includes responses characteristic of the hypersensitive response and the SA pathway in plants challenged with three different pseudomonads (Zdor and Anderson 1992).

Induction of systemic resistance through the JA/Et pathways by the biocontrol pseudomonads is triggered by their own secreted metabolites, in addition to the acyl homoserine lactones. Induced systemic resistance occurs with treatments of antimicrobials such as phenazines and pyrrolnitrin, as well as the volatile fermentation product from pyruvate, 2,3-butanediol, from *P. chlororaphis* O6 (Han et al. 2006; Kang et al. 2007; Park et al. 2011); DAPG (Iavicoli et al. 2003; Chae et al. 2020) and resorcinol (Calderón et al. 2019). Thus, although these pseudomonad compounds were identified initially because of their direct antimicrobial effects, these metabolites too are multitasking in the process of biocontrol through triggering systemic induced resistance. It is possible that the antifungals all trigger a second ROS burst that now initiates the systemic defense pathways. A similar ROS-based theory is debated for a common response to certain antibiotics in mammalian cells (Van Acker and Coenye 2017).

Production of these antimicrobials by the biocontrol pseudomonads is under nutritional regulation. For instance, with isolate *P. chlororaphis* O6, phenazine formation promoted under high glucose growth is accompanied by low pyrrolnitrin production (Park et al. 2011). Synthesis of many antimicrobials in the biocontrol pseudomonads is regulated by the Gac/Rsm system, meaning they are produced under the control of the acyl homoserine lactones once cell mass has reached a certain level and cell density. Thus, it is likely that the antimicrobials are produced after colonization is established at which time biofilms are in place and repression of innate immunity is no longer an issue. The biofilms also may act to concentrate these products by limiting ready diffusion for more effectiveness against a challenging fungal, nematode, or insect attack.

The activation of these protective defense pathways may perpetuate changes in the rhizosphere microbiome. For instance, Carvalhais et al. (2013, 2015) found root exudate composition differs between the wild type and mutants of Arabidopsis altered in the JA response; these metabolite changes correlate with altering the rhizosphere microbiome. Similarly, changes in root exudates and rhizosphere microbiome are found in plants with mutations in the SA defense pathway (Lebeis et al. 2015). Applications of exogenous SA also alters which members of a SycCom composed of endophytes and rhizosphere isolates dominate or became more minor species in the microbiome of the SA-treated plants (Lebeis et al. 2015). Some of the species that dominate use SA as a C source (Lebeis et al. 2015), which reiterates a primary role of the root exudates to nurture microbial growth with the feedback of a more-defended plant.

8.12 Summary

The studies discussed illustrate that plant and microbial metabolites engage in dynamic dialogs between plant and its rhizosphere microbiome. This review has focused on the key process of biofilm formation by the rhizosphere biocontrol pseudomonads in the regulation of plant health. Illustrated are the multiple roles of the simple metabolites, gluconate and polyamines, in the mechanisms that influence both plant and the supported microbiome. These metabolites are not just biochemicals in metabolic pathways but rather they are part of the dialog of successful communications between the plant and its selected microbiome. The interplay results in the biocontrol pseudomonads generating overlapping spheres of plant protection around the plant root, each of which has different modes of operation. There remains, however, the need for molecular understanding especially in the diversity of biocontrol processes initiated by the different pseudomonads. Each of the steps of the process of biocontrol through the sequence of root exudation, microbial root colonization, suppression of plant innate immunity, and activation of plant defenses still has unknowns.

References

Adams J, Wright M, Wagner H, Valiente J, Britt D, Anderson A (2017) Cu from dissolution of CuO nanoparticles signals changes in root morphology. Plant Physiol Biochem 110:108–117. https://doi.org/10.1016/j.plaphy.2016.08.005

Ahn SJ, Yang CH, Cooksey DA (2007) Pseudomonas putida 06909 genes expressed during colonization on mycelial surfaces and phenotypic characterization of mutants. J Appl Microbiol 103:120–132. https://doi.org/10.1111/j.1365-2672.2006.03232.x

Anderson AJ, Kim YC (2018) Biopesticides produced by plant-probiotic *Pseudomonas chlororaphis* isolates. Crop Protect 105:62–29

Anderson AJ, McLean JE, Jacobson AR, Britt DW (2018) CuO and ZnO nanoparticles modify interkingdom cell signaling processes relevant to crop production. J Agric Food Chem 66:6513–6524. https://doi.org/10.1021/acs.jafc.7b01302

Ardré M, Dufour D, Rainey PB (2010) Causes and biophysical consequences of cellulose production by *Pseudomonas fluorescens* SBW25 at the air-liquid interface. J Bacteriol 201:e00110–e00119. https://doi.org/10.1128/JB.00110-19

Armanhi JSL, de Souza RSC, Damasceno NB, de Araújo LM, Imperial J, Arruda P (2018) A community-based culture collection for targeting novel plant growth-promoting bacteria from the sugarcane microbiome. Front Plant Sci 8:2191. https://doi.org/10.3389/fpls.2017.02191

Aslam SN, Newman MA, Erbs G, Morrissey KL, Chinchilla D, Boller T, Jensen TT, De Castro C, Ierano T, Molinaro A, Jackson RW, Knight MR, Cooper RM (2008) Bacterial polysaccharides suppress induced innate immunity by calcium chelation. Curr Biol 18:1078–1083. https://doi.org/10.1016/j.cub.2008.06.061

Baetz U, Martinoia E (2014) Root exudates: the hidden part of plant defense. Trends Plant Sci 19: 90–98

Baker R (1968) Mechanisms of biocontrol of soil-borne pathogens. Annu Rev Phytopathol 6:263–294

Barret M, Frey-Klett P, Boutin M, Guillerm-Erckelboudt AY, Martin F, Guillot L, Sarniguet A (2009) The plant pathogenic fungus *Gaeumannomyces graminis* var. *tritici* improves bacterial growth and triggers early gene regulations in the biocontrol strain *Pseudomonas fluorescens* Pf29Arp. New Phytol 181:435–447. https://doi.org/10.1111/j.1469-8137.2008.02675.x

Basu Roy A, Sauer K (2014) Diguanylate cyclase NicD-based signalling mechanism of nutrient-induced dispersion by *Pseudomonas aeruginosa*. Mol Microbiol 94:771–793

Beckers GJ, Jaskiewicz M, Liu Y, Underwood WR, He SY, Zhang S, Conrath U (2009) Mitogen-activated protein kinases 3 and 6 are required for full priming of stress responses in *Arabidopsis thaliana*. Plant Cell 21:944–953. https://doi.org/10.1105/tpc.108.062158

Behrends V, Bell TJ, Liebeke M, Cordes-Blauert A, Ashraf SN, Nair C, Zlosnik JE, Williams HD, Bundy JG (2013) Metabolite profiling to characterize disease-related bacteria: gluconate excretion by *Pseudomonas aeruginosa* mutants and clinical isolates from cystic fibrosis patients. J Biol Chem 288:15098–15109. https://doi.org/10.1074/jbc.M112.442814

Berendsen RL, Pieterse CM, Bakker PA (2012) The rhizosphere microbiome and plant health. Trends Plant Sci 17:478–486. https://doi.org/10.1016/j.tplants.2012.04.001

Berens ML, Berry HM, Mine A, Argueso CT, Tsuda K (2017) Evolution of hormone signaling networks in plant defense. Annu Rev Phytopathol 55:401–425

Bergeau D, Mazurier S, Barbey C, Merieau A, Chane A, Goux D, Bernard S, Driouich A, Lemanceau P, Vicré M, Latour X (2019) Unusual extracellular appendages deployed by the model strain *Pseudomonas fluorescens* C7R12. PLoS One 14:e0221025. https://doi.org/10.1371/journal.pone.0221025

Bernier SP, Ha DG, Khan W, Merritt JH, O'Toole GA (2011) Modulation of *Pseudomonas aeruginosa* surface-associated group behaviors by individual amino acids through c-di-GMP signaling. Res Microbiol 162:680–688. https://doi.org/10.1016/j.resmic.2011.04.014

Biessy A, Filion M (2018) Phenazines in plant-beneficial *Pseudomonas* spp.: biosynthesis, regulation, function and genomics. Environ Microbiol 20:3905–3917

Biessy A, Novinscak A, Blom J, Léger G, Thomashow LS, Cazorla FM, Josic D, Filion M (2019) Diversity of phytobeneficial traits revealed by whole-genome analysis of worldwide-isolated phenazine-producing *Pseudomonas* spp. Environ Microbiol 21(1):437–455. https://doi.org/10.1111/1462-2920.14476

Bonebrake M, Anderson K, Valiente J, Jacobson A, McLean JE, Anderson A, Britt DW (2018) Biofilms benefiting plants exposed to ZnO and CuO nanoparticles studied with a root-mimetic Hollow Fiber Membrane. J Agric Food Chem 66:6619–6627. https://doi.org/10.1021/acs.jafc.7b02524

Briard B, Rasoldier V, Bomme P, ElAouad N, Guerreiro C, Chassagne P, Muszkeita L, Latge J-P, Mulard L, Beauvais A (2017) Dirhamnolipids secreted from *Pseudomonas aeruginosa* modify antifungal susceptibility of *Aspergillus fumigatus* by inhibiting β1,3 glucan synthase activity. ISME J 11:1578–1591

Cai YM, Hutchin A, Craddock J, Walsh MA, Webb JS, Tews I (2020) Differential impact on motility and biofilm dispersal of closely related phosphodiesterases in *Pseudomonas aeruginosa*. Sci Rep 10:6232. https://doi.org/10.1038/s41598-020-63008-5

Calderón CE, Tienda S, Heredia-Ponce Z, Arrebola E, Cárcamo-Oyarce G, Eberl L, Cazorla FM (2019) The compound 2-hexyl, 5-propyl resorcinol has a key role in biofilm formation by the biocontrol rhizobacterium *Pseudomonas chlororaphis* PCL1606. Front Microbiol 10:396. https://doi.org/10.3389/fmicb.2019.00396

Carvalhais LC, Dennis PG, Tyson GW, Schenk PM (2012) Application of metatranscriptomics to soil environments. J Microbiol Methods 91:246–251. https://doi.org/10.1016/j.mimet.2012.08.011

Carvalhais LC, Dennis PG, Badri DV, Kidd BN, Vivanco JM, Schenk PM (2015) Linking jasmonic acid signaling, root exudates, and rhizosphere microbiomes. Mol Plant-Microbe Interact 28:1049–1058. https://doi.org/10.1094/MPMI-01-15-0016-R

Castrillo G, Teixeira PJ, Paredes SH, Law TF, de Lorenzo L, Feltcher ME, Finkel OM, Breakfield NW, Mieczkowski P, Jones CD, Paz-Ares J, Dangl JL (2017) Root microbiota drive direct integration of phosphate stress and immunity. Nature 543:513–518. https://doi.org/10.1038/nature21417

Chae D-H, Kim D-R, Cheong MS, Lee YB, Kwak Y-S (2020) Investigating the induced systemic resistance mechanism of 2,4-diacetylphloroglucinol (DAPG) using DAPG hydrolase-transgenic *Arabidopsis*. Plant Pathol J 36:255–266. https://doi.org/10.5423/PPJ.OA.02.2020.0031

Chang WS, van de Mortel M, Nielsen L, Nino de Guzman G, Li X, Halverson LJ (2007) Alginate production by *Pseudomonas putida* creates a hydrated microenvironment and contributes to biofilm architecture and stress tolerance under water-limiting conditions. J Bacteriol 189:8290–8299. https://doi.org/10.1128/JB.00727-07

Chavarría M, Nikel PI, Pérez-Pantoja D, de Lorenzo V (2013) The Entner-Doudoroff pathway empowers *Pseudomonas putida* KT2440 with a high tolerance to oxidative stress. Environ Microbiol 15:1772–1785. https://doi.org/10.1111/1462-2920.12069

Chen D, Shao Q, Yin L, Younis A, Zheng B (2019) Polyamine function in plants: metabolism, regulation on development, and roles in abiotic stress responses. Front Plant Sci 9:1945. https://doi.org/10.3389/fpls.2018.01945

Cheng X, van der Voort M, Raaijmakers JM (2015) Gac-mediated changes in pyrroloquinoline quinone biosynthesis enhance the antimicrobial activity of *Pseudomonas fluorescens* SBW25. Environ Microbiol Rep 7:139–147. https://doi.org/10.1111/1758-2229.12231

Choi HW, Klessig DF (2016) DAMPs, MAMPs, and NAMPs in plant innate immunity. BMC Plant Biol 16(1):232. https://doi.org/10.1186/s12870-016-0921-2

Chung J-H, Song GC, Ryu CM (2016) Sweet scents from good bacteria: case studies on bacterial volatile compounds for plant growth and immunity. Plant Mol Biol 90:677–687

Ciancio A, Pieterse CMJ, Mercado-Blanco J (2019) Editorial: Harnessing useful rhizosphere microorganisms for pathogen and pest biocontrol. Front Microbiol 10:1935. https://doi.org/10.3389/fmicb.2019.01935

Collins AJ, Pastoras AB, Jarrod Smiths T, Dahlstrom K, O'Toole GA (2020) MapA, a second large RTX adhesin, contributes to biofilm formation by *Pseudomonas fluorescens*. J Bacteriol. https://doi.org/10.1101/2020.05.11.089839

Colvin KM, Gordon VD, Murakami K, Borlee BR, Wozniak DJ, Wong GC, Parsek MR (2011) The Pel polysaccharide can serve a structural and protective role in the biofilm matrix of *Pseudomonas aeruginosa*. PLoS Pathog 7:e1001264. https://doi.org/10.1371/journal.ppat.1001264

Colvin KM, Irie Y, Tart CS, Urbano R, Whitney JC, Ryder C, Howell PL, Wozniak DJ, Parsek MR (2012) The Pel and Psl polysaccharides provide *Pseudomonas aeruginosa* structural redundancy within the biofilm matrix. Environ Microbiol 14:1913–1928. https://doi.org/10.1111/j.1462-2920.2011.02657.x

Costerton JW, Stewart PS, Greenberg EP (1999) Bacterial biofilms: a common cause of persistent infections. Science 284:1318–1322. https://doi.org/10.1126/science.284.5418.1318

Cusano AM, Burlinson P, Deveau A, Vion P, Uroz S, Preston GM, Frey-Klett P (2011) *Pseudomonas fluorescens* BBc6R8 type III secretion mutants no longer promote ectomycorrhizal symbiosis. Environ Microbiol Rep 3:203–210. https://doi.org/10.1111/j.1758-2229.2010.00209.x

Cutruzzolà F, Frankenberg-Dinkel N (2016) Origin and impact of nitric oxide in *Pseudomonas aeruginosa* biofilms. J Bacteriol 198:55–65. https://doi.org/10.1128/JB.00371-15

Dakora FD, Phillips DA (2002) Root exudates as mediators of mineral acquisition in low-nutrient environments. Plant Soil 245:35–47

de Souza RSC, Armanhi JSL, Damasceno NB, Imperial J, Arruda P (2019) Genome sequences of a plant beneficial synthetic bacterial community reveal genetic features for successful plant colonization. Front Microbiol 10:1779. https://doi.org/10.3389/fmicb.2019.01779

Dennis PG, Miller AJ, Hirsch PR (2010) Are root exudates more important than other sources of rhizodeposits in structuring rhizosphere bacterial communities? FEMS Microbiol Ecol 72:313–327. https://doi.org/10.1111/j.1574-6941.2010.00860.x

Dimkpa CO, Zeng J, McLean JE, Britt DW, Zhan J, Anderson AJ (2012) Production of indole-3-acetic acid via the indole-3-acetamide pathway in the plant-beneficial bacterium *Pseudomonas chlororaphis* O6 is inhibited by ZnO nanoparticles but enhanced by CuO nanoparticles. Appl Environ Microbiol 78:1404–1410. https://doi.org/10.1128/AEM.07424-11

Dorosky RJ, Pierson LS III, Pierson EA (2018) *Pseudomonas chlororaphis* produces multiple R-tailocin particles that broaden the killing spectrum and contribute to persistence in rhizosphere communities. Appl Environ Microbiol 84:e01230–e01218. https://doi.org/10.1128/AEM.01230-18

Duca D, Lorv J, Patten CL, Rose D, Glick BR (2014) Indole-3-acetic acid in plant-microbe interactions. Antonie Van Leeuwenhoek 106:85–125. https://doi.org/10.1007/s10482-013-0095-y

Edwards J, Johnson C, Santos-Medellín C, Lurie E, Podishetty NK, Bhatnagar S, Eisen JA, Sundaresan V (2015) Structure, variation, and assembly of the root-associated microbiomes of rice. Proc Natl Acad Sci U S A 112:E911–E920. https://doi.org/10.1073/pnas.1414592112

Elad Y, Baker R (1985) The role of competition for iron and carbon in suppression of chlamydospore germination of *Fusarium* spp. by *Pseudomonas* spp. Phytopathology 75:1053–1059

Emami S, Alikhani HA, Pourbabaei AA, Etesami H, Sarmadian F, Motessharezadeh B (2019) Effect of rhizospheric and endophytic bacteria with multiple plant growth promoting traits on wheat growth. Environ Sci Pollut Res Int 26:19804–19813. https://doi.org/10.1007/s11356-019-05284-x

Farag MA, Zhang H, Ryu CM (2013) Dynamic chemical communication between plants and bacteria through airborne signals: induced resistance by bacterial volatiles. J Chem Ecol 39:1007–1018. https://doi.org/10.1007/s10886-013-0317-9

Fincheira P, Quiroz A (2018) Microbial volatiles as plant growth inducers. Microbiol Res 208:63–75

Frey-Klett P, Garbaye J, Tarkka M (2007) The mycorrhiza helper bacteria revisited. New Phytol 176:22–36. https://doi.org/10.1111/j.1469-8137.2007.02191.x

Fulaz S, Hiebner D, Barros CHN, Devlin H, Vitale S, Quinn L, Casey E (2019) Ratiometric imaging of the in situ pH distribution of biofilms by use of fluorescent mesoporous silica nanosensors. ACS Appl Mater Interfaces 11:32679–32688. https://doi.org/10.1021/acsami.9b09978

Fuqua C (2010) Passing the baton between laps: adhesion and cohesion in *Pseudomonas putida* biofilms. Mol Microbiol 77:533–536. https://doi.org/10.1111/j.1365-2958.2010.07250.x

Galet J, Deveau A, Hôtel L, Leblond P, Frey-Klett P, Aigle B (2014) Gluconic acid-producing *Pseudomonas* sp. prevent γ-actinorhodin biosynthesis by *Streptomyces coelicolor* A3(2). Arch Microbiol 196:619–627. https://doi.org/10.1007/s00203-014-1000-4

Garrett SD (1965) Toward biological control of soil-borne plant pathogens. In: Baker KF, Snyder WC (eds) Ecology of soil-borne plant pathogens. University of California Press, Berkeley, CA, pp 4–17. 571 pp

Giles CD, Hsu PC, Richardson AE, Hurst MR, Hill JE (2015) The role of gluconate production by *Pseudomonas* spp. in the mineralization and bioavailability of calcium-phytate to *Nicotiana tabacum*. Can J Microbiol 61:885–897. https://doi.org/10.1139/cjm-2015-0206

Giovannini L, Palla M, Agnolucci M, Avio L, Sbrana C, Turrini A, Giovannetti M (2020) Arbuscular mycorrhizal fungi and associated microbiota as plant biostimulants: research strategies for the selection of the best performing inocula. Agronomy 10:106. https://doi.org/10.3390/agronomy10010106

Girard IJ, Mcloughlin AG, de Kievit TR, Fernando DW, Belmonte MF (2016) Integrating large-scale data and RNA technology to protect crops from fungal pathogens. Front Plant Sci 7:631. https://doi.org/10.3389/fpls.2016.00631

Glick BR (2004) Bacterial ACC deaminase and the alleviation of plant stress. Adv Appl Microbiol 56:291–312. https://doi.org/10.1016/S0065-2164(04)56009-4

Gu Y, Wei ZM, Wang X, Friman V-P, Huang J, Wang X, Mei X, Xu Y, Shen Q, Jousset A (2016) Pathogen invasion indirectly changes the composition of soil microbiome via shifts in root exudation profile. Biol Fertil Soils 52:997–1005. https://doi.org/10.1007/s00374-016-1136-2

Gu S, Wei Z, Shao Z, Friman VP, Cao K, Yang T, Kramer J, Wang X, Li M, Mei X, Xu Y, Shen Q, Kümmerli R, Jousset A (2020) Competition for iron drives phytopathogen control by natural rhizosphere microbiomes. Nat Microbiol. https://doi.org/10.1038/s41564-020-0719-8

Hacquard S, Spaepen S, Garrido-Oter R, Schulze-Lefert P (2017) Interplay between innate immunity and the plant microbiota. Annu Rev Phytopathol 55:565–589. https://doi.org/10.1146/annurev-phyto-080516-035623

Han SH, Lee SJ, Moon JH, Park KH, Yang KY, Cho BH, Kim KY, Kim YW, Lee MC, Anderson AJ, Kim YC (2006) GacS-dependent production of 2R,3R-butanediol by *Pseudomonas chlororaphis* O6 is a major determinant for eliciting systemic resistance against *Erwinia carotovora* but not against *Pseudomonas syringae* pv. *tabaci* in tobacco. Mol Plant-Microbe Interact 19:924–930. https://doi.org/10.1094/MPMI-19-0924

Herrera Paredes S, Gao T, Law TF, Finkel OM, Mucyn T, Teixeira PJPL, Salas González I, Feltcher ME, Powers MJ, Shank EA, Jones CD, Jojic V, Dangl JL, Castrillo G (2018) Design of synthetic bacterial communities for predictable plant phenotypes. PLoS Biol 16:e2003962. https://doi.org/10.1371/journal.pbio.2003962

Hinsa SM, Espinosa-Urgel M, Ramos JL, O'Toole GA (2003) Transition from reversible to irreversible attachment during biofilm formation by *Pseudomonas fluorescens* WCS365 requires an ABC transporter and a large secreted protein. Mol Microbiol 49:905–918. https://doi.org/10.1046/j.1365-2958.2003.03615

Hortin JM, Anderson AJ, Britt DW, Jacobson AR, McLean JE (2019) Soil-derived fulvic acid and root exudates, modified by soil bacteria, alter CuO nanoparticle induced root stunting of wheat via Cu complexation. Environ Sci Nano 6:3638

Iavicoli A, Boutet E, Buchala A, Métraux JP (2003) Induced systemic resistance in *Arabidopsis thaliana* in response to root inoculation with *Pseudomonas fluorescens* CHA0. Mol Plant-Microbe Interact 16:851–858. https://doi.org/10.1094/MPMI.2003.16.10.851

Imperiali N, Dennert F, Schneider J, Laessle T, Velatta C, Fesselet M, Michele Wyler M, Mascher F, Mavrodi O, Mavrodi D, Maurhofer M, Keel C (2017) Relationships between root pathogen resistance, abundance and expression of *Pseudomonas* antimicrobial genes, and soil properties in representative Swiss agricultural soil. Front Plant Sci. https://doi.org/10.3389/fpls.2017.00427

Jain R, Behrens AJ, Kaever V, Kazmierczak BI (2012) Type IV pilus assembly in *Pseudomonas aeruginosa* over a broad range of cyclic di-GMP concentrations. J Bacteriol 194:4285–4294. https://doi.org/10.1128/JB.00803-12

Kang BR, Han SH, Zdor RE, Anderson AJ, Spencer M, Yang KY, Kim YH, Lee MC, Cho BH, Kim YC (2007) Inhibition of seed germination and induction of systemic disease resistance by

Pseudomonas chlororaphis O6 requires phenazine production regulated by the global regulator, GacS. J Microbiol Biotechnol 17(4):586–593

Katsuwon J, Anderson AJ (1992) Characterization of catalase activities in a root-colonizing isolate of *Pseudomonas putida*. Can J Microbiol 38:1026–1032. https://doi.org/10.1139/m92-169

Kim YC, Anderson AJ (2020) Integration of bacterial volatile organic compounds with plant health. In: Ryu C-M, Wiesskof L, Piechulla B (eds) Bacterial volatile compounds as mediators of airborne interactions. Springer, New York

Kim HJ, Nam HS, Anderson AJ, Yang KY, Cho BH, Kim YC (2007) Mutation in the *edd* gene encoding the 6-phosphogluconate dehydratase of *Pseudomonas chlororaphis* O6 impairs root colonization and is correlated with reduced induction of systemic resistance. Lett Appl Microbiol 44:56–61. https://doi.org/10.1111/j.1472-765X.2006.02029.x

Kimura S, Hunter K, Vaahtera L, Tran HC, Citterico M, Vaattovaara A, Rokka A, Stolze SC, Harzen A, Meißner L, Wilkens MMT, Hamann T, Toyota M, Nakagami H, Wrzaczek M (2020) CRK2 and C-terminal phosphorylation of NADPH oxidase RBOHD regulate reactive oxygen species production in Arabidopsis. Plant Cell 32:1063–1080

Kloepper JW, Leog J, Teintze M, Schroth MN (1980) *Pseudomonas* siderophores: a mechanism explaining disease-suppressive soils. Curr Microbiol 4:317–320

Kuc J (1987) Translocated signals for plant immunization. Ann Acad Sci 494:221–223

Lahesaare A, Ainelo H, Teppo A, Kivisaar M, Heipieper HJ, Teras R (2016) LapF and its regulation by Fis affect the cell surface hydrophobicity of *Pseudomonas putida*. PLoS One 11(11): e0166078. https://doi.org/10.1371/journal.pone.0166078

Lebeis SL, Paredes SH, Lundberg DS, Breakfield N, Gehring J, McDonald M, Malfatti S, Glavina del Rio T, Jones CD, Tringe SG, Dangl JL (2015) PLANT MICROBIOME. Salicylic acid modulates colonization of the root microbiome by specific bacterial taxa. Science 349:860–864. https://doi.org/10.1126/science.aaa8764

Levy A, Salas Gonzalez I, Mittelviefhaus M, Clingenpeel S, Herrera Paredes S, Miao J, Wang K, Devescovi G, Stillman K, Monteiro F, Rangel Alvarez B, Lundberg DS, Lu TY, Lebeis S, Jin Z, McDonald M, Klein AP, Feltcher ME, Rio TG, Grant SR, Doty SL, Ley RE, Zhao B, Venturi V, Pelletier DA, Vorholt JA, Tringe SG, Woyke T, Dangl JL (2017) Genomic features of bacterial adaptation to plants. Nat Genet 50(1):138–150. https://doi.org/10.1038/s41588-017-0012-9

Li T, Zhang J, Shen C, Li H, Qiu L (2019) 1-Aminocyclopropane-1-carboxylate: a novel and strong chemoattractant for the plant beneficial rhizobacterium *Pseudomonas putida* UW4. Mol Plant-Microbe Interact 32:750–759. https://doi.org/10.1094/MPMI-11-18-0317-R

Li E, Zhang H, Jiang H, Pieterse CMJ, Jousset A, Bakker PAHM, DeJonge R (2020) Experimental evolution-driven identification of *Arabidopsis* rhizosphere competence genes in *Pseudomonas protegens*. bioRxiv. 2020.12.01.407551. https://doi.org/10.1101/2020.12.01.407551

Lim GH, Liu H, Yu K, Liu R, Shine MB, Fernandez J, Burch-Smith T, Mobley JK, McLetchie N, Kachroo A, Kachroo P (2020) The plant cuticle regulates apoplastic transport of salicylic acid during systemic acquired resistance. Sci Adv 6:eaaz0478. https://doi.org/10.1126/sciadv.aaz0478

Lin YC, Sekedat MD, Cornell WC, Silva GM, Okegbe C, Price-Whelan A, Vogel C, Dietrich LEP (2018) Phenazines regulate Nap-dependent denitrification in *Pseudomonas aeruginosa* biofilms. J Bacteriol 200(9):e00031–e00018. https://doi.org/10.1128/JB.00031-18

Liu Y, Gokhale CS, Rainey PB, Zhang XX (2017) Unravelling the complexity and redundancy of carbon catabolic repression in *Pseudomonas fluorescens* SBW25. Mol Microbiol 105:589–605. https://doi.org/10.1111/mmi.13720

Liu Z, Beskrovnaya P, Melnyk RA, Hossain SS, Khorasani S, O'Sullivan LR, Wiesmann CL, Bush J, Richard JD, Haney CH (2018) A genome-wide screen identifies genes in rhizosphere-associated *Pseudomonas* required to evade plant defenses. mBio 9:e00433–e00418. https://doi.org/10.1128/mBio.00433-18

Liu C, Atanasov KE, Tiburcio AF, Alcázar R (2019) The polyamine putrescine contributes to H_2O_2 and RbohD/F-dependent positive feedback loop in *Arabidopsis* PAMP-triggered immunity. Front Plant Sci 10:894. https://doi.org/10.3389/fpls.2019.00894

Loper JE, Hassan KA, Mavrodi DV, Davis EW II, Lim CK, Shaffer BT, Elbourne LD, Stockwell VO, Hartney SL, Breakwell K, Henkels MD, Tetu SG, Rangel LI, Kidarsa TA, Wilson NL, van de Mortel JE, Song C, Blumhagen R, Radune D, Hostetler JB, Brinkac LM, Durkin AS, Kluepfel DA, Wechter WP, Anderson AJ, Kim YC, Pierson LS III, Pierson EA, Lindow SE, Kobayashi DY, Raaijmakers JM, Weller DM, Thomashow LS, Allen AE, Paulsen IT (2012) Comparative genomics of plant-associated *Pseudomonas* spp.: insights into diversity and inheritance of traits involved in multitrophic interactions. PLoS Genet 8:e1002784. https://doi.org/10.1371/journal.pgen.1002784

López-Baena FJ, Vinardell JM, Medina C (2019) Regulation of protein secretion systems mediated by cyclic diguanylate in plant-interacting bacteria. Front Microbiol 10:1289. https://doi.org/10.3389/fmicb.2019.01289

Lundberg DS, Lebeis SL, Paredes SH, Yourstone S, Gehring J, Malfatti S, Tremblay J, Engelbrektson A, Kunin V, Del Rio TG, Edgar RC, Eickhorst T, Ley RE, Hugenholtz P, Tringe SG, Dangl JL (2012) Defining the core *Arabidopsis thaliana* root microbiome. Nature 488:86–90. https://doi.org/10.1038/nature11237

Ma L, Zheng SC, Zhang TK, Liu ZY, Wang XJ, Zhou XK, Yang CG, Duo JL, Mo MH (2018) Effect of nicotine from tobacco root exudates on chemotaxis, growth, biocontrol efficiency, and colonization by *Pseudomonas aeruginosa* NXHG29. Antonie Van Leeuwenhoek 111:1237–1257. https://doi.org/10.1007/s10482-018-1035-7

Maleki S, Mærk M, Hrudikova R, Valla S, Ertesvåg H (2017) New insights into *Pseudomonas fluorescens* alginate biosynthesis relevant for the establishment of an efficient production process for microbial alginates. New Biotechnol 37(Pt A):2–8. https://doi.org/10.1016/j.nbt.2016.08.005

Marchi M, Boutin M, Gazengel K, Rispe C, Gauthier JP, Guillerm-Erckelboudt AY, Lebreton L, Barret M, Daval S, Sarniguet A (2013) Genomic analysis of the biocontrol strain *Pseudomonas fluorescens* Pf29Arp with evidence of T3SS and T6SS gene expression on plant roots. Environ Microbiol Rep 5:393–403. https://doi.org/10.1111/1758-2229.12048

Marco F, Busó E, Lafuente T, Carrasco P (2019) Spermine confers stress resilience by modulating abscisic acid biosynthesis and stress responses in *Arabidopsis* plants. Front Plant Sci 10:972. https://doi.org/10.3389/fpls.2019.00972

Marshall DC, Arruda BE, Silby MW (2019) Alginate genes are required for optimal soil colonization and persistence by *Pseudomonas fluorescens* Pf0-1. Access Microbiol. https://doi.org/10.1099/acmi.0.000021

Martínez-Gil M, Ramos-González MI, Espinosa-Urgel M (2014) Roles of cyclic Di-GMP and the Gac system in transcriptional control of the genes coding for the *Pseudomonas putida* adhesins LapA and LapF. J Bacteriol 196:1484–1495. https://doi.org/10.1128/JB.01287-13

Martins SJ, Flausino A, Marcio F, Pedroso P, Cunha MG, Rocha MR, Medeiros FHV (2019) Microbial volatiles organic compounds control anthracnose (*Colletotrichum lindemuthianum*) in common bean (*Phaseolus vulgaris L.*). Biol Contr 131:36–42

Mauch-Mani B, Baccelli I, Luna E, Flors V (2017) Defense priming: an adaptive part of induced resistance. Annu Rev Plant Biol 28:485–512. https://doi.org/10.1146/annurev-arplant-042916-041132

Mavrodi OV, McWilliams JR, Peter JO, Berim A, Hassan KA, Elbourne LDH, LeTourneau MK, Gang DR, Paulsen IT, Weller DM, Thomashow LS, Flynt AS, Mavrodi DV (2021) The effect of root exudates on the transcriptome of rhizosphere *Pseudomonas* spp. bioRxiv. 2021.01.08.425997. https://doi.org/10.1101/2021.01.08.425997

McManus P, Hortin J, Anderson AJ, Jacobson AR, Britt DW, Stewart J, McLean JE (2018) Rhizosphere interactions between copper oxide nanoparticles and wheat root exudates in a sand matrix: influences on copper bioavailability and uptake. Environ Toxicol Chem 37:2619–2632. https://doi.org/10.1002/etc.4226

Mermigka G, Amprazi M, Mentzelopoulou A, Amartolou A, Sarris PF (2020) Plant and animal innate immunity complexes: fighting different enemies with similar weapons. Trends Plant Sci 25:80–91. https://doi.org/10.1016/j.tplants.2019.09.008

Miller SH, Browne P, Prigent-Combaret C, Combes-Meynet E, Morrissey JP, O'Gara F (2010) Biochemical and genomic comparison of inorganic phosphate solubilization in *Pseudomonas* species. Environ Microbiol Rep 2:403–411. https://doi.org/10.1111/j.1758-2229.2009.00105.x

Molina MA, Ramos JL, Espinosa-Urgel M (2006) A two-partner secretion system is involved in seed and root colonization and iron uptake by *Pseudomonas putida* KT2440. Environ Microbiol 8:639–647. https://doi.org/10.1111/j.1462-2920.2005.00940.x

Molina L, Rosa R, Nogales J, Rojo F (2019) *Pseudomonas putida* KT2440 metabolism undergoes sequential modifications during exponential growth in a complete medium as compounds are gradually consumed. Environ Microbiol 21:2375–2390. https://doi.org/10.1111/1462-2920.14622

Monds RD, Silby MW, Mahanty HK (2001) Expression of the Pho regulon negatively regulates biofilm formation by *Pseudomonas aureofaciens* PA147-2. Mol Microbiol 42:415–426. https://doi.org/10.1046/j.1365-2958.2001.02641.x

Monds RD, Newell PD, Gross RH, O'Toole GA (2007) Phosphate-dependent modulation of c-di-GMP levels regulates *Pseudomonas fluorescens* Pf0-1 biofilm formation by controlling secretion of the adhesin LapA. Mol Microbiol 63:656–679. https://doi.org/10.1111/j.1365-2958.2006.05539.x

Nam HS, Spencer M, Anderson AJ, Cho BH, Kim YC (2003) Transcriptional regulation and mutational analysis of a *dctA* gene encoding an organic acid transporter protein from *Pseudomonas chlororaphis* O6. Gene 323:125–131. https://doi.org/10.1016/j.gene.2003.09.012

Nam HS, Anderson AJ, Yang KY, Cho BH, Kim YC (2006) The *dctA* gene of *Pseudomonas chlororaphis* O6 is under RpoN control and is required for effective root colonization and induction of systemic resistance. FEMS Microbiol Lett 256:98–104. https://doi.org/10.1111/j.1574-6968.2006.00092.x

Nandi M, Berry C, Brassinga AK, Belmonte MF, Fernando WG, Loewen PC, de Kievit TR (2016) *Pseudomonas brassicacearum* strain DF41 kills *Caenorhabditis elegans* through biofilm-dependent and biofilm-independent mechanisms. Appl Environ Microbiol 82:6889–6898. https://doi.org/10.1128/AEM.02199-16

Neal AL, Ahmad S, Gordon-Weeks R, Ton J (2012) Benzoxazinoids in root exudates of maize attract *Pseudomonas putida* to the rhizosphere. PLoS One 7:e35498. https://doi.org/10.1371/journal.pone.0035498

Newman MA, Sundelin T, Nielsen JT, Erbs G (2013) MAMP (microbe-associated molecular pattern) triggered immunity in plants. Front Plant Sci 4:139. https://doi.org/10.3389/fpls.2013.00139

Nicastro GG, Kaihami GH, Pulschen AA, Hernandez-Montelongo J, Boechat AL, de Oliveira Pereira T, Rosa CGT, Stefanello E, Colepicolo P, Bordi C, Baldini RL (2020) c-di-GMP-related phenotypes are modulated by the interaction between a diguanylate cyclase and a polar hub protein. Sci Rep 10:3077. https://doi.org/10.1038/s41598-020-59536-9

Noirot-Gros MF, Forrester S, Malato G, Larsen PE, Noirot P (2019) CRISPR interference to interrogate genes that control biofilm formation in *Pseudomonas fluorescens*. Sci Rep 9:15954. https://doi.org/10.1038/s41598-019-52400-5

O'Neill EM, Mucyn TS, Patteson JB, Finkel OM, Chung EH, Baccile JA, Massolo E, Schroeder FC, Dangl JL, Li B (2018) Phevamine A, a small molecule that suppresses plant immune responses. Proc Natl Acad Sci U S A 115:E9514–E9522. https://doi.org/10.1073/pnas.1803779115

O'Toole GA, Kolter R (1998) Flagellar and twitching motility are necessary for *Pseudomonas aeruginosa* biofilm development. Mol Microbiol 30:295–304. https://doi.org/10.1046/j.1365-2958.1998.01062.x

Okegbe C, Fields BL, Cole SJ, Beierschmitt C, Morgan CJ, Price-Whelan A, Stewart RC, Lee VT, Dietrich LEP (2017) Electron-shuttling antibiotics structure bacterial communities by modulating cellular levels of c-di-GMP. Proc Natl Acad Sci U S A 114:E5236–E5245. https://doi.org/10.1073/pnas.1700264114

Oku S, Komatsu A, Tajima T, Nakashimada Y, Kato J (2012) Identification of chemotaxis sensory proteins for amino acids in *Pseudomonas fluorescens* Pf0-1 and their involvement in chemotaxis to tomato root exudate and root colonization. Microbes Environ 27:462–629. https://doi.org/10.1264/jsme2.me12005

Oku S, Komatsu A, Nakashimada Y, Tajima T, Kato J (2014) Identification of *Pseudomonas fluorescens* chemotaxis sensory proteins for malate, succinate, and fumarate, and their involvement in root colonization. Microbes Environ 29:413–419. https://doi.org/10.1264/jsme2.ME14128

Oota M, Tsai AY, Aoki D, Matsushita Y, Toyoda S, Fukushima K, Saeki K, Toda K, Perfus-Barbeoch L, Favery B, Ishikawa H, Sawa S (2020) Identification of naturally occurring polyamines as root-knot nematode attractants. Mol Plant 13:658–665. https://doi.org/10.1016/j.molp.2019.12.010

Oteino N, Lally RD, Kiwanuka S, Lloyd A, Ryan D, Germaine KJ, Dowling DN (2015) Plant growth promotion induced by phosphate solubilizing endophytic *Pseudomonas* isolates. Front Microbiol 6:745. https://doi.org/10.3389/fmicb.2015.00745

Park JY, Oh SA, Anderson AJ, Neiswender J, Kim JC, Kim YC (2011) Production of the antifungal compounds phenazine and pyrrolnitrin from *Pseudomonas chlororaphis* O6 is differentially regulated by glucose. Lett Appl Microbiol 52:532–537. https://doi.org/10.1111/j.1472-765X.2011.03036.x

Park JY, Kang BR, Ryu CM, Anderson AJ, Kim YC (2018) Polyamine is a critical determinant of *Pseudomonas chlororaphis* O6 for GacS-dependent bacterial cell growth and biocontrol capacity. Mol Plant Pathol 19:1257–1266. https://doi.org/10.1111/mpp.12610

Park H, McGill SL, Arnold AD, Carlson RP (2020) Pseudomonad reverse carbon catabolite repression, interspecies metabolite exchange, and consortial division of labor. Cell Mol Life Sci 77:395–413. https://doi.org/10.1007/s00018-019-03377-x

Paulin MM, Novinscak A, Lanteigne C, Gadkar VJ, Filion M (2017) Interaction between 2,4-diacetylphloroglucinol- and hydrogen cyanide-producing *Pseudomonas brassicacearum* LBUM300 and *Clavibacter michiganensis* subsp. *michiganensis* in the tomato rhizosphere. Appl Environ Microbiol 83(13):e00073–e00017. https://doi.org/10.1128/AEM.00073-17

Pfeilmeier S, Saur IM-L, Rathjen JP, Zipfel C, Malone JG (2016) High levels of cyclic-di-GMP in plant-associated *Pseudomonas* correlate with evasion of plant immunity. Mol Plant Pathol 17: 521–531. https://doi.org/10.1111/mpp.12297

Pieterse CMJ, Van Pelt JA, Van Wees SCM, Ton J, Léon-Kloosterziel KM, Keurentjes JJB, Verhagen BWM, Knoester M, Van der Sluis J, Bakker PAHM, Van Loon LC (2001) Rhizobacteria-mediated induced systemic resistance: triggering, signalling and expression. Eur J Plant Pathol 107:51–61

Pishchany G, Kolter R (2020) On the possible ecological roles of antimicrobials. Mol Microbiol 113:580–587. https://doi.org/10.1111/mmi.14471

Pliego C, de Weert S, Lamers G, de Vicente A, Bloemberg G, Cazorla FM, Ramos C (2008) Two similar enhanced root-colonizing *Pseudomonas* strains differ largely in their colonization strategies of avocado roots and *Rosellinia necatrix* hyphae. Environ Microbiol 10:3295–3304. https://doi.org/10.1111/j.1462-2920.2008.01721.x

Pliego C, Crespo-Gómez JI, Pintado A, Pérez-Martínez I, de Vicente A, Cazorla FM, Ramos C (2019) Response of the biocontrol agent *Pseudomonas pseudoalcaligenes* AVO110 to Rosellinia necatrix exudate. Appl Environ Microbiol 85:e01741–e01718. https://doi.org/10.1128/AEM.01741-18

Polonio A, Vida C, de Vicente A, Cazorla FM (2017) Impact of motility and chemotaxis features of the rhizobacterium *Pseudomonas chlororaphis* PCL1606 on its biocontrol of avocado white root rot. Int Microbiol 20:95–104. https://doi.org/10.2436/20.1501.01.289

Quay SC, Friedman SB, Eisenberg RC (1972) Gluconate regulation of glucose catabolism in Pseudomonas fluorescens. J Bacteriol 112:291–298

Quecine MC, Kidarsa TA, Goebel NC, Shaffer BT, Henkels MD, Zabriskie TM, Loper JE (2015) An interspecies signaling system mediated by fusaric acid has parallel effects on antifungal metabolite production by *Pseudomonas protegens* strain Pf-5 and antibiosis of *Fusarium* spp. Appl Environ Microbiol 82:1372–1382. https://doi.org/10.1128/AEM.02574-15

Quintana-Rodriguez E, Duran-Flores D, Heil M, Camacho-Coronel X (2018) Damage-associated molecular patterns (DAMPs) as future plant vaccines that protect crops from pests. Sci Hortic 237:207–220

Raaijmakers JM, Mazzola M (2012) Diversity and natural functions of antibiotics produced by beneficial and plant pathogenic bacteria. Annu Rev Phytopathol 50:403–424. https://doi.org/10.1146/annurev-phyto-081211-172908

Rinaldo S, Giardina G, Mantoni F, Paone A, Cutruzzolà F (2018) Beyond nitrogen metabolism: nitric oxide, cyclic-di-GMP and bacterial biofilms. FEMS Microbiol Lett 365. https://doi.org/10.1093/femsle/fny029

Römling U, Galperin MY, Gomelsky M (2013) Cyclic di-GMP: the first 25 years of a universal bacterial second messenger. Microbiol Mol Biol Rev 77:1–52. https://doi.org/10.1128/MMBR.00043-12

Rouse SL, Matthews SJ, Dueholm MS (2018) Ecology and biogenesis of functional amyloids in *Pseudomonas*. J Mol Biol 430:3685–3695. https://doi.org/10.1016/j.jmb.2018.05.004

Ryan CA, Pearce G (2003) Systemins: a functionally defined family of peptide signals that regulate defensive genes in *Solanaceae* species. Proc Natl Acad Sci U S A 100(Suppl 2):14577–14580. https://doi.org/10.1073/pnas

Saijo Y, Loo EP (2020) Plant immunity in signal integration between biotic and abiotic stress responses. New Phytol 225:87–104. https://doi.org/10.1111/nph.15989

Salas-González I, Reyt G, Flis P, Custódio V, Gopaulchan D, Bakhoum N, Dew TP, Suresh K, Franke RB, Dangl JL, Salt DE, Castrillo G (2021) Coordination between microbiota and root endodermis supports plant mineral nutrient homeostasis. Science 371(6525):eabd0695. https://doi.org/10.1126/science.abd0695

Sasnow SS, Wei H, Aristilde L (2016) Bypasses in intracellular glucose metabolism in iron-limited *Pseudomonas putida*. Microbiologyopen 5(1):3–20. https://doi.org/10.1002/mbo3.287

Sasse J, Martinoia E, Northen T (2018) Feed your friends: do plant exudates shape the root microbiome? Trends Plant Sci 23:25–41. https://doi.org/10.1016/j.tplants.2017.09.003

Sauer K, Cullen MC, Rickard AH, Zeef LAH, Davies DG, Gilbert P (2004) Characterization of nutrient-induced dispersion in *Pseudomonas aeruginosa* PAO1 biofilm. J Bacteriol 186:7312–7326

Schenk ST, Stein E, Kogel KH, Schikora A (2012) *Arabidopsis* growth and defense are modulated by bacterial quorum sensing molecules. Plant Signal Behav 7:178–181. https://doi.org/10.4161/psb.18789

Schenk ST, Hernández-Reyes C, Samans B, Stein E, Neumann C, Schikora M, Reichelt M, Mithöfer A, Becker A, Kogel KH, Schikora A (2014) N-Acyl-homoserine lactone primes plants for cell wall reinforcement and induces resistance to bacterial pathogens via the salicylic acid/oxylipin pathway. Plant Cell 26:2708–2723. https://doi.org/10.1105/tpc.114.126763

Shah N, Gislason AS, Becker M, Belmonte MF, Fernando WGD, de Kievit TR (2020) Investigation of the quorum-sensing regulon of the biocontrol bacterium *Pseudomonas chlororaphis* strain PA23. PLoS One 15(2):e0226232. https://doi.org/10.1371/journal.pone.0226232

Shen L, Wu XQ, Zeng QW, Liu HB (2016) Regulation of soluble phosphate on the ability of phytate mineralization and β-propeller phytase gene expression of *Pseudomonas fluorescens* JZ-DZ1, a phytate-mineralizing rhizobacterium. Curr Microbiol 73:915–923. https://doi.org/10.1007/s00284-016-1139-0

Shinde S, Zerbs S, Collart FR, Cumming JR, Noirot P, Larsen PE (2019) *Pseudomonas fluorescens* increases mycorrhization and modulates expression of antifungal defense response genes in roots of aspen seedlings. BMC Plant Biol 19:4. https://doi.org/10.1186/s12870-018-1610-0

Smirnoff N, Arnaud D (2019) Hydrogen peroxide metabolism and functions in plants. New Phytol 221:1197–1214. https://doi.org/10.1111/nph.15488

Sobrero PM, Valverde C (2020) Comparative genomics and evolutionary analysis of RNA-binding proteins of the CsrA Family in the genus *Pseudomonas*. Front Mol Biosci 7:127

Spencer M, Ryu C-M, Yang KY, Kim YC, Kloepper JW, Anderson AJ (2003) Induced defence in tobacco by *Pseudomonas chlororaphis* O6 involves at least the ethylene pathway. Physiol Mol Plant Pathol 63:27–34

Spiers AJ, Rainey PB (2005) The *Pseudomonas fluorescens* SBW25 wrinkly spreader biofilm requires attachment factor, cellulose fibre and LPS interactions to maintain strength and integrity. Microbiology 151:2829–2839. https://doi.org/10.1099/mic.0.27984-0

Spiers AJ, Buckling A, Rainey PB (2000) The causes of *Pseudomonas* diversity. Microbiology 146: 2345–2350. https://doi.org/10.1099/00221287-146-10-2345

Stassen MJJ, Hsu SH, Pieterse CMJ, Stringlis IA (2021) Coumarin communication along microbiome-root-shoot axis. Trends Plant Sci 26:169–183. https://doi.org/10.1016/j.tplants.2020.09.008

Stringlis IA, Yu K, Feussner K, de Jonge R, Van Bentum S, Van Verk MC, Berendsen RL, Bakker PAHM, Feussner I, Pieterse CMJ (2018a) MYB72-dependent coumarin exudation shapes root microbiome assembly to promote plant health. Proc Natl Acad Sci U S A 115:E5213–E5222. https://doi.org/10.1073/pnas.1722335115

Stringlis IA, Proietti S, Hickman R, Van Verk MC, Zamioudis C, Pieterse CMJ (2018b) Root transcriptional dynamics induced by beneficial rhizobacteria and microbial immune elicitors reveal signatures of adaptation to mutualists. Plant J 93:166–180. https://doi.org/10.1111/tpj.13741

Stringlis IA, de Jonge R, Pieterse CMJ (2019) The age of coumarins in plant-microbe interactions. Plant Cell Physiol 60:1405–1419. https://doi.org/10.1093/pcp/pcz076

Svenningsen NB, Martínez-García E, Nicolaisen MH, de Lorenzo V, Nybroe O (2018) The biofilm matrix polysaccharides cellulose and alginate both protect *Pseudomonas putida* mt-2 against reactive oxygen species generated under matric stress and copper exposure. Microbiology 164: 883–888. https://doi.org/10.1099/mic.0.000667

Takishita Y, Charron JB, Smith DL (2018) Biocontrol rhizobacterium *Pseudomonas* sp. 23S induces systemic resistance in tomato (*Solanum lycopersicum* L.) against bacterial canker *Clavibacter michiganensis* subsp. *michiganensis*. Front Microbiol 9:2119. https://doi.org/10.3389/fmicb.2018.02119

Thiele B, Füllner K, Stein N, Oldiges M, Kuhn AJ, Hofmann D (2008) Analysis of amino acids without derivatization in barley extracts by LC-MS-MS. Anal Bioanal Chem 391:2663–2672. https://doi.org/10.1007/s00216-008-2167-9

Thomashow LS, Kwak Y-S, Weller DM (2019) Root-associated microbes in sustainable agriculture: models, metabolites and mechanisms. Pest Manag Sci 75:2360–2367

Tian T, Reverdy A, She Q, Sun B, Chai Y (2020) The role of rhizodeposits in shaping rhizomicrobiome. Environ Microbiol Rep 12:160–172. https://doi.org/10.1111/1758-2229.12816

Timmusk S, Behers L, Muthoni J, Muraya A, Aronsson A-C (2017) Perspectives and challenges of microbial application for crop improvement. Front Plant Sci 8:49

Tovi N, Frenk S, Hadar Y, Minz D (2019) Host specificity and spatial distribution preference of three *Pseudomonas* isolates. Front Microbiol 9:3263. https://doi.org/10.3389/fmicb.2018.03263

Ude S, Arnold DL, Moon CD, Timms-Wilson T, Spiers AJ (2006) Biofilm formation and cellulose expression among diverse environmental *Pseudomonas* isolates. Environ Microbiol 8:1997–2011. https://doi.org/10.1111/j.1462-2920.2006.01080.x

Upadhyay A, Kochar M, Rajam MV, Srivastava S (2017) Players over the surface: unraveling the role of exopolysaccharides in zinc biosorption by fluorescent *Pseudomonas* strain Psd. Front Microbiol 8:284. https://doi.org/10.3389/fmicb.2017.00284

Uroz S, Calvaruso C, Turpault MP, Frey-Klett P (2009) Mineral weathering by bacteria: ecology, actors and mechanisms. Trends Microbiol 17:378–387. https://doi.org/10.1016/j.tim.2009.05.004

Valentini M, Filloux A (2016) Biofilms and cyclic di-GMP (c-di-GMP) signaling: lessons from *Pseudomonas aeruginosa* and other bacteria. J Biol Chem 291:12547–12555. https://doi.org/10.1074/jbc.R115.711507

Van Acker H, Coenye T (2017) The role of reactive oxygen species in antibiotic-mediated killing of bacteria. Trends Microbiol 25:456–466. https://doi.org/10.1016/j.tim.2016.12.008

van Butselaar T, Van den Ackerveken G (2020) Salicylic acid steers the growth-immunity tradeoff. Trends Plant Sci 25:566–576. https://doi.org/10.1016/j.tplants.2020.02.002

van der Ent S, Koorneef A, Ton J, Pieterse CMJ (2018) Induced resistance – orchestrating defence mechanisms through crosstalk and priming. Annu Plant Rev 34:334–370. https://doi.org/10.1002/9781119312994.apr0371

van Rij ET, Girard G, Lugtenberg BJJ, Bloemberg GV (2005) Influence of fusaric acid on phenazine-1-carboxamide synthesis and gene expression of *Pseudomonas chlororaphis* strain PCL1391. Microbiology 151:2805–2814. https://doi.org/10.1099/mic.0.28063-0

Vives-Peris V, Molina L, Segura A, Gómez-Cadenas A, Pérez-Clemente RM (2018) Root exudates from citrus plants subjected to abiotic stress conditions have a positive effect on rhizobacteria. J Plant Physiol 228:208–217. https://doi.org/10.1016/j.jplph.2018.06.003

Vives-Peris V, de Ollas C, Gómez-Cadenas A, Pérez-Clemente RM (2020) Root exudates: from plant to rhizosphere and beyond. Plant Cell Rep 39:3–17. https://doi.org/10.1007/s00299-019-02447-5

Wagner MR, Lundberg DS, Del Rio TG, Tringe SG, Dangl JL, Mitchell-Olds T (2016) Host genotype and age shape the leaf and root microbiomes of a wild perennial plant. Nat Commun 7:12151. https://doi.org/10.1038/ncomms12151

Wang D, Yu JM, Dorosky RJ, Pierson LS III, Pierson EA (2016) The phenazine 2-hydroxy-phenazine-1-carboxylic acid promotes extracellular DNA release and has broad transcriptomic consequences in *Pseudomonas chlororaphis* 30-84. PLoS One 11(1):e0148003. https://doi.org/10.1371/journal.pone.0148003

Weisskopf L, Ryu C-M, Raaijmakers JM, Garbeva P (2016) Editorial: Smelly fumes: volatile-mediated communication between bacteria and other organisms. Front Microbiol. https://doi.org/10.3389/fmicb.2016.02031

Wilton M, Charron-Mazenod L, Moore R, Lewenza S (2015) Extracellular DNA acidifies biofilms and induces aminoglycoside resistance in *Pseudomonas aeruginosa*. Antimicrob Agents Chemother 60:544–553. https://doi.org/10.1128/AAC.01650-15

Wright M, Adams J, Yang K, McManus P, Jacobson A, Gade A, McLean J, Britt D, Anderson A (2016) A root-colonizing pseudomonad lessens stress responses in wheat imposed by CuO nanoparticles. PLoS One 11:e0164635. https://doi.org/10.1371/journal.pone.0164635

Wu G, Shortt BJ, Lawrence EB, Levine EB, Fitzsimmons KC, Shah DM (1995) Disease resistance conferred by expression of a gene encoding H_2O_2-generating glucose oxidase in transgenic potato plants. Plant Cell 7:1357–1368. https://doi.org/10.1105/tpc.7.9.1357

Wu L, Wang Z, Guan Y, Huang X, Shi H, Liu Y, Zhang X (2020) The (p)ppGpp-mediated stringent response regulatory system globally inhibits primary metabolism and activates secondary metabolism in *Pseudomonas protegens* H78. Appl Microbiol Biotechnol 104:3061–3079. https://doi.org/10.1007/s00253-020-10421-5

Xiao Y, Zhu W, He M, Nie H, Chen W, Huang Q (2019) High c-di-GMP promotes expression of *fpr-1* and *katE* involved in oxidative stress resistance in *Pseudomonas putida* KT2440. Appl Microbiol Biotechnol 103:9077–9089. https://doi.org/10.1007/s00253-019-10178-6

Yang F, Cao Y (2012) Biosynthesis of phloroglucinol compounds in microorganisms—review. Appl Microbiol Biotechnol 93:487–495. https://doi.org/10.1007/s00253-011-3712-

Yang CH, Menge JA, Cooksey DA (1994) Mutations affecting hyphal colonization and pyoverdine production in pseudomonads antagonistic toward *Phytophthora parasitica*. Appl Environ Microbiol 60:473–481

Yasmin S, Hafeez FY, Mirza MS, Rasul M, Arshad HMI, Zubair M, Iqbal M (2017) Biocontrol of bacterial leaf blight of rice and profiling of secondary metabolites produced by rhizospheric *Pseudomonas aeruginosa* BRp3. Front Microbiol 8:1895. https://doi.org/10.3389/fmicb.2017.01895

Yousef-Coronado F, Travieso ML, Espinosa-Urgel M (2008) Different, overlapping mechanisms for colonization of abiotic and plant surfaces by *Pseudomonas putida*. FEMS Microbiol Lett 288:118–124. https://doi.org/10.1111/j.1574-6968.2008.01339.x

Yu JM, Wang D, Pierson LS, Pierson EA (2017) Disruption of MiaA provides insights into the regulation of phenazine biosynthesis under suboptimal growth conditions in *Pseudomonas chlororaphis* 30-84. Microbiology 163:94–108. https://doi.org/10.1099/mic.0.000409

Yu K, Liu Y, Tichelaar R, Savant N, Lagendijk E, van Kuijk SJL, Stringlis IA, van Dijken AJH, Pieterse CMJ, Bakker PAHM, Haney CH, Berendsen RL (2019a) Rhizosphere-associated *Pseudomonas* suppress local root immune responses by gluconic acid-mediated lowering of environmental pH. Curr Biol 29:3913–3920.e4. https://doi.org/10.1016/j.cub.2019.09.015

Yu K, Pieterse CMJ, Bakker PAHM, Berendsen RL (2019b) Beneficial microbes going underground of root immunity. Plant Cell Environ 42(10):2860–2870. https://doi.org/10.1111/pce.13632

Yu XQ, Yan X, Zhang MY, Zhang LQ, He YX (2020) Flavonoids repress the production of antifungal 2,4-DAPG but potentially facilitate root colonization of the rhizobacterium *Pseudomonas fluorescens*. Environ Microbiol. https://doi.org/10.1111/1462-2920.15052.1111/pce.13632

Yuan J, Zhao J, Wen T, Zhao M, Li R, Goossens P, Huang Q, Bai Y, Vivanco JM, Kowalchuk GA, Berendsen RL, Shen Q (2018) Root exudates drive the soil-borne legacy of aboveground pathogen infection. Microbiome 6:156. https://doi.org/10.1186/s40168-018-0537-x

Zboralski A, Biessy A, Savoie MC, Novinscak A, Filion M (2020) Metabolic and genomic traits of phytobeneficial phenazine-producing *Pseudomonas* spp. are linked to rhizosphere colonization in *Arabidopsis thaliana* and *Solanum tuberosum*. Appl Environ Microbiol 86:e02443–e02419. https://doi.org/10.1128/AEM.02443-19

Zdor RE, Anderson AJ (1992) Influence of root colonizing bacteria on the defense responses of bean. Plant Soil 140:99–107

Zeng G, Vad BS, Dueholm MS, Christiansen G, Nilsson M, Tolker-Nielsen T, Nielsen PH, Meyer RL, Otzen DE (2015) Functional bacterial amyloid increases *Pseudomonas* biofilm hydrophobicity and stiffness. Front Microbiol 6:1099. https://doi.org/10.3389/fmicb.2015.01099

Zhou JM, Zhang Y (2020) Plant immunity: danger perception and signaling. Cell 181:978–989. https://doi.org/10.1016/j.cell.2020.04.028

Chapter 9
Plant Microbiome Modulation Through Seed Coating: A Novel Approach for a Smart and Efficient Microbial Delivery

Oumaima Zaher, Manal Mhada, Marwa El Graoui, Augustine T. Zvinavashe, Lamfeddal Kouisni, and Benedetto Marelli

Abstract Seeds represent the most vulnerable and critical stages of a plant's life, making them susceptible to several factors like genotype, environment, and seed quality, affecting germination, emergence, and vigor. Seed coating provides a (1) protective layer that improves plant early-stages development through nutrients, (2) plant growth-promoting microbes, and (3) biocontrol agents. In addition, its uniform application and precise supply contribute to making it a cost-effective technique for product delivery, germination homogeneity, seedling establishment, and preventing deficiencies in later growth stages. As only the needed amount of fertilization is provided to the seed, seed coating is considered a targeted application method for microorganism's supply and a cost-effective and sustainable one for macro- and micronutrient delivery.

Within this context, beneficial bacteria are of primary importance in alleviating environmental stresses and enabling a specific resistance to the plant. Bilateral interaction is the core of the synergetic relationship between a plant and its associated microbiome and helps shape this latter and modulate it for optimal performance. However, when it comes to using plant growth-promoting bacteria (PGPB), combined with an efficient seed coating technology, the challenge lies in the viability of the bacteria under this protective layer. Therefore, the actual market needs more adaption for both seed coating material and bacterial inoculum despite the rising

---

O. Zaher · M. Mhada · M. El Graoui
AgroBiosciences Department (AgBS), Mohammed VI Polytechnic University (UM6P), Ben-Guerir, Morocco

A. T. Zvinavashe · B. Marelli (✉)
Department of Civil and Environmental Engineering, Massachusetts Institute of Technology, Cambridge, MA, USA
e-mail: bmarelli@mit.edu

L. Kouisni
AgroBiosciences Department (AgBS), Mohammed VI Polytechnic University (UM6P), Ben-Guerir, Morocco

African Sustainable Agriculture Research Institute, Mohammed VI Polytechnic University (ASARI-UM6P), Laayoune, Morocco

B. A. Horwitz, P. K. Mukherjee (eds.), *Microbial Cross-talk in the Rhizosphere*, Rhizosphere Biology, https://doi.org/10.1007/978-981-16-9507-0_9

demand. Thus, the present chapter aims to discuss the opportunities and challenges of microbial seed coating technology and the mechanisms behind seed bacterial treatments.

Keywords Seed · Coating · Biopolymers · Salinity · Drought · Bioformulation

9.1 Introduction

In the post-green revolution era, the agricultural sector has reached its peak with the formulation of novel fertilizers and appropriate crop management. However, the hidden side—and disadvantageous one—of this green revolution is the excessive use of agricultural inputs, as it is challenging to have a uniform application over the soil surface, inducing environmental losses and constraints (eutrophication, alteration of soil microbial diversity, an increase of soil salinity, etc.) (Tuğrul 2019). Moreover, with the emergence of various severe challenges, we have witnessed a decrease in crop production, mainly due to climatic fluctuations, resulting in the emergence of new diseases, water scarcity, and soil salinization.

Cutting-edge strategies such as seed coating with "smart delivery" respond to today's agricultural needs, integrating basic research to advanced technologies. An example is the embodiment of rhizobacteria in a sustainable coating layer, to improve seed germination and seedlings development in harsh environments. The main challenges of an effective coating are its ability to keep the bacteria alive, the biofertilizer active, and keep nutrients in the seed microenvironment for efficient use (Zvinavashe et al. 2021b).

In addition to the environmental and ecological aspects covered, socio-economic values are equally important. Indeed, a precise seed placement enables a reduction of up to 50% of the applied formulations (fertilizers, PGPBs, etc.) (Bashan et al. 2014). Besides, the addition of this protective layer may prevent seed alteration and loss during storage and transportation and boost the germination of recalcitrant and valuable seeds (Afzal et al. 2020; Parisi et al. 2016). This technology will reduce the expenses of small-holder farmers, seed production companies, and research institutions.

Seed coating has proven its efficiency and has been used by seed companies. Layers composition was adapted to the targeted challenge. A mechanical protective layer was used to improve seed to soil contact and to protect seeds during storage. A biological protective layer contains living cells known for promoting growth by improving the seed microenvironment to facilitate access to nutrients and water. The same type of layer is used to protect the seed from insect predation and pathogens. In the absence of biotic or abiotic stress, seed coating can help in improving seedling survival and establishment.

Through this review, we aim to summarize the importance of the application of suitable seed coatings and to highlight that by adopting seed coating, instead of liquid or dry formulations, many practical challenges can be solved.

9.2 General Overview on the Seed Coating

9.2.1 Seed Coating Definition

Seed coating is a technique developed through the mid-nineteenth century used to improve sowing through increasing seed size and uniformity. This technique evolved in the mid-twentieth century from changing seed size and shape to producing coating formulas that influence the seed microenvironment. Coating formulas can contain pesticides, fertilizers, polymers, and growth stimulators (Afzal et al. 2020; Leinauer et al. 2010).

Seed coating is considered a tool for achieving accurate sowing, improving germination and seed emergence, and increasing seed performance. Through the application of exogenous materials on the surface of the seed coat, seed coating is used by the horticultural and crop industries worldwide for modifying the seed physical properties, applying tracer, colors, protectants, soil adjuvants, stimulating compounds (germination, growth, and stress resistance), nutrients, and plant beneficial microbes' inoculants (PBM) (Pedrini et al. 2017).

9.2.2 Seed Coating Formulation

Seed coating formulations contain active ingredients and carrier materials (polymeric or hygroscopic). These carriers are categorized according to their function as nutrients (P, K, Cu, Mn, and Zn) or fillers (cellulose, chitosan, talc, biochar, vermiculite, silk) used to increase seed shape, size, and weight. Most of the carriers are either organic, inorganic, or synthetic compounds. Carrier choice depends on the safety, availability of nutrients and micropores for PGPM inhabitancy, effect on the environment, and cost. For microbial seed coating, carriers should be chosen depending on their adherence to the seeds, effect on germination, seedling development, PGPM survival, and shelf life. Seed coating binders are adhesives added for adherence of the treatments to seeds to maintain coating integrity during and after drying by preventing cracking and dusting-off during handling and sowing. Binders can be used to extend the survival of microbial inoculants, for structural support, and for retention of active ingredients (Ma 2019). Most used binders include Gum Arabic, plant starches, methyl cellulose, and polyvinyl acetate (Table 9.1).

Table 9.1 Different biopolymers used in seed coating with proprieties and uses

Coating agents	Uses	Properties	Examples	References
Silica	Carrier for microorganism encapsulation.	Electrostatic Structure, degradability biosecurity and biocompatibility.	*Rhizophagus irregularis* and silicon dioxide in combination with *Pseudomonas libanensis* for cowpea production.	Ma et al. (2019), Ma et al. (2019)
Biochar	Seed coating, microorganism carrier.	Porous structure, large surface area, and negatively charged surface area increased water holding capacity and plant-nutrient retention.	Biochar seed coating on mountain brome (Bromus marginatus)	Williams et al. (2016)
Carboxymethyl cellulose (CMC)	Binder, and carrier in film coating.	Survival of seed inoculants, and high-water retention.	Carboxymethylcellulose coating with *Pseudomonas putida inoculation* on *Helianthus annuus* seeds.	Ma et al. (2019)
Polyvinyl alcohol (PVA)	Carrier	Biocompatibility, high water solubility, and chemical resistance	Polyvinyl alcohol (PVA) for Rhizobia survival on soybean seeds.	Damasceno et al. (2013)
Gum Arabic	Binding, film-forming carrier.	Abundant, cost-effective, biocompatibility, and water retention.	Effect of gum Arabic coatings with an antimicrobial agent on postharvest quality of strawberry.	Wani et al. (2021)
Bentonite	Carrier, adhesive, filler.	High swelling rate, adsorption, and colloidal properties	Evaluating the physical and physiological quality of *Brassica napus L.* seeds coated with bentonite.	de Melo et al. (2019)
Silk fibroin	Carrier in seed coating.	Stability, and controlled degradation.	Silk fibroin coating with *Rhizobium tropici* on *Phaseolus vulgaris*	Zvinavashe et al. (2019)
Xanthan gum	Microorganism carrier	Survival of seed inoculants, fast-hydrating, water-soluble hydrocolloid	Seeds coated with a K-165 bacterial formulation in 10% xanthan gum-talc	Ikpe et al. (2018), Ma et al. (2019), Schoina et al. (2011)
Alginate	Carrier, filler, and binder.	Biodegradable polymer, nontoxic, and easy to handle.	Seeds coated with K-165 encapsulated in sodium alginate-Pyrax	He et al. (2015), Rocha et al. (2019), Schoina et al. (2011)

9.2.3 *Active Ingredients*

9.2.3.1 Fertilizers–Nutrients

Through fertilizer delivery, adequate nutrient availability for the plant at the early growth stages could be achieved. Seed coating with appropriate amounts of macronutrients, micronutrients (copper, manganese, and zinc), or plant substances like biochar can reduce nutrient losses by placement on the seed, reducing competition from weeds. It has been proven that seed coating with nutrients improves germination, growth, and yield. However, macronutrient fertilization causes deleterious effects on seed germination and emergence due to phytotoxicity. To avoid toxicity, adding an initial layer or boundary layer before nutrients during seed coating is advised. For seed coating to be impactful, it should be done through a tailored study of plant requirements, soil types, and nutrient availability or deficiency (Afzal et al. 2020; Ma 2019; Pedrini et al. 2017).

9.2.3.2 Protectants

Protectant compounds include fungicides, pesticides, insecticides, nematicides, predator deterrents, and herbicides. They are used primarily for reducing predation, pathogen infection while they only slightly promote germination and emergence. For example, chitosan (biopolymer) could be used as an antifungal agent to improve the germination and quality of *Cynara scolymus* seeds. However, despite these benefits, some protectants have negative off-target environmental impacts. Fungicidal and insecticidal products indirectly affect the soil seed bank and wild bee diversity and distribution, interfering with agroecosystem processes (Ma 2019; Pedrini et al. 2017).

9.2.3.3 Plant and Algae Extracts

The valorization of plant and algae extract with growth stimulant functions through seed coating improves plant growth and physiological functions. For example, the combination of soy flour, cellulose, and diatomaceous earth with *Brassica oleracea* seeds significantly enhanced plant growth and biomass (Amirkhani et al. 2016; Madsen et al. 2016). Thus, plant and algae extract serve as sustainable, inexpensive, and green sources for the seed coating process. However, the type of growth stimulant used in seed coating can hinder germination, root growth, functions of biocontrol, and fertilization, and bind the seeds to the surrounding soils (Ma 2019).

9.2.3.4 Microorganisms

Rhizobia is the most used PGPM in seed coating for pulse crops due to its benefits on seedling growth and germination. It is an effective alternative to growth stimulants. PGPM represents the most efficient solution as it possesses a large range of plant growth-promoting traits (phytohormones and siderophores production) (Ma 2019).

An artificial carrier is a hostile environment for rhizobia due to osmotic and desiccation stress, and the presence of protectant compounds. The integration of inoculum in a coating carrier results in losses of microbial viability resulting in shorter shelf lives. To improve symbiotic organism survival, the choice of desiccation-resistant bacteria and a coating formulation that provides a high cell density of microbial inoculants are crucial (Pedrini et al. 2017). To ensure good shelf life and a controlled release, formulation of a bacterial coating should guarantee that the bacterial cells can either be trapped in a polyelectrolyte complex alternating charge through ionic interactions between acidic and basic component (e.g., chitosan polymer) or a porous inorganic carrier (e.g., calcium alginate hydrogels). For a successful PGPB delivery, many methods were developed. We find alginate microbeads as a substrate, nanofiber immobilized PGPB, a formulation containing liquid carriers, and osmo-protectants (trehalose) as an effective combination (Gregorio et al. 2017; Zvinavashe et al. 2021a). Microbial seed coating has become a promising tool for delivering minor amounts of microbial inoculants into the soil for crop production in precision agriculture scenarios (Ma 2019; Rocha et al. 2019).

9.2.3.5 Markers

Marker substances have been integrated into seed coating for tracing seed batches within the crop supply chain. Among all the markers (visible or invisible dyes, fluorescent, and chip markers), dyes are the most used markers in coating to help users discriminate the origin, variety, or treatment of the seeds and facilitate their sowing operations in the field. The choice of dyes used for labeling should benefit the plant in germination and seedling growth (Ma 2019). According to Tian et al. (2014), dyed coating with rhodamine B in *Vicia faba* seeds improved seedling growth, and the fluorescent dye can be observed in the plant shoot.

9.3 Coating Methods and Equipment

9.3.1 Coating Methods

9.3.1.1 Dry Powder Coating

Dry powder coating is a method used for mixing seeds with a dry powder used for the control of pests, antifungal or antibacterial treatments. Talc and graphite are largely used as dry powders or dusts. Carriers such as soy-based protein work as lubricants for the dry coating to improve seed-to-seed flowability. The dosage of dry powder applied on the seeds ranges from 0.06% to 1% with an amount proportional to seed size (Afzal et al. 2020).

9.3.1.2 Seed Dressing

Seed dressing is a low dosage application method for active materials, mostly chemical plant protectants. The liquid is atomized through a rotary disc onto the seed with a loading rate of 0.05–1% by weight. To optimize loading rate, finishing powders are applied during or after the treatments to absorb excess liquids (Afzal et al. 2020).

9.3.1.3 Film Coating

Film coating is a method that is considered an improved version of the slurry coating (a suspension applied on the seeds in a less firm and uniform layer) (O'Callaghan, 2016). Film coating is an application of a thin layer of coating with limited modification of the shape, size, and weight of the seed. It has lower interference with seed germination and facilitates the release of active ingredients or microbial inoculum (Rocha et al. 2019).

9.3.1.4 Pelleting

Pelleting is a seed coating technique that increases the weight and volume of the seed, as it allows the application of a higher amount of active ingredients, thus modifying the seed morphology into an ovoid or spherical shape (Rocha et al. 2019). It is usually used on small, thin, and irregular seeds, making them larger and easier to handle during the sowing (Mei et al. 2017).

9.3.1.5 Encrusting

Encrusting is a lighter form of pelleting, a more economical technique recommended for singularized seeds, where the seed is coated with a precise amount of active ingredients, causing a significant increase in its weight without changing the original shape of the seed (Zvinavashe et al. 2021b).

9.3.2 Coating Application

Seed coating equipment systems are divided to two types: Batch system and continuous flow system. A batch system applies the treatment on the seeds by batches while the flow system treats the seeds at a given flow rate. These equipment systems are: dry powder applicator, rotary pan, and pelleting pan. Dry powder applicator and rotary coater could be either a batch or continuous flow while most pelleting pan coating is performed on a batch basis. To achieve a good application uniformity and adherence for the five seed coating methods discussed above, three types of seed coating equipment are used separately or in tandem. To deliver the targeted dosage for each seed coating method, a computer metering technology (proportion control) is used to monitor seed flow and seed treatment application (Pedrini et al. 2017).

9.3.3 Challenges

9.3.3.1 Shelf Life

Extended shelf life is the most critical criterion in the commercialization of a specific seed coating. Inoculant survival on seeds depends on adjusting conditions favoring high cell initial loading (age, moisture, density, purity) and those favoring long-term stability. For rhizobia inoculation, it is recommended to test each ingredient in the seed coating treatment for compatibility (Pedrini et al. 2017).

9.3.3.2 Field Performance

Field performance inconsistency is one of the main restraints for the application of PBM-coated seeds. Research on PBM inoculation through seed coating has shown that it negatively affected plant performance or had a short-term effect. Inoculated coated seeds have been reported to have a reduced effect on crop productivity, nodulation, N-fixation, and biocontrol. Microbial seed coating studies are limited to either laboratory or greenhouse tests. However, a minor number of reports have studied all levels, from laboratory to field conditions (Rocha et al. 2019).

9.3.3.3 Cost Efficiency

Despite the positive and exciting results of seed coating, cost, scalability, and benefits on employment have rarely been investigated. To provide more complete proof, the possibility of commercialization of advanced seed coating technologies, a cost/ benefits evaluation should be done. For example, examining cost efficiency by comparing established plants from treated and untreated seed, accounting seeding time, the equipment uses, and the number of work hours for personnel (Pedrini et al. 2020).

9.3.4 Opportunities

Advancement in the seed coating industry has been made through research done by private companies and impact and efficiency evaluation done by the scientific community. It is an evolving field that offers a highly beneficial solution for agricultural and environmental problems (Pedrini et al. 2020; Pedrini et al. 2017).

Seed coating technologies could help achieve a cost-effective ecosystem recovery and food security. They can accelerate or delay a germination response, narrow or broaden the variability in seed germination rate within a population, or compensate for environmental and edaphic variability resulting from ecological disturbance and degradation factors. Recent studies have proven the efficiency of seed coating for wild seeds, and crop farming by directing seed-coating solutions for maximizing seedling establishment, plant growth, and yield. The use of biostimulants can also improve seed germination, seedling establishment, stress resistance, potentially reducing the need for environmentally harmful protectants and increasing food security (Pedrini et al. 2020, Pedrini et al. 2017).

9.4 Interaction Between the Plant and Its Microbiota

In the last few years of research, importance has been given predominantly to the two-way plant-microbe interactions, leaving behind third parties participating in this bilateral communication.

While "direct" plant growth and health promoting properties—as phytohormone production, increased nutrient availability, plant pathogen inhibition, and regulation of stress hormonal status—are still being studied (Oleńska et al. 2020), in-depth interactions between plants and their associated microbiome are largely overlooked.

The two examples discussed below, though recognized by the scientific community, have been understudied in comparison with what have become "standard" PGPB properties.

9.4.1 miRNAs as Key Mediators

Horizontal gene transfer (HGT) refers to the movement of genetic material to a "host" organism and is considered a significant contributor when it comes to microbial evolution. It leads to the gain of adaptive functions and, therefore, plays a key role in evolutionary shifts of a species (Van Etten and Bhattacharya 2020).

Interest in gene transfer (GT) has spiraled in the last decade. Although little attention has been given to GT when it comes to shaping the microbiota, evidence of this phenomenon can be traced back as far as 1999 (Nelson et al. 1999).

In this scope, we are focusing on miRNAs, which are small noncoding eukaryotic RNAs that regulate gene expression in almost all cellular processes: plant immune responses, host-pathogen interaction, and communication between plants and their associated microbes (Huang et al. 2019).

miRNAs are mobile mediators, allowing cross-kingdom conversations and trafficking between the host plant and its associated microbiome (Weiberg et al. 2015). Their mobility through plant tissues mediates communication between plant's shoot and root, therefore coordinating internal and external responses of the two compartments, and beyond, modulating microbial communities in a coevolutive way.

Such communication allows these small RNAs to participate in plant growth and innate immunity. However, as important as their contribution is, the translocation of the miRNAs from the hosts to the interacting microorganisms has been understudied due to limitations and lack of affordable technical tools.

Most findings have been orientated towards plant-pathogen interaction and plant immune responses. Studies revealed the role of miRNAs in the orchestration of plant innate immunity, by either being induced or suppressed when invaded by pathogenic microorganisms (Wang and Galili 2019).

On the other hand, very few but relevant studies addressed the bidirectional connection occurring inside the complex dynamic of the holobiont. It has been suggested that miRNAs constitute a communication channel between plants and rhizospheric microbial communities and, therefore, help shaping the soil microbiota (Middleton et al. 2021).

Subramanian (2019) focused on unraveling the mechanisms behind cell-to-cell movement of miRNAs involved in plant nutrient deficiency (Subramanian 2019). Furthermore, several researchers suggested that miRNAs play a key role in nodule formation and functioning in legume plants (Hoang et al. 2020). Indeed, Tsikou et al. recently demonstrated that these small RNAs translocate from shoot to root where they regulate a symbiosis repressor (TOO MUCH LOVE), thus controlling rhizobial infection (Tsikou et al. 2018).

9.4.2 Root Exudates

Root exudates are known for their role in facilitating plant-microbe interactions (Wilhelm et al., 2021). However, little is known on metabolites acting in the backstage of these interfaces, commonly named root exudates.

Root exudates refer to substances secreted from plant roots into the soil. They work as mediators of rhizospheric synchronization by providing a suitable environment for soil microorganisms. By feeding microbes, which in turn feed the plant, they participate in a chain reaction resulting in growth promotion, improved root architecture, stress resistance, etc.

Indeed, root exudates were found to be involved in several biological processes and their role in microbial recruitment. They have been documented to influence microbial ecosystem structures as a defense mechanism in drought conditions (Williams and de Vries 2020), as well as regulate rhizospheric bacterial concentrations (Makarova et al. 2021). Evidence has also been found on their role in legume-rhizobia symbiosis (Wang et al., 2021).

Further investigations revealed some of the underlying mechanisms and genes involved in root exudation (Vives-Peris et al. 2020). Haichar et al. also identified exuded molecules involved in the host-symbiont communication pipe (flavonoids, non-flavonoids, strigolactones, etc.) (Haichar et al. 2014).

Finally, yet importantly, Zhao et al. hypothesized that plants may optimize their nutrient use by adjusting their exudation patterns at different growth stages, based on nutrient demand of each phase (Zhao et al. 2021).

9.5 How Can Modulating the Plant's Microbiome Improve Crop Productivity?

The simultaneous occurrence of several abiotic and biotic stresses is threatening both food security and farmers' livelihood. Being associated with global climate change, this will likely be more evident in the future. Drought, soil salinization and disease and pest occurrence are the primary causes of crop losses worldwide.

9.5.1 Abiotic Factors

Plant adaptation to abiotic stresses depends upon intrinsic factors, from the activation of molecular networks involved in stress perception to the expression of specific stress-related genes. Plants are part of an integrative consortium bringing together plant's mechanisms and associated microbes (Wang et al. 2003). The plant-associated microbes are now becoming part of "*Plant Engineering Strategies*" for abiotic stress tolerance. Many studies focused on the effect of PGPB on plants, starting from bacteria isolation, going through their characterization, screening, and identification until the evaluation of their abilities to promote growth. This approach is widely adopted and has been the source of many PGPB that are commercialized in different forms. Under the circumstances where bacteria and their derived compounds can be used, the main challenges that remain are cost and use efficiency (Naamala and Smith 2021).

Besides PGPB, the exogenous application of osmolytes helps in response to osmotic stress. The principal function of osmolytes is to maintain cell turgor that drives the gradient for water uptake. Recent studies indicate that osmolytes can act as chemical chaperones by directly stabilizing membranes and proteins (Rydeen et al. 2018). Osmolytes are divided into three groups: amino acids (e.g., proline), quaternary amines (e.g., glycine betaine), and Pyrols/sugars (e.g., mannitol, trehalose). Phytohormones are also fascinating due to their instantaneous action (e.g., auxin and gibberellic acid). Micro-dose placement of osmolytes and phytohormones can result in improved stress tolerance by reinforcing the PGBP action, improving the seed microenvironment (e.g., pH, nutrients, microbiome), and hence improving seedling vigor and the functioning of the plant immune system.

The major threat of adding foreign bacteria to the indigenous ones found in the local soil is maintaining the already established balance. Many studies showed that by adding PGPB, microbial community composition in open field is not influenced (Shekhawat et al. 2021), indicating that beneficial microbes can be a powerful tool to enhance stress tolerance of crops in a sustainable manner.

Biopriming or direct application of biofertilizers (including PGPB) could be applied in agriculture to make crops more stress resistant and productive (Chua et al. 2020; Panuccio et al. 2018; Paul and Rakshit 2021; Yadav et al. 2018), but the application is not feasible under field conditions (Mitter et al. 2021). In this context, the use of beneficial endophytes and rhizobacteria through seed coating might be a more reliable method to promote plant growth under abiotic stress conditions.

Recently many studies adopted seed coating as an application method for PGPB immobilization and embodiment (Tables 9.1, 9.2, 9.3). The success of this method relies on the PGPB properties, seed characteristics, and the material used (Tables 9.1, 9.2, 9.3).

9.5.2 *Biotic Factors*

Seed coating is widely used to alleviate abiotic stress, but when it comes to biotic stress, few studies have tested, at the same time, the microorganism's effect and the coating material. Therefore, other inoculation techniques were used, like seed biopriming. It is found that the inoculation of Durum wheat with four *Bacillus* strains with proven plant growth-promoting ability and antimicrobial activity enhanced tolerance to Fusarium head blight (FHB) and reduced the disease incidence. In addition, as was already displayed by in vitro inhibition study, the strains showed an antifungal activity (Ibrahim 2019). In another study, *Trichoderma atroviride* was identified as a fungal strain showing potential biocontrol abilities against two Fusarium damping-off agents (*Fusarium avenaceum* or *F. culmorum*) in maize under greenhouse and field conditions. Coating seed with *Trichoderma atroviride* significantly increased the emergence rate of infected seedlings compared to untreated seeds (Coninck et al. 2020). More examples are presented in Table 9.3.

Years ago, researchers raised the importance of biomaterials in agriculture. Like the medical field, the uses in agriculture are numerous, going from polymeric hydrogels used as suspension (Mishra and Mishra 2016), antimicrobial polymeric

Table 9.2 Examples of applications of seed coating to alleviate abiotic stress in different crops

Type of stress	Crop	Type of PGPB	Coating material	Targeted trait	References
Drought stress	*Helianthus annuus*	*Pseudomonas putida*	Carboxy methyl cellulose	Plant biomass and root architecture	Sandhya et al. (2009)
	Cicer arietinum	*Paenibacillus lentimorbus*	Sodium alginate and CaCl2	Germination	Khan et al. (2011)
	Vigna unguiculata	*Bacillus* sp.	Jaggery slurry	Plant growth, yield and nutrient uptake	Minaxi et al. (2012)
	Vigna mungo L. and *Pisum sativum* L.	*Ochrobactrum pseudogrignonense*, *Pseudomonas* sp., and *Bacillus subtilis*	Carboxymethyl cellulose (CMC)	Shoot and root length, and dry weight. The root water content, Total chlorophyll content and root vigor	Saikia et al. (2018)
	Triticum Wheat (Triticum aestivum L.)	*Variovorax paradoxus*	Charcoal	N, P, K, Ca^{2+} and Na^{+} content, plant height, spike length, total number of grains/spike, 1000 grain weight, grain yield, straw yield and harvest index.	Chandra et al. (2019)
	Vigna unguiculata L.	*Pseudomonas libanensis*	Silicon dioxide and starch	Biomass and physiological traits.	Ma et al. (2019)
	Cowpea (*Vigna unguiculata* L. Walp)	*Pseudomonas libanensis*	Silicon dioxide and starch	Shout and root biomass, number of pods and seeds, seed weight and seed yield per plant.	Ma et al. (2019)
	Phaseolus vulgaris	*Rhizobium tropici*	Silk, trehalose, and hydrogel	Plant biomass, stomatal conductance, and nodulation.	Zvinavashe et al. (2021a)
Saline stress	*Gossypium hirsutum*	*Pseudomonas, Azotobacter, Rhizobium,*	Sucrose	Growth, physiological and ionic (K+/Na+) Characteristics.	Amjad et al. (2015)
	Zea mays L.	*Burkholderia phytofirmans* and *Enterobacter* sp.	Biochar	Nutrient balance.	Akhtar et al. (2015)

(continued)

Table 9.2 (continued)

Type of stress	Crop	Type of PGPB	Coating material	Targeted trait	References
	Triticum aestivum	*Bacillus* sp.	Press mud, compost, slurry and peat	Biochemical, growth and yield.	Shahzad et al. (2017)
	Phaseolus vulgaris	*Rhizobium tropici*	Silk and trehalose	Germination, shout and root biomass, rout architecture.	Mhada et al. (2021)
	Arachis hypogaea	*Bacillus* sp. REN51N	Charcoal	Pod and haulm yield, relative water content, production of enzymes, and uptake of potassium.	Pal et al. (2021)
Metallic stress	*Zea mays L.*	*Proteus mirabilis*	Salicylic acid (SA)	Chromium tolerance in maize	Islam et al. (2016)

Table 9.3 Examples of application of seed coating to activate plant immune system and application in disease biocontrol

Crop	Type of PGPB: Mode of action	Coating material	Targeted disease	References
Rice	*Pseudomonas fluorescens*: ACC deaminase production	Polymer	Rice fallow black gram diseases	Raja et al. (2017)
Okra and sunflower	*Trichoderma harzianum*: the ability of *Trichoderma* to produce antibiotics or hydrolytic enzymes	2% of glucose, gum arabic, sugar and mollases	Root rot fungi viz., *Macrophomina phaseolina, Rhizoctonia solani* and *Fusarium* spp.	Dawar et al. (2008)
Chickpea	*Micrococcus* sp.: The ability to produce fungal cell wall degrading enzymes namely, chitinase and glucanase	Talc-based bio-formulation	*Fusarium* sp.	Patel et al. (2021)

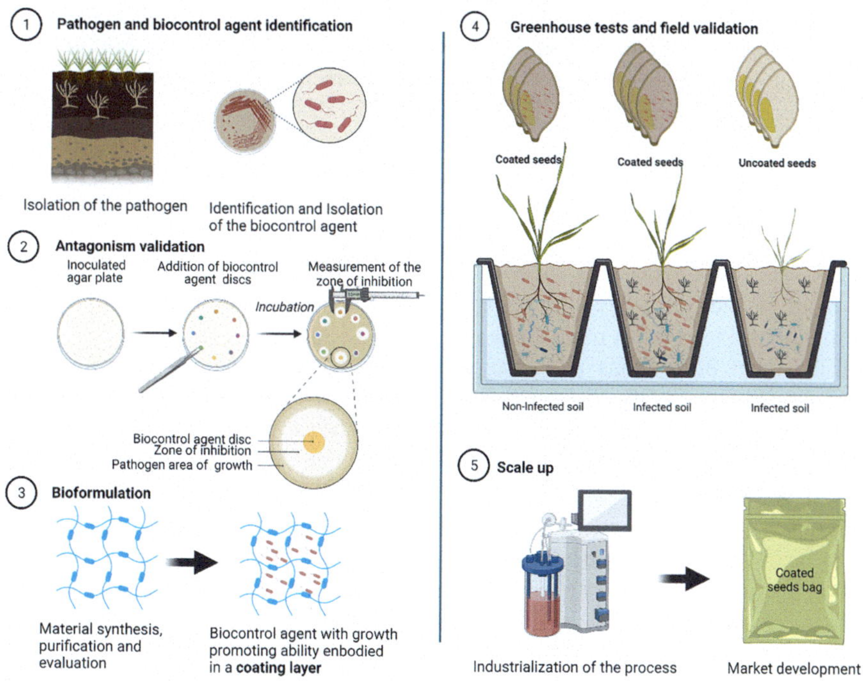

Fig. 9.1 Illustration of the actual state of the seed coating technology combining biocontrol agent and coating material. Figure created with BioRender.com

hydrogels (Cabral 2016), to fertilizers and seed coating (Brigham 2017; Freepons 1997). This review identified an existing gap in seed coating using both findings in PGPB applied for biotic stress and biomaterials. Few studies were found combining results and building on successful experiments from laboratory to field test (Fig. 9.1).

In 2017, Pseudomonas polymer coated black gram seeds were tested in a greenhouse and field to evaluate the efficiency of the endophytic *Pseudomonas fluorescens* formulation for the management of black gram diseases cultivated under rice fallow system (Raja et al. 2017). The application of such bioformulations in the fields can reduce the use of harmful chemicals, protect the environment and biological resources, and be an important component of integrated pest management (IPM) that can help the growers achieve a sustainable yield.

9.6 Market Development and Regulatory Considerations: New Markets and Niches

Seed improvement technologies have been developed in recent years due to strong demand from the agriculture industry for achieving a uniform and robust plant stand. Precision seeding has proven to be a suitable solution for reducing seed costs per unit area, improving seed quality, and increasing production flexibility. In parallel to that, there is an increase in awareness of consumers, farmers, and landowners about the importance of protecting water, land, and microbial resources to ensure sustainable production systems. In India, many farmers are supporting a national program known as zero-budget natural farming (ZBNF). This strategy involves using soil microbes and mulch rather than synthetic fertilizers to enrich lands (Eisenstein 2020).

All the actions undertaken by scientific communities, farmers, and policymakers were accompanied, and reinforced, by the rise of organic farming industries that formulate and implement regulations to control the inputs.

Like other innovations in the agricultural field, seed coating was subject to regulations, where synthetic materials are banned (Table 9.4). Thus the importance of adopting new findings in material sciences from the use of agro-industrial residues to produce cellulose (Bahloul et al. 2021; Kassab et al. 2020).

On the other hand, manipulating soil microbiota has proven to improve organic farming systems (De Corato 2020). Endophytic nitrogen-fixing bacteria are a rising trend in the plant-microbe interaction research area due to their ability to fix nitrogen in nonlegume plants. However, the limitations of conventional inoculation processes lie in microbial survival and unsuitable conditions during storage (Zvinavashe et al. 2021b). These are challenges that could be solved by designing tailored seed coating using appropriate bacteria or consortia and suitable biomaterials.

Since 2017, the seed coating materials market has witnessed increased demand, particularly in developing countries, such as India, China, Mexico, and Brazil. They have also set up manufacturing facilities and strong distribution networks across these regions (Seed Coating Market 2020), thus creating a growing market for the polymer segment and boosting creative startups and spin-offs to provide an easy-to-use application method for small farmers and seed producers.

Table 9.4 Seed coating regulations in the United States of America and Europe

Regulation	United States of America	Europe
Organic farming	For the crop to be considered organic, coating material should be: Synthetic substances listed for appropriate uses in (§ 205.601), or non-synthetic substances that are not prohibited are allowed as seed or planting stock treatments. E.g.: – Pesticides used must be compatible with organic production, including inert and active ingredients. – Ingredients used in pelleting must be non-synthetic or included on the National List (§ 205.601). – noculants used cannot be genetically modified.	For production to be considered organic, substances used should be pre-approved by the European Commission. E.g., Fertilizers, pesticides, and food additives should be listed as approved in European legislation (European Commission 2020).
Coating agents	Protectant products used on seeds should contain an EPA-approved dye (unnatural color to the seed) (Cornell Law School 2020).	Legislation banning added microplastics in seed coatings is expected to become effective around 2027 in Europe (Mitrano and Wohlleben 2020).
Labeling	Seed treatment should be reflected in the lot number (USDA 2017). Labels should contain the name of active ingredients, and a warning "Do not use for food, feed or oil purposes" (Label Manual 2013). For tracing, a complete record of each lot should be kept for 3 years (USDA 2017).	Following Article 49 of Regulation (EC) No 1107/2009: Packs must be marked with a label mentioning: active substance and dose (FAO 2021). The use of extensive best practice recommendations and pictograms is advised (Euroseeds 2020).

9.7 Conclusions and Future Perspectives

Seed coating is considered as an innovative solution to apply exogenous materials to enhance seed health and plant performance, without affecting the seed viability and characteristics. To improve the plant response, many studies focused on the design of new, precise, and smart coating that tend to be cost-effective, eco-friendly, and environmentally sustainable.

Research in plant-microbe interaction and in material sciences has recently focused on biofertilizers and their applications due to an increase in the awareness of the international community about sustainability and the expansion of the Organic Farming market. In relation with that, various microorganisms and appropriate materials have been under investigation, and in this review, we displayed some interesting results that need to be scaled up to the industrialization level.

The market is promising, and key drivers include three levels, (1) margin increase of farmers and agricultural companies, where the seed coating technology based on combining suitable levels of fertilizers and plant growth-promoting bacteria (PGPB) inoculation would be a good strategy to reduce the use of chemical fertilizers and their environmental impact, (2) seed coatings would support agricultural productivity on marginal lands by improving seed microbial microenvironment to restore the ecological system or reinforce with synergic strains, and (3) increase in the seed replacement rate to accelerate the adoption of commercialized coated seeds.

Microbiome modulations as a novel and efficient strategy should be a modular design and a continuous concept that includes bioprospecting and evaluation tests, formulation, industrialization of the process, and commercialization. Microbiome modulation strategy will then depend on the climatic predictions, crop, soil type, and agricultural practices. Research efforts must be accompanied by the encouragement of regulatory agencies and policy makers supporting sustainable practices in agriculture. The elucidation of the importance of PGPB to enhance plant tolerance and immunity has encouraged the scientific community and funders to invest in understanding the mechanisms and the application.

Acknowledgments We acknowledge funds from Université Mohammed VI Polytechique—MIT Research Program (UMRP) and OCP S.A. B.M. also acknowledges funds from the US Office of Naval Research (N N000141812258).

References

Afzal I, Javed T, Amirkhani M, Taylor AG (2020) Modern seed technology: seed coating delivery systems for enhancing seed and crop performance. Agriculture 10. https://doi.org/10.3390/agriculture10110526

Akhtar SS, Andersen MN, Naveed M, Zahir ZA, Liu F (2015) Interactive effect of biochar and plant growth-promoting bacterial endophytes on ameliorating salinity stress in maize. Funct Plant Biol 42:770–781

Amirkhani M, Netravali AN, Huang W, Taylor AG (2016) Investigation of soy protein–based biostimulant seed coating for broccoli seedling and plant growth enhancement. HortScience 51: 1121–1126. https://doi.org/10.21273/HORTSCI10913-16

Amjad M, Akhtar J, Rashid MS (2015) Evaluating the effectiveness of biofertilizer on salt tolerance of cotton (Gossypium hirsutum L.). Arch Agron Soil Sci 61:1165–1177. https://doi.org/10.1080/03650340.2014.990006

Bahloul A, Kassab Z, Aziz F, Hannache H, Bouhfid R, Qaiss AEK, Oumam M, El Achaby M (2021) Characteristics of cellulose microfibers and nanocrystals isolated from doum tree (Chamaerops humilis var. argentea). Cellulose 28:4089–4103. https://doi.org/10.1007/s10570-021-03793-y

Bashan Y, de-Bashan LE, Prabhu SR, Hernandez J-P (2014) Advances in plant growth-promoting bacterial inoculant technology: formulations and practical perspectives (1998–2013). Plant Soil 378:1–33. https://doi.org/10.1007/s11104-013-1956-x

Brigham CJ (2017) Chitin and chitosan: sustainable, medically relevant biomaterials. Int J Biotechnol Wellness Indust 6:41–47. https://doi.org/10.6000/1927-3037.2017.06.02.1

Cabral JD (2016) Antimicrobial polymeric hydrogels. In: Kalia S (ed) Polymeric hydrogels as smart biomaterials, Springer series on polymer and composite materials. Springer International Publishing, Cham, pp 153–170. https://doi.org/10.1007/978-3-319-25,322-0_6

Chandra D, Srivastava R, Gupta VVSR, Franco CMM, Paasricha N, Saifi SK, Tuteja N, Sharma AK (2019) Field performance of bacterial inoculants to alleviate water stress effects in wheat (Triticum aestivum L.). Plant Soil 441:261–281. https://doi.org/10.1007/s11104-019-04115-9

Chua M, Erickson TE, Merritt DJ, Chilton AM, Ooi MKJ, Muñoz-Rojas M (2020) Bio-priming seeds with cyanobacteria: effects on native plant growth and soil properties. Restoration Ecol 28:S168–S176. https://doi.org/10.1111/rec.13040

Coninck E, Scauflaire J, Gollier M, Liénard C, Foucart G, Manssens G, Munaut F, Legrève A (2020) Trichoderma atroviride as a promising biocontrol agent in seed coating for reducing Fusarium damping-off on maize. J Appl Microbiol 129:637–651. https://doi.org/10.1111/jam.14641

Cornell Law School. Seed treatment products [WWW Document]. Legal information Institute; 2020. https://www.law.cornell.edu/cfr/text/40/153.155

Damasceno R, Roggia I, Pereira C, de Sá E (2013) Rhizobia survival in seeds coated with polyvinyl alcohol (PVA) electrospun nanofibres. Can J Microbiol 59:716–719. https://doi.org/10.1139/cjm-2013-0508

Dawar S, Hayat S, Anis M, Zakki MJ (2008) Effect of seed coating material in the efficacy of microbial antagonists for the control of root rot fungi on Okra and Sunflower. Pak J Bot 40: 1269–1278

De Corato U (2020) Soil microbiota manipulation and its role in suppressing soil-borne plant pathogens in organic farming systems under the light of microbiome-assisted strategies. Chem Biol Technol Agric 7:17. https://doi.org/10.1186/s40538-020-00183-7

Eisenstein M (2020) Natural solutions for agricultural productivity. Nature 588:S58–S59. https://doi.org/10.1038/d41586-020-03445-4

European Commission. Organic production and products [WWW Document]; 2020. https://ec.europa.eu/info/food-farming-fisheries/farming/organic-farming/organic-production-and-products_en

Euroseeds. Euroseeds labelling recommendations [WWW Document]; 2020. https://euroseeds.eu/esta-the-european-seed-treatment-assurance-industry-scheme/euroseeds-labelling-recommendations/

FAO. Regulation supplementing the regulation on the marketing of oil and fibre plant seed. [WWW Document]. Base de données FAOLEX; 2021. http://www.fao.org/faolex/results/details/fr/c/LEX-FAOC130624/

Freepons D (1997) Enhancing food production with chitosan seed-coating technology. In: Applications of chitin and chitosan. CRC Press

Gregorio PRD, Michavila G, Muller LR, de Souza Borges C, Pomares MF, de Sá ELS, Pereira C, Vincent PA (2017) Beneficial rhizobacteria immobilized in nanofibers for potential application as soybean seed bioinoculants. PLoS One 12:e0176930. https://doi.org/10.1371/journal.pone.0176930

Haichar F e Z, Santaella C, Heulin T, Achouak W (2014) Root exudates mediated interactions belowground. Soil Biol Biochem 77:69–80. https://doi.org/10.1016/j.soilbio.2014.06.017

He Y, Wu Z, Tu L, Han Y, Zhang G, Li C (2015) Encapsulation and characterization of slow-release microbial fertilizer from the composites of bentonite and alginate. Appl Clay Sci 109–110:68–75. https://doi.org/10.1016/j.clay.2015.02.001

Hoang NT, Tóth K, Stacey G (2020) The role of microRNAs in the legume-Rhizobium nitrogen-fixing symbiosis. J Exp Bot 71:1668–1680. https://doi.org/10.1093/jxb/eraa018

Huang CY, Wang H, Hu P, Hamby R, Jin H (2019) Small RNAs—big players in plant-microbe interactions. Cell Host Microbe 26:173–182. https://doi.org/10.1016/j.chom.2019.07.021

Ibrahim EA-A (2019) Fundamental processes involved in seed priming. In: Hasanuzzaman M, Fotopoulos V (eds) Priming and pretreatment of seeds and seedlings: implication in plant stress tolerance and enhancing productivity in crop plants. Springer, Singapore, pp 63–115. https://doi.org/10.1007/978-981-13-8625-1_4

Ikpe A, Ebunilo P, Okovido J (2018) Geotechnical evaluation of bentonite clay for municipal solid waste landfill lining membrane. Appl J Environ Eng Sci 4:337–351

Kassab Z, Abdellaoui Y, Salim MH, El Achaby M (2020) Cellulosic materials from pea (Pisum Sativum) and broad beans (Vicia Faba) pods agro-industrial residues. Mater Lett 280(128):539. https://doi.org/10.1016/j.matlet.2020.128539

Khan N, Mishra A, Chauhan PS, Nautiyal CS (2011) Induction of Paenibacillus lentimorbus biofilm by sodium alginate and CaCl2 alleviates drought stress in chickpea. Ann Appl Biol 159:372–386. https://doi.org/10.1111/j.1744-7348.2011.00502.x

Label Manual. Unique product labeling [WWW Document]. environmental protection agency; 2013. https://www.epa.gov/sites/default/files/2014-06/documents/chap-18_0.pdf

Leinauer B, Serena M, Singh D (2010) Seed coating and seeding rate effects on turfgrass germination and establishment. HortTechnology 20:179–185. https://doi.org/10.21273/HORTTECH.20.1.179

Ma Y (2019) Seed coating with beneficial microorganisms for precision agriculture. Biotechnol Adv 37(107):423. https://doi.org/10.1016/j.biotechadv.2019.107423

Ma Y, Látr A, Rocha I, Freitas H, Vosátka M, Oliveira RS (2019) Delivery of inoculum of Rhizophagus irregularis via seed coating in combination with Pseudomonas libanensis for cowpea production. Agronomy. https://doi.org/10.3390/agronomy9010033

Madsen MD, Fidanza MA, Barney NS, Kostka SJ, Badrakh T, McMillan MF (2016) Low-dose application of nonionic alkyl terminated block copolymer surfactant enhances turfgrass seed germination and plant growth. HortTechnology 26:379–385. https://doi.org/10.21273/HORTTECH.26.4.379

Makarova LE, Markova YA, Morits AS, Karepova MS, Sidorov AV, Sokolova NA (2021) Interaction of Rhizospheric bacteria and nonrhizobial endophytic bacteria moving from the roots to the rhizosphere of pea plants (Pisum sativum). Appl Biochem Microbiol 57:514–520. https://doi.org/10.1134/s0003683821040104

Mei J, Wang W, Peng S, Nie L (2017) Seed pelleting with calcium peroxide improves crop establishment of direct-seeded rice under waterlogging conditions. Sci Rep 7:4878. https://doi.org/10.1038/s41598-017-04966-1

de Melo B, Assis F, Almeida C, Gomes J, Silva W, Dos I, Moreira S, Gomes Y, Santos, Lisboa J, Silva P, Silva S, Silva N, Primo D (2019) Physiological response of colza (Brassica napus L.) seeds coated and treated with alternative materials. Afr J Agric Res 14:943–948. https://doi.org/10.5897/AJAR2019.14044

Mhada M, Zvinavashe AT, Hazzoumi Z, Zeroual Y, Marelli B, Kouisni L (2021) Bioformulation of silk-based coating to preserve and deliver Rhizobium tropici to Phaseolus vulgaris under saline environments. Front Plant Sci 12:1260. https://doi.org/10.3389/fpls.2021.700273

Middleton H, Yergeau É, Monard C, Combier JP, El Amrani A (2021) Rhizospheric plant–microbe interactions: miRNAs as a key mediator. Trends Plant Sci 26:132–141. https://doi.org/10.1016/j.tplants.2020.09.005

Minaxi, Nain L., L., Yadav, R.C., Saxena, J., (2012). Characterization of multifaceted Bacillus sp. RM-2 for its use as plant growth promoting bioinoculant for crops grown in semi arid deserts. Appl Soil Ecol 59. doi: https://doi.org/10.1016/j.apsoil.2011.08.001.

Mishra SB, Mishra AK (2016) Polymeric hydrogels: a review of recent developments. In: Kalia S (ed) Polymeric hydrogels as smart biomaterials, Springer Series on Polymer and Composite Materials. Springer International Publishing, Cham, pp 1–17. https://doi.org/10.1007/978-3-319-25,322-0_1

Mitrano DM, Wohlleben W (2020) Microplastic regulation should be more precise to incentivize both innovation and environmental safety. Nat Commun 11:5324. https://doi.org/10.1038/s41467-020-19,069-1

Mitter EK, Tosi M, Obregón D, Dunfield KE, Germida JJ (2021) Rethinking crop nutrition in times of modern microbiology: innovative biofertilizer technologies. Front Sustain Food Syst 5:29. https://doi.org/10.3389/fsufs.2021.606815

Naamala J, Smith DL (2021) Microbial derived compounds, a step toward enhancing microbial inoculants technology for sustainable agriculture. Front Microbiol 12:294. https://doi.org/10.3389/fmicb.2021.634807

Nelson KE, Clayton RA, Gill SR, Gwinn ML, Dodson RJ, Haft DH, Hickey EK, Peterson JD, Nelson WC, Ketchum KA, McDonald L, Utterback TR, Malek JA, Linher KD, Garrett MM, Stewart AM, Cotton MD, Pratt MS, Phillips CA, Richardson D, Heidelberg J, Sutton GG,

Fleischmann RD, Eisen JA, White O, Salzberg SL, Smith HO, Venter JC, Fraser CM (1999) Evidence for lateral gene transfer between archaea and bacteria from genome sequence of Thermotoga maritima. Nature 399:323–329. https://doi.org/10.1038/20601

Oleńska E, Małek W, Wójcik M, Swiecicka I, Thijs S, Vangronsveld J (2020) Beneficial features of plant growth-promoting rhizobacteria for improving plant growth and health in challenging conditions: a methodical review. Sci Total Environ. https://doi.org/10.1016/j.scitotenv.2020.140682

Pal KK, Dey R, Sherathia DN, Devidayal, Mangalassery S, Kumar A, Rupapara RB, Mandaliya M, Rawal P, Bhadania RA, Thomas M, Patel MB, Maida P, Nawade BD, Ahmad S, Dash P, Radhakrishnan T (2021) Alleviation of salinity stress in peanut by application of endophytic bacteria. Front Microbiol 12:791. https://doi.org/10.3389/fmicb.2021.650771

Panuccio MR, Chaabani S, Roula R, Muscolo A (2018) Bio-priming mitigates detrimental effects of salinity on maize improving antioxidant defense and preserving photosynthetic efficiency. Plant Physiol Biochem 132:465–474. https://doi.org/10.1016/j.plaphy.2018.09.033

Parisi JJD, Biagi JD, Medina PF, Barbedo CJ (2016) Fungicide and drying effects on the viability of recalcitrant seeds of Inga vera subsp. affinis. Trop Plant Pathol 41:177–182. https://doi.org/10.1007/s40858-016-0084-6

Patel P, Patel K, Dhandhukia P, Thakker JN (2021) Plant growth promoting traits of marine Micrococcus sp. with bio-control ability against Fusarium in chickpea plant. Vegetos 34:94–101. https://doi.org/10.1007/s42535-021-00191-4

Paul S, Rakshit A (2021) Effect of seed bio-priming with Trichoderma viride strain BHU-2953 for enhancing soil phosphorus solubilization and uptake in soybean (Glycine max). J Soil Sci Plant Nutr 21:1041–1052. https://doi.org/10.1007/s42729-021-00420-4

Pedrini S, Merritt DJ, Stevens J, Dixon K (2017) Seed coating: science or marketing spin? Trends Plant Sci 22:106–116. https://doi.org/10.1016/j.tplants.2016.11.002

Pedrini S, Balestrazzi A, Madsen MD, Bhalsing K, Hardegree SP, Dixon KW, Kildisheva OA (2020) Seed enhancement: getting seeds restoration-ready. Restoration Ecol 28:S266–S275. https://doi.org/10.1111/rec.13184

Raja K, Karthikeyan M, Johnson I, Latha P, Saravanakumar D (2017) Antagonistic ACC deaminase producing pseudomonas fluorescens with polymer seed coating for the management of rice fallow black gram diseases. Adv Res 1–12. https://doi.org/10.9734/AIR/2017/34227

Rocha I, Ma Y, Souza-Alonso P, Vosátka M, Freitas H, Oliveira RS (2019) Seed coating: a tool for delivering beneficial microbes to agricultural crops. Front Plant Sci 10. https://doi.org/10.3389/fpls.2019.01357

Rydeen AE, Brustad EM, Pielak GJ (2018) Osmolytes and protein–protein interactions. J Am Chem Soc 140:7441–7444. https://doi.org/10.1021/jacs.8b03903

Saikia J, Sarma RK, Dhandia R, Yadav A, Bharali R, Gupta VK, Saikia R (2018) Alleviation of drought stress in pulse crops with ACC deaminase producing rhizobacteria isolated from acidic soil of Northeast India. Sci Rep 8:3560. https://doi.org/10.1038/s41598-018-21,921-w

Sandhya V, Ali SKZ, Grover M, Reddy G, Venkateswarlu B (2009) Alleviation of drought stress effects in sunflower seedlings by the exopolysaccharides producing Pseudomonas putida strain GAP-P45. Biol Fertil Soils 46:17–26. https://doi.org/10.1007/s00374-009-0401-z

Schoina C, Stringlis IA, Pantelides IS, Tjamos SE, Paplomatas EJ (2011) Evaluation of application methods and biocontrol efficacy of Paenibacillus alvei strain K-165, against the cotton black root rot pathogen Thielaviopsis basicola. Biol Control 58:68–73. https://doi.org/10.1016/j.biocontrol.2011.04.002

Seed Coating Market. Market research reports, marketing research company, business research by MarketsandMarkets [WWW Document]; 2020. https://www.marketsandmarkets.com/. Accessed 18 Aug 2021.

Shahzad S, Yahya Khan M, Zahir Z, Asghar H, Chaudhry U (2017) Comparative effectiveness of different carriers to improve the efficacy of bacterial consortium for enhancing wheat production under salt affected field conditions. Pak J Bot 49:1523–1530

Shekhawat K, Saad MM, Sheikh A, Mariappan K, Al-Mahmoudi H, Abdulhakim F, Eida AA, Jalal R, Masmoudi K, Hirt H (2021) Root endophyte induced plant thermotolerance by constitutive chromatin modification at heat stress memory gene loci. EMBO Rep 22:e51049. https://doi.org/10.15252/embr.202051049

Subramanian S (2019) Little RNAs go a long way: long-distance signaling by microRNAs. Mol Plant 12:18–20. https://doi.org/10.1016/j.molp.2018.12.001

Tsikou D, Yan Z, Holt DB, Abel NB, Reid DE, Madsen LH, Bhasin H, Sexauer M, Stougaard J, Markmann K (2018) Systemic control of legume susceptibility to rhizobial infection by a mobile microRNA. Science 236:233–236. https://doi.org/10.1126/science.aat6907(2018)

Tuğrul KM (2019) Soil management in sustainable agriculture, sustainable crop production. IntechOpen. https://doi.org/10.5772/intechopen.88319

USDA. Seed company records and the federal seed act [WWW Document]. Seed Regulatory and Testing Division; 2017. https://www.ams.usda.gov/sites/default/files/media/SeedCompanyRecordsandtheFederalSeedAct.pdf. Accessed 16 Aug 2021.

Van Etten J, Bhattacharya D (2020) Horizontal gene transfer in eukaryotes: not if, but how much? Trends Genet 36:915–925. https://doi.org/10.1016/j.tig.2020.08.006

Vives-Peris V, de Ollas C, Gómez-Cadenas A, Pérez-Clemente RM (2020) Root exudates: from plant to rhizosphere and beyond. Plant Cell Rep 39:3–17. https://doi.org/10.1007/s00299-019-02447-5

Wang W, Galili G (2019) Tuning the orchestra: miRNAs in plant immunity. Trends Plant Sci 24: 189–191. https://doi.org/10.1016/j.tplants.2019.01.009

Wang W, Vinocur B, Altman A (2003) Plant responses to drought, salinity and extreme temperatures: towards genetic engineering for stress tolerance. Planta 218:1–14. https://doi.org/10.1007/s00425-003-1105-5

Wani SM, Gull A, Ahad T, Malik AR, Ganaie TA, Masoodi FA, Gani A (2021) Effect of gum Arabic, xanthan and carrageenan coatings containing antimicrobial agent on postharvest quality of strawberry: assessing the physicochemical, enzyme activity and bioactive properties. Int J Biol Macromol 183:2100–2108. https://doi.org/10.1016/j.ijbiomac.2021.06.008

Weiberg A, Bellinger M, Jin H (2015) Conversations between kingdoms: small RNAs. Curr Opin Biotechnol 32:207–215. https://doi.org/10.1016/j.copbio.2014.12.025

Williams A, de Vries FT (2020) Plant root exudation under drought: implications for ecosystem functioning. New Phytol 225:1899–1905. https://doi.org/10.1111/nph.16223

Williams M, Dumroese RK, Page-Dumroese D, Hardegree S (2016) Can biochar be used as a seed coating to improve native plant germination and growth in arid conditions? J Arid Environ 125: 8–15. https://doi.org/10.1016/j.jaridenv.2015.09.011

Yadav RS, Singh V, Pal S, Meena SK, Meena VS, Sarma BK, Singh HB, Rakshit A (2018) Seed bio-priming of baby corn emerged as a viable strategy for reducing mineral fertilizer use and increasing productivity. Sci Hortic 241:93–99. https://doi.org/10.1016/j.scienta.2018.06.096

Zhao M, Zhao J, Yuan J, Hale L, Wen T, Huang Q, Vivanco JM, Zhou J, Kowalchuk GA, Shen Q (2021) Root exudates drive soil-microbe-nutrient feedbacks in response to plant growth. Plant Cell Environ 44:613–628. https://doi.org/10.1111/pce.13928

Zvinavashe AT, Lim E, Sun H, Marelli B (2019) A bioinspired approach to engineer seed microenvironment to boost germination and mitigate soil salinity. Proc Natl Acad Sci U S A 116:25,555–25,561. https://doi.org/10.1073/pnas.1915902116

Zvinavashe AT, Laurent J, Mhada M, Sun H, Fouda HME, Kim D, Mouhib S, Kouisni L, Marelli B (2021a) Programmable design of seed coating function induces water-stress tolerance in semi-arid regions. Nat Food 2:485–493. https://doi.org/10.1038/s43016-021-00315-8

Zvinavashe AT, Mardad I, Mhada M, Kouisni L, Marelli B (2021b) Engineering the plant microenvironment to facilitate plant-growth-promoting microbe association. J Agric Food Chem. https://doi.org/10.1021/acs.jafc.1c00138

Chapter 10
Trichoderma Rhizosphere Competence, Suppression of Diseases, and Biotic Associations

Valter Cruz-Magalhães, Fabiola Padilla-Arizmendi, John Hampton, and Artemio Mendoza-Mendoza

Abstract *Trichoderma* fungi are facultative plant symbionts that colonize multiple plant species. In natural environments, the process of root colonization by *Trichoderma* involves several stages. The first is the capacity of *Trichoderma* to colonize the rhizosphere by sensing molecules secreted by the roots, and the ability to compete or cooperate with the microbial community present in the rhizosphere. Once *Trichoderma* reaches the rhizoplane, the fungus must next overcome/manipulate the innate plant defenses by releasing secreted molecules which will allow its establishment inside the root. During this process of rhizosphere and root colonization, *Trichoderma* release specialized metabolites (commonly known as secondary metabolites) that manipulate the roots and also restrict bacteria, nematodes, and filamentous pathogens. The impact of *Trichoderma* in agriculture relies on the capacity to combat the negative effects of plant pathogens, and the ability to induce resistance in plant tissues far from the site of colonization (induced systemic resistance, ISR). ISR is activated by the sensing of *Trichoderma* elicitors and does not require root colonization. Here, we describe the different components as well as the impact of *Trichoderma* on the rhizosphere microbiome, the mechanisms of root colonization, and ISR activation.

Keywords *Trichoderma* · Microbiome · Rhizosphere · Rhizosphere competence · Disease suppressive soils · Endophytes · Specialised metabolites · Induced-systemic resistance · Endophytic continuum · Plant defense · Mycoparasitism

---

Fabiola Padilla-Arizmendi and Valter Cruz-Magalhães contributed equally with all other contributors.

V. Cruz-Magalhães
Department of Phytopathology, Federal University of Lavras, Lavras, MG, Brazil

F. Padilla-Arizmendi · J. Hampton · A. Mendoza-Mendoza (✉)
Bio-Protection Research Centre, Lincoln University, Lincoln, New Zealand
e-mail: artemio.mendoza@lincoln.ac.nz

B. A. Horwitz, P. K. Mukherjee (eds.), *Microbial Cross-talk in the Rhizosphere*, Rhizosphere Biology, https://doi.org/10.1007/978-981-16-9507-0_10

10.1 Introduction

The *Trichoderma* genus includes more than 370 different species (Cai and Druzhinina 2021). The genus inhabits many diverse ecosystems, including deserts, forests, grasslands, and oceans (Kim et al. 2020; Montoya and Quijano 2016; Su et al. 2018). Although *Trichoderma* spp. are considered cosmopolitan fungi, the species distribution is quite variable (Braithwaite et al. 2017; Jiang et al. 2016; Ma et al. 2020; Zachow et al. 2016). Some species are only present in specific geographic areas; for example, *Trichoderma* sp. *atroviride* B, which has been identified only in the Southern hemisphere (Braithwaite et al. 2017).

Trichoderma fungi interact with multiple organisms, including plants, bacteria, nematodes, and other fungi, establishing different lifestyles (Mukherjee et al. 2013). This interaction ability involves the secretion of numerous molecules, including cell wall degrading enzymes, non-enzymatic secreted proteins, and specialized metabolites (SMs) (Fig. 10.1) (Hermosa et al. 2014; Mendoza-Mendoza et al. 2003; Morán-Diez et al. 2009; Vinale et al. 2006; Woo et al. 2006). These secreted molecules are critical for survival in different environments, including during colonization of new niches, plant colonization, mycoparasitism, competition with other microbes, and nutrient assimilation (Druzhinina et al. 2011; Woo et al. 2006). These cell wall lytic enzymes, other secreted proteins, and SMs have a role in the induction of systemic protection in plant tissues far from the site of fungal colonization (e.g., leaves and shoots). The secreted molecules are an essential source of bioactive compounds used in agriculture and industry (Romano et al. 2017).

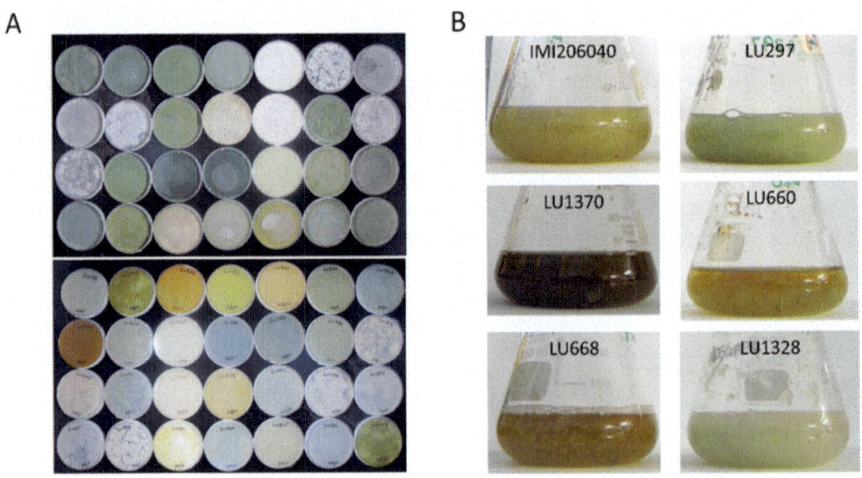

Fig. 10.1 Diversity of *Trichoderma* spp. (**a**) *Trichoderma* isolates grown on Petri dishes. The upper panel illustrates the sporulation of *Trichoderma* and the diversity of color pigmentation of the reproductive structures. The lower panel illustrates the secretion of metabolites into the media. (**b**) Pigmentation of potato dextrose broth medium from different *Trichoderma* isolates grown under the same conditions

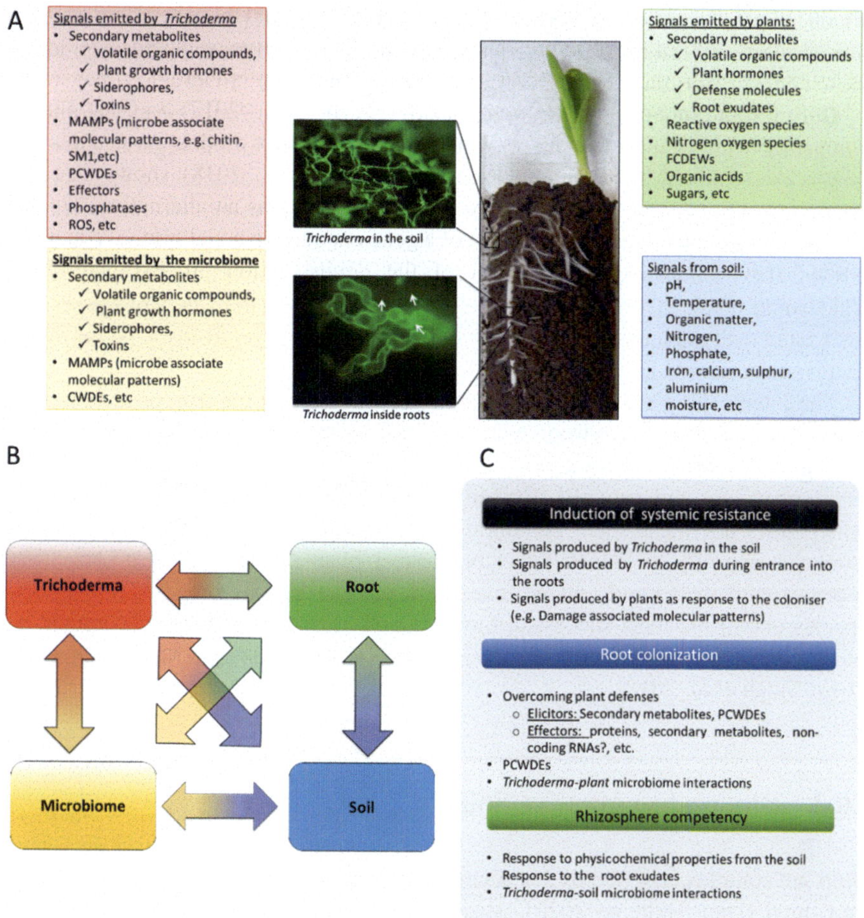

Fig. 10.2 ***Trichoderma* interaction with roots.** (**a**) The interaction involves signals emitted by the plant's root (root exudates), the plants and soil biomes signals, and the signals emitted by *Trichoderma*. *Trichoderma* in green was stained with WGA-Alexa Fluor. Dr Cripps-Guazzone kindly provided the picture of the maize-soil association. (**b**) Multifactorial association of *Trichoderma* with the roots in the soil. These interaction signals are influenced by the physico-chemical properties in the soil as well as the rhizosphere and endosphere microbiome. (**c**) Stages during *Trichoderma* root colonization and the mechanism of induction of plant defense. Induced systemic resistance is activated by signals emitted in the rhizosphere and during internal tissue colonization

Trichoderma spp. are considered facultative plant symbionts and are the fungal species applied most successfully in agriculture (Bae et al. 2009; Druzhinina et al. 2011; Harman et al. 2004; Mukherjee et al. 2013; Ownley et al. 2008; Rodriguez et al. 2009; Woo et al. 2014). This association is influenced by physicochemical conditions in the soil, the genetic background of the fungal and plant partners, and the presence of plant and soil microbiomes (Fig. 10.2a). The latter generates signals which are perceived by and modified by each other. The soil properties influence the interconnection between *Trichoderma*, roots, the soil, and the microbiomes; all of these factors determine the physiological fate of plants (Fig. 10.2b).

During root colonization, *Trichoderma* spp. must first colonize the rhizosphere, competing with members of the microbiome community in this region. Once the fungus crosses the rhizosphere, *Trichoderma* may or may not colonize the roots by producing cell wall degrading enzymes (CWDEs) and releasing effector molecules which suppress plant defenses (Guzmán-Guzmán et al. 2017, 2019; Hermosa et al. 2014, 2013; Mendoza-Mendoza et al. 2018; Nogueira-Lopez et al. 2018; Ramírez-Valdespino et al. 2019, 2018; Schweiger et al. 2021). These two processes are associated with the induction of systemic resistance in sites far from the colonization (Saravanakumar et al. 2016) (Fig. 10.2c).

The intense long-standing application of agrochemicals for crop production has resulted in significant environmental and soil health problems. Therefore, more ecologically friendly alternatives, such as the use of biocontrol and biofertilizer microorganisms, are required (Eltlbany et al. 2019). Under the right conditions, which does not mean ideal conditions for the plant, *Trichoderma* spp. have an extensive capacity to induce direct or indirect plant protection; these capacities, together with their rapid and profuse spore production, have made *Trichoderma* species one of the most successful bioinoculants, with almost 60% of commercial biocontrol agents containing *Trichoderma* in their formulation (Mukherjee et al. 2013; Singh et al. 2018).

10.2 Disease Suppressive Soils

Soils are complex ecosystems that accommodate an astonishing diversity of bacteria and fungi (Dini-Andreote 2020). A single gram of soil can contain around 10^{10} bacterial cells with an estimated biodiversity of 10^3 to 10^4 species (Raynaud and Nunan 2014). Plant roots, however, have molecular strategies that overcome the constant challenges posed by this diversity of microorganisms. Roots recruit and nurture soil microbes by the secretion of exudates, building a dynamic ecosystem in the region between the soil and root, called the rhizosphere. It is well documented that plants are functionally dependent on the extraordinary diversity of microorganisms from the rhizosphere and those able to colonize the plant tissue. They can protect plants against biotic and abiotic stresses and thereby enhance growth, productivity, biodiversity, and ecosystem function (Baedke et al. 2020; Berg et al. 2017; Brundrett et al. 2006; Clay and Holah 1999; Gibert et al. 2019; Gilbert 2014; Hardoim et al. 2015; Khare et al. 2018; Lareen et al. 2016; Poole 2017; Toju et al. 2019; Van Der Heijden et al. 2008; Zilber-Rosenberg and Rosenberg 2008). These associations support the theory that symbiosis is more the rule rather than the exception (Gilbert 2014).

The concept of disease suppressive soil is attributed to microorganisms that suppress plant pathogens and protect plants against diseases (Dini-Andreote 2020).

The disease suppressive soil depends on various factors, including the population dynamics of the pathogens, the genetic backgrounds of both the pathogen and host, biotic and abiotic conditions, and the composition and diversity of the plant and soil microbiome (Dini-Andreote 2020; Mendes et al. 2011; Santhanam et al. 2015; Trivedi et al. 2017). Disease suppressive soils can be classified as general, where the collective action of soil microbial communities restricts plant pathogens (Mazzola 2002), or specific, where one or more microbial taxa antagonize the pathogen (Dini-Andreote 2020; Gómez Expósito et al. 2017; Schlatter et al. 2017). While most research on microbial communities in disease suppressive soils has been focused on bacteria, highly disease suppressive soils exhibit high fungal diversity (Hadar and Papadopoulou 2012). In specific suppressive soils involving *Trichoderma*, the control mechanisms include (i) competition, (ii) parasitism, (iii) antibiosis, (iv) activation of disease-resistance genes, and (v) enhanced plant growth, all of which allow plants to become more resilient to biotic and abiotic stresses (Hadar and Papadopoulou 2012; Vyas and Mathur 2002).

Disease suppressive soils provide a natural microbe-based plant defense (Schlatter et al. 2017). However, when pathogens bypass this first barrier and colonize the plant roots, both the plant immune system mechanism and the endophytic microbiome response are triggered, as demonstrated recently by (Carrión et al. 2019). These authors suggested that secondary metabolism and cell wall degrading enzymes (CWDEs) from microorganisms might be critical elements of disease suppression. When *T. virens* colonizes maize roots, there is a significant increase in the production of SMs by the roots and the fungus (Schweiger et al. 2021) but also a significant increase in the gene expression of CWDEs from *Trichoderma* ((Malinich et al. 2019); Lawry et al., *in preparation*).

Trichoderma spp. are well-recognized fungal antagonists of filamentous pathogens (fungi and oomycetes, Table 10.1), bacteria (Table 10.2), and nematodes (Table 10.3) using mycoparasitism, antibiosis, and induced systemic resistance. These mechanisms may be synergistic with each other, but some act alone to protect plants (Druzhinina et al. 2011; Harman et al. 2004; Poveda et al. 2020).

10.2.1 Establishment in the Rhizosphere

The establishment of root symbiosis is one of the critical drivers of biocontrol success for *Trichoderma* spp. (Druzhinina et al. 2011; Harman et al. 2004; Hermosa et al. 2014, 2013). This root symbiosis is described as a three-step process, whereby (i) *Trichoderma* spp. colonize the rhizosphere and interact with the microbial community; (ii) when this succeeds, penetrate the root tissue by producing CWDEs which allow the crossing the plant cell wall and suppression or overcoming of plant defenses (Mendoza-Mendoza et al. 2017, 2018; Morán-Diez et al. 2009; Nogueira-López et al. 2020; Yedidia et al. 1999) and (iii) manipulate the host for the establishment of an endophytic relationship. Root colonization is regulated by the secretion of effector molecules, including those with hydrogen peroxide scavenging

Table 10.1 Representative examples of bioactivity of *Trichoderma* on soil-borne filamentous pathogens[a]

Trichoderma strain	Mode of action[b]	Host pathogen	Reference
T. asperellum T76-14	VOCs	*Fusarium incanatum*	Intana et al. (2021)
T. atroviride IMI206040	VOCs	*R. solani, Sclerotinia sclerotiorum*	Cruz-Magalhaes et al. (2018)
T. asperelloides TSU1	VOCs	*Corynespora cassiicola, F. incarnatum, Neopestalotiopsis clavispora, N. cubana, Sclerotium rolfsii*	Ruangwong et al. (2021)
T. virens, T. viride	VOCs	*F. oxysporum*	Li et al. (2018)
T. asperelloides PSU-P1	VOCs	*Colletotrichum* sp., *C. cassiicola, C. lunata, Ganoderma* sp., *P. oxalicum, N. clavispora, S. rolfsii, S. cucurbitacearum*	Phoka et al. (2020)
T. longibrachiatum MK425639, *T. longibrachiatum* MK751759, *T. harzianum* MK751758, *T. pleuroti* MK751757	Antagonism, VOCs	*S. sclerotiorum, S. rolfsii, F. oxysporum*	Rajani et al. (2021)
T. longibrachiatum MK425639, *T. longibrachiatum* MK751759, *T. harzianum* MK751758, *T. pleuroti* MK751757	Antagonism	*Macrophomina phaseolina*	Rajani et al. (2021)
T. longibrachiatum EF5	Antagonism, VOCs	*S. rolfsii, M. phaseolina*	Sridharan et al. (2020)
T. harzianum, T. ghanense, T. asperellum,	Antagonism, VOCs	*S. sclerotiorum, F. solani, R. solani*	Qualhato et al. (2013)
T. tomentosum	Antagonism	*S. sclerotiorum*	Qualhato et al. (2013)
T. tomentosum	VOCs	*F. solani, R. solani*	Qualhato et al. (2013)
T. virens GV29-8	VOCs	*Botrytis cinerea*	Velázquez-Robledo et al. (2011)
T. harzianum T-E5	VOCs, Antagonism	*F. oxysporum f. sp. cucumerinum*	Zhang et al. (2014a)
T. asperellum	VOCs	*Magnaporthiopsis maydis*	Degani et al. (2021)
T. koningii SMF2	Antagonism	*Pyrenophora tritici-repentis, S. sclerotiorum*	Xiao-Yan et al. (2006)

(continued)

Table 10.1 (continued)

Trichoderma strain	Mode of action[b]	Host pathogen	Reference
T. koningii Tr5	Mycoparasitism (CWDEs)	*S. cepivorum*	Metcalf and Wilson (2001)
Mixture of *Trichoderma koningii, T. aureoviride* and *T. longibrachiatum*	Antagonism	*S. cepivorum*	Escande et al. (2002)
T. atroviride KACC40552, *T. atroviride* KAAC40557, *T. virens* KACC40929	Antibiosis	*P. melonis, P. cactorum, P. drechsleri, P. sojae, P. capsici, P. nicotianae, P. infestans*	Bae et al. (2016)
T. gamsii KAACC40553	Antibiosis	*P. melonis, P. cactorum, P. drechsleri, P. capsici, P. infestans*	Bae et al. (2016)
T. harzianum KACC40871	Antibiosis	*P. sojae, P. nicotianae, P. infestans*	Bae et al. (2016)
T. brevicompactum KACC40931	Antibiosis	*P. melonis, P. cactorum, P. drechsleri, P. sojae, P. capsici, P. infestans*	Bae et al. (2016)
T. brevicompactum KACC41707	Antibiosis	*P. melonis, P. capsici, P. infestans*	Bae et al. (2016)
T. virens KACC40717	Antibiosis	*P. drechsleri, P. melonis, P. capsici, P. infestans*	Bae et al. (2016)

[a]Only a few representative studies are indicated
[b]Antagonism includes the role of CWDEs and antibiosis

activity, peroxidases and catalases, and SMs (Guzmán-Guzmán et al. 2017, 2019; Mendoza-Mendoza et al. 2018; Nogueira-Lopez et al. 2018; Ramírez-Valdespino et al. 2019, 2018).

10.2.2 Rhizosphere Competence

The rhizosphere can be described as the area around a plant root that is inhabited by a unique population of microorganisms influenced by the chemicals released from plant roots. Rhizosphere chemistry involves the root exudates, their breakdown products, and the microbiome products of soil-derived chemicals (Pétriacq et al. 2017). The ability of a microbe to colonize and proliferate over time within the rhizosphere is termed rhizosphere competence (RC). The molecular mechanisms of RC in *Trichoderma* include, among others: (i) the production of SMs with antibiotic, antifungal, and fungistatic activity, which inhibit potential microbial competitors (Cruz-Magalhaes et al. 2018; Nieto-Jacobo et al. 2017); (ii) *Trichoderma* spp.

Table 10.2 Representative examples of bioactivity of *Trichoderma* on soil-borne nematode pathogens[a]

Trichoderma strain[a]	Mode of action[b]	Host pathogen	Reference
Trichoderma spp.	Antagonistic	*M. incognita*	Khan et al. (2020)
Trichoderma spp.	Antagonistic	*M. incognita*	Pocurull et al. (2019)
Trichoderma spp.	Antagonistic	*Pratylenchus brachyurus*	Oliveira et al. (2019)
Trichoderma spp.	Antagonistic	*M. javanica*	Kiriga et al. (2018)
Trichoderma spp.	Antagonistic	*M. hapla*	Braithwaite et al. (2016)
Trichoderma spp.	Induce resistance	*P. brachyurus*	Miamoto et al. (2017)
Trichoderma spp.	Induce resistance	*M. javanica*	Miamoto et al. (2017)
T. afroharzianum	Toxic metabolites	*Globodera rostochiensis*	Benttoumi et al. (2020)
T. album	Nematostatic metabolite against J2	*Tylenchulus semipenetrans*	Elzawahry et al. (2015)
T. atroviride IMI352941	Direct parasitism, Induction of systemic resistance	*M. incognita*	De Medeiros et al. (2017)
T. citrinoviride Snef1910	Toxic metabolite to J2 and eggs	*M. incognita*	Fan et al. (2020)
T. harzianum	Antagonistic	*M. incognita*	Feyisa et al. (2016)
T. atroviride IMI206040	Metabolites, direct parasitism of eggs	*M. javanica*	Sharon et al. (2007)
T. harzianum BI	Direct parasitism of eggs, Induction of systemic resistance	*M. javanica*	Sahebani and Hadavi (2008)
T. harzianum T-78	Induction of resistance by JA and SA pathways	*M. Incognita*	Martínez-Medina et al. (2017a)
T. harzianum SZMC 1647	Parasitism of eggs, chitinolytic enzyme production	*Caenorhabditis elegans*	Szabó et al. (2012)
T. harzianum T908	Induction of SAR	*M. Incognita*	Leonetti et al. (2017)
T. harzianum ThzID1-M3	Direct parasitism by colonization of juveniles (J2) and cysts	*Globodera pallida*	Contina et al. (2017)
T. harzianum	Toxic metabolites	*G. rostochiensis*	Benttoumi et al. (2020)
T. hirsutum	Toxic metabolites	*G. rostochiensis*	Benttoumi et al. (2020)

(continued)

Table 10.2 (continued)

Trichoderma strain[a]	Mode of action[b]	Host pathogen	Reference
T. longibrachiatum T6	Direct parasitism of J2 and eggs, Induction of resistance	*Heterodera avenae*	Zhang et al. (2017)
T. viride	Antagonistic	*M. incognita*	Kumar and Chand (2015)
T. harzianum ESALQ-1306	Adhere and immobilize eggs and juveniles (J2)	*M. incognita*	Mascarin et al. (2012)
T. harzianum (isolate-27) *and T. viride* (isolate-08)	Suppression of nematode reproduction and root galling	*M. incognita*	Al-Hazmi and Tariqjaveed (2016)

[a]Only a few representative studies are indicated here
[b]Antagonistic might includes the role of CWDEs and antibiosis

perceiving the sugars and other metabolites present in the root exudates and activating chemo-attraction signaling pathways to direct the fungus to the roots; at the same time the fungus must avoid / overcome potential antifungal products secreted in the root exudates (Moreno-Ruiz et al. 2020); (iii) the production of molecules like siderophores and other SMs able to solubilize, and consequently make available metal ions at the root surface (Cai et al. 2015). These capacities may be "sensed" by the plant and allow the persistence of the fungus in the rhizosphere. *Trichoderma* spp. chelate and reduce Fe, Cu, Mn, and Zn and also solubilize insoluble phosphate, improving their uptake by plant roots (Altomare et al. 1999; De Santiago et al. 2009; Yedidia et al. 2001).

Fungal SMs modulate plant growth, inhibit plant pathogens, and induce protection against abiotic stresses (Brookes 2017; Contreras-Cornejo et al. 2014; Druzhinina et al. 2011; Garnica-Vergara et al. 2016; Harman et al. 2004; Kuchár 2019; Makiola 2018; Nieto-Jacobo et al. 2017; Patil et al. 2016; Salazar-Badillo et al. 2015; Werner et al. 2016; Zeilinger et al. 2016). In pairwise interaction experiments (one microbe and one plant), it has been observed that SMs emitted by *Trichoderma* are potent inducers of plant growth (Cruz-Magalhães et al. 2019; Garnica-Vergara et al. 2016; Nieto-Jacobo et al. 2017; Zeilinger et al. 2016), provide abiotic stress protection (Salazar-Badillo et al. 2015), induce plant disease resistance (Contreras-Cornejo et al. 2014), and directly inhibit the growth of plant pathogens (Cruz-Magalhães et al. 2019; Patil et al. 2016; Werner et al. 2016). Multipartite interactions in field experiments have also demonstrated that *Trichoderma* spp. increase the bacteria and mycorrhizal fungi population in the rhizosphere (Eltlbany et al. 2019; Illescas et al. 2020; Pang et al. 2017; Qiao et al. 2019; Saravanakumar et al. 2017; Zhang et al. 2018, 2013) (Table 10.4).

10.2.2.1 *Trichoderma*–Soil-Borne Fungi Interactions

Trichoderma spp. interact with other microorganisms by using different mechanisms; the best-known association is mycoparasitism when *Trichoderma* kills other

Table 10.3 Effect of SMs from *Trichoderma* spp. on bacterial pathogens[a]

Trichoderma strain[a]	Mode of action[b]	Bacterial pathogens	References
T. harzianum	Antibiosis	*Xanthomonas campestris*	Anwar and Iqbal (2017)
T. harzianum	Antibiosis	*Clavibacter michiganensis*	Anwar and Iqbal (2017)
T. harzianum	Antibiosis	*Escherichia coli*	Anwar and Iqbal (2017)
T. harzianum	Antibiosis	*Pseudomonas aeruginosa*	Anwar and Iqbal (2017)
T. harzianum	Antibiosis	*Staphylococcus aureus*	Anwar and Iqbal (2017)
T. asperelloides T136, *T. pseudoharzianum* T113, *T. pseudoharzianum* T129, *T. pseudoharzianum* T160, *T. viridae*	Antibiosis	*Ralstonia solanacearum*	Khan et al. (2020)
T. asperelloides T136, *T. pseudoharzianum* T113, *T. pseudoharzianum* T129, *T. pseudoharzianum* T160, *T. viride*	Antibiosis	*X. campestris*	Khan et al. (2020)
T. harzianum T-22, *T. virens* G-41	Antibiosis, VOCs	Diverse *Bacillus* sp. strains	Li et al. (2020)
T. harzianum T-22, *T. virens* G-41	Antibiosis, VOCs	*Terribacillus* sp. LR2	Li et al. (2020)
T. harzianum T-22, *T. virens* G-41	Antibiosis, VOCs	Diverse *Fictibacillus* sp. strains	Li et al. (2020)
T. harzianum T-22, *T. virens* G-41	Antibiosis, VOCs	Diverse *Exiguobacterium* sp. strains	Li et al. (2020)
T. harzianum T-22, *T. virens* G-41	Antibiosis	*Rhizobium* sp. *LR18*	Li et al. (2020)
T. harzianum T-22, *T. virens* G-41	Antibiosis	*Ensifer* sp. *TS7*	Li et al. (2020)
T. harzianum T-22, *T. virens* G-41	Antibiosis	*Stenotrophomonas* sp. *TS2*	Li et al. (2020)
T. harzianum T-22, *T. virens* G-41	Antibiosis	*Thermomonas* sp. *TS6*	Li et al. (2020)
T. harzianum T-22, *T. virens* G-41	Antibiosis	*Kosakonia* sp. *LS11*	Li et al. (2020)
T. harzianum T-22, *T. virens* G-41	Antibiosis	*Pantoea* sp. *LR8*	Li et al. (2020)
T. harzianum T-22, *T. virens* G-41	Antibiosis	*Acinetobacter* sp. *LR6*	Li et al. (2020)
T. harzianum T-22, *T. virens* G-41	Antibiosis, VOCs	Diverse *Pseudomonas* sp. strains	Li et al. (2020)
T. harzianum T-22, *T. virens* G-41	Antibiosis, VOCs	Diverse *Chryseobacterium* strains	Li et al. (2020)
T. harzianum T-22, *T. virens* G-41	Antibiosis	Diverse *Curtobacterium* sp. strains	Li et al. (2020)

(continued)

Table 10.3 (continued)

Trichoderma strain[a]	Mode of action[b]	Bacterial pathogens	References
T. harzianum T-22, *T. virens* G-41	Antibiosis, VOCs	Diverse *Microbacterium* sp. strains	Li et al. (2020)
T. harzianum T-22, *T. virens* G-41	Antibiosis	*Paenarthrobacter* sp. LR20	Li et al. (2020)
T. harzianum T-22, *T. virens* G-41	Antibiosis, VOCs	Diverse *Pseudoarthrobacter* sp. strains	Li et al. (2020)
T. harzianum T-22, *T. virens* G-41	Antibiosis	Diverse *Micrococcaceae* sp. strains	Li et al. (2020)
T. harzianum T-22, *T. virens* G-41	Antibiosis	*Arthrobacter* sp. TR3	(Li et al. 2020)
T. harzianum T-22, *T. virens* G-41	Antibiosis	Diverse *Deinococcus* sp. strains	Li et al. (2020)
T. harzianum T-22, *T. virens* G-41	Antibiosis	*Xanthomonas* sp. *LR14*	Li et al. (2020)
T. harzianum T-22, *T. virens* G-41	Antibiosis	*Burkholderiaceae* TS1	Li et al. (2020)
T. harzianum T-22, *T. virens* G-41	Antibiosis, VOCs	*Moraxella* sp. TS14	Li et al. (2020)
T. harzianum T-22, *T. virens* G-41	Antibiosis	*Escherichia coli*	Li et al. (2020)
T. virens G-41	Antibiosis, VOCs	Diverse *Paenarthrobacter* sp. strains	Li et al. (2020)
T. harzianum T9	Antibiosis	*Salmonella typhi* ATCC5784	Phupiewkham et al. (2015)
T. harzianum T9	Antibiosis	Diverse *B. subtilis* strains	Phupiewkham et al. (2015)
T. harzianum T9	Antibiosis	*B. cereus* ATCC11778	Phupiewkham et al. (2015)
T. harzianum T9	Antibiosis	*B. amyloliquefaciens* TISTR1014	Phupiewkham et al. (2015)
T. harzianum T9	Antibiosis	*B. licheniformis* TISTR1010	Phupiewkham et al. (2015)
T. harzianum T9	Antibiosis	Diverse *Staphylococcus aureus* strains	Phupiewkham et al. (2015)
T. harzianum T9	Antibiosis	*E. coli* O157:H7	Phupiewkham et al. (2015)
T. harzianum T9	Antibiosis	*Vibrio cholerae*	Phupiewkham et al. (2015)
T. longibrachiatum MD33	Antibiosis	*B. subtilis*	Sarsaiya et al. (2020)

(continued)

Table 10.3 (continued)

Trichoderma strain[a]	Mode of action[b]	Bacterial pathogens	References
T. longibrachiatum MD33	Antibiosis	*B. mycoides*	Sarsaiya et al. (2020)
T. longibrachiatum MD33	Antibiosis	*Staphylococcus* spp.	Sarsaiya et al. (2020)

[a]Only a few representative studies are indicated here
[b]Antibiosis indicates soluble secondary metabolites from *Trichoderma*; VOCs means Volatile organic compounds

Table 10.4 Rhizosphere and endosphere microbiome affected by *Trichoderma*

Trichoderma	Rhizosphere	Microorganisms affected	**References**
T. rossicum NAU-18	Grassland plant rhizosphere	Increased: Archaeorhizomyces. Decreased: *Ophiosphaerella.*	Zhang et al. (2018)
T. harzianum T22	Tomato rhizosphere	Increased: Proteobacteria (*Azospirillum, Acidovorax, Pseudomonas, Sphingobium,* and *Arthrobacter*), Fermicutes (*Paenibacillus*), Acidobacteria *Gp3, Gp4, Gp6, and Gp16*, and *Bacteroidetes* (*Pedobacter*)	Eltlbany et al. (2019)
The combined application of top dressing and *T. harzianum* T-34	Wheat rhizosphere	Increased: *Kaistobacter, Gemmatimonas, Luteolibacter, Flavisolibacter, Janthinobacterium.* Mycorrhizal fungus *Claroideoglomus* Increased also in the endosphere: *Lysobacter. Devosia, Rhizobium,* and *Sphingomonas* Decreased: *Williamsia, Haliangium,* and *Steroidobacter.*	Illescas et al. (2020)
T. harzianum	Maize rhizosphere	Increased: Acidobacteria	Saravanakumar et al. (2017)
T. harzianum T-E5	Cucumber rhizosphere	Increased: *Rhodotarzetta rosea, Melastiza cornubiensis, Uncultured soil basidiomycete*	Zhang et al. (2013)
T. harzianum SQR-T037 -enriched organic fertilize	Tomato rhizoasphere	Increased: Actinobacteria, Firmicutes, Proteobacteria, *Chytridiomycota*	Pang et al. (2017)
T. guizhouense NJAU 4742 with Organic fertilizer	Maize-cabbage cropping system	Increased: the genera *Cladorrhinum* and *Massilia* Decreased: *Zavarzinella, Rubritepida,* and *Bdellovibrio*	Qiao et al. (2019)

fungi and uses their nutrients (Mukherjee et al. 2022). During this process, *Trichoderma* recognizes the prey by chemosensing molecules secreted by the host fungi (Moreno-Ruiz et al. 2020), followed by the coiling of the hyphae and secretion of CWDEs and SMs (Harman et al. 2004). The comparison of the first three sequenced *Trichoderma* genomes (*T. reesei, T. virens,* and *T. atroviride*) suggested that mycoparasitism is the ancestral lifestyle of the genus (Karlsson et al. 2017; Kubicek et al. 2011; Schmoll et al. 2016). Mycoparasitism and antibiosis are the mechanisms that *Trichoderma* uses to control soil-borne filamentous pathogens (including fungi and oomycetes) and bacteria and contribute to specific suppressive soil effect.

The biocontrol activity of filamentous pathogens by *Trichoderma* involves the production of SMs (volatile and non-volatile compounds), which restrict the pathogen (Cruz-Magalhães et al. 2019; Li et al. 2018; Mukherjee et al. 2022), and fungal host cell wall degradation by the action of *Trichoderma'*s CWDEs. These include chitinases, β-glucanases, and proteases (Carsolio et al. 1999; Cortés et al. 1998; Donzelli and Harman 2001; Flores et al. 1997; Geremia et al. 1993; Pozo et al. 2004; Vázquez-Garcidueñas et al. 1998; Viterbo et al. 2004). The role of CWDEs in biocontrol has been demonstrated by genetic manipulation of CWDEs encoding genes, *prb1* (encoding a protease), *ech42* (encoding an endochitinase 42) in *T. atroviride* IMI206040 (Carsolio et al. 1999; Flores et al. 1997) and *Tvbgn*3 (encoding a β-1, 6-glucanase), *tvsp1* (encoding a protease) from *T. virens* GV29-8 (Djonović et al. 2006; Pozo et al. 2004).

The mechanisms involved in the perception of the prey by *Trichoderma* spp. are not well understood. Many of the genes encoding CWDEs, however, are already induced before contact with the fungal host and might involve the perception of diffusible signals (Cortés et al. 1998). Moreover, the production of CWDEs is catabolically repressed by either carbon or nitrogen and induced by the fungal cell walls or by chitin and its derivatives (Cortés et al. 1998; Donzelli and Harman 2001).

The role of pH in mycoparasitism was demonstrated by gene deletion of PAC1, a pH-regulatory gene in *T. harzianum* CECT 2413 and *T. virens* IMI 304061. The absence of PAC1 generated *T. harzianum* strains unable to overgrow *R. solani, Rhizoctonia meloni* and *Phytophthora citrophthora* (Moreno-Mateos et al. 2007) and *T. virens* strains with decreased ability to compete with or overgrow *R. solani* and *S. rolfsii* (Trushina et al. 2013).

Trichoderma spp. SMs release is typically combined with other mechanisms like the production of CWDEs and competition for space and nutrients in the rhizosphere. These, together, have a strong impact on the ability of *Trichoderma* to control plant pathogens, stimulate plant defense, and enhance plant growth (Zeilinger and Schumacher 2013). *Trichoderma* spp. produce more than 1,000 different SMs with an astonishing chemical diversity (Hermosa et al. 2014; Zeilinger et al. 2016). These molecules include terpenes, polyketides, non-ribosomal peptides, peptaibols, diketopiperazine-like compounds, and lactones (Hermosa et al. 2014). SMs produced by *Trichoderma* have antagonistic activity against bacteria, yeasts, and fungi (Zeilinger et al. 2016), as well as promoting plant growth and inducing systemic resistance (Garnica-Vergara et al. 2016; Salazar-Badillo et al. 2015; Velázquez-Robledo et al. 2011; Zeilinger et al. 2016).

In *T. reesei,* the methyltransferase LaeA, a global regulator of SMs, regulates the production of multiple SMs, including polyketides and non-ribosomal peptides (Karimi-Aghcheh et al. 2013). Nutrients in the medium influence the production of SMs, which also vary between species and isolates (Guo et al. 2019; Kottb et al. 2015a; Nieto-Jacobo et al. 2017; Speckbacher et al. 2020). For example, the production of the auxin indole-3-acetic acid and volatile organic compounds (VOCs) varies depending on the medium and the strain used (Nieto-Jacobo et al. 2017). The production of SM also depends on light and other environmental factors (Speckbacher et al. 2020). These findings suggest that, depending on the environmental conditions in the soil, *Trichoderma* might or might not produce specific SMs, which may impact biocontrol.

Like other microbial SMs, VOCs biosynthesis is regulated by different factors during growth, such as media composition (nutrient availability), temperature, pH, light-dark status, and culture age (Speckbacher et al. 2020). In addition, the perception of fungal VOCs by *Trichoderma* spp. induces the secretion of VOCs (Zhang et al. 2014a) and soluble metabolites with fungicidal activity (Li et al. 2018). VOCs are chemically very diverse, and include alcohols, aldehydes, alkenes, acids, esters, lactones, terpenoids, and ketones (Guo et al. 2020). Currently, there are more than 470 different VOCs identified in *Trichoderma* spp. (Siddiquee 2014). VOCs released by *Trichoderma* spp. inhibit or kill soil-borne plant pathogens, including fungi, oomycetes, and bacteria, representing an effective mechanism of biocontrol (Elsherbiny et al. 2020; Intana et al. 2021).

The activity of mycoparasitism and the inhibition of filamentous pathogen growth by VOCs are independent of each other, but there are strains that both produce VOCs that inhibit the pathogens and have mycoparasitic activity. Other *Trichoderma* strains, whose VOCs do not inhibit specific pathogens, have mycoparasitic activity which has a significant role in controlling those pathogens (Rajani et al. 2021). (Table 10.1).

10.2.2.2 *Trichoderma*-Nematodes Interactions

Trichoderma spp. are able to control nematodes by different mechanisms, including inhibition at different nematode developmental stages, production of lytic enzymes, and competition for space and nutrients. Furthermore, *Trichoderma* spp. activate hormone-mediated (salicylic and jasmonic acid, strigolactones, among others) plant-defense mechanisms (De Medeiros et al. 2017; Martínez-Medina et al. 2017a; Morán-Diez et al. 2021; Poveda et al. 2020). As for microbial pathogens, priming plant defense helps control nematodes.

Trichoderma affects the establishment, development, and reproduction of the nematode *Meloidogyne javanica* (Martínez-Medina et al. 2017a; Poveda et al. 2020). This is reflected in a reduction in the number of galls and egg masses per plant, and a reduction in egg hatching (Martínez-Medina et al. 2017a). It has been demonstrated that the SMs secreted by *T. harzianum* control *M. incognita* on tomato plants (Khan et al. 2018). These metabolites have a significant effect on egg hatching and increase the mortality of the juvenile (J2) stage. The direct application of spores of *T. harzianum* has been demonstrated to be an effective mechanism to reduce the population of J2s (Khan et al. 2018).

T. longibrachiatum directly penetrates the eggs of the cyst-forming nematode *Heterodera avenae,* most likely with the help of extracellular chitinases and proteases, which affect the structural components of the eggshell. This reduces the number of eggs capable of hatching and, therefore, the number of infective J2 (Zhang et al. 2017, 2014b). The cyst-forming nematode *Globodera pallida* is also affected by *T. harzianum* at both the cyst and J2s stages (Contina et al. 2017). Further examples of *Trichoderma* effects on soil-borne nematode pathogens are provided in Table 10.2. The fungus may directly affect the nematode, or indirectly inhibit its ability to damage the host, by inducing systemic resistance.

10.2.2.3 *Trichoderma*–Bacteria Interactions

Trichoderma spp. and bacteria interact with each other by releasing SMs (Li et al. 2020). These associations can be negative (Table 10.3) or positive for the bacteria (Table 10.4). SMs, including VOCs released by *Trichoderma* and bacteria, can be either inhibitory or beneficial for the bacteria and *Trichoderma* (Li et al. 2020).

Trichoderma strains affect the soil bacterial and fungal communities in a pH- and N-supply dependent manner, respectively (Zhang et al. 2018). *Trichoderma* spp. in combination or not with organic fertilizers, induce changes in the rhizosphere microbial community in grassland plants, cucumber, tomato, maize, black pepper, potato, and wheat (Illescas et al. 2020; Pang et al. 2017; Qiao et al. 2019; Saravanakumar et al. 2017; Zhang et al. 2018, 2013) (Table 10.4). *Trichoderma guizhouense* NJAU 4742 amendment with organic fertilizer increased the rhizosphere microbial community in terms of diversity, richness, and abundances, far beyond the direct impact of adding the *Trichoderma* inoculum (Qiao et al. 2019).

The use of consortia inoculum, including bacteria and *Trichoderma,* can result in additive plant protection against biotic and abiotic stresses (Eltlbany et al. 2019; Wang et al. 2021, 2019). For example, the use of a rhizosphere-derived consortium of *B. subtilis* and *T. harzianum* provided a higher abundance of beneficial rhizosphere bacterium in potato plants, suppression of common scab of potato, and an increase in yield of tubers (Wang et al. 2021, 2019). In moderate P-limited soils under greenhouse conditions, *T. harzianum* increases plant growth and the relative representation of Acidobacteria in the maize rhizosphere (Saravanakumar et al. 2017). In the tomato rhizosphere, *T. harzianum* T22 also increased the relative abundance of Proteobacteria (*Azospirillum, Acidovorax, Pseudomonas, Sphingobium,* and *Arthrobacter*), Fermicutes (*Paenibacillus*), Acidobacteria Gp3, Gp4, Gp6, and Gp16, and Bacteroidetes (*Pedobacter*) (Eltlbany et al. 2019) (Table 10.4).

Nitrogen addition to the soil has a significant impact on the microbial community, and it is a stronger determinant in the establishment of microbial communities than *Trichoderma harzianum* T-34. However, *T. harzianum* T-34 increased the rhizosphere levels of the mycorrhizal fungus *Claroideoglomus* (Illescas et al. 2020) (Table 10.4).

10.3 *Trichoderma*–Plant Interaction

10.3.1 Theory of the Endophytic Continuum

Early stages of root colonization by *Trichoderma* include fungal attachment to the root surface, formation of specialized penetration structures, synthesis of plant cell wall degrading enzymes, and evasion/suppression of plant immune reactions (Nogueira-Lopez et al. 2018; Yedidia et al. 1999). During root colonization, *Trichoderma* suppresses the expression of pectinase and xylanase inhibitors, which directly protect the plant from cell wall degradation, mediated by the enzymes released by fungi (Lawry et al., in preparation). *Trichoderma's* mechanisms to suppress the expression of these genes is unknown, but most likely involve effector-like molecules produced during the interaction (Guzmán-Guzmán et al. 2019; Mendoza-Mendoza et al. 2018; Ramírez-Valdespino et al. 2019).

However, diverse *Trichoderma* strains have shown positive, neutral, and sometimes detrimental effects in plants (Hoyos-Carvajal et al. 2009; Kottb et al. 2015a; Nieto-Jacobo et al. 2017) (Fig. 10.3). Moreover, plant benefits depend on the plant genotype (Tucci et al. 2011) and on root colonization, as shown recently in maize (Lawry et al., in preparation), which suggests the presence of still unknown molecules from both the plants and the fungus.

10.3.2 Metabolomics

Within the complex interactions in the rhizosphere, the interkingdom communication between plants and microorganisms is regulated by the production of specialized metabolites that permit the establishment of beneficial or pathogenic interactions. The use of high-throughput omics technologies (transcriptomics, proteomics and metabolomics) has allowed, to some extent, the characterization of the interactions between beneficial microorganisms and plants, including the interactions between *Trichoderma* with their hosts. Overall the *Trichoderma*-plant interaction involves the modulation of (i) gene expression from both organisms (Brotman et al. 2013; De Palma et al. 2019; Moran-Diez et al. 2012; Morán-Diez et al. 2015), (ii) proteins secretion (Nogueira-Lopez et al. 2018; Shoresh and Harman 2008a), and (iii) plant metabolome (Brotman et al. 2013, 2012; Coppola et al. 2019; Vinci et al. 2018; Yedidia et al. 2003). Within omics, the most widely used technique for

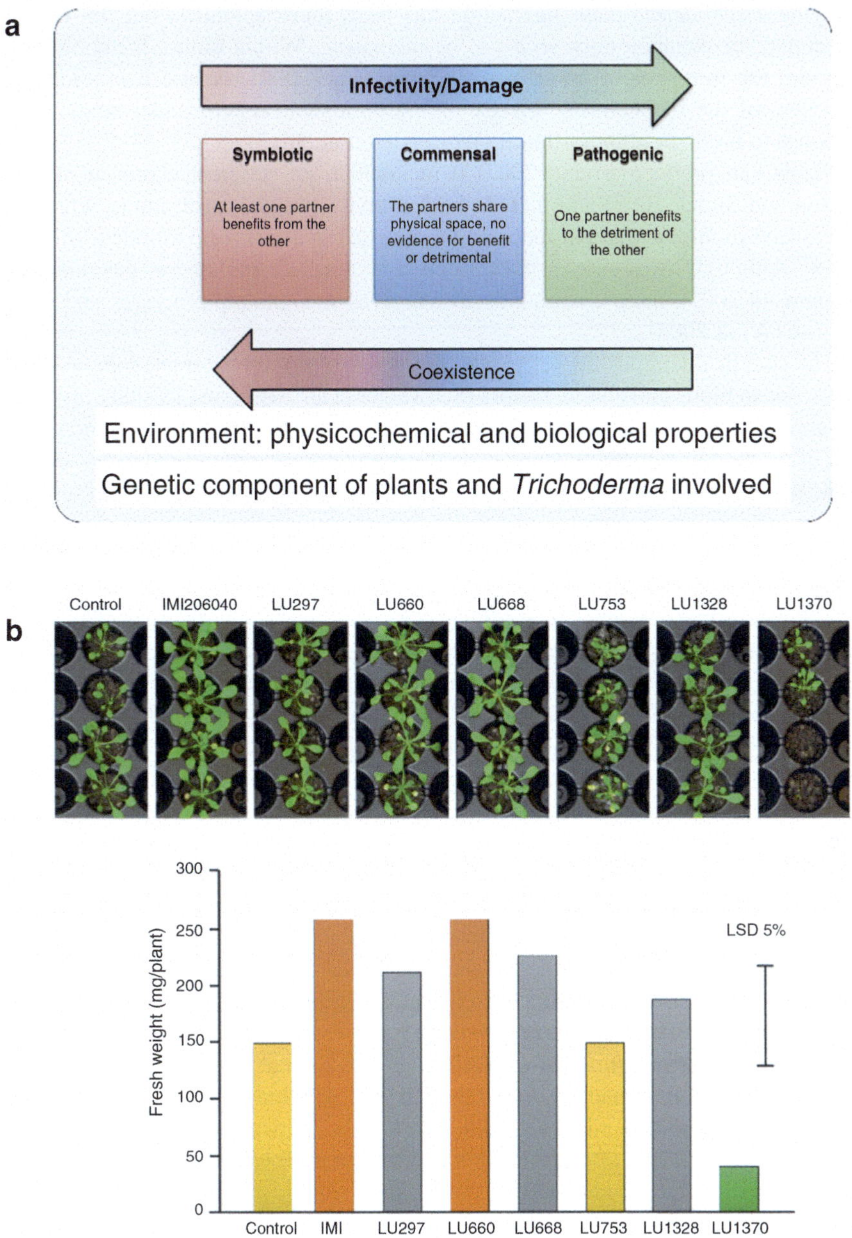

Fig. 10.3 (**a**) The endophytic continuum hypothesis. The outcome of the association of *Trichoderma* with plants depends on the environment and the genetics of the organisms in the association. These also include the microbiomes from the soil, rhizosphere, and endosphere. (**b**) An example of the same plant genotype and different *Trichoderma* genetic backgrounds; *Arabidopsis thaliana* Col0 in interaction with different species of *Trichoderma* in gamma radiated soil. (From Nieto-Jacobo et al., 2017)

studying *Trichoderma*-plant interaction has been transcriptomics, but the use of proteomics or metabolomics seems to be increasing. Within omics, metabolomics provides the most helpful information (Fiehn et al. 2000) because it provides an overview of the biochemical activity of the biological system under study, as a response to biotic or abiotic stimuli (Wolfender et al. 2013).

Plants can produce around 200,000 metabolites of different chemical classes (Dixon and Strack 2003) and include both primary (crucial to sustaining life) and specialized (non-essential but required for survival in a given environment) metabolites. Although SMs are non-essential for normal growth or development (Grotewold 2005), these compounds are involved in signaling processes and plant defense (Wink 2003).

The chemical composition of plants, also known as the phytometabolome, fluctuates due to biotic and abiotic factors (Schweiger et al. 2014), and its study involves the application of targeted and untargeted approaches. In metabolomics, the selection of an analytical method depends on the research question desired to be solved; while targeted metabolomics aims at the identification and quantification of already known metabolites, untargeted metabolic fingerprinting allows the generation and comparison of metabolic phenotypes that fluctuate in a given biological system, rather than metabolite identification (Wolfender et al. 2013).

The use of metabolomics to study the beneficial interaction between plants and microorganisms has increased. Studies of the beneficial interaction between arbuscular mycorrhizal fungi (AMF) and plants have been mainly performed using targeted metabolomics approaches and have unraveled systemic plant responses mediated by AMF colonization (Schweiger and Müller 2015). The use of metabolomics to study *Trichoderma*-plant interactions has used both targeted and untargeted approaches. Overall, the *Trichoderma* spp. root colonization directly impacts the content of phytohormones, soluble sugars, amino acids, phenolic compounds, citrate cycle intermediates, and polyamines in the host (Brotman et al. 2012; Vinci et al. 2018; Yedidia et al. 2003). The direct effects of *Trichoderma* SMs in the host metabolome have also been tested (Mazzei et al. 2016; Schweiger et al. 2021). Mazzei et al. (2016) analyzed the metabolome modulation of tomato-systemic tissue after it was exposed to the specialized metabolites 6-pentyl-α-pyrone (6PP) and harzianic acid isolated from *T. atroviride* P1 and *T. harzianum* M10, respectively. The results revealed significant differences in the content of amino acids, sugars, GABA, acetylcholine, and other metabolites and concluded that the metabolome modulation is dose-dependent for both compounds. Schweiger et al. (2021) tested the role of two mutants affected in SMs production in *T. virens* and their impact in the root metabolome. These authors found that metabolic composition of *T. virens*-colonized roots differed profoundly from that of non-colonized roots, with the effects depending on the fungal genotype. In particular, the concentrations of several metabolites derived from the shikimate pathway, including amino acids and several flavonoids, were modulated.

The modulation of the maize systemic tissue metabolome following *Trichoderma* root colonization and the influence of fertilizers applied in the soil has been evaluated by Vinci et al. (2018). These authors demonstrated that the maize leaf primary metabolism and the total phosphorus (P) and nitrogen (N) content are modulated by two factors, (i) the presence of *T. harzianum* in the rhizosphere and (ii) the application of phosphate-fertilizers. The study concluded that combining the application of organic phosphate-fertilizers to the soil and the inoculation of maize roots with *T. harzianum* significantly increased plant P, N, and chlorophyll content.

The application of untargeted and targeted metabolomics approaches coupled with transcriptome analyses of the tripartite interaction between tomato plants, *T. harzianum,* and aphids were recently assessed by Coppola et al. (2019), who demonstrated that *T. harzianum* modulated the gene expression and the phytometabolome of tomato, by promoting endogenous plant defenses after aphid infestation. Overall, the presence of *T. harzianum* in the tomato rhizosphere modified the phloem sap metabolome, the regulation of phytohormones balance and more likely improved the cross talk with natural aphids enemies via salicylate derivates and terpenes. Despite the *Trichoderma*-plant metabolomics analyses carried out, the metabolic events activated in plants as a response to *Trichoderma* colonization are mostly unknown. For further discussion on metabolomics see Chap. 6.

10.3.3 Endophytic Trichoderma*: Plant Immune System, Host Recognition, and Colonization*

In nature, chemical signals secreted by plants and microorganisms mediate beneficial or pathogenic interactions (Bais et al. 2006). For the establishment of beneficial interactions, microorganisms have developed the ability to modulate plant defense systems (Hassani et al. 2018; Jayaraman et al. 2014). In response to microorganisms, plants employ a conserved two-tier innate immune system described by the following steps. Upon microbial colonization, the first line of defense provided by plant immunity (basal defense) is activated, and it is characterized by the recognition of self from non-self structures. Thus, the plant responds to (i) specific microbial molecules known as microbe- or pathogen-associated molecular patterns (MAMPs or PAMPs), such as fungal chitin, or flagellin (Jones and Dangl 2006) and/or (ii) endogenous elicitors such as cell wall fragments, known as damage-associated molecular patterns (DAMPs) (Ibrahim et al. 2013). MAMPs and DAMPs are sensed by the host via specialized pattern recognition receptors (PRRs) that activate the MAMP/DAMP-triggered immunity (MTI). In response, beneficial or pathogenic microorganisms produce effector-like proteins to overcome MTI and establish a successful infection resulting in effector-triggered susceptibility (ETS). The second plant-defense line leads to a robust and accelerated response known as effector-triggered immunity (ETI), as a result of the recognition of microbial effectors (Cook et al. 2015; De Wit et al. 2009). ETI is more specific than PTI and usually leads to a hypersensitive response (HR) at the infection site causing disease resistance (Boller and Felix 2009; Cui et al. 2015).

During the establishment of beneficial plant-microbial interactions, the secretion of effector molecules facilitates plant colonization. By definition, any molecule that alters the physiology, structure, or function of another organism, facilitating the infection and triggering defense responses, can be considered as an effector (Kamoun 2006). Effector molecules are characterized by their ability to suppress plant defense, thus promoting pathogens' virulence or symbionts' compatibility (Ramírez-Valdespino et al. 2019). During plant colonization by *Trichoderma,* there is a transient suppression of the host defenses (Coppola et al. 2019; Mendoza-Mendoza et al. 2018; Morán-Diez et al. 2009). It may be possible that plant receptors distinguish the proteins and other molecules secreted by *Trichoderma*, allowing colonization to occur (Mendoza-Mendoza et al. 2018; Ramírez-Valdespino et al. 2019). Proteins were the first molecules proposed as effectors in pathogens, and cerato-platinins (Brotman et al. 2008), hydrophobins (Guzmán-Guzmán et al. 2017), glycoside-hydrolases (Moran-Diez et al. 2009), and SSCPs (Small Secreted Cysteine-Rich Proteins) (Lamdan et al. 2015) have been identified as fungal effectors during *Trichoderma*-plant interactions.

Besides proteins, specialized metabolites such as lactones, peptaibols, polyketides, phytohormones, trichothecenes, VOCs, and non-volatile terpenes have received attention as putative effectors in the *Trichoderma*-plant interactions, as well as small RNAs (Ramírez-Valdespino et al. 2019). Thus, the identification and functional characterization of additional *Trichoderma* spp. effectors will improve the understanding of *Trichoderma* as a fungal symbiont. Colonization involves overcoming toxic compounds released by the host in response to the invasion and protecting the hyphal tips against defense compounds released by the plant. In this opportunistic/facultative interaction, *Trichoderma* uses sucrose or other plant nutrients and, in turn, increases the plant's immunity levels and improves photosynthetic abilities (Vargas et al. 2009). Besides their ability to produce effectors, plant-associated beneficial microorganisms such as mycorrhizal fungi and *Trichoderma* spp. are capable of metabolizing plant-derived carbohydrates (Vargas et al. 2009). These polysaccharides secreted by plant roots are ideal substrates of lytic enzymes produced by beneficial fungi. Hydrolysis of polysaccharides may initiate host colonization (Druzhinina et al. 2011). The enrichment of lytic enzymes biosynthesis in *Trichoderma* spp. indeed seems to regulate plant roots colonization. For example, the silencing of the *T. harzianum Thpg-1*, encoding an endopolygalacturonase (pectin hydrolase), has a negative effect on the ability to colonize tomato roots (Moran-Diez et al. 2009), suggesting that pectin degradation is a critical step for the establishment of a *Trichoderma*-host interaction. Only oxygenic photosynthetic organisms such as plants are capable of synthesizing sucrose, making them the only source of sucrose in the rhizosphere (Salerno and Curatti 2003). *Trichoderma* spp. have developed the capacity to use sucrose as a carbon source. In this regard, Vargas et al. (2009) reported that *T. virens* possesses an intracellular invertase for sucrose hydrolysis (TvInv) that is expressed in the presence of plant roots. Furthermore, unpublished data from the same authors suggest that *T. virens* harbors in its genome two putative membrane-associated sucrose transporters. Besides these findings, further analyses are required to understand carbohydrates metabolism in *Trichoderma*-plant interactions.

Trichoderma root colonization seems to be a multifactorial and tightly controlled process, where fungal growth is restricted to certain root tissues. Transmission electron and confocal microscopy have revealed that *Trichoderma* spp. penetrate the first and second layers of the roots epidermis, colonizing inter- and intracellular spaces (Nogueira-Lopez et al. 2018; Yedidia et al. 1999). Apart from the knowledge regarding colonization patterns of *Trichoderma* in plants such as maize and tomato, the exact mechanisms that regulate and limit fungal growth to non-vascular tissues are not fully understood. Since VOCs have been considered as important signaling molecules during inter- and intrakingdom communication, the role of *T. virens* VOCs synthesized by the terpene synthase Vir4 was characterized. Vir4 regulates, in addition to sesquiterpenes, multiple other SMs, which are produced in the presence of the maize root. Moreover Vir4 is required for maize root colonization (Schweiger et al. 2021).

10.3.4 Plant Growth Promotion Induced by **Trichoderma** *spp.*

The capability of *Trichoderma* spp. to enhance plant growth has been assessed repeatedly, and it can occur in either soil or axenic systems (Contreras-Cornejo et al. 2013). Growth promotion effects have been reported from different plants including radish, pepper, cucumber, tomato, and *A. thaliana* (Azarmi et al. 2011; Baker 1984; Chang 1986; Contreras-Cornejo et al. 2009; Gravel et al. 2007). Plant development and biomass production depend on diverse environmental factors, including nutrient availability. Plant-associated microorganisms, including *Trichoderma* spp., facilitate plant-nutrient uptake by enhancing the availability of micronutrients such as P and Fe (Yedidia et al. 2001), thus inducing plant growth. Also, *Trichoderma* spp. improve plant-nutrient uptake by modifying the architecture of the root system (Samolski et al. 2012), facilitating water and nutrient uptake. Additional mechanisms used by *Trichoderma* spp. to improve plant growth include the production of phytohormones such as auxins (Contreras-Cornejo et al. 2009), an increase in the tolerance to abiotic stress (Brotman et al. 2013; Donoso et al. 2008), the inhibition of plant pathogens, and the activation of the plant immune system (Keswani et al. 2014; Pieterse et al. 2009). The chemical diversity of specialized metabolites produced by *Trichoderma* spp. provide them with the ability to survive even in hostile environments, and to interact with the host and enhance plant health. Within the diversity of bioactive molecules produced by *Trichoderma* spp. VOCs have been associated with the ability to enhance plant fitness; hence, an overview of *Trichoderma* VOCs is presented in the following section.

10.3.5 **Trichoderma spp.** ***VOCs and Plant Fitness***

VOCs synthesized by plant-associated microorganisms are considered chemical signals that mediate intra- and interkingdom communication acting in antagonism, mutualism, and intra- and interspecies cross talk. These signals are species- or strain-specific and fundamental to the regulation of cellular and developmental processes (Bitas et al. 2013; Fiedler et al. 2001; Kai et al. 2016; Morath et al. 2012). *Trichoderma* spp. are rich sources of VOCs that comprise diverse classes of bioactive compounds including pyrones, C-8 compounds, sesquiterpenes, monoterpenes, diterpenes, alkane compounds, and many more (Zeilinger and Schuhmacher 2013).

Independent analyses of the influence of *Trichoderma* spp. VOCs on growth of *A. thaliana* have revealed that not all the species tested have the same outcome. While some increase the plant growth and biomass, others do not have any effect and a few have detrimental effects on plant growth (Lee et al. 2016; Nieto-Jacobo et al. 2017). *Trichoderma* VOCs have also been associated with plant protection, being capable of displaying antifungal and nematicidal activities (Kottb et al. 2015b; Martinez-Medina et al. 2017a, b; Yang et al. 2012). Also, *Trichoderma* spp. VOCs have been shown to induce salt tolerance in *A. thaliana* (Jalali et al. 2017), stimulate iron uptake, and induce systemic resistance (Martinez-Medina et al. 2017b) (Fig. 10.4).

10.4 Induction of Systemic Resistance (ISR) by *Trichoderma*

Biological control agents (BCAs) can suppress the activity of plant pathogens by different mechanisms, directly or indirectly (Köhl et al. 2019). Direct action against the pathogen can occur due to mycoparasitism and antibiosis, among other forms (see above). In the indirect form, BCAs can induce resistance or prepare the plant for increased resistance against infections by a pathogen without direct antagonistic interaction with the pathogen (Köhl et al. 2019; Mendoza-Mendoza et al. 2017). This priming effect is a viable strategy when the direct mechanisms are ineffective against pathogens (Poveda et al. 2020). The induction of systemic resistance (ISR) is a durable and effective response against a broad spectrum of pathogens (Hermosa et al. 2013). It is considered the most prominent biological control mechanism (Elsharkawy 2020). ISR in plants is the result of the interaction and activation of signaling pathways, which leads to physiological and biochemical alterations in plants (Nawrocka and Małolepsza 2013). ISR thus provides a strong reaction against pathogen attack (Bisen et al. 2016). In this process, plant surface pattern recognition receptors (PRR) recognize specific cell surface components of microorganisms (Hermosa et al. 2013; Mendoza-Mendoza et al. 2018). The initial basal defenses to suppress the growth of the invader are initiated on the outer surface of the host cell (Hermosa et al. 2013). In the case of beneficial microorganisms, these elicitors are

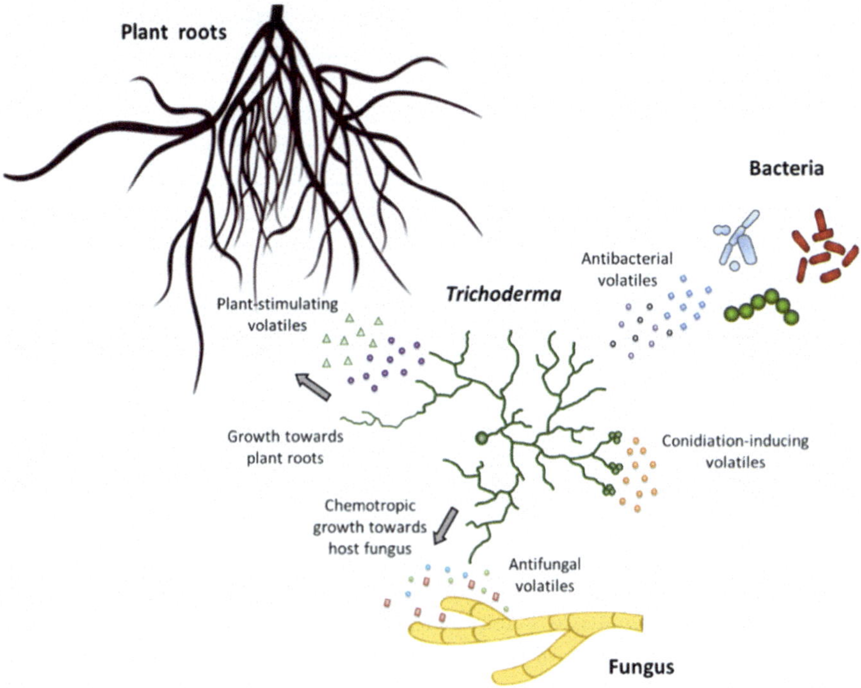

Fig. 10.4 Scheme representing the possible roles of *Trichoderma* VOCs in the rhizosphere during its interaction with plants and other microorganisms. Under non-sterile soil conditions, different organisms coexist and use VOCs as signaling molecules. *Trichoderma* spp. emit VOCs that are perceived by the surrounding organisms triggering different outcomes. Root exudates act as chemoattractants causing *Trichoderma* to grow towards plant roots. *Trichoderma* emitted VOCs increase tolerance to abiotic stress, stimulate iron uptake, stimulate root branching and nutrient and water uptake, induce systemic resistance, and promote plant growth. Also, *Trichoderma* VOCs not only act as inducers of conidiation in *Trichoderma* itself but also affect other soil microorganisms, displaying antimicrobial activities against fungi and bacteria, facilitating persistence in soil and plant protection against pathogens. VOCs secreted by *Trichoderma* are represented as circles, plant roots VOCs as triangles, bacterial VOCs as diamonds, and other fungal VOCs as rectangles. Image adapted from Zeilinger and Schumacher (2013)

known as MAMPs, while elicitors associated with harmful microbes are called pathogen-associated molecular patterns—PAMPs (Bisen et al. 2016; Hermosa et al. 2013). Systemic stimuli of the defense mechanisms can prepare the plant's immune response to act faster against subsequent attacks by any type of pathogen (Poveda et al. 2020). These responses can contribute to reducing the spread of the disease (Bisen et al. 2016; Mendoza-Mendoza et al. 2018). PRR-based signaling can also recognize endogenous components that result from microbial hydrolytic activity in plant tissues, such as cell wall oligomers or cuticle fragments. These molecules are known as damage-associated molecular patterns (DAMPs) (Hermosa et al. 2013).

The activation of PRRs leads to a cascade of signal transduction and the activation of immunity triggered by MAMP or PAMP (Hermosa et al. 2013;

Mendoza-Mendoza et al. 2018). The interaction between PAMP and the corresponding plant receptor activates defense responses in the host and is called PAMP triggered immunity (PTI) (Bisen et al. 2016). Like PAMPs, MAMPs from biological control agents are also associated with ISR, and their responses involve the generation of ionic fluxes, reactive oxygen species (ROS), nitric oxide and ethylene (ET) and, later, involve callose accumulation and the biosynthesis of antimicrobial substances (Bisen et al. 2016). MAMP responses are called MAMP-triggered immunity (MPI). This process triggers modifications that promote the fortification of the cell wall through the deposit of structural substances, such as callose and lignin (Yedidia et al. 1999). In this process, there is also the deposit of chitinases and glucanases—proteins related to pathogenesis (PR), which are capable of degrading the cell wall of pathogens (Hermosa et al. 2014, 2012; Morán-Diez et al. 2009). In addition, there is an accumulation of phytohormones such as SA, JA, and ET, whose signal is transmitted systemically through neighboring cells (Mendoza-Mendoza et al. 2018). The activation of priming involves transcription factors such as WRKYs, MYBs, and MYCs, which in many cases are transcriptionally upregulated in the roots during the interaction with *Trichoderma* (Coppola et al. 2019; Morán-Diez et al. 2021). The plant signals that regulate ISR in *T. virens* are oxylipins other than JA (Wang et al. 2020a). The accumulation of 12-oxophytodienoate and two γ-ketols in plants treated with *T. virens* correlates with ISR and the protection of maize leaves against *Colletotrichum graminicola* (Wang et al. 2020a, b). Interestingly, the systemic resistance induced by *Trichoderma* is transmitted through generations by unknown mechanisms that most likely involve epigenetic modifications in the seeds (Medeiros et al. 2017; Morán-Diez et al. 2021).

Trichoderma's ISR is not only restricted to protection against filamentous pathogens and bacteria, but also against nematodes (De Medeiros et al. 2017; Martínez-Medina et al. 2017a, b; Martinez-Medina et al. 2013; Morán-Diez et al. 2021), insect herbivores (Contreras-Cornejo et al. 2018), and attracting parasitoids on aphids (Coppola et al. 2019). The mechanisms of ISR are directly associated with the production of SMs in the plant as well as changes in the metabolome in the leaves (Contreras-Cornejo et al. 2018; Coppola et al. 2019). Inoculation of the rhizosphere of maize with *T. harzianum* increased the abundance of pest regulating arthropods and chewing herbivores, and decreased the number of piercing-sucking herbivores (Contreras-Cornejo et al. 2021).

Trichoderma spp. can live under adverse conditions in many ecosystems, promote changes in the rhizosphere that contribute to plant growth, and coordinate defense mechanisms against pathogens (Verma et al. 2007). *Trichoderma* spp. can colonize plant tissues above and below ground (Elsharkawy 2020; Nogueira-Lopez et al. 2018, 2019), and this interaction stimulates diverse defense signaling mechanisms (Pratap Singh et al. 2021). Many species that grow in the rhizosphere are able to colonize the internal tissues of plant roots (Mukherjee et al. 2013). This colonization has beneficial effects on plants and stimulates systemic resistance against attack by pathogens (Shoresh and Harman 2008b; Vinale et al. 2006).

10.5 Conclusions

Trichoderma spp. are one of the most successful biocontrol agents in agricultural production. They have the capacity to produce a broad diversity of CWDEs and SMs which have direct positive effects in plants, can inhibit plant pathogens, and can induce the accumulation of plant growth promoting bacteria and mycorrhizal fungi. However, the mode of action of *Trichoderma* depends on the environment, the genetics of the plant host, and the presence of the microbiome. To succeed in the rhizosphere *Trichoderma* must communicate with the microbiome, overcome the plant defenses, and be able to induce systemic resistance.

Acknowledgments The work from AMM was supported by the Tertiary Education Commission through the Bio-protection Research Centre-Lincoln University and The Massey-Lincoln and Agricultural Industry Trust (MLAIT).

References

Al-Hazmi AS, TariqJaveed M (2016) Effects of different inoculum densities of Trichoderma harzianum and Trichoderma viride against Meloidogyne javanica on tomato. Saudi J Biol Sci 23:288–292. https://doi.org/10.1016/j.sjbs.2015.04.007

Altomare C, Norvell WA, Bjorkman T, Harman GE (1999) Solubilization of phosphates and micronutrients by the plant-growth-promoting and biocontrol fungus trichoderma harzianum rifai 1295-22. Appl Environ Microbiol 65:2926–2933. https://doi.org/10.1128/AEM.65.7.2926-2933.1999

Anwar J, Iqbal Z (2017) Effect of growth conditions on antibacterial activity of Trichoderma harzianum against selected pathogenic bacteria Sarhad. J Agric 33:501–510

Azarmi R, Hajieghrari B, Giglou A (2011) Effect of Trichoderma isolates on tomato seedling growth response and nutrient uptake. Afr J Biotechnol 10. https://doi.org/10.5897/AJB10.1600

Bae H, Sicher RC, Kim MS, Kim SH, Strem MD, Melnick RL, Bailey BA (2009) The beneficial endophyte Trichoderma hamatum isolate DIS 219b promotes growth and delays the onset of the drought response in Theobroma cacao. J Exp Bot 60:3279–3295. https://doi.org/10.1093/jxb/erp165

Bae S-J, Mohanta TK, Chung JY, Ryu M, Park G, Shim S, Hong S-B, Seo H, Bae D-W, Bae I, Kim J-J, Bae H (2016) Trichoderma metabolites as biological control agents against Phytophthora pathogens. Biol Control 92:128–138. https://doi.org/10.1016/j.biocontrol.2015.10.005

Baedke J, Fábregas-Tejeda A, Nieves Delgado A (2020) The holobiont concept before Margulis. J Exp Zool Part B Mol Dev Evol 334:149–155. https://doi.org/10.1002/jez.b.22931

Bais HP, Weir TL, Perry LG, Gilroy S, Vivanco JM (2006) The role of root exudates in rhizosphere interactions with plants and other organisms. Annu Rev Plant Biol 57:233–266. https://doi.org/10.1146/annurev.arplant.57.032905.105159

Baker R (1984) The controlled experiment in the scientific method with special emphasis on biological control. Am Phytopathol Soc 74(9):1019–1021

Benttoumi N, Colagiero M, Sellami S, Boureghda H, Keddad A, Ciancio A (2020) Diversity of nematode microbial antagonists from algeria shows occurrence of nematotoxic trichoderma spp. Plants 9:941

Berg G, Köberl M, Rybakova D, Müller H, Grosch R, Smalla K (2017) Plant microbial diversity is suggested as the key to future biocontrol and health trends. FEMS Microbiol Ecol:93. https://doi.org/10.1093/femsec/fix050

Bisen K, Keswani C, Patel JS, Sarma BK, Singh HB (2016) Trichoderma spp.: efficient inducers of systemic resistance in plants. In: Choudhary DK, Varma A (eds) Microbial-mediated induced systemic resistance in plants. Springer Singapore, Singapore, pp 185–195

Bitas V, Kim HS, Bennett JW, Kang S (2013) Sniffing on microbes: diverse roles of microbial volatile organic compounds in plant health. Mol Plant Microbe Interact 26:835–843. https://doi.org/10.1094/mpmi-10-12-0249-cr

Boller T, Felix G (2009) A renaissance of elicitors: perception of microbe-associated molecular patterns and danger signals by pattern-recognition receptors. Annu Rev Plant Biol 60:379–406. https://doi.org/10.1146/annurev.arplant.57.032905.105346

Braithwaite M, Clouston A, Minchin R, Yardley J, Nieto-Jacobo MF, Mendoza-Mendoza A, Steyaert J, Hill R, Marshall J, Stewart A (2016) The density-dependent effect of initial nematode population levels on the efficacy of Trichoderma as a bio-nematicide against Meloidogyne hapla on tomato. Austr Plant Pathol 45:473–479

Braithwaite M, Johnston PR, Ball SL, Nourozi F, Hay AJ, Shoukouhi P, Chomic A, Lange C, Ohkura M, Nieto-Jacobo MF, Cummings NJ, Bienkowski D, Mendoza-Mendoza A, Hill RA, McLean KL, Stewart A, Steyaert JM, Bissett J (2017) Trichoderma down under: species diversity and occurrence of Trichoderma in New Zealand. Austr Plant Pathol 46:11–30. https://doi.org/10.1007/s13313-016-0457-9

Brookes JJ (2017) Endophytes in maize (Zea mays) in New Zealand. M.Sc. Thesis. Lincoln University

Brotman Y, Briff E, Viterbo A, Chet I (2008) Role of swollenin, an expansin-like protein from Trichoderma in plant root colonization. Plant Physiol 147:779–789. https://doi.org/10.1104/pp.108.116293

Brotman Y, Landau U, Cuadros-Inostroza Á, Takayuki T, Fernie AR, Chet I, Viterbo A, Willmitzer L (2013) Trichoderma-plant root colonization: escaping early plant defense responses and activation of the antioxidant machinery for saline stress tolerance. PLOS Pathogens 9: e1003221. https://doi.org/10.1371/journal.ppat.1003221

Brotman Y, Lisec J, Meret M, Chet I, Willmitzer L, Viterbo A (2012) Transcript and metabolite analysis of the Trichoderma-induced systemic resistance response to Pseudomonas syringae in Arabidopsis thaliana. Microbiology 158:139–146. https://doi.org/10.1099/mic.0.052621-0

Brundrett M, Schulz B, Boyle C, Sieber T (2006) Microbial root endophytes.

Cai F, Chen W, Wei Z, Pang G, Li R, Ran W, Shen Q (2015) Colonization of Trichoderma harzianum strain SQR-T037 on tomato roots and its relationship to plant growth, nutrient availability and soil microflora. Plant Soil 388:337–350. https://doi.org/10.1007/s11104-014-2326-z

Cai F, Druzhinina IS (2021) In honor of John Bissett: authoritative guidelines on molecular identification of Trichoderma. Fungal Diversity 107:1–69. https://doi.org/10.1007/s13225-020-00464-4

Carrión VJ, Perez-Jaramillo J, Cordovez V, Tracanna V, de Hollander M, Ruiz-Buck D, Mendes LW, van Ijcken WFJ, Gomez-Exposito R, Elsayed SS, Mohanraju P, Arifah A, van der Oost J, Paulson JN, Mendes R, van Wezel GP, Medema MH, Raaijmakers JM (2019) Pathogen-induced activation of disease-suppressive functions in the endophytic root microbiome. Science 366:606–612. https://doi.org/10.1126/science.aaw9285

Carsolio C, Benhamou N, Haran S, Cortés C, Gutiérrez A, Chet I, Herrera-Estrella A (1999) Role of the Trichoderma harzianum endochitinase gene, ech42, in mycoparasitism. Appl Environ Microbiol 65:929–935. https://doi.org/10.1128/AEM.65.3.929-935.1999

Chang Y-C (1986) Increased growth of plants in the presence of the biological control agent Trichoderma harzianum. Plant Dis 70(2)

Clay K, Holah J (1999) Fungal endophyte symbiosis and plant diversity in successional fields. Science 285:1742–1744

Contina J, Dandurand L, Knudsen G (2017) Use of GFP-tagged Trichoderma harzianum as a tool to study the biological control of the potato cyst nematode Globodera pallida. Appl Soil Ecol 115:31–37

Contreras-Cornejo HA, Macías-Rodríguez L, Cortés-Penagos C, López-Bucio J (2009) Trichoderma virens, a plant beneficial fungus, enhances biomass production and promotes lateral root growth through an auxin-dependent mechanism in Arabidopsis. Plant Physiol 149:1579–1592. https://doi.org/10.1104/pp.108.130369

Contreras-Cornejo HA, Macías-Rodríguez L, del-Val E, Larsen J (2018) The root endophytic fungus Trichoderma atroviride induces foliar herbivory resistance in maize plants. Appl Soil Ecol 124:45–53

Contreras-Cornejo HA, Macías-Rodríguez L, Herrera-Estrella A, López-Bucio J (2014) The 4-phosphopantetheinyl transferase of Trichoderma virens plays a role in plant protection against Botrytis cinerea through volatile organic compound emission. Plant Soil 379:261–274

Contreras-Cornejo HA, Viveros-Bremauntz F, del-Val E, Macías-Rodríguez L, López-Carmona DA, Alarcón A, González-Esquivel CE, Larsen J (2021) Alterations of foliar arthropod communities in a maize agroecosystem induced by the root-associated fungus Trichoderma harzianum. J Pest Sci 94:363–374. https://doi.org/10.1007/s10340-020-01261-3

Cook DE, Mesarich CH, Thomma BPHJ (2015) Understanding plant immunity as a surveillance system to detect invasion. Annu Rev Phytopathol 53:541–563. https://doi.org/10.1146/annurev-phyto-080614-120114

Coppola M, Diretto G, Digilio MC, Woo SL, Giuliano G, Molisso D, Pennacchio F, Lorito M, Rao R (2019) Transcriptome and Metabolome Reprogramming in Tomato Plants by Trichoderma harzianum strain T22 Primes and Enhances Defense Responses Against Aphids. Front Physiol 10:745–745. https://doi.org/10.3389/fphys.2019.00745

Cortés C, Gutiérrez A, Olmedo V, Inbar J, Chet I, Herrera-Estrella A (1998) The expression of genes involved in parasitism by Trichoderma harzianum is triggered by a diffusible factor. Mol Gen Genet 260:218–225. https://doi.org/10.1007/s004380050889

Cruz-Magalhaes V, Nieto-Jacobo MF, van Zijll de Jong E, Rostas M, Padilla-Arizmendi F, Kandula D, Kandula J, Hampton J, Herrera-Estrella A, Steyaert JM, Stewart A, Loguercio LL, Mendoza-Mendoza A (2018) The NADPH oxidases Nox1 and Nox2 differentially regulate volatile organic compounds, fungistatic activity, plant growth promotion and nutrient assimilation in Trichoderma atroviride. Front Microbiol 9:3271. https://doi.org/10.3389/fmicb.2018.03271

Cruz-Magalhães V, Nieto-Jacobo MF, van Zijll de Jong E, Rostás M, Padilla-Arizmendi F, Kandula D, Kandula J, Hampton J, Herrera-Estrella A, Steyaert JM, Stewart A, Loguercio LL, Mendoza-Mendoza A (2019) The NADPH oxidases Nox1 and Nox2 differentially regulate volatile organic compounds, fungistatic activity, plant growth promotion and nutrient assimilation in Trichoderma atroviride. Front Microbiol 9. https://doi.org/10.3389/fmicb.2018.03271

Cui H, Tsuda K, Parker JE (2015) Effector-triggered immunity: from pathogen perception to robust defense. Annu Rev Plant Biol 66:487–511. https://doi.org/10.1146/annurev-arplant-050213-040012

de Medeiros HA, de Araújo Filho JV, de Freitas LG, Castillo P, Rubio MB, Hermosa R, Monte E (2017) Tomato progeny inherit resistance to the nematode Meloidogyne javanica linked to plant growth induced by the biocontrol fungus Trichoderma atroviride. Sci Rep 7:1–13

De Palma M, Salzano M, Villano C, Aversano R, Lorito M, Ruocco M, Docimo T, Piccinelli AL, D'Agostino N, Tucci M (2019) Transcriptome reprogramming, epigenetic modifications and alternative splicing orchestrate the tomato root response to the beneficial fungus Trichoderma harzianum. Hortic Res 6:5. https://doi.org/10.1038/s41438-018-0079-1

de Santiago A, Quintero JM, Avilés M, Delgado A (2009) Effect of Trichoderma asperellum strain T34 on iron nutrition in white lupin. Soil Biol Biochem 41:2453–2459. https://doi.org/10.1016/j.soilbio.2009.07.033

De Wit PJ, Mehrabi R, Van den Burg HA, Stergiopoulos I (2009) Fungal effector proteins: past, present and future. Mol Plant Pathol 10:735–747. https://doi.org/10.1111/j.1364-3703.2009.00591.x

Degani O, Khatib S, Becher P, Gordani A, Harris R (2021) Trichoderma asperellum secreted 6-pentyl-α-pyrone to control Magnaporthiopsis maydis, the maize Late Wilt disease agent. Biology 10:897. https://doi.org/10.3390/biology10090897

Dini-Andreote F (2020) Endophytes: the second layer of plant defense. Trends Plant Sci 25:319–322. https://doi.org/10.1016/j.tplants.2020.01.007

Dixon R, Strack D (2003) Phytochemistry meets genome analysis, and beyond. Phytochemistry 62:815–816

Djonović S, Pozo MJ, Kenerley CM (2006) Tvbgn3, a beta-1,6-glucanase from the biocontrol fungus Trichoderma virens, is involved in mycoparasitism and control of Pythium ultimum. Appl Environ Microbiol 72:7661–7670. https://doi.org/10.1128/aem.01607-06

Donoso EP, Bustamante RO, Carú M, Niemeyer HM (2008) Water deficit as a driver of the mutualistic relationship between the fungus Trichoderma harzianum and two wheat genotypes. Appl Environ Microbiol 74:1412–1417. https://doi.org/10.1128/aem.02013-07

Donzelli BG, Harman GE (2001) Interaction of ammonium, glucose, and chitin regulates the expression of cell wall-degrading enzymes in Trichoderma atroviride strain P1. Appl Environ Microbiol 67:5643–5647. https://doi.org/10.1128/aem.67.12.5643-5647.2001

Druzhinina IS, Seidl-Seiboth V, Herrera-Estrella A, Horwitz BA, Kenerley CM, Monte E, Mukherjee PK, Zeilinger S, Grigoriev IV, Kubicek CP (2011) Trichoderma: the genomics of opportunistic success. Nat Rev Microbiol 9:749–759

Elsharkawy MM (2020) Chapter 7—Trichoderma genes for improving plant resistance to the pathogens. In: Gupta VK, Zeilinger S, Singh HB, Druzhinina I (eds) New and future developments in microbial biotechnology and bioengineering. Elsevier, pp 157–170

Elsherbiny EA, Amin BH, Aleem B, Kingsley KL, Bennett JW (2020) Trichoderma volatile organic compounds as a biofumigation tool against late blight pathogen phytophthora infestans in postharvest potato tubers. J Agric Food Chem 68:8163–8171. https://doi.org/10.1021/acs.jafc.0c03150

Eltlbany N, Baklawa M, Ding G-C, Nassal D, Weber N, Kandeler E, Neumann G, Ludewig U, van Overbeek L, Smalla K (2019) Enhanced tomato plant growth in soil under reduced P supply through microbial inoculants and microbiome shifts. FEMS Microbiol Ecol:95. https://doi.org/10.1093/femsec/fiz124

Elzawahry AM, Khalil A, Allam A, Mostafa RG (2015) Effect of the bio-agents (Bacillus megaterium and Trichoderma album) on citrus nematode (Tylenchulus semipenetrans) infecting baladi orange and. J Phytopathol Pest Manag 2(2):1–8

Escande AR, Laich FS, Pedraza MV (2002) Field testing of honeybee-dispersed Trichoderma spp. to manage sunflower head rot (Sclerotinia sclerotiorum). Plant Pathol 51:346–351. https://doi.org/10.1046/j.1365-3059.2002.00723.x

Fan H, Yao M, Wang H, Zhao D, Zhu X, Wang Y, Liu X, Duan Y, Chen L (2020) Isolation and effect of Trichoderma citrinoviride Snef1910 for the biological control of root-knot nematode, Meloidogyne incognita. BMC Microbiol 20:299. https://doi.org/10.1186/s12866-020-01984-4

Feyisa B, Lencho A, Selvaraj T, Getaneh G (2016) Evaluation of some botanicals and Trichoderma harzianum against root-knot nematode (Meloidogyne incognita (Kofoid and White) Chit wood) in tomato. J Entomol Nematol 8:11–18

Fiedler K, Schutz E, Geh S (2001) Detection of microbial volatile organic compounds (MVOCs) produced by moulds on various materials. Int J Hyg Environ Health 204:111–121. https://doi.org/10.1078/1438-4639-00094

Fiehn O, Kopka J, Dormann P, Altmann T, Trethewey RN, Willmitzer L (2000) Metabolite profiling for plant functional genomics. Nat Biotechnol 18:1157–1161. https://doi.org/10.1038/81137

Flores A, Chet I, Herrera-Estrella A (1997) Improved biocontrol activity of Trichoderma harzianum by over-expression of the proteinase-encoding gene prb1. Curr Genet 31:30–37. https://doi.org/10.1007/s002940050173

Garnica-Vergara A, Barrera-Ortiz S, Munoz-Parra E, Raya-Gonzalez J, Mendez-Bravo A, Macias-Rodriguez L, Ruiz-Herrera LF, Lopez-Bucio J (2016) The volatile 6-pentyl-2H-pyran-2-one from Trichoderma atroviride regulates Arabidopsis thaliana root morphogenesis via auxin signaling and ETHYLENE INSENSITIVE 2 functioning. New Phytol 209:1496–1512. https://doi.org/10.1111/nph.13725

Geremia RA, Goldman GH, Jacobs D, Ardiles W, Vila SB, Van Montagu M, Herrera-Estrella A (1993) Molecular characterization of the proteinase-encoding gene, prb1, related to mycoparasitism by Trichoderma harzianum. Mol Microbiol 8:603–613. https://doi.org/10.1111/j.1365-2958.1993.tb01604.x

Gibert A, Tozer W, Westoby M (2019) Plant performance response to eight different types of symbiosis. New Phytol. https://doi.org/10.1111/nph.15392

Gilbert SF (2014) Symbiosis as the way of eukaryotic life: the dependent co-origination of the body. J Biosci 39:201–209

Gómez Expósito R, de Bruijn I, Postma J, Raaijmakers JM (2017) Current insights into the role of rhizosphere bacteria in disease suppressive soils. Front Microbiol 8:2529–2529. https://doi.org/10.3389/fmicb.2017.02529

Gravel V, Antoun H, Tweddell RJ (2007) Growth stimulation and fruit yield improvement of greenhouse tomato plants by inoculation with Pseudomonas putida or Trichoderma atroviride: possible role of indole acetic acid (IAA). Soil Biol Biochem 39:1968–1977. https://doi.org/10.1016/j.soilbio.2007.02.015

Grotewold E (2005) Plant metabolic diversity: a regulatory perspective. Trends Plant Sci 10:57–62. https://doi.org/10.1016/j.tplants.2004.12.009

Guo Y, Ghirardo A, Weber B, Schnitzler J-P, Benz JP, Rosenkranz M (2019) Trichoderma species differ in their volatile profiles and in antagonism toward ectomycorrhiza Laccaria bicolor. Front Microbiol:10. https://doi.org/10.3389/fmicb.2019.00891

Guo Y, Jud W, Ghirardo A, Antritter F, Benz JP, Schnitzler JP, Rosenkranz M (2020) Sniffing fungi - phenotyping of volatile chemical diversity in Trichoderma species. New Phytol 227:244–259. https://doi.org/10.1111/nph.16530

Guzmán-Guzmán P, Alemán-Duarte MI, Delaye L, Herrera-Estrella A, Olmedo-Monfil V (2017) Identification of effector-like proteins in Trichoderma spp. and role of a hydrophobin in the plant-fungus interaction and mycoparasitism. BMC Genet 18:16. https://doi.org/10.1186/s12863-017-0481-y

Guzmán-Guzmán P, Porras-Troncoso MD, Olmedo-Monfil V, Herrera-Estrella A (2019) Trichoderma species: versatile plant symbionts. Phytopathology 109:6–16. https://doi.org/10.1094/phyto-07-18-0218-rvw

Hadar Y, Papadopoulou KK (2012) Suppressive composts: microbial ecology links between abiotic environments and healthy plants. Annu Rev Phytopathol 50:133–153. https://doi.org/10.1146/annurev-phyto-081211-172914

Hardoim PR, van Overbeek LS, Berg G, Pirttilä AM, Compant S, Campisano A, Döring M, Sessitsch A (2015) The hidden world within plants: ecological and evolutionary considerations for defining functioning of microbial endophytes. Microbiol Mol Biol Rev 79:293–320. https://doi.org/10.1128/mmbr.00050-14

Harman GE, Howell CR, Viterbo A, Chet I, Lorito M (2004) Trichoderma species—opportunistic, avirulent plant symbionts. Nat Rev Microbiol 2:43–56. https://doi.org/10.1038/nrmicro797

Hassani MA, Durán P, Hacquard S (2018) Microbial interactions within the plant holobiont. Microbiome 6:58. https://doi.org/10.1186/s40168-018-0445-0

Hermosa R, Cardoza R, Rubio M, Gutiérrez S, Monte E (2014) Secondary metabolism and antimicrobial metabolites of Trichoderma. In: Biotechnology and biology of Trichoderma. Elsevier, pp 125–137

Hermosa R, Rubio MB, Cardoza RE, Nicolás C, Monte E, Gutiérrez S (2013) The contribution of Trichoderma to balancing the costs of plant growth and defense. Int Microbiol 16:69–80. https://doi.org/10.2436/20.1501.01.181

Hermosa R, Viterbo A, Chet I, Monte E (2012) Plant-beneficial effects of Trichoderma and of its genes. Microbiology (Reading) 158:17–25. https://doi.org/10.1099/mic.0.052274-0

Contreras-Cornejo HA et al (2013) Promotion of plant growth and the induction of systemic defence by trichoderma: physiology, genetics and gene expression. In: Trichoderma: biology and applications. CABI, Wallingford, pp 173–194

Hoyos-Carvajal L, Orduz S, Bissett J (2009) Growth stimulation in bean (Phaseolus vulgaris L.) by Trichoderma. Biol Control 51:409–416

Ibrahim ZA, Armour CL, Phipps S, Sukkar MB (2013) RAGE and TLRs: relatives, friends or neighbours? Mol Immunol 56:739–744. https://doi.org/10.1016/j.molimm.2013.07.008

Illescas M, Rubio MB, Hernández-Ruiz V, Morán-Diez ME, Martínez de Alba AE, Nicolás C, Monte E, Hermosa R (2020) Effect of inorganic N top dressing and Trichoderma harzianum seed-inoculation on crop yield and the shaping of root microbial communities of wheat plants cultivated under high basal N fertilization. Front Plant Sci 11:575861. https://doi.org/10.3389/fpls.2020.575861

Intana W, Kheawleng S, Sunpapao A (2021) Trichoderma asperellum T76-14 released volatile organic compounds against postharvest fruit rot in muskmelons (Cucumis melo) caused by Fusarium incarnatum. J Fungi (Basel) 7. https://doi.org/10.3390/jof7010046

Jalali F, Zafari D, Salari H (2017) Volatile organic compounds of some Trichoderma spp. increase growth and induce salt tolerance in Arabidopsis thaliana. Fungal Ecol 29:67–75. https://doi.org/10.1016/j.funeco.2017.06.007

Jayaraman D, Gilroy S, Ané J-M (2014) Staying in touch: mechanical signals in plant–microbe interactions. Curr Opin Plant Biol 20:104–109. https://doi.org/10.1016/j.pbi.2014.05.003

Jiang Y, Wang J-L, Chen J, Mao L-J, Feng X-X, Zhang C-L, Lin F-C (2016) Trichoderma biodiversity of agricultural fields in east China reveals a gradient distribution of species. PLoS One 11:e0160613. https://doi.org/10.1371/journal.pone.0160613

Jones JD, Dangl JL (2006) The plant immune system. Nature 444:323–329. https://doi.org/10.1038/nature05286

Kai M, Effmert U, Piechulla B (2016) Bacterial-plant-interactions: approaches to unravel the biological function of bacterial volatiles in the rhizosphere. Front Microbiol 7:108. https://doi.org/10.3389/fmicb.2016.00108

Kamoun S (2006) A catalogue of the effector secretome of plant pathogenic oomycetes. Annu Rev Phytopathol 44:41–60. https://doi.org/10.1146/annurev.phyto.44.070505.143436

Karimi-Aghcheh R, Bok JW, Phatale PA, Smith KM, Baker SE, Lichius A, Omann M, Zeilinger S, Seiboth B, Rhee C, Keller NP, Freitag M, Kubicek CP (2013) Functional analyses of Trichoderma reesei LAE1 reveal conserved and contrasting roles of this regulator. G3 (Bethesda) 3:369–378. https://doi.org/10.1534/g3.112.005140

Karlsson M, Atanasova L, Jensen DF, Zeilinger S, Heitman J, James TY, Crous PW (2017) Necrotrophic mycoparasites and their genomes. Microbiol Spectr 5:5.2.08. https://doi.org/10.1128/microbiolspec.FUNK-0016-2016

Keswani C, Mishra S, Sarma BK, Singh SP, Singh HB (2014) Unraveling the efficient applications of secondary metabolites of various Trichoderma spp. Appl Microbiol Biotechnol 98:533–544. https://doi.org/10.1007/s00253-013-5344-5

Khan MR, Ahmad I, Ahamad F (2018) Effect of pure culture and culture filtrates of Trichoderma species on root-knot nematode, Meloidogyne incognita infesting tomato. Indian Phytopathol 71:265–274

Khan RAA, Najeeb S, Mao Z, Ling J, Yang Y, Li Y, Xie B (2020) Bioactive secondary metabolites from Trichoderma spp. against phytopathogenic bacteria and root-knot nematode. Microorganisms 8:401. https://doi.org/10.3390/microorganisms8030401

Khare E, Mishra J, Arora NK (2018) Multifaceted interactions between endophytes and plant: developments and prospects. Front Microbiol 9:2732. https://doi.org/10.3389/fmicb.2018.02732

Kim K, Heo YM, Jang S, Lee H, Kwon S-L, Park MS, Lim YW, Kim J-J (2020) Diversity of Trichoderma spp. in Marine Environments and Their Biological Potential for Sustainable Industrial Applications. Sustainability 12:4327

Kiriga AW, Haukeland S, Kariuki GM, Coyne DL, Beek NV (2018) Effect of Trichoderma spp. and Purpureocillium lilacinum on Meloidogyne javanica in commercial pineapple production in Kenya. Biol Control 119:27–32

Köhl J, Kolnaar R, Ravensberg WJ (2019) Mode of action of microbial biological control agents against plant diseases: relevance beyond efficacy. Front Plant Sci:10. https://doi.org/10.3389/fpls.2019.00845

Kottb M, Gigolashvili T, Großkinsky DK, Piechulla B (2015a) Trichoderma volatiles effecting Arabidopsis: from inhibition to protection against phytopathogenic fungi. Front Microbiol:6. https://doi.org/10.3389/fmicb.2015.00995

Kottb M, Gigolashvili T, Großkinsky DK, Piechulla B (2015b) Trichoderma volatiles effecting Arabidopsis: from inhibition to protection against phytopathogenic fungi. Front Microbiol 6:995. https://doi.org/10.3389/fmicb.2015.00995

Kubicek CP, Herrera-Estrella A, Seidl-Seiboth V, Martinez DA, Druzhinina IS, Thon M, Zeilinger S, Casas-Flores S, Horwitz BA, Mukherjee PK, Mukherjee M, Kredics L, Alcaraz LD, Aerts A, Antal Z, Atanasova L, Cervantes-Badillo MG, Challacombe J, Chertkov O, McCluskey K, Coulpier F, Deshpande N, von Döhren H, Ebbole DJ, Esquivel-Naranjo EU, Fekete E, Flipphi M, Glaser F, Gómez-Rodríguez EY, Gruber S, Han C, Henrissat B, Hermosa R, Hernández-Oñate M, Karaffa L, Kosti I, Le Crom S, Lindquist E, Lucas S, Lübeck M, Lübeck PS, Margeot A, Metz B, Misra M, Nevalainen H, Omann M, Packer N, Perrone G, Uresti-Rivera EE, Salamov A, Schmoll M, Seiboth B, Shapiro H, Sukno S, Tamayo-Ramos JA, Tisch D, Wiest A, Wilkinson HH, Zhang M, Coutinho PM, Kenerley CM, Monte E, Baker SE, Grigoriev IV (2011) Comparative genome sequence analysis underscores mycoparasitism as the ancestral life style of Trichoderma. Genome Biol 12:R40. https://doi.org/10.1186/gb-2011-12-4-r40

Kuchár M (2019) Endophytic fungi associated with cabbage in New Zealand and their potential for biological control. Ph.D. Thesis. Lincoln University

Kumar P, Chand R (2015) Bio-efficacy of Trichoderma harzianum against root knot nematode Meloidogyne incognita on Brinjal. Ann Plant Prot Sci 23:361–364

Lamdan N-L, Shalaby S, Ziv T, Kenerley CM, Horwitz BA (2015) Secretome of Trichoderma interacting with maize roots: role in induced systemic resistance. Mol Cell Proteomics 14:1054–1063. https://doi.org/10.1074/mcp.M114.046607

Lareen A, Burton F, Schafer P (2016) Plant root-microbe communication in shaping root microbiomes. Plant Mol Biol 90:575–587. https://doi.org/10.1007/s11103-015-0417-8

Lee S, Yap M, Behringer G, Hung R, Bennett JW (2016) Volatile organic compounds emitted by Trichoderma species mediate plant growth. Fungal Biol Biotechnol 3:7. https://doi.org/10.1186/s40694-016-0025-7

Leonetti P, Zonno MC, Molinari S, Altomare C (2017) Induction of SA-signaling pathway and ethylene biosynthesis in Trichoderma harzianum-treated tomato plants after infection of the root-knot nematode Meloidogyne incognita. Plant Cell Rep 36:621–631. https://doi.org/10.1007/s00299-017-2109-0

Li N, Alfiky A, Wang W, Islam M, Nourollahi K, Liu X, Kang S (2018) Volatile compound-mediated recognition and inhibition between trichoderma biocontrol agents and Fusarium oxysporum. Front Microbiol:9. https://doi.org/10.3389/fmicb.2018.02614

Li N, Islam MT, Kang S (2020) Secreted metabolite-mediated interactions between rhizosphere bacteria and Trichoderma biocontrol agents. PLoS One 14:e0227228. https://doi.org/10.1371/journal.pone.0227228

Ma J, Tsegaye E, Li M, Wu B, Jiang X (2020) Biodiversity of Trichoderma from grassland and forest ecosystems in Northern Xinjiang, China. 3 Biotech 10:362. https://doi.org/10.1007/s13205-020-02301-6

Makiola A (2018) Characterising plant pathogen communities and their environmental drivers at a national scale. Ph.D. Thesis. Lincoln University

Malinich EA, Wang K, Mukherjee PK, Kolomiets M, Kenerley CM (2019) Differential expression analysis of Trichoderma virens RNA reveals a dynamic transcriptome during colonization of Zea mays roots. BMC Genomics 20:280. https://doi.org/10.1186/s12864-019-5651-z

Martinez-Medina A, Fernandez I, Sánchez-Guzmán M, Jung S, Pascual J, Pozo M (2013) Deciphering the hormonal signalling network behind the systemic resistance induced by Trichoderma harzianum in tomato. Front Plant Sci:4. https://doi.org/10.3389/fpls.2013.00206

Martínez-Medina A, Fernandez I, Lok GB, Pozo MJ, Pieterse CMJ, Van Wees SCM (2017a) Shifting from priming of salicylic acid- to jasmonic acid-regulated defences by Trichoderma protects tomato against the root knot nematode Meloidogyne incognita. New Phytol 213:1363–1377. https://doi.org/10.1111/nph.14251

Martinez-Medina A, Van Wees SCM, Pieterse CMJ (2017b) Airborne signals from Trichoderma fungi stimulate iron uptake responses in roots resulting in priming of jasmonic acid-dependent defences in shoots of Arabidopsis thaliana and Solanum lycopersicum. Plant Cell Environ 40:2691–2705. https://doi.org/10.1111/pce.13016

Mascarin GM, Bonfim Junior M, de Araújo Filho J (2012) Trichoderma harzianum reduces population of Meloidogyne incognita in cucumber plants under greenhouse conditions Embrapa Arroz e Feijão-Artigo em periódico indexado (ALICE).

Mazzei P, Vinale F, Woo SL, Pascale A, Lorito M, Piccolo A (2016) Metabolomics by proton high-resolution magic-angle-spinning nuclear magnetic resonance of tomato plants treated with two secondary metabolites isolated from Trichoderma. J Agric Food Chem 64:3538–3545. https://doi.org/10.1021/acs.jafc.6b00801

Mazzola M (2002) Mechanisms of natural soil suppressiveness to soilborne diseases. Antonie van Leeuwenhoek 81:557–564. https://doi.org/10.1023/A:1020557523557

Medeiros HA, Araújo Filho JV, Freitas LG, Castillo P, Rubio MB, Hermosa R, Monte E (2017) Tomato progeny inherit resistance to the nematode Meloidogyne javanica linked to plant growth induced by the biocontrol fungus Trichoderma atroviride. Sci Rep 7:40,216. https://doi.org/10.1038/srep40216

Mendes R, Kruijt M, De Bruijn I, Dekkers E, van der Voort M, Schneider JH, Piceno YM, DeSantis TZ, Andersen GL, Bakker PA (2011) Deciphering the rhizosphere microbiome for disease-suppressive bacteria. Science 332:1097–1100

Mendoza-Mendoza A, Pozo MJ, Grzegorski D, Martínez P, García JM, Olmedo-Monfil V, Cortés C, Kenerley C, Herrera-Estrella A (2003) Enhanced biocontrol activity of Trichoderma through inactivation of a mitogen-activated protein kinase. Proc Natl Acad Sci U S A 100:15,965–15,970

Mendoza-Mendoza A, Nogueira-López G, Arizmendi F, Cripps-Guazzone N, Nieto-Jacobo M, Lawry R, Kandula D, Salazar-Badillo F, Salas-Muñoz S, Mauricio-Castillo J (2017) Mechanisms of growth promotion by members of the rhizosphere fungal genus Trichoderma. Adv PGPR Res:1–15

Mendoza-Mendoza A, Zaid R, Lawry R, Hermosa R, Monte E, Horwitz BA, Mukherjee PK (2018) Molecular dialogues between Trichoderma and roots: role of the fungal secretome. Fungal Biol Rev 32:62–85. https://doi.org/10.1016/j.fbr.2017.12.001

Metcalf DA, Wilson CR (2001) The process of antagonism of Sclerotium cepivorum in white rot affected onion roots by Trichoderma koningii. Plant Pathol 50:249–257. https://doi.org/10.1046/j.1365-3059.2001.00549.x

Miamoto A, Silva MTR, Dias-Arieira CR, Puerari HH (2017) Alternative products for Pratylenchus brachyurus and Meloidogyne javanica management in soya bean plants. J Phytopathol 165:635–640

Montoya A, Quijano G (2016) Isolation of Trichoderma Spp. from desert soil, biocontrol potential evaluation and liquid culture production of conidia using agricultural fertilizers. J Fertil Pestic 2016. https://doi.org/10.4172/2471-2728.1000163

Moran-Diez E, Hermosa R, Ambrosino P, Cardoza RE, Gutierrez S, Lorito M, Monte E (2009) The ThPG1 endopolygalacturonase is required for the trichoderma harzianum-plant beneficial interaction. Mol Plant Microbe Interact 22:1021–1031. https://doi.org/10.1094/mpmi-22-8-1021

Morán-Diez E, Hermosa R, Ambrosino P, Cardoza RE, Gutiérrez S, Lorito M, Monte E (2009) The ThPG1 endopolygalacturonase is required for the Trichoderma harzianum–plant beneficial interaction. Mol Plant Microbe Interact 22:1021–1031. https://doi.org/10.1094/mpmi-22-8-1021

Moran-Diez E, Rubio B, Dominguez S, Hermosa R, Monte E, Nicolas C (2012) Transcriptomic response of Arabidopsis thaliana after 24 h incubation with the biocontrol fungus Trichoderma harzianum. J Plant Physiol 169:614–620. https://doi.org/10.1016/j.jplph.2011.12.016

Morán-Diez ME, Martínez de Alba ÁE, Rubio MB, Hermosa R, Monte E (2021) Trichoderma and the plant heritable priming responses. J Fungi 7:318. https://www.mdpi.com/2309-608X/7/4/318

Morán-Diez ME, Trushina N, Lamdan NL, Rosenfelder L, Mukherjee PK, Kenerley CM, Horwitz BA (2015) Host-specific transcriptomic pattern of Trichoderma virens during interaction with maize or tomato roots. BMC Genomics:16. https://doi.org/10.1186/s12864-014-1208-3

Morath SU, Hung R, Bennett JW (2012) Fungal volatile organic compounds: a review with emphasis on their biotechnological potential. Fungal Biol Rev 26:73–83. https://doi.org/10.1016/j.fbr.2012.07.001

Moreno-Mateos MA, Delgado-Jarana J, Codon AC, Benitez T (2007) pH and Pac1 control development and antifungal activity in *Trichoderma harzianum*. Fungal Genet Biol 44:1355–1367

Moreno-Ruiz D, Lichius A, Turrà D, Di Pietro A, Zeilinger S (2020) Chemotropism assays for plant symbiosis and mycoparasitism related compound screening in Trichoderma atroviride. Front Microbiol 11:601251. https://doi.org/10.3389/fmicb.2020.601251

Mukherjee P, Mendoza-Mendoza A, Zeilinger S, Horwitz B (2022) Mycoparasitism as a mechanism of Trichoderma-mediated suppression of plant diseases. Fungal Biol Rev 39:15–33

Mukherjee PK, Horwitz BA, Herrera-Estrella A, Schmoll M, Kenerley CM (2013) Trichoderma research in the genome era. Annu Rev Phytopathol 51:105–129. https://doi.org/10.1146/annurev-phyto-082712-102353

Nawrocka J, Małolepsza U (2013) Diversity in plant systemic resistance induced by Trichoderma. Biol Control 67:149–156. https://doi.org/10.1016/j.biocontrol.2013.07.005

Nieto-Jacobo M, Steyaert J, Salazar F, Nguyen D, Rostás M, Braithwaite M, de Souza J, Jimenez Bremont J, Ohkura M, Stewart A, Mendoza-Mendoza A (2017) Environmental growth conditions of Trichoderma spp. affects indole acetic acid derivatives, volatile organic compounds and plant growth promotion. Front Plant Sci 8. https://doi.org/10.3389/fpls.2017.00102

Nogueira-Lopez G, Greenwood DR, Middleditch M, Winefield C, Eaton C, Steyaert JM, Mendoza-Mendoza A (2018) The apoplastic secretome of Trichoderma virens during interaction with maize roots shows an inhibition of plant defence and scavenging oxidative stress secreted proteins. Front Plant Sci 9:409. https://doi.org/10.3389/fpls.2018.00409

Nogueira-López G, Lawry R, Echaide-Aquino JF, Mendoza-Mendoza A (2020) Chapter 5—Chemical communication between Trichoderma and plants. In: Gupta VK, Zeilinger S, Singh HB, Druzhinina I (eds) New and future developments in microbial biotechnology and bioengineering. Elsevier, pp 109–139

Nogueira-Lopez G, Padilla-Arizmendi F, Inwood S, Lyne S, Steyaert JM, Nieto-Jacobo MF, Stewart A, Mendoza-Mendoza A (2019) TrichoGate: an improved vector system for a large scale of functional analysis of trichoderma genes. Front Microbiol 10:2794. https://doi.org/10.3389/fmicb.2019.02794

Oliveira KCL, Meneses AC, Silva JM, Tavares RLC (2019) Biological management of pratylenchus brachyurus in soybean crops. Revista Caatinga 32:041–051

Ownley BH, Griffin MR, Klingeman WE, Gwinn KD, Moulton JK, Pereira RM (2008) Beauveria bassiana: endophytic colonization and plant disease control. J Invertebr Pathol 98:267–270. https://doi.org/10.1016/j.jip.2008.01.010

Pang G, Cai F, Li R, Zhao Z, Li R, Gu X, Shen Q, Chen W (2017) Trichoderma-enriched organic fertilizer can mitigate microbiome degeneration of monocropped soil to maintain better plant growth. Plant Soil 416:181–192. https://doi.org/10.1007/s11104-017-3178-0

Patil AS, Patil SR, Paikrao HM (2016) Trichoderma secondary metabolites: their biochemistry and possible role in disease management. In: Choudhary DK, Varma A (eds) Microbial-mediated induced systemic resistance in plants. Springer Singapore, Singapore, pp 69–102

Pétriacq P, Williams A, Cotton A, McFarlane AE, Rolfe SA, Ton J (2017) Metabolite profiling of non-sterile rhizosphere soil. Plant J 92:147–162. https://doi.org/10.1111/tpj.13639

Phoka N, Suwannarach N, Lumyong S, Ito S-i, Matsui K, Arikit S, Sunpapao A (2020) Role of volatiles from the endophytic fungus Trichoderma asperelloides PSU-P1 in biocontrol potential and in promoting the plant growth of Arabidopsis thaliana. J Fungi 6:341. https://www.mdpi.com/2309-608X/6/4/341

Phupiewkham W, Sirithorn P, Saksirirat W, Thammasirirak S (2015) Antibacterial agents from Trichoderma harzianum strain T9 against pathogenic bacteria Chiang Mai. J. Sci 42:304–316

Pieterse CMJ, Leon-Reyes A, Van der Ent S, Van Wees SCM (2009) Networking by small-molecule hormones in plant immunity. Nat Chem Biol 5:308. https://doi.org/10.1038/nchembio.164

Pocurull M, Fullana AM, Ferro M, Valero P, Escudero N, Saus E, Gabaldón T, Sorribas FJ (2019) Commercial formulates of Trichoderma induce systemic plant resistance to meloidogyne incognita in tomato and the effect is additive to that of the Mi-1.2 resistance gene. Front Microbiol 10

Poole P (2017) Shining a light on the dark world of plant root–microbe interactions. Proc Natl Acad Sci 114:4281. https://doi.org/10.1073/pnas.1703800114

Poveda J, Abril-Urias P, Escobar C (2020) Biological control of plant-parasitic nematodes by filamentous fungi inducers of resistance: Trichoderma mycorrhizal and endophytic fungi. Front Microbiol 11:992. https://doi.org/10.3389/fmicb.2020.00992

Pozo MJ, Baek JM, García JM, Kenerley CM (2004) Functional analysis of tvsp1, a serine protease-encoding gene in the biocontrol agent Trichoderma virens. Fungal Genet Biol 41:336–348. https://doi.org/10.1016/j.fgb.2003.11.002

Pratap Singh S, Keswani C, Pratap Singh S, Sansinenea E, Xuan Hoat T (2021) Trichoderma spp. mediated induction of systemic defense response in brinjal against Sclerotinia sclerotiorum. Curr Res Microb Sci 2:100051. https://doi.org/10.1016/j.crmicr.2021.100051

Qiao C, Penton CR, Xiong W, Liu C, Wang R, Liu Z, Xu X, Li R, Shen Q (2019) Reshaping the rhizosphere microbiome by bio-organic amendment to enhance crop yield in a maize-cabbage rotation system. Appl Soil Ecol 142:136–146. https://doi.org/10.1016/j.apsoil.2019.04.014

Qualhato TF, Lopes FAC, Steindorff AS, Brandão RS, Jesuino RSA, Ulhoa CJ (2013) Mycoparasitism studies of Trichoderma species against three phytopathogenic fungi: evaluation of antagonism and hydrolytic enzyme production. Biotechnol Lett 35:1461–1468. https://doi.org/10.1007/s10529-013-1225-3

Rajani P, Rajasekaran C, Vasanthakumari MM, Olsson SB, Ravikanth G, Uma Shaanker R (2021) Inhibition of plant pathogenic fungi by endophytic Trichoderma spp. through mycoparasitism and volatile organic compounds. Microbiol Res 242:126,595. https://doi.org/10.1016/j.micres.2020.126595

Ramírez-Valdespino CA, Casas-Flores S, Olmedo-Monfil V (2019) Trichoderma as a model to study effector-like molecules. Front Microbiol 10:1030. https://doi.org/10.3389/fmicb.2019.01030

Ramírez-Valdespino CA, Porras-Troncoso MD, Corrales-Escobosa AR, Wrobel K, Martínez-Hernández P, Olmedo-Monfil V (2018) Functional characterization of TvCyt2, a member of the p450 monooxygenases from Trichoderma virens relevant during the association with plants and mycoparasitism. Mol Plant Microbe Interact 31:289–298. https://doi.org/10.1094/mpmi-01-17-0015-r

Raynaud X, Nunan N (2014) Spatial ecology of bacteria at the microscale in soil. PLoS One 9:–e87217. https://doi.org/10.1371/journal.pone.0087217

Rodriguez RJ, White JF Jr, Arnold AE, Redman RS (2009) Fungal endophytes: diversity and functional roles. New Phytol 182:314–330. https://doi.org/10.1111/j.1469-8137.2009.02773.x

Romano G, Costantini M, Sansone C, Lauritano C, Ruocco N, Ianora A (2017) Marine microorganisms as a promising and sustainable source of bioactive molecules. Mar Environ Res 128:58–69. https://doi.org/10.1016/j.marenvres.2016.05.002

Ruangwong OU, Wonglom P, Suwannarach N, Kumla J, Thaochan N, Chomnunti P, Pitija K, Sunpapao A (2021) Volatile organic compound from Trichoderma asperelloides TSU1: impact on plant pathogenic fungi. J Fungi (Basel) 7. https://doi.org/10.3390/jof7030187

Sahebani N, Hadavi N (2008) Biological control of the root-knot nematode Meloidogyne javanica by Trichoderma harzianum. Soil Biol Biochem 40:2016–2020

Salazar-Badillo FB, Sanchez-Rangel D, Becerra-Flora A, Lopez-Gomez M, Nieto-Jacobo F, Mendoza-Mendoza A, Jimenez-Bremont JF (2015) Arabidopsis thaliana polyamine content is modified by the interaction with different Trichoderma species. Plant Physiol Biochem 95:49–56. https://doi.org/10.1016/j.plaphy.2015.07.003

Salerno GL, Curatti L (2003) Origin of sucrose metabolism in higher plants: when, how and why? Trends Plant Sci 8:63–69. https://doi.org/10.1016/s1360-1385(02)00029-8

Samolski I, Rincon AM, Pinzon LM, Viterbo A, Monte E (2012) The qid74 gene from Trichoderma harzianum has a role in root architecture and plant biofertilization. Microbiology 158:129–138. https://doi.org/10.1099/mic.0.053140-0

Santhanam R, Weinhold A, Goldberg J, Oh Y, Baldwin IT (2015) Native root-associated bacteria rescue a plant from a sudden-wilt disease that emerged during continuous cropping. Proc Natl Acad Sci 112:E5013–E5020

Saravanakumar K, Fan L, Fu K, Yu C, Wang M, Xia H, Sun J, Li Y, Chen J (2016) Cellulase from Trichoderma harzianum interacts with roots and triggers induced systemic resistance to foliar disease in maize. Sci Rep 6:35,543. https://doi.org/10.1038/srep35543

Saravanakumar K, Li Y, Yu C, Wang Q-Q, Wang M, Sun J, Gao J-X, Chen J (2017) Effect of Trichoderma harzianum on maize rhizosphere microbiome and biocontrol of Fusarium Stalk rot. Sci Rep 7:1–13

Sarsaiya S, Jain A, Fan X, Jia Q, Xu Q, Shu F, Zhou Q, Shi J, Chen J (2020) New insights into detection of a dendrobine compound from a novel endophytic Trichoderma longibrachiatum strain and its toxicity against phytopathogenic bacteria. Front Microbiol 11:337

Schlatter D, Kinkel L, Thomashow L, Weller D, Paulitz T (2017) Disease suppressive soils: new insights from the soil microbiome. Phytopathology 107:1284–1297. https://doi.org/10.1094/phyto-03-17-0111-rvw

Schmoll M, Dattenbock C, Carreras-Villasenor N, Mendoza-Mendoza A, Tisch D, Aleman MI, Baker SE, Brown C, Cervantes-Badillo MG, Cetz-Chel J, Cristobal-Mondragon GR, Delaye L, Esquivel-Naranjo EU, Frischmann A, Gallardo-Negrete Jde J, Garcia-Esquivel M, Gomez-Rodriguez EY, Greenwood DR, Hernandez-Onate M, Kruszewska JS, Lawry R, Mora-Montes HM, Munoz-Centeno T, Nieto-Jacobo MF, Nogueira Lopez G, Olmedo-Monfil V, Osorio-Concepcion M, Pilsyk S, Pomraning KR, Rodriguez-Iglesias A, Rosales-Saavedra MT, Sanchez-Arreguin JA, Seidl-Seiboth V, Stewart A, Uresti-Rivera EE, Wang CL, Wang TF, Zeilinger S, Casas-Flores S, Herrera-Estrella A (2016) The genomes of three uneven siblings: footprints of the lifestyles of three Trichoderma species. Microbiol Mol Biol Rev 80:205–327. https://doi.org/10.1128/MMBR.00040-15

Schweiger R, Baier MC, Persicke M, Müller C (2014) High specificity in plant leaf metabolic responses to arbuscular mycorrhiza. Nat Commun 5:3886. https://doi.org/10.1038/ncomms4886. https://www.nature.com/articles/ncomms4886#supplementary-information

Schweiger R, Müller C (2015) Leaf metabolome in arbuscular mycorrhizal symbiosis. Curr Opin Plant Biol 26:120–126. https://doi.org/10.1016/j.pbi.2015.06.009

Schweiger R, Padilla-Arizmendi F, Nogueira-Lopez G, Rostas M, Lawry R, Brown C, Hampton J, Steyaert JM, Muller C, Mendoza-Mendoza A (2021) Insights into metabolic changes caused by the Trichoderma virens-maize root interaction. Mol Plant Microbe Interact 34:524–537. https://doi.org/10.1094/MPMI-04-20-0081-R

Sharon E, Chet I, Viterbo A, Bar-Eyal M, Nagan H, Samuels GJ, Spiegel Y (2007) Parasitism of Trichoderma on Meloidogyne javanica and role of the gelatinous matrix. Eur J Plant Pathol 118:247–258

Shoresh M, Harman GE (2008a) The molecular basis of shoot responses of maize seedlings to Trichoderma harzianum T22 inoculation of the root: a proteomic approach. Plant Physiol 147:2147–2163. https://doi.org/10.1104/pp.108.123810

Shoresh M, Harman GE (2008b) The relationship between increased growth and resistance induced in plants by root colonizing microbes. Plant Signal Behav. 3:737–739. https://doi.org/10.4161/psb.3.9.6605

Siddiquee S (2014) Recent advancements on the role and analysis of volatile compounds (VOCs) from Trichoderma. In: Biotechnology and biology of trichoderma. Elsevier, pp 139–175

Singh A, Shukla N, Kabadwal B, Tewari A, Kumar J (2018) Review on plant-Trichoderma-pathogen interaction. Int J Curr Microbiol Appl Sci 7:2382–2397

Speckbacher V, Ruzsanyi V, Wigger M, Zeilinger S (2020) The Trichoderma atroviride strains P1 and IMI 206040 differ in their light-response and VOC production. Molecules 25:208. Retrieved from https://www.mdpi.com/1420-3049/25/1/208

Sridharan AP, Sugitha T, Karthikeyan G, Uthandi S (2020) Comprehensive profiling of the VOCs of Trichoderma longibrachiatum EF5 while interacting with Sclerotium rolfsii and Macrophomina phaseolina. Microbiol Res 236:126436. https://doi.org/10.1016/j.micres.2020.126436

Su D, Ding L, He S (2018) Marine-derived Trichoderma species as a promising source of bioactive secondary metabolites. Mini Rev Med Chem 18:1702–1713

Szabó M, Csepregi K, Gálber M, Virányi F, Fekete C (2012) Control plant-parasitic nematodes with Trichoderma species and nematode-trapping fungi: the role of chi18-5 and chi18-12 genes in nematode egg-parasitism. Biol Control 63:121–128

Toju H, Okayasu K, Notaguchi M (2019) Leaf-associated microbiomes of grafted tomato plants. Sci Rep 9:1787. https://doi.org/10.1038/s41598-018-38344-2

Trivedi P, Delgado-Baquerizo M, Trivedi C, Hamonts K, Anderson IC, Singh BK (2017) Keystone microbial taxa regulate the invasion of a fungal pathogen in agro-ecosystems. Soil Biol Biochem 111:10–14

Trushina N, Levin M, Mukherjee PK, Horwitz BA (2013) PacC and pH–dependent transcriptome of the mycotrophic fungus *Trichoderma virens*. BMC Genomics 14:138. https://doi.org/10.1186/1471-2164-14-138

Tucci M, Ruocco M, De Masi L, De Palma M, Lorito M (2011) The beneficial effect of Trichoderma spp. on tomato is modulated by the plant genotype. Mol Plant Pathol 12:341–354. https://doi.org/10.1111/j.1364-3703.2010.00674.x

van der Heijden MG, Bardgett RD, van Straalen NM (2008) The unseen majority: soil microbes as drivers of plant diversity and productivity in terrestrial ecosystems. Ecol Lett 11:296–310. https://doi.org/10.1111/j.1461-0248.2007.01139.x

Vargas WA, Mandawe JC, Kenerley CM (2009) Plant-derived sucrose is a key element in the symbiotic association between Trichoderma virens and maize plants. Plant Physiol 151:792–808. https://doi.org/10.1104/pp.109.141291

Vázquez-Garcidueñas S, Leal-Morales CA, Herrera-Estrella A (1998) Analysis of the beta-1,3-glucanolytic system of the biocontrol agent Trichoderma harzianum. Appl Environ Microbiol 64:1442–1446. https://doi.org/10.1128/aem.64.4.1442-1446.1998

Velázquez-Robledo R, Contreras-Cornejo H, Macías-Rodríguez L, Hernández-Morales A, Aguirre J, Casas-Flores S, López-Bucio J, Herrera-Estrella A (2011) Role of the 4-phosphopantetheinyl transferase of Trichoderma virens in secondary metabolism and induction of plant defense responses. Mol Plant Microbe Interact 24:1459–1471

Verma M, Brar SK, Tyagi RD, Surampalli RY, Valéro JR (2007) Antagonistic fungi, Trichoderma spp.: panoply of biological control. Biochem Eng J 37:1–20. https://doi.org/10.1016/j.bej.2007.05.012

Vinale F, Marra R, Scala F, Ghisalberti EL, Lorito M, Sivasithamparam K (2006) Major secondary metabolites produced by two commercial Trichoderma strains active against different phytopathogens. Lett Appl Microbiol 43:143–148. https://doi.org/10.1111/j.1472-765X.2006.01939.x

Vinci G, Cozzolino V, Mazzei P, Monda H, Spaccini R, Piccolo A (2018) An alternative to mineral phosphorus fertilizers: the combined effects of Trichoderma harzianum and compost on Zea mays, as revealed by 1H NMR and GC-MS metabolomics. PLoS One 13:e0209664. https://doi.org/10.1371/journal.pone.0209664

Viterbo A, Harel M, Chet I (2004) Isolation of two aspartyl proteases from Trichoderma asperellum expressed during colonization of cucumber roots. FEMS Microbiol Lett 238:151–158

Vyas R, Mathur K (2002) Distribution of Trichoderma spp. in cumin rhizosphere and their potential in suppression of wilt. Indian Phytopathol 55:451–457

Wang KD, Borrego EJ, Kenerley CM, Kolomiets MV (2020a) Oxylipins other than jasmonic acid are xylem-resident signals regulating systemic resistance induced by Trichoderma virens in maize. Plant Cell 32:166–185. https://doi.org/10.1105/tpc.19.00487

Wang KD, Gorman Z, Huang PC, Kenerley CM, Kolomiets MV (2020b) Trichoderma virens colonization of maize roots triggers rapid accumulation of 12-oxophytodienoate and two γ-ketols in leaves as priming agents of induced systemic resistance. Plant Signal Behav 15: 1792187. https://doi.org/10.1080/15592324.2020.1792187

Wang Z, Li Y, Zhao Y, Zhuang L, Yu Y, Wang M, Liu J, Wang Q (2021) A microbial consortium-based product promotes potato yield by recruiting rhizosphere bacteria involved in nitrogen and carbon metabolisms. Microb Biotechnol 14:1961–1975. https://doi.org/10.1111/1751-7915.13876

Wang Z, Li Y, Zhuang L, Yu Y, Liu J, Zhang L, Gao Z, Wu Y, Gao W, Ding GC, Wang Q (2019) A rhizosphere-derived consortium of Bacillus subtilis and Trichoderma harzianum suppresses common scab of potato and increases yield. Comput Struct Biotechnol J 17:645–653. https://doi.org/10.1016/j.csbj.2019.05.003

Werner S, Polle A, Brinkmann N (2016) Belowground communication: impacts of volatile organic compounds (VOCs) from soil fungi on other soil-inhabiting organisms. Appl Microbiol Biotechnol 100:8651–8665. https://doi.org/10.1007/s00253-016-7792-1

Wink M (2003) Evolution of secondary metabolites from an ecological and molecular phylogenetic perspective. Phytochemistry 64:3–19. https://doi.org/10.1016/S0031-9422(03)00300-5

Wolfender JL, Rudaz S, Choi YH, Kim HK (2013) Plant metabolomics: from holistic data to relevant biomarkers. Curr Med Chem 20:1056–1090

Woo SL, Ruocco M, Vinale F, Nigro M, Marra R, Lombardi N, Pascale A, Lanzuise S, Manganiello G, Lorito M (2014) Trichoderma-based products and their widespread use in agriculture. Open Mycol J:8

Woo SL, Scala F, Ruocco M, Lorito M (2006) The molecular biology of the interactions between Trichoderma spp., phytopathogenic fungi, and plants. Phytopathology 96:181–185. https://doi.org/10.1094/phyto-96-0181

Xiao-Yan S, Qing-Tao S, Shu-Tao X, Xiu-Lan C, Cai-Yun S, Yu-Zhong Z (2006) Broad-spectrum antimicrobial activity and high stability of Trichokonins from Trichoderma koningii SMF2 against plant pathogens. FEMS Microbiol Lett 260:119–125. https://doi.org/10.1111/j.1574-6968.2006.00316.x

Yang C, Hamel C, Gan Y, Vujanovic V (2012) Bacterial endophytes mediate positive feedback effects of early legume termination times on the yield of subsequent durum wheat crops. Can J Microbiol 58:1368–1377. https://doi.org/10.1139/w2012-114

Yedidia I, Benhamou N, Chet I (1999) Induction of defense responses in cucumber plants (Cucumis sativus L.) by the biocontrol agent Trichoderma harzianum. Appl Environ Microbiol 65:1061–1070. https://aem.asm.org/content/aem/65/3/1061.full.pdf

Yedidia I, Shoresh M, Kerem Z, Benhamou N, Kapulnik Y, Chet I (2003) Concomitant induction of systemic resistance to Pseudomonas syringae pv. lachrymans in cucumber by Trichoderma

asperellum (T-203) and accumulation of phytoalexins. Appl Environ Microbiol 69:7343–7353. https://doi.org/10.1128/aem.69.12.7343-7353.2003
Yedidia I, Srivastva AK, Kapulnik Y, Chet I (2001) Effect of Trichoderma harzianum on micro-element concentrations and increased growth of cucumber plants. Plant Soil 235:235–242
Zachow C, Berg C, Müller H, Monk J, Berg G (2016) Endemic plants harbour specific Trichoderma communities with an exceptional potential for biocontrol of phytopathogens. J Biotechnol 235: 162–170. https://doi.org/10.1016/j.jbiotec.2016.03.049
Zeilinger S, Gruber S, Bansal R, Mukherjee PK (2016) Secondary metabolism in Trichoderma—chemistry meets genomics. Fungal Biol Rev 30:74–90. https://doi.org/10.1016/j.fbr.2016.05.001
Zeilinger S, Schuhmacher R. (2013). Volatile organic metabolites of Trichoderma spp.:biosynthesis, biology and analytics. Trichoderma—biology and applications. Wallingford, Oxfordshire: CAB International. In (pp. 110-127).
Zeilinger S, Schumacher R (2013) Volatile organic metabolites of Trichoderma spp.: biosynthesis, biology and analytics. In: Mukherjee PK, Horwitz BA, Singh US, Mukherjee M, Schmoll M (eds) Trichoderma: biology and applications. CAB International, pp 110–127
Zhang F, Huo Y, Cobb AB, Luo G, Zhou J, Yang G, Wilson GW, Zhang Y (2018) Trichoderma biofertilizer links to altered soil chemistry, altered microbial communities, and improved grassland biomass. Front Microbiol 9:848
Zhang F, Yang X, Ran W, Shen Q (2014a) Fusarium oxysporum induces the production of proteins and volatile organic compounds by Trichoderma harzianum T-E5. FEMS Microbiol Lett 359: 116–123. https://doi.org/10.1111/1574-6968.12582
Zhang F, Zhu Z, Yang X, Ran W, Shen Q (2013) Trichoderma harzianum T-E5 significantly affects cucumber root exudates and fungal community in the cucumber rhizosphere. Appl Soil Ecol 72: 41–48
Zhang S, Gan Y, Ji W, Xu B, Hou B, Liu J (2017) Mechanisms and characterization of Trichoderma longibrachiatum T6 in suppressing nematodes (Heterodera avenae) in wheat. Front Plant Sci 8: 1491
Zhang S, Gan Y, Xu B, Xue Y (2014b) The parasitic and lethal effects of Trichoderma longibrachiatum against Heterodera avenae. Biol Control 72:1–8
Zilber-Rosenberg I, Rosenberg E (2008) Role of microorganisms in the evolution of animals and plants: the hologenome theory of evolution. FEMS Microbiol Rev 32:723–735. https://doi.org/10.1111/j.1574-6976.2008.00123.x

Chapter 11
Ectomycorrhizal Symbiosis: From Genomics to Trans-Kingdom Molecular Communication and Signaling

José Eduardo Marqués-Gálvez, Claire Veneault-Fourrey, and Annegret Kohler

Abstract Ectomycorrhizal symbioses are among the most widespread associations between roots of trees and soil fungi in forest ecosystems and contribute significantly to the sustainability of forest ecosystems through nutrient cycling and carbon sequestration. Due to increasing genomic and transcriptomic resources and the development of new tools for functional characterization, significant progress has been made in understanding the establishment and functioning of this mutualistic interaction. Here we summarize our current knowledge about the evolution of ectomycorrhizal symbiosis, the signaling and communication necessary for the development of the plant-microbe interface and how nutrient exchange will affect the association, from both the plant and fungal perspective.

Keywords Ectomycorrhiza · Symbiosis · Tree-microbe interactions · Genomics · Signaling · Effector

11.1 Introduction

Around 6000 plant species belonging to Pinaceae and most Angiosperms (Fabaceae, Fagaceae, Myrtaceae, Dipterocarpaceae, Fagaceae, Salicaceae, etc.) establish ectomycorrhizal (ECM) symbiosis with around 20,000 ectomycorrhizal (ECM) fungal species, both Ascomycota and Basidiomycota (Brundrett and Tedersoo 2018). These ECM symbioses are among the most widespread associations between roots of trees and soil fungi in forest ecosystems and contribute significantly to the sustainability of the forest ecosystems through nutrient cycling and carbon sequestration (Brundrett and Tedersoo 2018). The beneficial association between trees and microbes is a good example of a complex system that is shaped by the organisms present and further by the environmental forces acting on them.

J. E. Marqués-Gálvez · C. Veneault-Fourrey · A. Kohler (✉)
Université de Lorraine, INRAE, UMR Interactions Arbres-Microorganismes, Centre INRAE Grand Est-Nancy, Champenoux, France
e-mail: annegret.kohler@inrae.fr

B. A. Horwitz, P. K. Mukherjee (eds.), *Microbial Cross-talk in the Rhizosphere*, Rhizosphere Biology, https://doi.org/10.1007/978-981-16-9507-0_11

In this review, we will discuss the recent advances in elucidating the nutritional, molecular, and hormonal determinants of ECM interactions from both the plant and fungal perspectives, with a particular focus on the *Populus* x *Laccaria bicolor* system, which is one of the most used models to study plant-ECM interactions (Cregger et al. 2021). Further, we will summarize the current status of genome sequencing of ectomycorrhizal species and their plant hosts and we will discuss the lessons we learned so far from large-scale genomics and how this changed our view on ectomycorrhiza evolution.

11.2 Tree Genomes

The first draft genome from a tree, the black cottonwood *Populus trichocarpa*, was published in 2006 (Tuskan et al. 2006). *P. trichocarpa* was selected for the genome sequencing because of its modest genome size (485 Mb), its rapid growth, relative ease of experimental manipulation, and the availability of genetic tools (Tuskan et al. 2006). These features make *Populus* an ideal model host for mycorrhizal studies on the molecular level, in addition to the fact that both Arbuscular Mycorrhiza (AM) and ECM fungi are able to colonize its roots. One major finding from the genome of the long-lived Poplar compared to the annual *Arabidopsis* was the overall expansion of genes coding for membrane transporters. Among them are two mycorrhizal-specific phosphate transporters, suggesting that mycorrhizal symbiosis could have an important impact for the mineral nutrition of this tree (Tuskan et al. 2006).

Because their genomes are among the largest of all organisms, genome-wide analyses of conifers are particularly challenging (Nystedt et al. 2013). The first gymnosperm genome was the 20-gigabase genome of Norway spruce (*Picea abies*), published in 2013 (Nystedt et al. 2013). With more high-throughput sequencing methods available and decreasing sequencing costs the number of plant and in particular tree genomes are constantly increasing. In 2018, 148 plant reference genomes were available including 52 tree species (Wegrzyn et al. 2020). Among them we find more and more genomes of trees with importance for ectomycorrhiza research like *Eucalyptus grandis* (Myburg et al. 2014), *Pinus taeda* (Neale et al. 2014), *Betula pendula* (Salojärvi et al. 2017), or *Quercus robur* (Plomion et al. 2018).

Due to their enormous genome size, sequence data for forest trees often originate from experiments without a reference genome using transcriptomic approaches (de novo assembly of RNA sequences) (Wegrzyn et al. 2020). A transcriptome-based phylogeny recently allowed tracking the macroevolution of *Pinus* with nearly complete species sampling (Jin et al. 2021). It revealed that about 90% of existing pine species originated in the Miocene, that topography played a primary role in pine diversification while the aridity index was decisive for the niche rate shift. Moreover, fire has forced diversification and adaptive evolution of Pines (Jin et al. 2021).

How have microbial interactions and in particular mycorrhizal interactions impacted this plant evolution? Comparison of genomes from ancient plants with recent ones, in particular bryophytes with tracheophytes, allows to unravel how existing plant life diversity evolved and showed that biotic interactions with plants have shaped plant diversity (Lutzoni et al. 2018; Delaux et al. 2019). It is possible to infer traits of ancestral plants on the basis of observations from today's bryophytes and tracheophytes and to trace their evolution (Delaux et al. 2019; Delaux and Schornack 2021). How have plants and in particular trees evolved recognition systems that can distinguish between friend and foe? To benefit from microbes, plants had to evolve genetic toolboxes for symbiosis. These modules, like for example the Common Symbiotic Pathway (CSP, see Sect. 11.4.2), have been developed by early land plants and have been maintained despite the risk of getting hijacked by pathogens (for review, see Delaux and Schornack 2021). Often the perception of microbes in plants relies on extracellular or membrane-anchored nucleotide-binding site leucine-rich repeat (NB–LRR)-related proteins and receptor-like kinases (RLK). Phylogenetic analyses have identified RLKs already present in both bryophytes and tracheophytes (Han 2019). Interestingly, in the genome of the long-lived oak tree expanded gene families are enriched in Gene Ontology terms related to biotic interactions, including NB-LRR and RLK encoding genes (Plomion et al. 2018). Comparison with genomes from other trees and herbaceous plant genomes showed that many gene families generally expanded in trees are related to plant immunity (Plomion et al. 2018). From the comparison of bryophyte and tracheophyte genomes we further know that the perception of chitin, a fungal cell wall component, and the subsequent signaling pathways are conserved and were likely present in a common ancestor (Galotto et al. 2020). Many plant hormones have also been linked to biotic and in particular symbiotic interactions (see Sect. 11.5). The main biosynthetic and signaling pathways for salicylic acid, jasmonic acid, and ethylene are ancient and predated or developed probably together with the transition of plants to land (Bowles et al. 2020). While intracellular fungal symbionts (AM fungi) all activate the CSP, the situation for ectomycorrhizal fungi is less clear. Poplar roots colonize less with *Laccaria bicolor* when two genes of the CSP (CCaMK and POLLUX) are silenced (Cope et al. 2019), but trees from the Pinaceae family lacking most genes of this pathway form ectomycorrhiza very well (Garcia et al. 2015), which demonstrates that it is not essential (see Sect. 11.4.2). The sequencing of more trees from different phylogenetic clades, angiosperms and gymnosperm, will hopefully provide an answer as to whether ectomycorrhiza formation relies on another symbiosis signaling pathway or whether several accommodation ways evolved or were recruited from existing pathways independently from each other.

11.3 Fungal Genomes

Within the *Populus* sequencing project numerous sequence reads that mapped to microorganisms were identified spanning three kingdoms, 37 genera, and 99 species (Tuskan et al. 2006; Wullschleger et al. 2013), emphasizing its large microbiome. Shortly after the release of the *Populus* genome, there was a call to sequence *Populus* associated microbes (Martin et al. 2004; Cheng and Tuskan 2009). As a result, the first genome of an ectomycorrhizal fungus, *Laccaria bicolor*, was published in 2008 (Martin et al. 2008), as well as a few years later a poplar pathogen, *Melampsora larici-populina* (Duplessis et al. 2011) and finally the endomycorrhizal (AM) fungus *Rhizophagus irregularis* (Tisserant et al. 2013), able to form arbuscules with young *Populus* roots. The genome of *Laccaria bicolor* revealed proliferation of repetitive elements, reduction of gene families coding for secreted plant cell wall degrading enzymes (PCWDE) and surprisingly, since so far only known for pathogenic fungi, many genes coding for effector-like small secreted proteins (SSPs) with unknown function, several only expressed in symbiotic tissues (Martin et al. 2008, see Sect. 11.5.1). To clarify whether these were fundamental features of ectomycorrhizal genomes and to trace the evolution of these gene families, a large-scale initiative to produce a much larger set of mycorrhizal genomes was started. Led by the Joint Genome Institute (JGI) and in the frame of their community sequencing program the Mycorrhizal Genomics Initiative (MGI) started in 2011 a project called "Exploring the Genome Diversity of Mycorrhizal Fungi to Understand the Evolution and Functioning of Symbiosis in Woody Shrubs and Trees." The mycorrhizal fungi targeted for sequencing were selected based on their phylogenetic diversity and their ability to establish different types of mycorrhizal symbiosis (ECM, ericoid mycorrhiza, orchid mycorrhiza). Among the selected fungi were some species already established as models for mycorrhizal symbiosis (*Pisolithus*, *Amanita*, *Piloderma*, *Hebeloma*, *Suillus*, *Cenococcum*, *Oidiodendron*, *Tulasnella*, *Sebacina*) with existing in vitro mycorrhization systems (with *Eucalyptus*, *Populus*, *Quercus*, *Pinus*, *Blueberry*, *Arabidopsis*). The first comparative analysis included a set of 13 genomes of mycorrhizal fungi and was published by Kohler et al. (2015). It confirmed that ectomycorrhizal genomes contain a reduced gene set encoding PCWDEs as compared to their ancestral wood decayers. Nevertheless, the different ECM fungi have retained unique arrays of PCWDEs, depending on their respective saprophytic ancestor (Kohler et al. 2015). *Hebeloma cylindrosporum*, nested in a group of white-rot decayers, still possesses three manganese peroxidase genes, while *Amanita muscaria* lacks lignin peroxidase genes, but their loss preceded the evolution of the ECM habit and was therefore probably not (directly) caused by it (Kohler et al. 2015). Surprisingly, orchid and ericoid mycorrhizal genomes do not show any reduction of PCWDE genes, probably reflecting a primitive mode of symbiotic life substantially relying on the saprotrophic ability or a switching lifestyle from saprotrophic to endophytic to endosymbiotic (Kohler et al. 2015; Martin et al. 2016; Martino et al. 2018). Since the ECM lifestyle evolved several times independently, the question arises whether already existing genes or new genes are necessary for symbiosis? Both prove true, since a large set of symbiosis-upregulated genes have orthologs in brown- and white-rot fungi, thus suggesting that they are not unique to mycorrhizal symbionts and tend to be associated with core metabolic pathways or transport (Kohler et al. 2015). But at the same time 7–38% of the symbiosis-induced genes are orphan genes, suggesting that the convergent evolution of the ECM habit occurred via the repeated evolution of lineage-specific gene sets, like the Mycorrhiza-induced small-secreted proteins (MiSSPs) (Kohler et al. 2015). This conservation of symbiosis-induced genes was examined more closely in Miyauchi et al. (2020), a combined analysis of 135 fungal genomes from 73 saprotrophic, endophytic, and pathogenic species and 62 mycorrhizal species, including ten transcriptomes from ECM associations. A phylostratigraphic approach was applied to characterize the evolutionary origins of ectomycorrhizal lineages on the basis of gene functions in extant organisms. In Ascomycota and Basidiomycota, an average of 74% and 67% of the ECM-induced genes predated the evolution of ectomycorrhizal symbiosis. It implies that symbiosis-induced genes have mostly been co-opted from saprotrophic ancestors for ectomycorrhiza development. However, the phylostratigraphic analysis also highlighted that 9–22% of ectomycorrhiza-induced genes are restricted to specific mycorrhizal lineages (Miyauchi et al. 2020).

Miyauchi et al. (2020) confirmed the general trend toward losses of PCWDE in ectomycorrhizal clades throughout the phylogenetic tree, but also that there is a wide diversity in the decomposing ability. Furthermore, some species like *Gautieria morchelliformis* or *Acephala macrosclerotium* are probably just in the transition to symbiosis and the evolution to a mutualist still in progress. As suggested by Martino et al. (2018) for ericoid mycorrhiza, a dual or intermediate lifestyle could give a greater flexibility, advantageous under certain environmental conditions. Miyauchi et al. (2020) showed that *A. macrosclerotium* is able to regulate its PCWDE genes at the transcriptional level during the symbiotic interaction with pine roots and thus probably able to avoid defense reactions during root colonization. This would suggest that loss of genes encoding PCWDEs is a consequence of, but not a pre-requisite for, the evolution of ectomycorrhizal mutualisms (Miyauchi et al. 2020).

A first intra-genus comparative genomic analysis with *Suillus*, a host specialist, was recently conducted in order to elucidate how host specificity may be encoded into ECM fungal genomes (Lofgren et al. 2021). The different *Suillus* had highly dynamic genomes with numerous rapidly evolving gene families and many domain expansions and contractions. Targeted analyses supported a role for secondary metabolites but not for SSPs or G-protein coupled receptors in *Suillus* host specificity. Both secondary metabolites and pathways involved in the deactivation of reactive oxygen species could be associated with enhanced host specificity (Lofgren et al. 2021). A phylogenomic-based approach identified *Larix* as the ancestral host of *Suillus*, with multiple independent switches between white and red pine hosts (Lofgren et al. 2021). More intra-genus or intra-species comparative genomics but also comparative transcriptomics using different host trees is needed to identify key modules linked to host specificity.

11.4 Molecular Dialogue Between Partners: What's New from the Plant Point of View

11.4.1 *Plant Small Secreted Proteins*

The interaction between roots of plants and microorganisms requires a continuous cross talk that causes alteration and adaptation from both partners. Plant exudates shape bacterial and fungal diversity surrounding their roots and, although every plant produces exudates, the amount and composition of root exudates varies depending on the genotype of the host, the developmental stage, and abiotic stresses (Hugoni et al. 2018; Sasse et al. 2018). Among the different exometabolites that can be released to the rhizosphere by plants, we can find several molecules such as flavonoids or strigolactones that are well known to play a role in the interaction between plant and microbes (Mierziak et al. 2014; Tsuzuki et al. 2016). Recent works also point out the importance of protein effectors that may play a role in plant-microbial interaction mechanisms. Plett et al. (2017) found a total of 417 putative SSPs from *P. trichocarpa* transcriptomic data, including 79 with predicted nuclear localization signals. Of these, four putative SSPs were found to be capable of entering the *L. bicolor* nucleus and some of them were capable of modifying fungal growth and hyphal branching. Ongoing functional analyses are currently being carried out to identify and characterize candidate poplar SSPs that could regulate fungal gene expression and promote symbiosis (Yang et al. 2019). It seems that poplar SSPs may play a role not only in the interaction with ECM, but with other fungal guilds, such as endophytes. 41% of all poplar extracellular proteins regulated by *P. trichocarpa x Mortierella elongata* interaction were considered SSPs, with upregulated genes encoding SSPs involved in lipid transfer and cell wall loosening, while SSPs related to plant defense and cell adhesion were downregulated (Liao et al. 2019).

11.4.2 *A New Perspective for the Common Symbiosis Pathway*

AM and legume-rhizobia symbiotic signaling pathways have been much more studied than ECM. Both symbiotic interactions are mediated by the activation in the host plant of the CSP, which is a signal transduction pathway from the plasma membrane to the nucleus that ultimately promotes mycorrhiza or nodule formation in legumes (reviewed in Oldroyd 2013; Genre and Russo 2016; Zipfel and Oldroyd 2017). Although the first studies were carried out in legumes due to their capacity to form AM and legume-rhizobia type of symbiotic interactions, CSP is conserved and required in all plants capable of forming AM (Martin et al. 2017). Briefly, host plants recognize fungal (Myc factors) or bacterial (Nod factors) signals though receptor-like kinases (RLKs, detailed below), which trigger downstream signals that go to the cell nucleus and ultimately results in Ca2+ spiking, activating a calcium and calmodulin dependent kinase (CCaMK) that phosphorylates CYCLOPS, a CCaMK interacting protein that can directly and/or indirectly regulate gene expression (Oldroyd 2013; Genre and Russo 2016; Zipfel and Oldroyd 2017).

Several genes of this pathway are known to be consistently lost in non-host plants while they are conserved in the genomes of AM plants, and a great proportion of them are supposed to be actively involved in symbiotic processes (Delaux et al. 2014; Bravo et al. 2016). The presence of CSP genes and their role in ECM interactions is not fully known yet. Radhakrishnan et al. (2020) recently found out that, with very few exceptions, a symbiosis related RLK gene called *SYMRK*, *CCaMK*, and *CYCLOPS*, the GRAS transcription factor *RAD1*, and two half ABCG transporters *STR* and *STR2* are consistently lost in non-mutualistic lineages. Of these, the CSP genes *SYMRK*, *CCaMK*, and *CYCLOPS* are retained in plants that are capable of forming both intracellular and intercellular symbioses, such as *Populus* spp., whose roots interact with a wide range of microorganisms including AM, ECM, and endophyte fungi. For several Pinaceae species and other plants hosting fungi and bacteria exclusively intercellularly, these genes are completely lost suggesting their dispensability for intercellular symbiosis (Garcia et al. 2015; Radhakrishnan et al. 2020). The fact that these genes may be dispensable, does not necessary mean that they do not play a role in plant-microbe interactions. In fact, there is recent evidence that CSP is not exclusive to AM and legume-rhizobia symbioses, but may also be involved in other interactions such as ECM (Cope et al. 2019) or endophytes (Skiada et al. 2020). We will not focus on endophytes, since this is not the purpose of this chapter, but it is worth mentioning that it has been recently proven that the recognition of *Fusarium* by model legumes requires the activation of the CSP at least partially, since the suppression of both CSP genes and LysM-RLK genes moderately impairs the endophytic colonization of this species (Skiada et al. 2020).

A role of the CSP in ECM interaction has been recently reported (Cope et al. 2019). As it occurs in AM symbioses, the ECM fungus *L. bicolor* also releases signals (lipochitooligosaccharides, LCOs) to the rhizobiome, and the host plant *P. trichocarpa* is able to recognize and process these signals by an as yet unknown mechanism, producing a nuclear Ca^{2+} spike comparable to that which occurs in AM symbiosis between the same host plant and *Rhizophagus irregularis*. In addition, *P. trichocarpa* RNAi silencing lines with reduced expression of *CCaMRK* showed a partial impairment in the ECM formation with *L. bicolor* and a total impairment of the AM formation with *R. irregularis* (Cope et al. 2019). The independent arise of ECM lifestyle in multiple fungal lineages (Tedersoo and Smith 2013; Martin et al. 2016; Hoeksema et al. 2018), the importance of species-specific fungal small secreted proteins in the establishment of ECM (see Sect. 11.5.1), and the fact that, although apparently dispensable, CSP could play a role in some ECM interactions (Garcia et al. 2015; Cope et al. 2019; Radhakrishnan et al. 2020) suggest the absence of a single common plant molecular mechanism for ECM interactions. Future research focused on a wider variety of different ECM symbioses should shed light on the commonalities and differences of ECM molecular mechanisms, among themselves and among other symbiotic interactions.

11.4.3 The Role of Plant Receptor-Like Kinases in ECM Interactions

RLKs are transmembrane proteins that possess an extracellular amino domain and an intracellular carboxyl domain with kinase activity. In plant genomes there are hundreds of genes encoding for these receptors and some of them are known to play a prominent role in plant development or interaction with its surroundings (Shiu and Bleecker 2001). As stated in the previous section, RLKs represent the first step in the molecular symbiosis pathway in AM and legume-rhizobia symbioses. For ECM symbiosis, a G-type lectin receptor-like kinase *PtLecRLK1* was found to have an important role in the formation of the ECM between *P. trichocarpa* and *L. bicolor* (Labbé et al. 2019). This receptor is consistently present in *P. trichocarpa* genotypes that preferentially form ECM with *L. bicolor* over *P. deltoides* genotypes which consistently lack the locus for this receptor. The transgene *PtLecRLK1* can mediate colonization of the non-host species *Arabidopsis* by *L. bicolor*, since *Arabidopsis 35S:PtLecRLK1* transgenic plants in co-cultivation with *L. bicolor* are capable to form a mantle and Hartig net-like structures, while this is not the case for Wildtype (WT)-*Arabidopsis* plants. Additionally, up to 24 genes related to defense are downregulated in *Arabidopsis 35S:PtLecRLK1* co-cultured with *L. bicolor* versus only one gene in WT *Arabidopsis* (Labbé et al. 2019).

Several implications arise from the discovery of the role of RLKs in ECM symbiosis. First of all, it points to an important role of these receptors in ECM host specificity, which is consistent with ECM interactions being more specific than AM (van der Heijden et al. 2015). It will also be crucial to understand the downstream signaling pathways that *PtLecRLK1* triggers. Put together, the works from Cope et al. (2019) and Labbé et al. (2019) may lead to the hypothesis that *PtLEcRLK1*, alone or forming a receptor complex, could act as the receptor of LCOs from *L. bicolor* and trigger the CSP. But taking into account that heterologous expression of *PtLecRLK1* in *A. thaliana* allows *L. bicolor* colonization and that this species has lost multiple genes of the symbiosis toolkit, including several from the CSP (Delaux et al. 2014; Radhakrishnan et al. 2020), this hypothesis does not seem to be the most likely. Another possible role of this plant receptor could be the recognition of effectors from the fungal partner. The downregulation of several defense-related genes in *A. thaliana* carrying the *PtLecRLK1* gene in the presence of *L. bicolor* reinforces this hypothesis since *L. bicolor* SSPs such as *MiSSP7* or *MiSSP7*.6 are known to interfere with plant immunity (Plett et al. 2011, 2014a, b; Kang et al. 2020). Overall, a deeper study of the downstream cascade of *PtLecRLK1* together with the search for more RLKs involved in partner signals recognition will significantly improve our understanding into the molecular mechanisms and plant-fungal dialogue that occurs in ECM symbiosis.

11.5 Beneficial Fungi at Work in Establishing an Ectomycorrhizal Relationship with Their Host Tree: Tools and Mode of Action

11.5.1 Mycorrhiza-Induced Small-Secreted Proteins (MiSSPs)

Effectors are SSPs (<250 aa) that are capable of altering another organism, mainly by modulating, suppressing, or simply modifying the immune system of the partner (Martin and Kamoun 2011). While effectors were first described and, thus, are more thoroughly studied, in pathogens such as bacteria, filamentous fungi, or oomycetes (Varden et al. 2017; Khan et al. 2018), later works have evidenced an important role of effectors in AM (Kloppholz et al. 2011) and ECM fungi (Plett et al. 2011, a; Pellegrin et al. 2019a; Kang et al. 2020). The duality of plant receptors to recognize microbial signals released by both mutualistic and pathogenic microbes and trigger immunity and symbiosis programs is now well established (Rey et al. 2013; Miyata et al. 2014). For instance, the LysM receptor-like protein kinases OsCERK1 (Zhang et al. 2015) or MtLYK9 (Gibelin-Viala et al. 2019) play dual roles in immunity and symbiosis through the recognition of LCOs (Feng et al. 2019). It was also recently shown that LCOs production is not restricted to mutualistic fungi but is a more ancestral fungal trait (Rush et al. 2020). Altogether these results suggest that in addition to the triggering of the symbiotic pathway, mutualistic fungi might release effectors to dampen the plant's innate immunity, as plant-pathogenic microbes do.

To dissect the fungal contribution to the molecular dialogue leading to mutualistic interactions, comparative genomics and transcriptomics enabled the identification of putative secreted fungal proteins among them MiSSPs (Martin et al. 2008; Pellegrin et al. 2015; Kohler et al. 2015; de Freitas Pereira et al. 2018). Using proteomics, the effective production of these SSPs has been confirmed in mycelium of *Hebeloma cylindrosporum* (Doré et al. 2015) and *L. bicolor*-poplar ectomycorrhiza (Villalobos Solis et al. 2020). This latter study provides evidence of newly identified peptides, stressing the need of new technique development to identify SSPs at the protein level. Since all these analyses were performed in vitro, one can wonder if such MiSSPs are synthesized in soil conditions. Liao and collaborators showed that in soil, the ectomycorrhizal fungus *Piloderma croceum* in contact with *Pinus taeda* roots expresses four SSPs with homology to known effector-like proteins, two of them being significantly upregulated in ECM tissues (Liao et al. 2014).

Several MiSSPs were functionally characterized, mainly in the model interaction between the ectomycorrhizal fungus *Laccaria bicolor* and poplar roots. The first mutualistic effectors discovered were MiSSP7 for the ectomycorrhizal fungus

Laccaria bicolor (Plett et al. 2011) and SP7 from the endomycorrhizal fungus Glomus (now *Rhizophagus irregularis*; Kloppholz et al. 2011). Both effectors are required for *in planta* fungal growth and interfere with hormonal-signaling pathways (jasmonate and ethylene, respectively). MiSSP7 blocks JA-signaling pathway through its interaction with the poplar co-repressor of JA-signaling PtJAZ6. MiSSP7 prevents JAZ degradation triggered by JA, which consequently blocks JA-signaling, allowing in planta fungal growth (Plett et al. 2014b). More precisely, MiSSP7 strengthens the binding of PtJAZ6 to the transcription factor PtMYC2.1, suggesting that MiSSP7 maintains the repression of PtMYC2.1-regulated genes (Daguerre et al. 2020). This study highlights that a mutualistic effector may promote the symbiotic interaction through altered dynamics of a JA-signaling-associated protein–protein interaction network, like effectors from plant-pathogenic microbes.

In addition, other *Laccaria bicolor* MiSSPs have been characterized. MiSSP7.6 is required for later stages of symbiosis development and interacts with a poplar transcription factor, whose Arabidopsis ortholog is involved in plant immunity (Kang et al. 2020). These results emphasize that plant-interacting fungi, either beneficial or detrimental, use the peptide (or small proteins)-mediated communication as a strategy to control plant hormonal signaling pathways. Another study shows that the secreted *Pisolithus albus* PaMiSSP10b protein interacts with and increases the enzymatic activity of the *Eucalyptus grandis* S-adenosyl methionine decarboxylase (AdoMetDC) (Plett et al. 2019), consequently inducing the N-acetyl spermine biosynthesis (Plett et al. 2019). Polyamines are nitrogen-containing compounds that (i) contribute to plant immunity as regulatory molecules and (ii) could serve as C and N-sources. Consequently, several hypotheses need to be tested to explain the exact role in immunity and/or metabolism of increased polyamines synthesis to sustain ectomycorrhizal symbiosis. The first one will be to confirm whether the ECM fungus is able to use putrescine, spermine, or spermidine as C and N-sources, as previously suggested and through the presence of the appropriate transporters within the genome (Lucic et al. 2008). The second hypothesis is that increasing polyamine biosynthesis diverts resources from other pathways, e.g., AdoMet is required for ethylene production. This is the first study showing manipulation of the host metabolism by a mutualistic fungal effector, which favors *in planta* root colonization of *P. albus*.

Altogether these studies highlight that pathogenic and mutualistic fungi evolve similar tools (diffusible SSPs called effectors) to manipulate their host plants. However, it remains to be determined if we can generalize the two observations showing that the ECM fungus *L. bicolor* inhibits JA-signaling pathway, while AM and biotrophic pathogens induce JA-responses during host colonization and that polyamines are detrimental to necrotrophic fungi, whereas they favor the plant colonization by mutualistic fungi. More studies on mutualistic effectors are thus required to better understand the multifaceted and adaptable immune system that allows them to restrict pathogenic microbes while allowing beneficial associations (Vannier et al. 2019). In particular, assessing, if the opposite direction for the control of plant immunity is only a matter of a strict spatio-temporal regulation, would be of great interest.

Several data strengthen the idea for a role of fungal SSPs not only as effectors in fungal-plant interactions, but also in the adaptation to their biotic environment and saprotrophic growth. For instance, proteomic analysis of the secretome of *Hebeloma cylindrosporum* free-living mycelium revealed that 17% of the secreted proteins were SSPs (Doré et al. 2015). Expression of some *Hebeloma* SSP-encoding genes is regulated by the environmental conditions and/or interaction with the host. Likewise, some *C. geophilum* MiSSPs were not only upregulated in the ECM, but also in mycelium and sclerotia in contact with the host plant. This suggests that these MiSSPs are induced by the host plant but are not likely to play a direct role in the fungal-plant communication at the symbiotic interface. Rather, they have a role in fungal biology (de Freitas Pereira et al. 2018). The study of the mutualistic effector LbMiSSP8 illustrates this point. This protein contains a DWRR repetitive motif which is the site of KEX2 protease-processing, leading to the release of three peptides of 4 amino-acid (DSDW). MiSSP8 or its derived peptides are necessary at early stage of ECM development for hyphal aggregation and pseudoparenchyma formation (Pellegrin et al. 2019a, b). Interestingly, KEX2-processed repeat proteins (KEPs) were identified in the whole fungal kingdom, with increased numbers in the genomes of plant-interacting fungi, whether decomposers, pathogens, symbionts, or endophytes (Le Marquer et al. 2019). This indicates that KEX2-processes repeat proteins and their derivate peptides might play a role in the mechanisms of plant-fungal interactions. Since KEPs proteins fall into different functional categories, e.g., sexual pheromones, toxins, structural role in fungal cell wall, or effectors promoting host-microbe interactions (for review Ma et al. 2018), the next step will be to decipher the exact function(s) in symbiosis of KEPs (and their derivate peptides) for ECM fungi, and to distinguish between signaling or structural molecules. Within the same line, another overlooked hypothesis would be that these SSPs are involved in microbe-microbes interactions, as showed in plant-pathogenic fungi (Snelders et al. 2018, 2020). Interestingly, the genome of the common root endophyte *Mortierrela elongata* has a reduced number of SSP-encoding genes. In addition, a similar number and pattern of *MeSSPs* are expressed in mycelium in natural soil (in which other microbes are naturally present) in the absence or presence of poplar (Liao et al. 2019). This suggests that *M. elongata* SSPs might play a role in microbes-microbes interaction, rather (or to a lesser extent) than in the host-plant-fungal interaction. Comparative genomics studies highlighted two types of SSPs: the lineage-specific orphan genes and the genes shared with saprotrophic fungi (Pellegrin et al. 2015; Miyauchi et al. 2020). It is more likely that the SSPs involved in microbes-microbes interactions belongs to this last group. Another function that could be attributed to fungal SSPs would be their involvement in host specificity. This hypothesis still needs to be addressed.

11.5.2 The Setup of the Biotrophic Interface Through the Strict Regulation of CAZymes

In root apices, hyphae grow in the apoplast between rhizodermal cells to differentiate an intraradicular and finger-like hyphal network, the so-called Hartig net (Balestrini and Kottke 2016). This labyrinthine-like hyphal structure is a key feature of a fully mature and functional ectomycorrhiza. The formation of the Hartig net is synonymous with the symbiotic interface promoting the bi-directional translocation of solutes (Smith and Read 2008). Differentiation of the Hartig net is the result of fungal and plant cell wall remodeling, through biochemical changes, allowing *in planta* fungal growth inside the apoplast. For instance, localized loosening and redistribution of un-esterified pectins within plant cell walls have been identified (Balestrini and Bonfante 2014; Sillo et al. 2016). Redistribution of fungal cell wall proteins (e.g., hydrophobins, the symbiosis-regulated acidic polypeptides of 32 kDa, in the ectomycorrhizal *Pisolithus microcarpus* interacting with *Eucalyptus globulus)* have been revealed through immunocytology (Laurent et al. 1999; Tagu et al. 2001). The apoplastic colonization in the middle lamella by the ectomycorrhizal hyphae is thought to rely on the mechanical force of hyphal tip growth (Peterson and Massicotte 2004). Fungal auxins were proposed to promote loosening or extensibility of the rhizodermic cell wall to facilitate hyphal growth (Gay et al. 1994). In recent years, progress has been achieved towards the characterization of the fungal molecular players, namely the fungal CAZymes acting on plant cell wall, required for cell wall remodeling. First, a sequential expression of distinct CAZymes occurred during ectomycorrhiza development between poplar-*L. bicolor* (Veneault-Fourrey et al. 2014). Among them, the symbiosis-induced ß-1,4-endoglucanase LbGH5-CBM1 is a secreted fungal endocellulase with a high activity against cellulose and galactomannans, acting on poplar cell walls. It is an important determinant for successful symbiotic fungal colonization (Zhang et al. 2018). In situ localization of LbGH5-CBM1 in ectomycorrhizal rootlets shows that it is located in cell walls of hyphae forming the Hartig net and mantle sheath. In addition to *LbGH5-CBM1*, three *L. bicolor* genes coding for pectinases (polygalacturonases) of the glycosyl hydrolase family 28 (GH28) are induced in *L. bicolor-Populus* ectomycorrhizal roots (Veneault-Fourrey et al. 2014), suggesting that Hartig net development may require pectin degradation. Pectins are present in plant middle lamella, primary cell and secondary walls and accumulate in the early stages of cell expansion. One of the Laccaria GH28 CAZyme family members, LbGH28A, has its highest activity towards pectin and polygalacturonic acid. In situ localization of LbGH28A indicates that this pectinase is located on both fungal and plant cell walls at the symbiotic hyphal front (Zhang et al. 2021). One *L. bicolor* gene encoding a lytic polysaccharide monooxygenase-like protein of the X325 family is upregulated in ECM symbiosis and fruiting-bodies. The corresponding protein is not an LPMO since it does not perform oxidative cleavage of plant cell wall polysaccharides nor fungal cell wall ones. LbW135 is localized in fungal cell wall or in the apoplast between fungal cells and rhizodermal cells, at the site of Hartig net suggesting a role of this protein in symbiosis (Labourel et al. 2020). In addition to the CAZymes, other fungal proteins are predicted to restructure the plant-fungal interface (e.g., lectins, hydrophobins, and expansins (Veneault-Fourrey et al. 2014). The two characterized hydrolytic enzymes would liberate oligomers from the plant cell wall which could be sensed as DAMPS (Damaged-Associated Molecular Patterns) by the plant cell, as is the case during plant pathogen interaction. Consequently, it will be of great interest, in the coming years, to better understand why the host cells do not react the same way when DAMPs are liberated through colonization by mutualistic or pathogenic fungi. Is it a question of DAMPs-threshold or a different DAMPs nature, or a more efficient control of the signaling pathway triggered by the mutualistic fungi? In addition, identification of the plant molecular players remodeling the plant cell wall should provide some answers to the previous questions.

11.6 How Plant and Fungal Hormones and Metabolites Regulate ECM Formation

Plant hormones are considered the master regulators of developmental and environmental responses, both biotic and abiotic. The involvement of phytohormones such as auxin, ethylene (ET), or jasmonic acid (JA) in modulating ECM symbioses directly or indirectly have been studied during the past years (Pellegrin et al. 2019b and references therein). Lateral root stimulation by auxin of either fungal and plant origin (Vayssières et al. 2015) or limitation within root tissue through impairment of the Hartig net formation by plant ET and JA (Plett et al. 2014b) are good examples of this. More recently, Basso et al. (2020) assessed the hormonal and transcriptional landscape of *P. tremula x alba* in the presence or absence of *L. bicolor* colonization, showing the accumulation of salicylic acid (SA) in mid-stage ECM in vitro. They also showed that exogenous treatment of JA, SA, ET, or gibberellic acid (GA) affected in different ways the development of ECM and the physiology of the plant host. Finally, by analyzing differentially expressed genes in all these conditions they put in focus the central role of JA and the synergistic effect of JA-SA and JA-ET signaling in poplar, highlighting the regulation of several genes putatively involved in cell wall remodeling such as pectin lyases, xyloglucan endotransglucosylases, or cellulose synthases. Other regulated genes include those involved in plant defense and communication with the surroundings, such as protease inhibitors or terpene synthases. These results underline the importance not only of single plant hormones, but also of their balance and cross talk to control and modulate the colonization of the roots by the fungal partner.

Several past studies have shown that ECM fungi are also able to produce hormones, mainly ethylene, auxins, cytokinins, absicic acid, and salicylic acid (Splivallo et al. 2009; Regvar et al. 1997; Kraigher et al. 1991; Kovac and Zel 1995; Basso et al. 2020). The cocktail of hormones secreted by the hyphae depend on fungal species. The roles of fungal auxins have been extensively reviewed

(Sukumar et al. 2013; Pellegrin et al. 2019b) and will not be addressed here (see Chap. 4). Briefly, fungal auxins are suspected to enhance lateral root formation and likely loosen pectin within the middle lamella. The role of other hormones produced by ECM fungi is not well known but they are likely to interact with plant-hormone signaling pathways. More recently, some studies have focused on fungal volatiles. For instance, the ectomycorrhizal fungus *L. bicolor* produces sesquiterpenes which promote lateral roots formation, enhancing the root surface area for (i) further fungal colonization and thus for nutrients exchanges through the biotrophic interface or (ii) a better plant nutrient uptake and an improved fungal access to plant-derived carbon via root exudates (Ditengou et al. 2015). *T. vaccinum* releases volatiles not usually associated with fungi, like limonene and β-barbatene, and geosmin (Abdulsalam et al. 2021). Geosmin biosynthesis genes are upregulated in ectomycorrhiza but the exact role of these fungal volatiles is not yet elucidated. The importance of fungal secondary metabolites in manipulating plant hormone signaling has been also highlighted in plant-pathogen interactions. In the model *Arabidopsis* and the plant-pathogenic fungus *Colletotrichum higginsianum*, the fungal terpenoid higginsianin B blocks JA-synthesis or signaling by preventing the degradation of JAZ-proteins, the repressor of jasmonates responses (Dallery et al. 2020). Fungal metabolites might also be the signals allowing interkingdom communication, and influencing the microbe-microbe interactions (for review Weisskopf et al. 2021). Further investigation on the cocktail of fungal terpenes produced by ECM fungi in the presence of host plants and other members of the microbiome (either pathogenic or other mutualistic) would provide new insight in the fungal mechanisms that shape root mycobiome composition and organization.

11.7 Nutrient Trading: Which Partner Takes the Control?

Nutrient exchange between host plant and ECM fungi is one of the better-known properties of ECM interactions, while fungi provide increased access to nutrients—mainly nitrogen and phosphorous—and water, they receive in exchange photo-assimilated carbon from the host plant (Smith and Read 2008). Several studies have focused on how nutrient exchanges could act as a control tool for the plant to determine the success of plant-microbe interactions. While the host plant prioritizes interactions with AM or rhizobia partners that provide more beneficial outcomes (i.e., more nitrogen acquisition) by allocating more carbon to them in comparison to other less-beneficial partners (Simms et al. 2006; Bever et al. 2009; Sachs et al. 2010; Kiers et al. 2011; Argüello et al. 2016), no or very little "reward" system occurs in ECM symbiosis. Corrêa et al. (2012) showed that carbon allocation to the ECM partner did not suppose a cost to the host plant. In the ECM interaction between *Eucalyptus grandis* and three different *Pisolithus* isolates, the host plant was able to discriminate the less beneficial isolate, in terms of nitrogen acquisition, and to limit its colonization in comparison to the others. This was, however, not explained by differential allocation of carbon, but it seemed to be related to upregulation of plant defense genes (Hortal et al. 2017). C/N exchange seems to be of importance in other ECM interactions. Bogar et al. (2019) showed that *Larix occidentalis* hosts can discriminate between different *Suillus* species, but this discrimination did not seem to be based on N transfer capacities from the different species. They showed, however, that substrates with more N supply resulted in more C allocation in ECM roots. The N received by *Eucalyptus grandis* seems to be dependent on the amount of C transferred to the fungal partner *Pisolithus albus*. Moreover, high access to plant-available inorganic N resulted in reduced mycorrhization, suggesting that the plant nutritional status drives the extent of fungal colonization (Plett et al. 2020). This experiment, however, was made with a single fungal partner, and therefore it could not be concluded that nutrient exchange could be used as a plant tool to select more beneficial ECM partners. On the other hand, when *Pisolithus albus* has access to less C, it transfers more N to the plant, suggesting that nitrogen transfer might be controlled by the ECM fungus, in particular with its C needs (Plett et al. 2020). This is in accordance with the dependance on C availability for the decomposition and uptake of soil-N by ECM fungi (Rineau et al. 2013). Overall, evidence up to now suggests that plant C allocation does not suppose a generalized mechanism of control from the plant to select fungal partners in ECM interactions. Whether nutrient exchange plays a direct or indirect role in this cannot yet be discarded and further research will be needed to unveil this issue.

Despite the importance of nutrient exchanges in ECM symbioses, the main changes in metabolite pools in ECM root tips and leaves of mycorrhizal plants are not related to nutrients (Tschaplinski et al. 2014; Kaling et al. 2018). For example, the ECM root tips of *Populus trichocarpa–L. bicolor* displayed an increased turnover of metabolites associated with the benzoate degradation pathway (Tschaplinski et al. 2014), including a decrease in benzyl alcohol containing phenolic glycosides, and the accumulation of benzoic acid and many hydroxylated benzoic acid metabolites. Benzoate detoxification has been described in several other symbiotic relationships (Mornico et al. 2011; Cheng et al. 2013; Liu et al. 2013); nevertheless, its role is currently unknown. One hypothesis is that this pathway may be required for the detoxification of host-produced defensive xenobiotics.

Kaling et al. (2018) demonstrated metabolomic changes in leaves of mycorrhizal poplar trees. In particular, changes in nitrogen allocation associated with downregulation of phenolics and an upregulation of defensive metabolites suggest a shift of the resources from constitutive phenol-based to more specialized defensive and protective compounds. Further investigation is required to identify the systemic signal(s) leading to these changes. Is it from plant or fungal origin? Are these modifications the sole reflection of the profound nutritional changes in the ectomycorrhizae?

Active research was also performed to assess at the metabolomic levels what differentiates a "receptive" host plant from a "recalcitrant" one. For example, *L. bicolor* is able to form mycorrhiza with the host *P. trichocarpa* but less with *P. deltoides*. This ECM fungus equally induced defense-associated metabolites in the roots of both type poplar plants undergoing colonization. However, the incompatible relationship was characterized by less significantly regulated metabolites in plant tissues in contact with fungal hyphae (Tschaplinski et al. 2014). These results would suggest that trees easily colonized by ECM fungi are receptive due to a lack of defensive metabolites when compared to recalcitrant hosts. The authors propose that compatible versus incompatible interactions could be discriminated by both the diversity and half-life of the host's defensive metabolites as well as its metabolic responsiveness at the first steps of fungal colonization (Tschaplinski et al. 2014). In the same line and using *E. grandis*, Wong et al. showed that the repression of several unidentified metabolites (likely plant origin) appears to be necessary, during the pre-symbiotic interaction, for the fungus to establish successful ECM colonization in later stages (Wong et al. 2019). This led to the hypothesis that ECM fungi isolates that establish a high level of colonization might be more successful to inactivate the plant immune system. Further investigations are required in the future.

11.8 Conclusions and Open Questions

Figure 11.1 summarizes our current understanding of the processes at the plant-microbe interface. Despite the recent advances in ectomycorrhizal research due to increasing genomic and transcriptomic resources as well as in functional characterization of key modules (e.g., MiSSPs, RLKs), many questions still remain unanswered.

Why are some ECM species highly specific and form mycorrhiza only with one or a few host trees? Is there a similar principle, as for pathogenic microbes? Incompatibility factors, which are still unidentified? The first intra-genus comparative genome studies (Lofgren et al. 2021) point to secondary metabolites and pathways involved in the deactivation of reactive oxygen species, but more model systems and comparative studies are needed. Another open question concerns an "ectomycorrhizal symbiosis pathway" similar to the CSP identified for AM fungi. Are there any minimum requirements, certain core genes that both fungal and plant partner must possess in order to successfully establish ectomycorrhizal symbiosis? Also very little is known about the role of small RNAs or changes in DNA methylation during mutualistic interactions. Plett et al. (2019) showed for *E. grandis* that posttranslational modifications like arginine methylation via protein arginine methyltransferase have important effects on ectomycorrhizal symbiosis and the regulation of hormone signaling pathways. Wong-Bajracharya et al. (2022) demonstrated that *Pisolithus microcarpus* encodes a miRNA that enters plant cells and stabilizes ECM interaction. Are these general mechanisms?

The current knowledge on the molecular signaling in ECM interactions was obtained with a few in vitro model systems. These sterile systems do not take into account the influence of other microorganisms present in natural conditions. It is essential to address the question how other microbes, how the tree microbiome influences the way ectomycorrhizal fungi interact with their host tree and how several ectomycorrhizal fungi communicate and interact while colonizing the same host root system. Can symbiotic fungi with redundant functions colonize at the

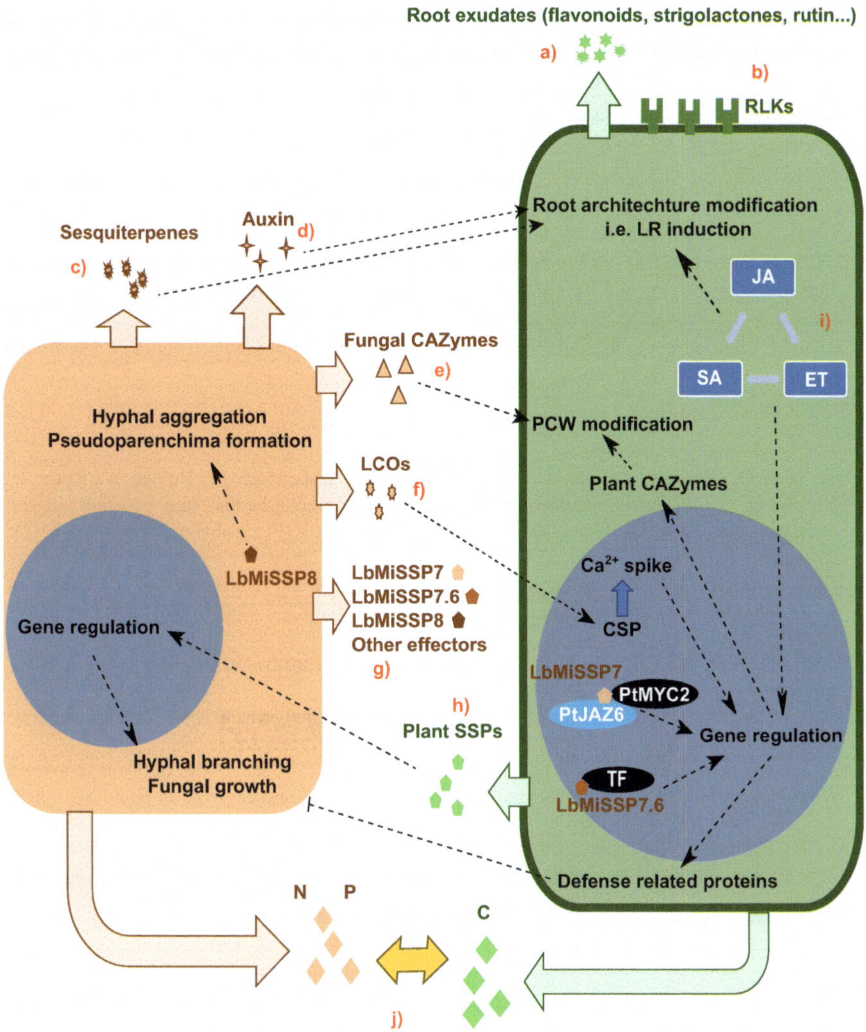

Fig. 11.1 Molecular dialogue between partners in Ectomycorrhizal (ECM) symbiosis. (**a**) Root exudates contain several compounds, including flavonoids, rutin, or strigolactones that shape microbial community of the rhizosphere (Hugoni et al. 2018; Sasse et al. 2018). (**b**) One *Populus trichocarpa* G-type lectin receptor-like kinase (PtLecRLK1) has been found to play a role in ECM formation and fungal-host specificity (Labbé et al. 2019). This and other RLKs may play a fundamental role in signal perception from fungal origin, such as lipochitooligosaccharides (LCOs) or Mycorrhiza-induced small-secreted proteins (MiSSPs). (**c**) *Laccaria bicolor* volatile sesquiterpenes are known to induce lateral root formation of plant species. (Ditengou et al. 2015). (**d**) Fungal auxin is also known to promote lateral root formation and to participate in the loosening of the pectin in the middle lamella (Sukumar et al. 2013; Pellegrin et al. 2019b). (**e**) Several fungal CAZYmes such as LbGH5-CBM1, LbGH28A, or LbW135 are known to play a role in plant cell wall remodeling (Veneault-Fourrey et al. 2014; Zhang et al. 2018, 2021; Labourel et al. 2020). (**f**) *L. bicolor* secretes LCOs and induces, by a yet unknown mechanism that could involve the Common Symbiosis Pathway, a nuclear Ca2+spike, comparable to what occurs in Arbuscular Mycorrhiza symbiosis (Cope et al. 2019) (**g**) Several *L. bicolor* MiSSPs have been characterized

same time? Or do the fungi need to offer different functions to the plant? Can they coexist together or only successively? In the future, various stress factors could also be tested to see how they disturb the balance of the system and to make predictions how abiotic stress, such as drought or higher temperature, would impact symbiotic interactions.

Acknowledgments This research was supported by the Laboratory of Excellence ARBRE (ANR-11-LABX-0002-01), the Region Lorraine, the European Regional Development Fund, and the Plant–Microbe Interfaces Scientific Focus Area in the Genomic Science Program, the Office of Biological and Environmental Research in the US DOE Office of Science.

References

Abdulsalam O, Wagner K, Wirth S, Kunert M, David A, Kallenbach M, Boland W, Kothe E, Krause K (2021) Phytohormones and volatile organic compounds, like geosmin, in the ectomycorrhiza of Tricholoma vaccinum and Norway spruce (Picea abies). Mycorrhiza 31(2): 173–188

Argüello A, O'Brien MJ, van der Heijden MG, Wiemken A, Schmid B, Niklaus PA (2016) Options of partners improve carbon for phosphorus trade in the arbuscular mycorrhizal mutualism. Ecol Lett 19(6):648–656

Balestrini R, Bonfante P (2014) Cell wall remodeling in mycorrhizal symbiosis: a way towards biotrophism. Front Plant Sci 5:237–237

Balestrini R, Kottke I (2016) Structure and development of ectomycorrhizal roots. Mol Mycorrhizal Symbiosis:47–61

Basso V, Kohler A, Miyauchi S, Singan V, Guinet F, Šimura J et al (2020) An ectomycorrhizal fungus alters sensitivity to jasmonate, salicylate, gibberellin, and ethylene in host roots. Plant Cell Environ 43(4):1047–1068

Bever JD, Richardson SC, Lawrence BM, Holmes J, Watson M (2009) Preferential allocation to beneficial symbiont with spatial structure maintains mycorrhizal mutualism. Ecol Lett 12(1): 13–21

Fig. 11.1 (continued) as important agents in the molecular dialogue that takes place during ECM formation. LbMiSSP7 interacts with the poplar co-repressor of jasmonate signaling PtJAZ6, impairing its degradation and maintaining its repression to PtMYC2.1 transcription factor and the genes under its regulation (Plett et al. 2011, 2014a, b; Daguerre et al. 2020). LbMiSSP7.6 interacts with a poplar transcription factor (TF), whose *Arabidopsis* homolog is known to be involved in plant immunity (Kang et al. 2020). LbMiSSP8 and/or its derivate peptides play a role in hyphal aggregation and pseudoparenchyma formation (Pellegrin et al. 2019a). Other MiSSPs from different organisms are known to be also important for ECM establishment, not only in vitro but also in soil conditions (Liao et al. 2014; Plett et al. 2019). (**h**) Plant SSPs that are capable of entering *L. bicolor* nucleus and modify fungal branching and growth have been identified as secreted proteins expressed in *Populus trichocarpa*–*L. bicolor* ECM (Plett et al. 2017). (**i**) Hormonal cross talk between jasmonate (JA), salicylic acid (SA), and ethylene (ET) regulates plant responses to colonization and fungal accommodation by influencing root architecture, plant cell wall remodeling and plant defense, among other processes (Basso et al. 2020). (**j**) Nutrient exchanges between partners are the main trait of ECM symbiosis (Smith and Read 2008). Whether nutrient trading is a mechanism of control for partner selection is yet to be determined in ECM interactions

Bogar L, Peay K, Kornfeld A, Huggins J, Hortal S, Anderson I, Kennedy P (2019) Plant-mediated partner discrimination in ectomycorrhizal mutualisms. Mycorrhiza 29(2):97–111
Bowles AM, Bechtold U, Paps J (2020) The origin of land plants is rooted in two bursts of genomic novelty. Curr Biol 30(3):530–536
Bravo A, York T, Pumplin N, Mueller LA, Harrison MJ (2016) Genes conserved for arbuscular mycorrhizal symbiosis identified through phylogenomics. Nature Plants 2(2):1–6
Brundrett MC, Tedersoo L (2018) Evolutionary history of mycorrhizal symbioses and global host plant diversity. New Phytol 220(4):1108–1115
Cheng ZM, Tuskan GA (2009) Populus community mega-genomics: coming of age. Crit Rev Plant Sci 28(5):282–284
Cheng XY, Tian XL, Wang YS, Lin RM, Mao ZC, Chen N, Xie BY (2013) Metagenomic analysis of the pinewood nematode microbiome reveals a symbiotic relationship critical for xenobiotics degradation. Sci Rep 3:1869
Cope KR, Bascaules A, Irving TB, Venkateshwaran M, Maeda J, Garcia K et al (2019) The ectomycorrhizal fungus Laccaria bicolor produces lipochitooligosaccharides and uses the common symbiosis pathway to colonize Populus roots. Plant Cell 31(10):2386–2410
Corrêa A, Gurevitch J, Martins-Loução MA, Cruz C (2012) C allocation to the fungus is not a cost to the plant in ectomycorrhizae. Oikos 121:449–463
Cregger MA, Carper DL, Christel S, Doktycz MJ, Labbe J, Michener JK et al (2021) Plant–microbe interactions: from genes to ecosystems using populus as a model system. Phytobiomes Journal 5(1):29–38
Daguerre Y, Basso V, Hartmann-Wittulski S, Schellenberger R, Meyer L, Bailly J, Kohler A, Plett JM, Martin F, Veneault-Fourrey C (2020) The mutualism effector MiSSP7 of Laccaria bicolor alters the interactions between the poplar JAZ6 protein and its associated proteins. Sci Rep 10(1):20,362
Dallery J-F, Zimmer M, Halder V, Suliman M, Pigné S, Le Goff G, Gianniou DD, Trougakos IP, Ouazzani J, Gasperini D, O'Connell RJ (2020) Inhibition of jasmonate-mediated plant defences by the fungal metabolite higginsianin B. J Exp Bot 71(10):2910–2921
Delaux PM, Schornack S (2021) Plant evolution driven by interactions with symbiotic and pathogenic microbes. Science 371(6531)
Delaux PM, Varala K, Edger PP, Coruzzi GM, Pires JC, Ané JM (2014) Comparative phylogenomics uncovers the impact of symbiotic associations on host genome evolution. PLoS Genet 10(7):e1004487
Delaux PM, Hetherington AJ, Coudert Y, Delwiche C, Dunand C, Gould S et al (2019) Reconstructing trait evolution in plant evo–devo studies. Curr Biol 29(21):R1110–R1118
Ditengou FA, Müller A, Rosenkranz M, Felten J, Lasok H, Van Doorn MM, Legué V, Palme K, Schnitzler JP, Polle A (2015) Volatile signalling by sesquiterpenes from ectomycorrhizal fungi reprogrammes root architecture. Nat Commun 6. https://doi.org/10.1038/ncomms7279
Doré J, Perraud M, Dieryckx C, Kohler A, Morin E, Henrissat B, Lindquist E, Zimmermann SD, Girard V, Kuo A, Grigoriev IV, Martin F, Marmeisse R, Gay G (2015) Comparative genomics, proteomics and transcriptomics give new insight into the exoproteome of the basidiomycete Hebeloma cylindrosporum and its involvement in ectomycorrhizal symbiosis. New Phytol 208(4):1169–1187
Duplessis S, Cuomo CA, Lin Y-C et al (2011) Obligate biotrophy features unraveled by the genomic analysis of rust fungi. Proc Natl Acad Sci U S A 108:9166–9171
Feng F, Sun J, Radhakrishnan GV, Lee T, Bozsóki Z, Fort S, Gavrin A, Gysel K, Thygesen MB, Andersen KR, Radutoiu S, Stougaard J, Oldroyd GED (2019) A combination of chitooligosaccharide and lipochitooligosaccharide recognition promotes arbuscular mycorrhizal associations in Medicago truncatula. Nat Commun 10(1):5047–5047
de Freitas Pereira M, Veneault-Fourrey C, Vion P, Guinet F, Morin E, Barry KW, Lipzen A, Singan V, Pfister S, Na H, Kennedy M, Egli S, Grigoriev I, Martin F, Kohler A, Peter M (2018) Secretome analysis from the ectomycorrhizal ascomycete Cenococcum geophilum. Front Microbiol 9(141)

Galotto G, Abreu I, Sherman C, Liu B, Gonzalez-Guerrero M, Vidali L (2020) Chitin triggers calcium-mediated immune response in the plant model Physcomitrella patens. Mol Plant-Microbe Interact 33(7):911–920
Garcia K, Delaux PM, Cope KR, Ané JM (2015) Molecular signals required for the establishment and maintenance of ectomycorrhizal symbioses. New Phytol 208(1):79–87
Gay G, Normand L, Marmeisse R, Sotta B, Debaud JC (1994) Auxin overproducer mutants of *Hebeloma cylindrosporum* romagnesi have increased mycorrhizal activity. New Phytol 128: 645–657
Genre A, Russo G (2016) Does a common pathway transduce symbiotic signals in plant–microbe interactions? Front Plant Sci 7:96
Gibelin-Viala C et al (2019) The Medicago truncatula LysM receptor-like kinase LYK9 plays a dual role in immunity and the arbuscular mycorrhizal symbiosis. New Phytol 223(3):1516–1529
Han GZ (2019) Origin and evolution of the plant immune system. New Phytol 222(1):70–83
van der Heijden MG, Martin FM, Selosse MA, Sanders IR (2015) Mycorrhizal ecology and evolution: the past, the present, and the future. New Phytol 205(4):1406–1423
Hoeksema JD, Bever JD, Chakraborty S, Chaudhary VB, Gardes M, Gehring CA et al (2018) Evolutionary history of plant hosts and fungal symbionts predicts the strength of mycorrhizal mutualism. Commun Biol 1(1):1–10
Hortal S, Plett KL, Plett JM, Cresswell T, Johansen M, Pendall E, Anderson IC (2017) Role of plant–fungal nutrient trading and host control in determining the competitive success of ectomycorrhizal fungi. ISME J 11(12):2666–2676
Hugoni M, Luis P, Guyonnet J, el Zahar Haichar F (2018) Plant host habitat and root exudates shape fungal diversity. Mycorrhiza 28(5):451–463
Jin WT, Gernandt DS, Wehenkel C, Xia XM, Wei XX, Wang XQ (2021) Phylogenomic and ecological analyses reveal the spatiotemporal evolution of global pines. Proc Natl Acad Sci 118(20)
Kaling M, Schmidt A, Moritz F, Rosenkranz M, Witting M, Kasper K, Janz D, Schmitt-Kopplin P, Schnitzler JP, Polle A (2018) Mycorrhiza-triggered transcriptomic and metabolomic networks impinge on herbivore fitness. Plant Physiol 176(4):2639–2656
Kang H, Chen X, Kemppainen M, Pardo AG, Veneault-Fourrey C, Kohler A, Martin FM (2020) The small secreted effector protein MiSSP7. 6 of Laccaria bicolor is required for the establishment of ectomycorrhizal symbiosis. Environ Microbiol 22(4):1435–1446
Khan M, Seto D, Subramaniam R, Desveaux D (2018) Oh, the places they'll go! A survey of phytopathogen effectors and their host targets. Plant J 93(4):651–663
Kiers ET, Duhamel M, Beesetty Y, Mensah JA, Franken O, Verbruggen E et al (2011) Reciprocal rewards stabilize cooperation in the mycorrhizal symbiosis. Science 333(6044):880–882
Kloppholz S, Kuhn H, Requena N (2011) A secreted fungal effector of Glomus intraradices promotes symbiotic biotrophy. Curr Biol 21(14):1204–1209
Kohler A, Kuo A, Nagy LG, Morin E, Barry KW, Buscot F, Canback B, Choi C, Cichocki N, Clum A, Colpaert J, Copeland A, Costa MD, Dore J, Floudas D, Gay G, Girlanda M, Henrissat B, Herrmann S, Hess J, Hogberg N, Johansson T, Khouja HR, LaButti K, Lahrmann U, Levasseur A, Lindquist EA, Lipzen A, Marmeisse R, Martino E, Murat C, Ngan CY, Nehls U, Plett JM, Pringle A, Ohm RA, Perotto S, Peter M, Riley R, Rineau F, Ruytinx J, Salamov A, Shah F, Sun H, Tarkka M, Tritt A, Veneault-Fourrey C, Zuccaro A, Tunlid A, Grigoriev IV, Hibbett DS, Martin F, C. Mycorrhizal Genomics Initiative (2015) Convergent losses of decay mechanisms and rapid turnover of symbiosis genes in mycorrhizal mutualists. Nat Genet 47(4):410–U176
Kovac M, Zel J (1995) The effect of aluminum on cytokinins in the mycelia of Amanita muscaria. J Plant Growth Regul 14:117–120
Kraigher H, Grayling A, Wang TL, Hanke DE (1991) Cytokinin production by two ectomycorrhizal fungi in liquid culture. Phytochemistry 30:2249–2254
Labbé J, Muchero W, Czarnecki O, Wang J, Wang X, Bryan AC et al (2019) Mediation of plant–mycorrhizal interaction by a lectin receptor-like kinase. Nature Plants 5(7):676–680

Labourel A, Frandsen KEH, Zhang F, Brouilly N, Grisel S, Haon M, Ciano L, Ropartz D, Fanuel M, Martin F, Navarro D, Rosso MN, Tandrup T, Bissaro B, Johansen KS, Zerva A, Walton PH, Henrissat B, Leggio LL, Berrin JG (2020) A fungal family of lytic polysaccharide monooxygenase-like copper proteins. Nat Chem Biol 16(3):345–350

Laurent P, Voiblet C, Tagu D, de Carvalho D, Nehls U, De Bellis R, Balestrini R, Bauw G, Bonfante P, Martin F (1999) A novel class of ectomycorrhiza-regulated cell wall polypeptides in Pisolithus tinctorius. Mol Plant-Microbe Interact 12(10):862–871

Le Marquer M, San Clemente H, Roux C, Savelli B, Frey NF (2019) Identification of new signalling peptides through a genome-wide survey of 250 fungal secretomes. BMC Genomics 20(1):64

Liao HL, Chen Y, Bruns TD, Peay KG, Taylor JW, Branco S, Talbot JM, Vilgalys R (2014) Metatranscriptomic analysis of ectomycorrhizal roots reveals genes associated with Piloderma-Pinus symbiosis: improved methodologies for assessing gene expression in situ. Environ Microbiol 16(12):3730–3742

Liao HL, Bonito G, Rojas JA, Hameed K, Wu S, Schadt CW et al (2019) Fungal endophytes of Populus trichocarpa alter host phenotype, gene expression, and rhizobiome composition. Mol Plant-Microbe Interact 32(7):853–864

Liu N, Zhang L, Zhou H, Zhang M, Yan X, Wang Q, Long Y, Xie L, Wang S, Huang Y, Zhou Z (2013) Metagenomic insights into metabolic capacities of the gut microbiota in a fungus-cultivating termite (Odontotermes yunnanensis). PLoS One 8:e69184

Lofgren LA, Nguyen NH, Vilgalys R, Ruytinx J, Liao HL, Branco S et al (2021) Comparative genomics reveals dynamic genome evolution in host specialist ectomycorrhizal fungi. New Phytol 230(2):774–792

Lucic E, Fourrey C, Kohler A, Martin F, Chalot M, Brun-Jacob A (2008) A gene repertoire for nitrogen transporters in Laccaria bicolor. New Phytol 180(2):343–364

Lutzoni F, Nowak MD, Alfaro ME, Reeb V, Miadlikowska J, Krug M et al (2018) Contemporaneous radiations of fungi and plants linked to symbiosis. Nat Commun 9(1):1–11

Ma L-S, Pellegrin C, Kahmann R (2018) Repeat-containing effectors of filamentous pathogens and symbionts. Curr Opin Microbiol

Martin F, Kamoun S (2011) Effectors in plant-microbe interactions. John Wiley & Sons

Martin F, Tuskan GA, DiFazio SP, Lammers P, Newcombe G, Podila GK (2004) Symbiotic sequencing for the Populus mesocosm. New Phytol 161(2):330–335

Martin F, Aerts A, Ahrén D, Brun A, Danchin EGJ, Duchaussoy F, Gibon J, Kohler A, Lindquist E, Pereda V, Salamov A, Shapiro HJ, Wuyts J, Blaudez D, Buée M, Brokstein P, Canbäck B, Cohen D, Courty PE, Coutinho PM, Delaruelle C, Detter JC, Deveau A, DiFazio S, Duplessis S, Fraissinet-Tachet L, Lucic E, Frey-Klett P, Fourrey C, Feussner I, Gay G, Grimwood J, Hoegger PJ, Jain P, Kilaru S, Labbé J, Lin YC, Legué V, Le Tacon F, Marmeisse R, Melayah D, Montanini B, Muratet M, Nehls U, Niculita-Hirzel H, Secq MPO-L, Peter M, Quesneville H, Rajashekar B, Reich M, Rouhier N, Schmutz J, Yin T, Chalot M, Henrissat B, Kües U, Lucas S, Van de Peer Y, Podila GK, Polle A, Pukkila PJ, Richardson PM, Rouzé P, Sanders IR, Stajich JE, Tunlid A, Tuskan G, Grigoriev IV (2008) The genome of Laccaria bicolor provides insights into mycorrhizal symbiosis. Nature 452(7183):88–92

Martin F, Kohler A, Murat C, Veneault-Fourrey C, Hibbett DS (2016) Unearthing the roots of ectomycorrhizal symbioses. Nat Rev Microbiol 14(12):760–773

Martin FM, Uroz S, Barker DG (2017) Ancestral alliances: plant mutualistic symbioses with fungi and bacteria. Science 356(6340)

Martino E, Morin E, Grelet GA, Kuo A, Kohler A, Daghino S et al (2018) Comparative genomics and transcriptomics depict ericoid mycorrhizal fungi as versatile saprotrophs and plant mutualists. New Phytol 217(3):1213–1229

Mierziak J, Kostyn K, Kulma A (2014) Flavonoids as important molecules of plant interactions with the environment. Molecules 19(10):16,240–16,265

Miyata K, Kozaki T, Kouzai Y, Ozawa K, Ishii K, Asamizu E, Okabe Y, Umehara Y, Miyamoto A, Kobae Y et al (2014) The bifunctional plant receptor, Os CERK1, regulates both chitin-triggered immunity and arbuscular mycorrhizal symbiosis in rice. Plant Cell Physiol 55:1864–1872

Miyauchi S, Kiss E, Kuo A, Drula E, Kohler A, Sánchez-García M et al (2020) Large-scale genome sequencing of mycorrhizal fungi provides insights into the early evolution of symbiotic traits. Nat Commun 11(1):1–17

Mornico D, Miché L, Béna G, Nouwen N, Verméglio A, Vallenet D, Smith AAT, Giraud E, Medigue C, Moulin L (2011) Comparative genomics of Aeschynomene symbionts: insights into the ecological lifestyle of nod-independent photosynthetic bradyrhizobia. Genes 3:35–61

Myburg AA, Grattapaglia D, Tuskan GA, Hellsten U, Hayes RD, Grimwood J et al (2014) The genome of Eucalyptus grandis. Nature 510(7505):356–362

Neale DB, Wegrzyn JL, Stevens KA, Zimin AV, Puiu D, Crepeau MW et al (2014) Decoding the massive genome of loblolly pine using haploid DNA and novel assembly strategies. Genome Biol 15(3):1–13

Nystedt B, Street NR, Wetterbom A, Zuccolo A, Lin YC, Scofield DG et al (2013) The Norway spruce genome sequence and conifer genome evolution. Nature 497(7451):579–584

Oldroyd GE (2013) Speak, friend, and enter: signalling systems that promote beneficial symbiotic associations in plants. Nat Rev Microbiol 11(4):252–263

Pellegrin C, Morin E, Martin FM, Veneault-Fourrey C (2015) Comparative analysis of secretomes from ectomycorrhizal fungi with an emphasis on small-secreted proteins. Front Microbiol 6(1278)

Pellegrin C, Daguerre Y, Ruytinx J, Guinet F, Kemppainen M, Frey NFD et al (2019a) Laccaria bicolor MiSSP8 is a small-secreted protein decisive for the establishment of the ectomycorrhizal symbiosis. Environ Microbiol 21(10):3765–3779

Pellegrin C, Martin F, Veneault-Fourrey C (2019b) Molecular signalling during the ectomycorrhizal symbiosis. In: Gressler M, Hoffmeister D (eds) Biology of the fungal cell. The mycota (a comprehensive treatise on fungi as experimental systems for basic and applied research), vol 8. Springer, Cham, pp 95–109

Peterson RL, Massicotte HB (2004) Exploring structural definitions of mycorrhizas, with emphasis on nutrient-exchange interfaces. Can J Bot 82(8):1074–1088

Plett JM, Kemppainen M, Kale SD, Kohler A, Legué V, Brun A et al (2011) A secreted effector protein of Laccaria bicolor is required for symbiosis development. Curr Biol 21(14):1197–1203

Plett JM, Daguerre Y, Wittulsky S, Vayssières A, Deveau A, Melton SJ et al (2014a) Effector MiSSP7 of the mutualistic fungus Laccaria bicolor stabilizes the Populus JAZ6 protein and represses jasmonic acid (JA) responsive genes. Proc Natl Acad Sci 111(22):8299–8304

Plett JM, Khachane A, Ouassou M, Sundberg B, Kohler A, Martin F (2014b) Ethylene and jasmonic acid act as negative modulators during mutualistic symbiosis between L accaria bicolor and P opulus roots. New Phytol 202(1):270–286

Plett JM, Yin H, Mewalal R, Hu R, Li T, Ranjan P et al (2017) Populus trichocarpa encodes small, effector-like secreted proteins that are highly induced during mutualistic symbiosis. Sci Rep 7(1):1–13

Plett KL, Raposo AE, Anderson IC, Piller SC, Plett JM (2019) Protein arginine methyltransferase expression affects ectomycorrhizal symbiosis and the regulation of hormone signaling pathways. Mol Plant-Microbe Interact 32(10):1291–1302

Plett KL, Singan VR, Wang M, Ng V, Grigoriev IV, Martin F, Plett JM, Anderson IC (2020) Inorganic nitrogen availability alters Eucalyptus grandis receptivity to the ectomycorrhizal fungus Pisolithus albus but not symbiotic nitrogen transfer. New Phytol 226(1):221–231

Plomion C, Aury JM, Amselem J, Leroy T, Murat F, Duplessis S et al (2018) Oak genome reveals facets of long lifespan. Nature Plants 4(7):440–452

Radhakrishnan GV, Keller J, Rich MK, Vernié T, Mbadinga DLM, Vigneron N et al (2020) An ancestral signalling pathway is conserved in intracellular symbioses-forming plant lineages. Nature Plants 6(3):280–289

Regvar M, Gogala N, Znidarsic N (1997) Jasmonic acid affects mycorrhization of spruce seedlings with Laccaria laccata. Trees Struct Funct 11:511–514

Rey T, Nars A, Bonhomme M, Bottin A, Huguet S, Balzergue S, Jardinaud M-F, Bono J-J, Cullimore J, Dumas B et al (2013) NFP, a LysM protein controlling Nod factor perception, also intervenes in Medicago truncatula resistance to pathogens. New Phytol 198:875–886

Rineau F, Shah F, Smits MM, Persson P, Johansson T, Carleer R, Troein C, Tunlid A (2013) Carbon availability triggers the decomposition of plant litter and assimilation of nitrogen by an ectomycorrhizal fungus. ISME J 7:2010–2022

Rush TA, Puech-Pagès V, Bascaules A, Jargeat P, Maillet F, Haouy A, Maës AQ, Carriel CC, Khokhani D, Keller-Pearson M, Tannous J, Cope KR, Garcia K, Maeda J, Johnson C, Kleven B, Choudhury QJ, Labbé J, Swift C, O'Malley MA, Bok JW, Cottaz S, Fort S, Poinsot V, Sussman MR, Lefort C, Nett J, Keller NP, Bécard G, Ané J-M (2020) Lipo-chitooligosaccharides as regulatory signals of fungal growth and development. Nat Commun 11(1):3897

Sachs JL, Russell JE, Lii YE, Black KC, Lopez G, Patil AS (2010) Host control over infection and proliferation of a cheater symbiont. J Evol Biol 23:1919–1927

Salojärvi J, Smolander OP, Nieminen K, Rajaraman S, Safronov O, Safdari P et al (2017) Genome sequencing and population genomic analyses provide insights into the adaptive landscape of silver birch. Nat Genet 49(6):904–912

Sasse J, Martinoia E, Northen T (2018) Feed your friends: do plant exudates shape the root microbiome? Trends Plant Sci 23(1):25–41

Shiu SH, Bleecker AB (2001) Receptor-like kinases from Arabidopsis form a monophyletic gene family related to animal receptor kinases. Proc Natl Acad Sci 98(19):10,763–10,768

Sillo F, Fangel JU, Henrissat B, Faccio A, Bonfante P, Martin F, Willats WG, Balestrini R (2016) Understanding plant cell-wall remodelling during the symbiotic interaction between Tuber melanosporum and Corylus avellana using a carbohydrate microarray. Planta 244(2):347–359

Simms EL, Taylor DL, Povich J, Shefferson RP, Sachs JL, Urbina M, Tausczik Y (2006) An empirical test of partner choice mechanisms in a wild legume–rhizobium interaction. Proc R Soc B Biol Sci 273(1582):77–81

Skiada V, Avramidou M, Bonfante P, Genre A, Papadopoulou KK (2020) An endophytic fusarium–legume association is partially dependent on the common symbiotic signalling pathway. New Phytol 226(5):1429–1444

Smith SE, Read DJ (2008) Mycorrhizal symbiosis, 3rd edn. Academic Press, London, pp 1–9

Snelders NC, Kettles GJ, Rudd JJ, Thomma BPHJ (2018) Plant pathogen effector proteins as manipulators of host microbiomes? Mol Plant Pathol 19(2):257–259

Snelders NC, Rovenich H, Petti GC, Rocafort M, van den Berg GCM, Vorholt JA, Mesters JR, Seidl MF, Nijland R, Thomma B (2020) Microbiome manipulation by a soil-borne fungal plant pathogen using effector proteins. Nat Plants 6(11):1365–1374

Splivallo R, Fischer U, Gobel C, Feussner I, Karlovsky P (2009) Truffles regulate plant root morphogenesis via the production of auxin and ethylene. Plant Physiol 150:2018–2029

Sukumar P, Legué V, Vayssières A, Martin F, Tuskan GA, Kalluri UC (2013) Involvement of auxin pathways in modulating root architecture during beneficial plant-microorganism interactions. Plant Cell Environ 36(5):909–919

Tagu D, De Bellis R, Balestrini R, De Vries OMH, Piccoli G, Stocchi V, Bonfante P, Martin F (2001) Immunolocalization of hydrophobin HYDPt-1 from the ectomycorrhizal basidiomycete Pisolithus tinctorius during colonization of Eucalyptus globulus roots. New Phytol 149(1): 127–135

Tedersoo L, Smith ME (2013) Lineages of ectomycorrhizal fungi revisited: foraging strategies and novel lineages revealed by sequences from belowground. Fungal Biol Rev 27(3–4):83–99

Tisserant E, Malbreil M, Kuo A, Kohler A, Symeonidi A, Balestrini R et al (2013) Genome of an arbuscular mycorrhizal fungus provides insight into the oldest plant symbiosis. Proc Natl Acad Sci 110(50):20,117–20,122

Tschaplinski TJ, Plett JM, Engle NL, Deveau A, Cushman KC, Martin MZ, Doktycz MJ, Tuskan GA, Brun A, Kohler A, Martin F (2014) Populus trichocarpa and Populus deltoides exhibit

different metabolomic responses to colonization by the symbiotic fungus Laccaria bicolor. Mol Plant-Microbe Interact 27(6):546–556
Tsuzuki S, Handa Y, Takeda N, Kawaguchi M (2016) Strigolactone-induced putative secreted protein 1 is required for the establishment of symbiosis by the arbuscular mycorrhizal fungus Rhizophagus irregularis. Mol Plant-Microbe Interact 29(4):277–286
Tuskan GA, DiFazio S, Jansson S et al (2006) The genome of black cottonwood, Populus trichocarpa (Torr & Gray). Science 313:1596–1604
Vannier N, Agler M, Hacquard S (2019) Microbiota-mediated disease resistance in plants. PLoS Pathog 15(6):e1007740
Varden FA, De la Concepcion JC, Maidment JH, Banfield MJ (2017) Taking the stage: effectors in the spotlight. Curr Opin Plant Biol 38:25–33
Vayssières A, Pěnčík A, Felten J, Kohler A, Ljung K, Martin F, Legué V (2015) Development of the poplar-Laccaria bicolor ectomycorrhiza modifies root auxin metabolism, signaling, and response. Plant Physiol 169(1):890–902
Veneault-Fourrey C, Commun C, Kohler A, Morin E, Balestrini R, Plett J, Danchin E, Coutinho P, Wiebenga A, de Vries RP, Henrissat B, Martin F (2014) Genomic and transcriptomic analysis of Laccaria bicolor CAZome reveals insights into polysaccharides remodelling during symbiosis establishment. Fungal Genet Biol 72:168–181
Villalobos Solis MI, Poudel S, Bonnot C, Shrestha HK, Hettich RL, Veneault-Fourrey C, Martin F, Abraham PE (2020) A viable new strategy for the discovery of peptide proteolytic cleavage products in plant-microbe interactions. Mol Plant-Microbe Interact 33(10):1177–1188
Wegrzyn JL, Falk T, Grau E, Buehler S, Ramnath R, Herndon N (2020) Cyberinfrastructure and resources to enable an integrative approach to studying forest trees. Evol Appl 13(1):228–241
Weisskopf L, Schulz S, Garbeva P (2021) Microbial volatile organic compounds in intra-kingdom and inter-kingdom interactions. Nat Rev Microbiol 19(6):391–404
Wong JWH, Lutz A, Natera S, Wang M, Ng V, Grigoriev I, Martin F, Roessner U, Anderson IC, Plett JM (2019) The influence of contrasting microbial lifestyles on the pre-symbiotic metabolite responses of Eucalyptus grandis roots. Front Ecol Evol 7(10)
Wong-Bajracharya J, Singan VR, Monti R, Plett KL, Ng V, Grigoriev IV et al (2022) The ectomycorrhizal fungus Pisolithus microcarpus encodes a microRNA involved in cross-kingdom gene silencing during symbiosis. Proc Natl Acad Sci 119(3)
Wullschleger SD, Weston DJ, DiFazio SP, Tuskan GA (2013) Revisiting the sequencing of the first tree genome: Populus trichocarpa. Tree Physiol 33(4):357–364
Yang X, Cheng H, Li L, Hu R, Liu D, Morrel-Falvey J, Labbé J, Doktycz M (2019) Plant-microbe interfaces: functional characterization of small secreted proteins in poplar in relation to Populus-Laccaria symbiosis. Genomic Sciences Program 2019 Annual Principal Investigator(PI) Meeting. Abstract book, pp 226–227
Zhang X, Dong W, Sun J, Feng F, Deng Y, He Z, Oldroyd GED, Wang E (2015) The receptor kinase CERK1 has dual functions in symbiosis and immunity signalling. Plant J 81:258–267
Zhang F, Anasontzis GE, Labourel A, Champion C, Haon M, Kemppainen M, Commun C, Deveau A, Pardo A, Veneault-Fourrey C, Kohler A, Rosso M-N, Henrissat B, Berrin J-G, Martin F (2018) The ectomycorrhizal basidiomycete Laccaria bicolor releases a secreted β-1,4 endoglucanase that plays a key role in symbiosis development. New Phytol 220(4):1309–1321
Zhang F, Labourel A, Haon M, Kemppainen M, Da Silva ME, Brouilly N, Veneault-Fourrey C, Kohler A, Rosso MN, Pardo A et al (2021) The ectomycorrhizal basidiomycete Laccaria bicolor releases a GH28 polygalacturonase that plays a key role in symbiosis establishment. New Phytol. https://doi.org/10.1111/nph.17940
Zipfel C, Oldroyd GE (2017) Plant signalling in symbiosis and immunity. Nature 543(7645): 328–336

Chapter 12
Fungal Effector Proteins: Molecular Mediators of Fungal Symbionts of Plants

Lauren B. L. Stone, Israel Enrique Padilla-Guerrero, and Michael J. Bidochka

Abstract Plant-fungal symbioses are of great importance to agriculture. Phytopathogenic fungi, which cause disease in plants as they consume host tissues, are a major threat to global food security. Conversely, endophytes and mycorrhizal fungi, which can engage in mutually beneficial relationships with host plants, are capable of promoting plant health, growth, and development. Such plant-fungal symbioses, whether they be harmful or beneficial to host plants, require complex and continuous molecular cross talk, including the secretion of fungal effector proteins into the plant apoplast and cytoplasm. Fungal effectors are broadly defined as proteins which modulate plant physiology in ways which facilitate fungal colonization and growth within the host plant. These proteins typically share three common features: (1) an amino(N)-terminal signal peptide to allow for secretion, (2) high cysteine content that facilitates structural stability, and (3) small molecular mass (typically $\leq$30 kDa). However, fungi often possess a large number of genes which fit these criteria—only a subset of which truly act as effector proteins. Progress in identifying effector proteins has been hindered by the fact that most apparently do not share significant sequence similarities with effectors of other genera. This is thought to be an adaptive trait within the context of the co-evolutionary arms race between plant-associated fungi and their host plants, in which plants are continually evolving novel variations in their resistance (R) proteins to allow for improved detection of fungal effector proteins, while fungi diversify their effectors to evade detection. Despite this fact, there are a limited number of domains and motifs which have been found in effector proteins from a variety of plant-associated fungi. Some of these conserved sequence features, such as CFEM and LysM domains, are shared among effector proteins of phytopathogenic and mutualistic fungi, while others may be tailored to the unique lifestyles of individual species or genera. Relying upon the common features and few

L. B. L. Stone · M. J. Bidochka (✉)
Department of Biological Sciences, Brock University, St. Catharines, ON, Canada
e-mail: lstone22@uwo.ca; mbidochka@brocku.ca

I. E. Padilla-Guerrero
Division of Natural and Exact Sciences, Department of Biology, University of Guanajuato, Campus Guanajuato, Guanajuato, Mexico

B. A. Horwitz, P. K. Mukherjee (eds.), *Microbial Cross-talk in the Rhizosphere*, Rhizosphere Biology, https://doi.org/10.1007/978-981-16-9507-0_12

conserved motifs and domains, in silico tools are often relied upon to mine fungal genomes for candidate effector proteins. Although this process is valuable in curating lists of potential effectors, experimental validation of effector status is still required. To that end, many studies have utilized knockout or overexpression of putative effector protein-coding genes in order to elucidate the importance of putative effectors in the establishment and maintenance of fungal-plant symbioses—a method which is susceptible to issues of gene redundancy. The vast majority of literature on fungal effectors has focused on fungal pathogens of above-ground plant tissue; however this chapter also emphasizes the limited state of understanding on effector proteins from rhizospheric fungi.

Keywords Metarhizium · Endophyte · Effector proteins · Plant symbiosis

12.1 Plant-Fungal Symbioses

Throughout their evolution, some fungi have acquired the capacity to establish symbiotic associations with plants. These symbioses exist on a scale from pathogenic to mutualistic, depending on the fungal species.

Phytopathogenic fungi obtain their nourishment from hosts, providing nothing beneficial in return, and thereby reducing plant fitness and ultimately leading to disease. Such pathogens are traditionally categorized into three groups: (1) biotrophs, which extract nutrients from living plant tissue, (2) necrotrophs, which kill and feed on plant tissues, and (3) hemibiotrophs, which begin as biotrophs, but progress to a necrotrophic lifestyle. Fungal diseases of plants are a major threat to global food security, with fungi responsible for 64% of local plant extinctions (Fisher et al. 2012). Based on harvest statistics from the five most agriculturally important crops (rice, wheat, maize, potatoes, and soybean), a study by Fisher et al. (2012) estimated that even low-grade persistence of fungal disease leads to a level of crop loss equating to enough food for nearly 600 million people. If severe epidemics were to occur concurrently in these five crops, 61% of the world's population would be without a sufficient supply of food (Fisher et al. 2012).

Mutualistic fungi, including mycorrhizae and endophytes, also obtain nourishment from plants. However, unlike phytopathogens, these fungi reciprocally convey benefits to their host plant, making this relationship mutually beneficial. Such mutualistic relationships can promote plant health, growth, and development by, for example, providing nutrients, improving stress tolerance, and defending the host against phytopathogenic microbes via competitive exclusion, stimulation of the plant immune system, or direct parasitism of other organisms. Given these benefits, mutualistic fungi are often used in agriculture as soil amendments which can increase crop yield, improve soil quality, and reduce the need for chemical fertilizers, fungicides, and pesticides. Underground networks of mycorrhizal fungi can even mediate inter-plant cross talk between the same or different species via the transfer of resources, stress signals, and allelochemicals (Gorzelak et al. 2015). For example, in what may be a form of reciprocal altruism, mycorrhizal networks can allow nutrient-

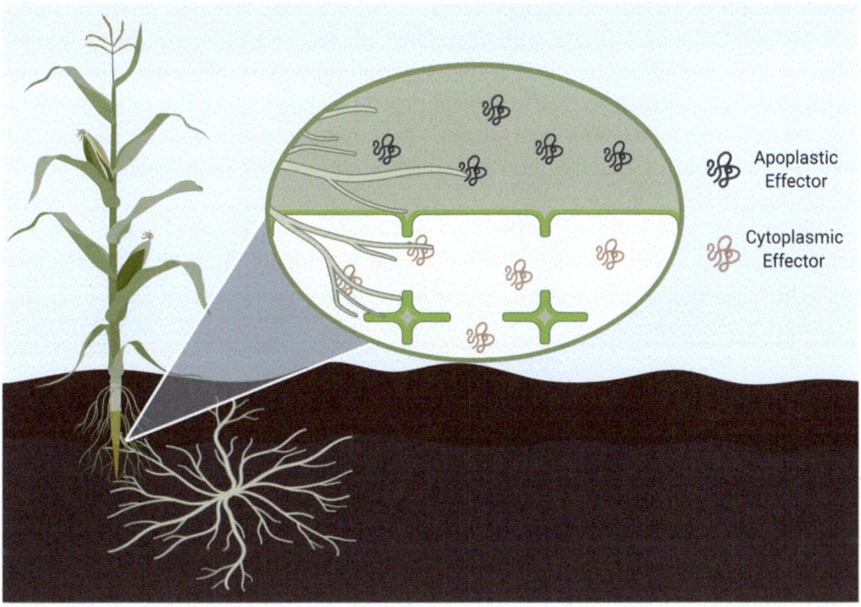

Fig. 12.1 Secretion of effector proteins by a soil-borne fungus. Given their amino(N)-terminal signal peptide, fungal effector proteins can be targeted for secretion via the ER-Golgi apparatus route. Secreted effectors may act in the host apoplast or cytoplasm, allowing for targeting of extracellular or intracellular plant processes, respectively. Image created with BioRender

rich plants to donate resources to worse-off neighbors, with the expectation that resources will be returned in the future (Gorzelak et al. 2015).

Cross talk between fungi and their host plants dictates several aspects of the symbiotic relationship, including the extent of fungal colonization, nutrient fixation and exchange, and defense responses. In rhizospheric systems, plant roots produce a variety of substances that mediate their interactions with microbial invaders and the greater rhizosphere microbiome, such as immune proteins and secondary metabolites, including antimicrobial compounds. In turn, fungi secrete substances, including small metabolites and effector proteins, which can influence plant physiology. This chapter focuses on such fungal mediators of such interactions—specifically, on fungal effector proteins.

Effector proteins can be broadly defined as any fungal protein which interferes with or modifies plant physiology (i.e., structure and/ or function) in ways which modulate fungal colonization or growth within the host plant. Such proteins may be secreted by the fungi into the cytoplasm of plant cells or simply into the extracellular space, allowing these proteins to be broadly categorized into cytoplasmic and apoplastic effectors, respectively (Fig. 12.1) (Sonah et al. 2016). The definition of fungal effectors is necessarily vague, as the precise effects of such proteins on host plants varies greatly between species and between individual effectors from the same species. This is both due to evolutionary pressure for rapid and large-scale diversification of fungal effector proteins, as well as the fact that different fungi cultivate unique relationships with their host plants, and thus require their own unique suite of proteins to modify plant physiology. Overall, effectors are not characterized by a distinct set of chemical properties nor targets, but rather by their function in interactions with host plants—that is, to interfere with or manipulate plant structure and/ or function in ways which accommodate fungal invasion (Uhse and Djamei 2018).

This chapter will summarize the current literature on fungal effector proteins, including common features, examples of some previously characterized effectors from rhizospheric fungi, as well as methods that have been used to find and characterize effectors.

12.2 Role of Effector Proteins During Fungal Colonization of Plants

Jones and Dangl (2006) proposed a "zig-zag" model of the interaction between the plant immune system and plant-associated microbes, in which plants are believed to possess two lines of innate immune defense, the first of which involves recognition of conserved molecular features found in microbial invaders (both pathogenic and non-pathogenic) referred to as microorganism-associated molecular patterns (MAMPs) (Newman et al. 2013; Sonah et al. 2016). Plant pattern recognition receptors (PPRs) localized to the plasma membrane of plant cells can recognize particular MAMPs, inducing an immune response, termed MAMP-triggered immunity (MTI) (Sonah et al. 2016; Trdá et al. 2015). This MTI may involve changes to the plant cell wall, as well as expression of a variety of antimicrobial compounds, including reactive oxygen species (ROS), reactive nitrogen species (RNS), proteases, protease inhibitors, and chitinases (Newman et al. 2013; Sonah et al. 2016; Trdá et al. 2015). Apparently induced by this MTI, some plant-associated fungi have been found to alter their expression and secretion of effector proteins in ways which apparently aid in establishment and maintenance of their interaction with their host plants (Guzmán-Guzmán et al. 2017; Ramírez-Valdespino et al. 2019; Sonah et al. 2016). For fungal pathogens, this may involve proteins which assist in overcoming the plant defense response in order to facilitate pathogenesis, whereas, for plant mutualists, effector proteins may simply modulate an ongoing symbiosis in which the fungus is able to remain associated with host plants despite their defense responses (Guzmán-Guzmán et al. 2017). In either case, there may be overlap in the function of some of their effector proteins. For instance, plants possess PRRs which can recognize and bind to chitin, a major component of the fungal cell wall and thus an important MAMP in both plant pathogenic and mutualistic fungi, which then triggers an immune response (Fig. 12.2) (Guzmán-Guzmán et al. 2017; Rafiqi et al. 2013). This MAMP-triggered immunity (MTI) may include the production of chitinases which act to degrade chitin in the cell wall of the invading fungi

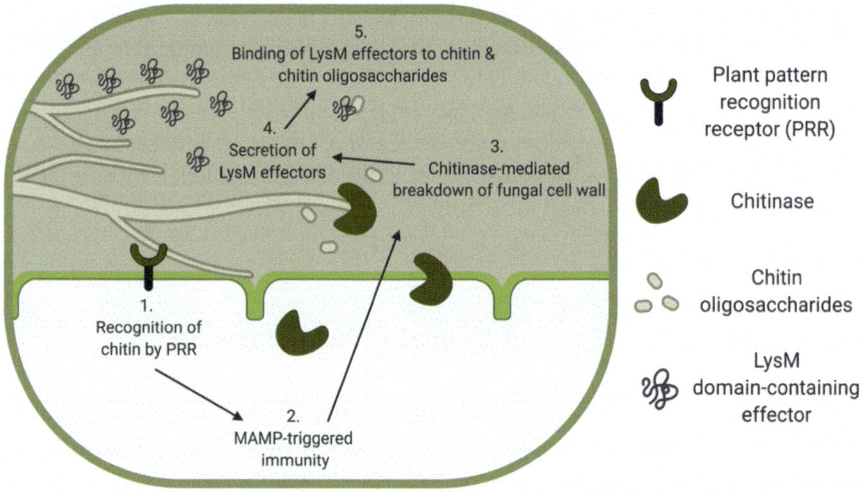

Fig. 12.2 Interaction between plant host and LysM domain-containing effector proteins secreted by an invading fungi. Given its presence in fungal cell walls, chitin represents a major microorganism-associated molecular pattern (MAMP) for all fungi. Plant pattern recognition receptors (PRRs) can recognize and bind to chitin from invading fungi (1), triggering the first line of innate immune defense (2). This MAMP-triggered immunity (MTI) may include production of chitinases, which can degrade the fungal cell wall (3), releasing chitin oligosaccharides. In response, the fungus may secrete LysM domain-containing effectors (4), which can bind to chitin in the fungal cell wall, thereby minimizing additional chitinase degradation, and/or bind to chitin oligosaccharides, thereby preventing further host immune stimulation. Image created with BioRender

(Fig. 12.2) (Rafiqi et al. 2013; Ramírez-Valdespino et al. 2019). In response, some fungi secrete effector proteins containing a LysM domain, which allows for binding of the effector to chitin in the fungal cell wall, thereby preventing host plant hydrolases (e.g., chitinase) from binding and degrading chitin (Fig. 12.2). For instance, Avr4, a LysM effector from tomato pathogen *Cladosporium fulvum,* binds to polymeric chitin in the fungal cell wall, thereby protecting the cell wall from degradation (Van Den Burg et al. 2006). Additionally, binding of LysM effectors to chitin or degraded chitin products (i.e., chitin oligosaccharides) prevents their detection by plant LysM receptors (e.g., LysM receptor-like kinases) which can trigger defense responses (Fig. 12.2). For example, Ecp6, another LysM effector from *C. fulvum*, binds to chitin oligosaccharides released after degradation of chitin via host plant chitinases, thereby preventing their detection by plant receptors (De Jonge et al. 2010).

Regardless of whether these effector proteins are coming from a plant pathogen or a plant-beneficial fungus, host plants may recognize the fungus and mount a second line of plant immunity—effector-triggered immunity (ETI) (Sonah et al. 2016). ETI, another form of innate immunity, utilizes so-called resistance (R) proteins, which can recognize certain effectors and induce an additional defense response (Sonah et al. 2016). Particularly in the case of plant pathogens, ETI often includes a hypersensitive response (HR) in which rapid death of plant cells, localized to the site of infection, is induced in order to restrict the spread of the invading microbe (Selin et al. 2016). This dynamic relationship between fungus and host plant sets up a co-evolutionary arms-race, in which plants are continually evolving novel variations in their R proteins to allow for improved detection of fungal effector proteins, while fungi are adapting their arsenal of effector proteins to evade detection by host plants (Newman et al. 2013; Sonah et al. 2016; Trdá et al. 2015).

12.3 Common Features of Fungal Effector Proteins

Table 12.1 shows a representative list of fungal effector proteins identified in soil-borne fungi, including their known or suspected roles in plant interactions. Although several effectors in Table 12.1 contain functional domains, most fungal effectors lack conserved domains and do not share any significant sequence similarity with effectors of other genera (Hassing et al. 2019; Sonah et al. 2016). However, there is likely a bias in the literature towards effectors which do possess conserved sequence features, as they may be easier to identify. The lack of overall commonality in the various effector proteins may be related to the previously mentioned co-evolutionary arms race between plant-associated fungi and their plant hosts, in which there is a strong evolutionary pressure for diversification of effector proteins in order to avoid detection by plant hosts. This rapid evolution required in effector gene sequences, likely requiring numerous insertions, deletions, and single-nucleotide polymorphisms, could, in part, explain the lack of sequence similarity between effector proteins of different fungal species (Selin et al. 2016).

However, there are some features common to most known effector proteins, which can be used to aid in the identification of candidate effector proteins (CEPs), as will be discussed later. Most obviously, effector proteins typically require an amino(N)-terminal signal peptide to target them for the extracellular secretory pathway from the rough endoplasmic reticulum of the fungus and into the plant, where they will exert their effects (Guzmán-Guzmán et al. 2017; Hassing et al. 2019; Sonah et al. 2016).

Another feature common to many effectors is high cysteine content. Cysteine enrichment may allow the proteins to form an abundance of disulfide bridges—a characteristic which may be selected in the evolution of effector proteins as it would facilitate maintenance of structural integrity and proper folding during secretion and *in planta* (Hassing et al. 2019). This may be especially important for effectors which operate in the protease-rich apoplastic space (Hassing et al. 2019). Additionally, many effector proteins are relatively small in size and thus, in screening for potential effector proteins, many studies use a cutoff size of $\leq$300 amino acids (Hassing et al. 2019; Rafiqi et al. 2013). However, these features do not apply to all known effectors, with several large effectors with few or no cysteine residues having been found. For instance, the cytoplasmic effector AvrM from the plant pathogen

Table 12.1 Characterized effectors from selected soil-borne fungi. Many known effectors do not possess conserved domains

Species	Effector	Domain/ family	Function(s) in interaction with host plants	References
Clonostachys rosea	LYSM2	LysM	Protection of fungal cell wall against host hydrolytic enzymes	Dubey et al. (2020)
Colletotrichum tofieldiae	CtNIS1	NIS1	Suppression of MAMP-triggered ROS production	Irieda et al. (2019)
Fusarium oxysporum f. sp. *lycopersici*	Mep1	Fungalysin metallopeptidase (M36)	Cleavage of host chitinases	Jashni et al. (2015a)
	Sep1	Subtilisin-like serine protease	Cleavage of host chitinases	Jashni et al. (2015a)
	Six family (Six1, Six2, Six3, Six4, Six5, Six6, Six7)	None	Virulence and/or avirulence factors	Thatcher et al. (2012)
Fusarium graminearum	FGL1	Lipase (class 3)	Inhibition of host immune-related callose formation	Blümke et al. (2014)
	XylA	Glycosyl hydrolase 11	Plant cell wall degradation	Tini et al. (2020)
Laccaria bicolor	MiSSP7	None	Blocking of host jasmonic acid-dependent transcription of defense-related genes	Plett et al. (2014)
Leptosphaeria maculans	AvrLm family effectors (AvrLm1, AvrLm2, AvrLm3, AvrLm4-7, AvrLm11, AvrLmJ1)	None	Avirulence	van de Wouw et al. (2010)
Piriformospora indica	FGB1	None	Suppression of MAMP-triggered ROS production	Wawra et al. (2016)
	PIIN_08944	None	Interference with host salicylic acid-mediate immune responses	Akum et al. (2015)
Rhizoctonia solani	RsLysM	LysM	Suppression of chitin-triggered immunity	Dölfors et al. (2019)
	AGLIP1	Lipase (class 3)	Suppression of MAMP-triggered immunity, induction of host cell death	Li et al. (2019)

(continued)

Table 12.1 (continued)

Species	Effector	Domain/ family	Function(s) in interaction with host plants	References
Rhizophagus irregularis	RiCRN1	None	Arbuscule development	Voß et al. (2018)
	RiSLM	LysM	Suppression of chitin-triggered immunity	Zeng et al. (2020)
	SIS1	None	Regulation of colonization	Tsuzuki et al. (2016)
	SP7	None	Suppression of host defense responses	Kloppholz et al. (2011)
Sclerotinia sclerotiorum	SsSSVP1	None	Manipulation of host energy metabolism	Lyu et al. (2016)
Sporisorium reilianum	SAD1	None	Suppression of apical dominance	Ghareeb et al. (2015)
Trichoderma atroviride	Ep1	Cerato-platanin	Induction of host defense-related genes, accumulation of ROS & phenolic compounds	Salas-Marina et al. (2015)
	Tal6	LysM	Protection from host chitinases, evasion from host defense responses	Romero-Contreras et al. (2019)
Trichoderma virens	Sm1	Cerato-platanin	Induction of defense-related genes	Djonović et al., 2006 (2007), Salas-Marina et al. (2015)
	Sm2	Cerato-platanin	Root colonization, host defense activation	Crutcher et al. (2015), Gaderer et al. (2015)
	TVHYDII1	Hydrophobin (class II)	Root colonization	Guzmán-Guzmán et al. (2017)
Verticillium dahliae	Ave1	RlpA-like DPBB, Expansin-like EG45	Induction of host defense-related genes	Castroverde et al. (2016)
	VdICSH1	Isochorismatase	Interference with host salicylate & jasmonate signaling	Zhu et al. (2017)
	VdSCP7	None	Modulation of host immunity	Zhang et al. (2017)

(continued)

Table 12.1 (continued)

Species	Effector	Domain/ family	Function(s) in interaction with host plants	References
	PevD1	Alternaria alternata allergen 1	Induction of host cell death, inhibition of antifungal proteins	Zhang et al. (2019)
	Vd2LysM	LysM	Suppression of chitin-triggered immunity, protection from host chitinases	Kombrink et al. (2017)
Verticillium nonalfalfae	VnaChtBP	Carbohydrate-binding module	Suppression of chitin-triggered immunity, protection from host chitinases	Volk et al. (2019)

Melampsora lini is 314 amino acids in size and contains only a single cysteine residue (Sperschneider et al. 2018).

12.4 Types and Examples of Known Effectors from Rhizospheric Fungi

Although most effector proteins apparently lack conserved domains and do not share any significant sequence similarity with effectors of other genera, there are a limited number of conserved sequence motifs and domains that have been found in effectors of several species.

This section will provide examples of such common families of effectors, with an emphasis on those found in rhizospheric fungi. Some common effector domains and motifs, such as those involved in colonization and evasion or inhibition of the plant defense response, may be shared by fungal pathogens and mutualists of plants, while others are tailored to particular lifestyles.

12.4.1 Effectors from Plant Pathogenic Fungi

Phytopathogenic fungi secrete a variety of effector proteins which contribute to their virulence. Common to effectors of pathogenic fungi are domains and motifs involved in degradation and colonization of plant tissues, as well as evasion and manipulation of the plant immune system.

The infection process begins with adhesion of the fungus to a potential host plant. Hydrophobins—highly hydrophobic secreted proteins containing a domain with eight cysteine residues—may aid filamentous fungi in this process, allowing for adhesion to hydrophobic plant surfaces (Quarantin et al. 2019). Such proteins, sometimes considered to act as effectors, may also aid in hyphal development, formation and dispersal of spores, interactions with the host defense system, and additional environmental interactions (Quarantin et al. 2019). Similarly, effectors with a so-called Common in Fungal Extracellular Membranes (CFEM) domain, containing eight conserved cysteine residues, may act, in part, as cell surface receptors or adhesion molecules, potentially mediating interactions between the fungus and its host or other organisms (Guzmán-Guzmán et al. 2017; Kou et al. 2017). However, the exact function of this domain, particularly in the context of effector proteins, remains unclear. Hydrophobins and CFEM domain-containing proteins, including known and suspected effectors, have been identified in the genome of a variety of phytopathogens, including soil-borne *Fusarium* and *Verticillium* spp. (Marton et al. 2018; Quarantin et al. 2019).

In order to colonize hosts and cause disease, phytopathogenic fungi must also evade and manipulate plant defense responses—a process which depends on a multitude of effector proteins. Such effectors may suppress plant innate immunity or interfere with host cell signaling, MAMP perception, execution of defense responses, and similar evasive processes. For instance, the previously discussed LysM domain, which binds to chitin in the fungal cell wall and/ or degraded chitin products, has been found in proteins known to act as effectors from a variety of pathogenic species, such as the RsLysM effector from soil-borne *Rhizoctonia solani* (Dölfors et al. 2019). Another function of several known effector proteins is protease inhibition. Protease inhibitors (PIs), when acting as fungal effectors, are capable of interfering with the action of proteases secreted as part of the plant defense response, thereby protecting fungal proteins from cleavage by plant proteases (Jashni et al. 2015a). One example of this type of effector is Pit2 from maize pathogen *Ustilago maydis* which inhibits three separate cysteine proteases important to the maize defense response (Jashni et al. 2015b). A homolog of Pit2 has been identified in the genome of soil-borne maize pathogen *Sporisorium reilianum,* deletion of which reduced *S. reilianum* virulence (Schweizer et al. 2018). In addition to PIs, some effectors are known to have protease activity, potentially working, in part, to cleave host plant immune proteins. For instance, in the soil pathogen *Fusarium oxysporum* f. sp. *lycopersici,* metalloprotease effector Mep1 works with Sep1, a serine protease effector, to cooperatively cleave host plant chitinases, thereby protecting the fungal cell wall (Franceschetti et al. 2017; Jashni et al. 2015a). However, some metalloproteases may act as avirulence proteins, triggering plant defense responses. The AVR-Pita1 metalloprotease effector from *Magnaporthe oryzae,* for example, is recognized by rice R protein Pita, which triggers an ETI response in host plants, including initiation of the hypersensitive response, thereby leading to host cell death (Khang et al. 2008). Fungi may also secrete enzymatic effectors, such as hydrolases, targeting non-proteinaceous components of the plant defense response. For instance, VdICSH1, an effector produced by soil pathogen *Verticillium dahliae,* has isochorismatase hydrolase activity, thereby allowing it to suppress the isochorismate pathway for salicylic acid synthesis (Zhu et al. 2017). Isochorismatase domain-containing proteins, which may act as effectors, have been identified in the genomes of additional phytopathgenic fungi, including *F. oxysporum* (Thatcher et al. 2016).

To achieve their ultimate goal of host death, necrotrophs and hemibiotrophs also possess effectors, such as the previously described avirulence proteins, which trigger plant cell death. To this effect, a variety of phytopathogenic fungi possess effectors which act as necrosis-inducing proteins (NLPs), which trigger cell death, while often eliciting strong defense responses. NLPs were first purified from the soil-borne hemibiotroph *F. oxysporum* f. sp. *erythroxyli*, but have since been identified in numerous additional phytopathogens (Bailey 1995).

In addition to the aforementioned families of proteins containing domains common to several effectors, there are a few motifs which have been found to be highly conserved in several effectors. Arguably the most notable of these motifs is Y/F/WxC, which has been identified in over 300 effectors from a variety of powdery mildew and rust fungi (Godfrey et al. 2010). Although the function of this motif is unclear, its presence among only haustoria-producing pathogens suggests a potential role in haustoria formation and/ or transfer of effectors bearing this motif across the extrahaustorial membrane (Godfrey et al. 2010). Similarly, an RxLR motif, thought to be involved in translocation from the plant apoplast into the cytoplasm, has been identified in a variety of phytopathogenic effectors (Dou et al. 2008). Although commonly found in oomycete effectors, known or suspected effectors containing an RxLR motif have also been identified in the genomes of a variety of fungi, including the soil-borne pathogens *F. oxysporum* and *R. solani* (Taylor et al. 2016; Yamamoto et al. 2019).

The aforementioned examples—hydrophobins, CFEM domain-containing proteins, LysM domain-containing proteins, proteases, protease inhibitors, necrosis-inducing proteins, as well as the Y/F/WxC and RxLR motifs—are by no means an exhaustive list of all of the functional domains and motifs found in effector proteins of more than one species of phytopathogenic fungi. It is important to reiterate, though, that many or most fungal effectors actually do not contain any recognizable domains or motifs.

12.4.2 Effectors from Plant Mutualists

The current literature on effector proteins in plant mutualists is limited. Much of the current understanding of effectors from plant-beneficial fungi seems to come from studies of these proteins in *Trichoderma* spp., many of which exist in the rhizosphere and are capable of mutualistically colonizing plant roots. Additional studies examining effectors in endophytes, such as *Piriformospora indica* and *Colletotrichum tofieldiae*, as well as mycorrhizal fungi, including *Rhizophagus* spp. and *Laccaria bicolor*, have been undertaken, though they have primarily identified candidate effectors, with only a few CEPs appearing to have been experimentally confirmed to act as effector proteins (Hacquard et al. 2016; Plett et al. 2014; Rafiqi et al. 2013; Sędzielewska Toro and Brachmann 2016).

A limited number of known and suspected effectors with similar functional domains to those found in plant pathogens have been identified in the genomes of mutualistic fungi. Given that the initial stages of colonization are similar in all plant-colonizing fungi, effector domains and motifs involved in fungal adhesion to the host plant, differentiation of colonization structures, and similar processes may be similar between phytopathogenic and mutualistic fungi (Lo Presti et al. 2015). For instance, of the <20 proteins experimentally confirmed to act as effectors in *Trichoderma* spp., four were hydrophobins (Ramírez-Valdespino et al. 2019). As previously discussed, among other roles, hydrophobins aid filamentous fungi in adhesion to plant surfaces, thereby facilitating colonization. Hydrophobin effectors from *Trichoderma* may also activate plant defense-related genes/ pathways in certain species (Ramírez-Valdespino et al. 2019). Known or suspected hydrophobin domain-containing proteins have been identified in *L. bicolor* and endophytic *Clonostachys rosea* (Dubey et al. 2014; Plett et al. 2012). Similarly, CFEM domain-containing proteins, which may act as effectors during colonization of host plants, have been found in the genome of *Trichoderma* spp., *P. indica*, and *L. bicolor* (Guzmán-Guzmán et al. 2017; Liu et al. 2019; Martin et al. 2008). Guzmán-Guzmán et al. (2017) showed that *tacfem1*, a candidate CFEM domain-containing effector-coding gene from endophytic *T. atroviride*, was upregulated when the fungus was co-cultivated with host *Arabidopsis* seedlings, suggesting it may act as an effector during host colonization.

Although beneficial to their host, mutualistic fungi must, like phytopathogens, modulate plant defense responses to facilitate ongoing symbioses. To this end, known or suspected effectors containing a LysM domain have been identified among endophytes and mycorrhizal fungi, including *C. rosea*, *P. indica*, *Rhizophagus* spp., and *Trichoderma* spp. (Dubey et al. 2020; Rafiqi et al. 2013; Romero-Contreras et al. 2019; Zeng et al. 2020). In addition to utilizing some techniques common to plant pathogenic effectors, plant mutualists may also possess effector proteins with strategies quite different from those found in plant pathogens. For instance, suppressing the host plant immune response may not always be ideal for plant mutualists, which benefit from their hosts' ability to protect themselves from invading pathogens. In fact, effector proteins which apparently boost the plant defense response without causing host death or disease have been found in mutualistic species. As an example, certain effector proteins characterized as cerato-platanins from *Trichoderma* spp., such as Sm1 from *T. virens* and Epl1 from *T. atroviride*, are apparently capable of bringing about changes in the host plant that allow it to better protect itself, such as induction of defense-related genes and production of ROS via yet-unknown mechanisms (Djonović et al. 2006; Ramírez-Valdespino et al. 2019; Seidl et al. 2006). However, immune-suppressing effector proteins have also been identified in plant mutualists. For instance, Fgb1 and PIIN_08944, effectors of endophytic *P. indica,* appear to suppress host plant MAMP-triggered ROS production in response to certain MAMPs (β-glucan and chitin, respectively), with PIIN_08944 also interfering with the salicylic acid-mediated immune response which, in part, is important for regulation of the hypersensitive response (Akum et al. 2015; Betsuyaku et al. 2018; Wawra et al. 2016). Furthermore, during colonization of host *Arabidopsis* plants, *C. tofieldiae* secretes an immune-suppressing homolog of a necrosis-inducing secreted protein 1 (NIS1). Necrosis-inducing proteins act as effectors in several phytopathogens (Irieda et al. 2019). Although the effects of NIS1 on host plants may vary between fungal species, it often suppresses MTI, including the production of ROS in response to certain MAMPs (bacterial flagellin and fungal chitin), as well as ETI, including the hypersensitive response (Irieda et al. 2019). For both *P. indica* and *C. tofieldiae,* the presence of immune-boosting effectors is unknown and, in general, the interplay between immune-supporting and immune-suppressing effectors in the establishment and maintenance of mutualistic relationships remains unclear. It is possible that effectors which interfere with plant defense responses are of more importance during initial colonization of hosts, with immune-boosting effectors then becoming more highly expressed, allowing for the maintenance of plant health once the mutualistic fungus has effectively established itself.

In addition to conserved domains, mutualistic fungi may possess motifs common to multiple effectors. For instance, 25 candidate effector proteins from *P. indica* have been found to possess a conserved 7-amino acid RSIDELD motif near their carbon (C)-terminal, whose functions are unknown (Rafiqi et al. 2013).

Overall, effector motifs and domains which play roles important to both mutualists and their phytopathogenic counterparts, such as adhesion to and colonization of plant tissues, may be shared between the groups. However, the effector repertoire of plant-beneficial fungi is likely highly tailored to their lifestyle, with effectors capable of mediating the maintenance of their ongoing mutualistic symbioses. Additional research into the unique effector arsenals of mutualists is, thus, warranted.

12.4.2.1 Endophytic Insect Pathogenic Fungi

There exists a unique group of endophytic fungi which are capable of parasitizing soil insects. The most well-studied of these endophytic insect pathogenic fungi (EIPF) are rhizosphere competent *Beauveria bassiana* and *Metarhizium* spp.

The relationship between *Metarhizium* spp. and their host plants involves a symbiotic exchange of nutrients that is intimately related to the ability of this genus to parasitize insects (Behie et al. 2017). Nitrogen, a limiting nutrient for plant growth, is found in abundance in insects (~10% of biomass) (Behie et al. 2017; Fagan et al. 2002). Thus, soil insects represent a critical reservoir of nitrogen which *Metarhizium* can capture via insect parasitism and subsequently translocate via mycelia to host plants, thereby promoting plant growth and productivity (Behie and Bidochka 2014). In exchange for this insect-derived nitrogen, *Metarhizium* receives fixed carbon in the form of plant photosynthate (Behie et al. 2017). The relationship between *B. bassiana* and its host plants may involve a similar tripartite exchange of nutrients. Like *Metarhizium* spp., *B. bassiana* has been shown to be capable of transferring insect-derived nitrogen to plants, as well as possess sugar transporters which may allow for assimilation of plant photosynthate (Behie and Bidochka 2014; Behie et al. 2017).

Given their ability to simultaneously promote plant growth and parasitize insects, including a number of common pests, agricultural applications of these EIPF (endophytic insect pathogenic fungi) have been studied, including their use as alternatives to chemical insecticides. *B. bassiana* and *Metarhizium* spp. are considered good candidates for such applications as they are found naturally in soil worldwide and, unlike chemical insecticides, appear to present minimal risk to the environment and the health of humans and other vertebrates (Brunner-Mendoza et al. 2019; Zimmermann 2007). The insect hosts of EIPF vary according to the fungal species, with some, such as *M. acridum*, acting as specialists against a narrow range of host insects, and others, such as *M. robertsii* and *B. bassiana*, being generalists which can infect a broad range of hosts. In the context of their use as biological control agents, this can allow for very controlled or relatively broad targeting of soil insects, depending on the species chosen. In order to regulate and optimize the utility of these fungi as biological control agents and soil amendments, it is critical to understand how they interact with their plant hosts, including the potential use of effector proteins in mediating plant colonization. However, the literature on this topic is very limited.

Although they are capable of insect pathogenesis, there is phylogenetic evidence that EIPF evolved from a lineage of plant symbionts which subsequently evolved insect pathogenesis capabilities. Phylogenetic analyses have shown that *Metarhizium* and *Beauveria* are related to fungal grass endophytes, with the lineage leading to *Metarhizium* having diverged from the grass endophyte *Epichloë festucae* ca. 100 million years ago (Barelli et al. 2016; Gao et al. 2011; Spatafora et al. 2007). Furthermore, whole-genome analyses of *Metarhizium* spp. indicated they are more closely related to plant symbionts than animal pathogenic fungi, again suggesting they evolved from an endophytic lineage which later gained insect pathogenesis capabilities (Barelli et al. 2016; Gao et al. 2011). Thus, although effector-coding genes require rapid evolution, those produced by EIPF may still bear some resemblance to those in other endophytes.

The secretomes of *Metarhizium* spp. have previously been examined; however the literature on effectors in this genus appears to have only gone so far as identifying CEPs in a singular species. Pattemore et al. (2014) annotated the whole-genome sequence of *M. anisopliae* and, in doing so, identified 242 “candidate effectors,” which the authors defined as proteins containing a secretion signal and a motif which is known to be associated with pathogenicity in fungal pathogens. However, the study did not experimentally confirm whether any of the identified candidate effectors were true effector proteins (Pattemore et al. 2014). Beyond that study, the only literature pertaining to effectors in *Metarhizium* appears to be those which have examined proteins with functional domains or motifs similar to those found in effectors, though the proteins were not necessarily investigated in the context of effectors. Huang et al. (2020) recently described an M35 metalloprotease, *MrM35-4*, from *M. robertsii* which was required for fungal virulence against insect hosts. However, deletion of a closely related metalloprotease, *MrM35-2,* apparently did not exhibit this effect on virulence against insects (Huang et al. 2020). Additionally the authors found that phylogenetic analysis of M35 metalloproteases from *M. robertsii* and other fungal and bacterial species placed *MrM35-2* in a lineage of metalloproteases from plant pathogenic fungi, including the previously described *AVR-Pita1* metalloprotease effector from phytopathogen *Magnaporthe oryzae* (Huang et al. 2020). However, the effector status of this CEP has not yet been experimentally confirmed. Furthermore, Moonjely et al. (2019) investigated the role of a hydrophobin (hyd3) and a serine protease (subtilisin-like serine protease Pr1A) in *M. robertsii* colonization of barley. Although RNA sequencing of *M. robertsii* on 7-day-old germinating seedlings showed these genes were upregulated during colonization of bean roots, no direct correlation between the genes and plant root association was found, with gene knockouts having no apparent effect on barley root colonization (Moonjely et al. 2019). However, this may have been due to redundancy, with similar proteins compensating for the loss of the deleted genes. Additionally, given that plant host preferences of *M. robertsii* have been established in the literature, it is possible that these proteins, if acting as effectors, are important during colonization of particular plants, such as monocots, but less so in dicots (Moonjely and Bidochka 2019). Overall, there does not appear to be any currently published literature confirming the effector status of CEPs in *Metarhizium*.

Similarly, there has been minimal research on effectors in *B. bassiana*. Moonjely et al. (2018) identified two hydrophobins, *hyd1* and *hyd2*, important for *B. bassiana* colonization of bean roots. Furthermore, preliminary bioinformatic analyses conducted by Mei et al. (2020) identified putative effector-coding genes; however the effector status of these CEPs has not yet been experimentally confirmed. As with the previously discussed M35 metalloprotease from *Metarhizium*, a study by Cen et al. (2017) established that LysM domain-containing proteins are important to *B. bassiana* insect pathogenesis. The authors found that deletion mutants of two LysM domain-containing proteins impaired the ability of *B. bassiana* to infect host insects (Cen et al. 2017). Moreover, complementation of the deletion mutants with Slp1, a LysM effector protein from plant pathogenic fungus *Magnaporthe oryzae*, allowed for restoration of full virulence (Cen et al. 2017). This ability for a LysM effector from *M. oryzae*, which itself is not an insect pathogen, to rescue insect pathogenesis capabilities in *B. bassiana* deletion mutants suggests a possible dual role of some EIPF effector proteins in the colonization of plant roots, as well as insect pathogenesis. This is supported by the fact that EIPF, including both *Beauveria* and *Metarhizium* spp., apparently utilize similar mechanisms to infect insects as they do to colonize plant hosts, possibly mediated, in some cases, by gene duplication events (Branine et al. 2019).

12.5 Methods for Identifying and Characterizing Effector Proteins

A variety of experimental and computational methods have been utilized to find and characterize fungal effector proteins. Recently, there has been an increased reliance on in silico analyses as a foundation in this process, with several bioinformatics tools available. For instance, there are a number of computational programs designed to predict the presence of signal peptides, including the aforementioned N-terminal signal peptide required by most effectors to target them for secretion (Sonah et al. 2016). Although these programs, such as SignalP and TargetP, are only predictive, requiring experimental validation, they are considered to have high accuracy (Sonah et al. 2016). Using such tools allows for rapid, high-throughput screening of large numbers of proteins for the potential to be secreted extracellularly, often altogether replacing experimental secretion analysis in recent studies pertaining to effector proteins (Guzmán-Guzmán et al. 2017; Neu and Debener 2019; Rafiqi et al. 2013). However, given that the signal peptides which target proteins for extracellular secretion can be quite similar in sequence to transmembrane domains, accurate prediction of secretion status typically requires these programs to be used in conjunction with those which can predict the presence of transmembrane helices, such as TMHMM (Sonah et al. 2016).

Several computational tools can also be used to screen CEPs for motifs and functional domains common to some known effectors. For instance, Motif Scan, a program which combines several databases pertaining to motif patterns and profiles, such as PROSITE, HAMAP, and Pfam, can be used to detect sequence motifs based on amino acid sequence (El-Gebali et al. 2019; Pedruzzi et al. 2015; Sigrist et al. 2013). However, it should be noted that many motifs apparently associated with effector proteins are relatively short in length (often ~4 amino acids), and thus the presence of these motifs may merely be a coincidence. Additionally, programs such as SUPERFAMILY and Pfam, both of which can operate through the HMMER server, can be used to classify proteins based on their domain(s) and/ or family classification(s), including those families previously described, such as LysM and CFEM domain-containing proteins, cerato-platanins, hydrophobins, proteases, and protease inhibitors (El-Gebali et al. 2019; Gough et al. 2001).

There also exists an in silico tool, EffectorP, designed specifically to predict whether a given protein is an effector based on its amino acid sequence (Sperschneider et al. 2018). The program does not establish definitive criteria or thresholds for predicting effector status, such as having $\leq$300 amino acids, given that effectors with a wide variety of characteristics have been found (Sperschneider et al. 2018). Rather, EffectorP utilizes a machine learning method to identify patterns in the features of known effectors, most notably protein size and net charge, as well as the abundance of cysteine and serine residues (Sperschneider et al. 2018). The program is trained with data sets containing known effectors and secreted proteins believed not to be effectors, allowing it to better learn how to identify potential effectors based on their predicted features (Sperschneider et al. 2018). The most recent version of the program, EffectorP 2.0, claims to have an accuracy of 89% (Sperschneider et al. 2018). It is important to note, though, that the program is trained specifically with effectors from plant pathogenic fungi, and thus its accuracy in predicting the effector status of proteins from plant mutualists is uncertain (Sperschneider et al. 2018). Furthermore, in the context of endophytic insect-pathogenic fungi, such as *M. robertsii* and *B. bassiana*, the negative data sets used to train the program (i.e., the data sets the program is told are not effector proteins) included secreted animal pathogen proteins which may be of concern in the case of fungal effectors that also serve a role in insect pathogenesis, as was previously suggested by the importance of LysM proteins in *B. bassiana* for infecting host insects (Cen et al. 2017; Sperschneider et al. 2018). Additionally, the program does not screen for the presence of a signal peptide, and thus secretion analysis by other means, whether it be experimental or in silico, is necessary.

Although valuable in screening fungal genomes for candidate effectors, results garnered from in silico tools such as EffectorP still require experimental validation. As of yet, there is not a standardized method for confirming the effector status of CEPs. One method of beginning experimentation is to simply analyze the expression of a suspected effector in the presence of a host plant. For instance, Guzmán-Guzmán et al. (2017) set up interaction assays in which *Trichoderma* was inoculated on one end of a culture plate, and germinating host plant *Arabidopsis* seedlings were placed on the other end. Quantitative reverse transcription PCR (RT-qPCR) was then utilized to analyze the expression of their candidate effectors at three time points: (1) before the fungus made contact with the root; (2) at contact with the root; and (3) once the *Trichoderma* spp. overgrew the *Arabidopsis* root (Guzmán-Guzmán et al. 2017). Expression at these time points was then compared to the expression of the candidate effectors in control axenic cultures (Guzmán-Guzmán et al. 2017). This allowed for determination of changes in gene expression, and, if changes in expression were observed, provided some hints as to the stage(s) of colonization a potential effector may be particularly important to (i.e., preparing for colonization, actual establishment of colonization, and/ or maintenance of colonization) (Guzmán-Guzmán et al. 2017). Changes in gene expression in the presence of a plant is a characteristic of effector proteins but can also occur with non-effectors. Thus, additional experimentation must be undertaken to confirm whether a CEP truly functions as an effector by modifying plant physiology in ways which aid in the establishment and maintenance of symbiosis.

To that end, many studies have utilized gene knockouts/silencing and/or overexpression to investigate the effects of experimentally disrupting expression of suspected effector proteins, with the expectation that, if the CEP is truly an effector, doing so would affect the ability of the fungus to colonize its host plant. Interpretation of the results of such experiments may be simpler for phytopathogenic fungi, for which plant disease state may be taken as an indicator of the effect and importance of a candidate effector. Nevertheless, gene knockout and overexpression experiments have also been conducted during investigations of CEPs in plant mutualists. For instance, the previously described immune-suppressing *PIIN_08944* gene from endophytic *P. indica* was established as an effector by way of gene deletion (Akum et al. 2015). By utilizing qPCR and fungal-specific primers to investigate the relative fungal biomass in host *A. thaliana* roots, Akum et al. (2015) determined that the ability of a *PIIN_08944* knockout strain of *P. indica* to colonize host roots was significantly delayed as compared to a wild-type strain over a 21-day period. As a further example, in attempting to identify effectors in the endophyte *Epichloë festucae*, Hassing et al. (2019) created knockout and overexpressing strains for four suspected effector proteins. They compared the phenotype of the mutant strains to wild-type strains both in axenic culture, analyzed by way of colony and hyphal morphology, and in their interaction with a host plant, *Lolium perenne* (Hassing et al. 2019). The host plants infected with mutant strains were compared to those infected with wild-type *E. festucae* in terms of the effects of gene deletion or overexpression on the length and number of tillers (stems) growing from the plant (Hassing et al. 2019). Additionally, pseudostem cross-sections of infected plants were stained to examine whether the number of hyphae per intercellular space differed in wild-type and mutant strains (Hassing et al. 2019). The authors found no significant differences in the culture nor *in planta* phenotype of any of their four knockout or overexpressing strains as compared to wild-type strains (Hassing et al. 2019). However, these results do not necessarily indicate that the suspected effectors were not *E. festucae* effector proteins. A critical issue with the knockout method is its susceptibility to issues of redundancy, with effector proteins with the same or a similar function potentially being capable of making up for the function of those which have been knocked out (Hassing et al. 2019). Similarly, lack of change in the plant-fungal interaction with overexpression strains may not necessarily indicate that a given protein is not an effector, as the *in planta* expression levels of the protein (as well as redundant proteins) in wild-type strains may already be sufficient to maximize the ability of the fungus to colonize its host plant in that regard, with overexpression not providing any additional benefit. Despite these issues, knockout and overexpression methods have proven quite effective in establishing effector status of candidate effectors from several species, including one or more species from the genera *Trichoderma, Fusarium,* and *Verticillium* (Guzmán-Guzmán et al. 2017; Marton et al. 2018; Selin et al. 2016).

12.6 Conclusion

In order to mediate interactions with host plants, plant-associated fungi secrete effector proteins into the plant apoplast and cytoplasm, which modulate host plant physiology in ways which facilitate fungal colonization and growth. These proteins typically require an N-terminal signal peptide for fungal secretion and are often small in size and high in cysteine content. A limited number of effector families with similar domains and motifs have been identified in several fungal genera. These include hydrophobins, CFEM domain-containing proteins, LysM domain-containing proteins, proteases, protease inhibitors, necrosis-inducing proteins, cerato-platanins, as well as the Y/F/WxC and RxLR motifs. Despite a few such conserved sequence features, though, the effector repertoire of fungi is presumably highly tailored to their individual lifestyles. In fact, the effector proteins of a given species typically have little to no sequence similarity with those found in other species or genera, likely due to the unique lifestyles of different fungi, as well as rapid evolution required of effectors to avoid detection by host plants in the co-evolutionary arms race. This fact has hindered progress in the identification of novel effectors, particularly in species in which such proteins have not been previously characterized.

Identifying effector proteins typically begins with utilizing in silico tools to mine fungal genomes for CEPs based on the aforementioned common features, followed by experimental validation of a select number of putative effectors. Currently, there is no standardized pipeline for experimental validation of the effector status of CEPs, though many studies take a gene knockout or overexpression approach. However, these approaches may provide false negatives due to issues of gene redundancy.

The literature on effector proteins from rhizospheric fungi, especially plant mutualists, is limited, with most knowledge stemming from investigations of pathogens of above-ground tissue. Nonetheless, rhizospheric fungi are of great agricultural importance. Soil-borne phytopathogens are often capable of causing disease in a wide range of crop plants. Conversely, plant root mutualists, particularly endophytic insect pathogenic fungi, can improve soil quality and promote plant growth and health, and thus have been utilized as alternatives to chemical fertilizers and pesticides. Thus, continued research into understanding of how rhizospheric fungi mediate their interactions with host plants, including their use of effector proteins, will be vital to global food security and ongoing agricultural advancements.

Acknowledgments This research was conducted with the assistance of a Natural Sciences and Engineering Research Council of Canada to MJB (RGPIN-2014-04496).

References

Akum FN, Steinbrenner J, Biedenkopf D, Imani J, Kogel KH (2015) The *Piriformospora indica* effector PIIN_08944 promotes the mutualistic Sebacinalean symbiosis. Front Plant Sci 6:906. https://doi.org/10.3389/fpls.2015.00906

Bailey BA (1995) Purification of a protein from culture filtrates of *Fusarium oxysporum* that induces ethylene and necrosis in leaves of *Erythroxylum coca*. Phytopathology 85:1250–1255. https://doi.org/10.1094/Phyto-85-1250

Barelli L, Moonjely S, Behie SW, Bidochka MJ (2016) Fungi with multifunctional lifestyles: endophytic insect pathogenic fungi. Plant Mol Biol 90:657–664. https://doi.org/10.1007/s11103-015-0413-z

Behie SW, Bidochka MJ (2014) Ubiquity of insect-derived nitrogen transfer to plants by endophytic insect-pathogenic fungi: an additional branch of the soil nitrogen cycle. Appl Environ Microbiol 80:1553–1560. https://doi.org/10.1128/AEM.03338-13

Behie SW, Moreira CC, Sementchoukova I, Barelli L, Zelisko PM, Bidochka MJ (2017) Carbon translocation from a plant to an insect-pathogenic endophytic fungus. Nat Commun 8:14,245. https://doi.org/10.1038/ncomms14245

Betsuyaku S, Katou S, Takebayashi Y, Sakakibara H, Nomura N, Fukuda H (2018) Salicylic acid and jasmonic acid pathways are activated in spatially different domains around the infection site during effector-triggered immunity in *Arabidopsis thaliana*. Plant Cell Physiol 59:8–16. https://doi.org/10.1093/pcp/pcx181

Blümke A, Falter C, Herrfurth C, Sode B, Bode R, Schäfer W et al (2014) Secreted fungal effector lipase releases free fatty acids to inhibit innate immunity-related callose formation during wheat head infection. Plant Physiol 165:346–358. https://doi.org/10.1104/pp.114.236737

Branine M, Bazzicalupo A, Branco S (2019) Biology and applications of endophytic insect-pathogenic fungi. PLoS Pathog 15:e1007831

Brunner-Mendoza C, del Reyes-Montes M, Moonjely S, Bidochka MJ, Toriello C (2019) A review on the genus *Metarhizium* as an entomopathogenic microbial biocontrol agent with emphasis on its use and utility in Mexico. Biocontrol Sci Tech 29:83–102. https://doi.org/10.1080/09583157.2018.1531111

Castroverde CDM, Nazar RN, Robb J (2016) *Verticillium* Ave1 effector induces tomato defense gene expression independent of Ve1 protein. Plant Signal Behav 11:e1245254. https://doi.org/10.1080/15592324.2016.1245254

Cen K, Li B, Lu Y, Zhang S, Wang C (2017) Divergent LysM effectors contribute to the virulence of *Beauveria bassiana* by evasion of insect immune defenses. PLoS Pathog 13:e1006604. https://doi.org/10.1371/journal.ppat.1006604

Crutcher FK, Moran-Diez ME, Ding S, Liu J, Horwitz BA, Mukherjee PK, Kenerley CM (2015) A paralog of the proteinaceous elicitor SM1 is involved in colonization of maize roots by *Trichoderma virens*. Fungal Biol 119:476–486. https://doi.org/10.1016/j.funbio.2015.01.004

De Jonge R, Van Esse HP, Kombrink A, Shinya T, Desaki Y, Bours R et al (2010) Conserved fungal LysM effector Ecp6 prevents chitin-triggered immunity in plants. Science 329:953–955. https://doi.org/10.1126/science.1190859

Djonović S, Pozo MJ, Dangott LJ, Howell CR, Kenerley CM (2006) Sm1, a proteinaceous elicitor secreted by the biocontrol fungus *Trichoderma virens* induces plant defense responses and systemic resistance. Mol Plant-Microbe Interact 19:838–853. https://doi.org/10.1094/MPMI-19-0838

Djonović S, Vargas WA, Kolomiets MV, Horndeski M, Wiest A, Kenerley CM (2007) A proteinaceous elicitor Sm1 from the beneficial fungus *Trichoderma virens* is required for induced systemic resistance in maize. Plant Physiol 145:875–889. https://doi.org/10.1104/pp.107.103689

Dölfors F, Holmquist L, Dixelius C, Tzelepis G (2019) A LysM effector protein from the basidiomycete *Rhizoctonia solani* contributes to virulence through suppression of chitin-triggered immunity. Mol Gen Genomics 294:1211–1218. https://doi.org/10.1007/s00438-019-01573-9

Dou D, Kale SD, Wang X, Chen Y, Wang Q, Wang X et al (2008) Conserved C-terminal motifs required for avirulence and suppression of cell death by *Phytophthora sojae* effector Avr1b. Plant Cell 20:1118–1133. https://doi.org/10.1105/tpc.107.057067

Dubey MK, Jensen DF, Karlsson M (2014) Hydrophobins are required for conidial hydrophobicity and plant root colonization in the fungal biocontrol agent *Clonostachys rosea*. BMC Microbiol 14:18. https://doi.org/10.1186/1471-2180-14-18

Dubey M, Vélëz H, Broberg M, Jensen DF, Karlsson M (2020) LysM proteins regulate fungal development and contribute to hyphal protection and biocontrol traits in *Clonostachys rosea*. Front Microbiol 11:679. https://doi.org/10.3389/fmicb.2020.00679

El-Gebali S, Mistry J, Bateman A, Eddy SR, Luciani A, Potter SC et al (2019) The Pfam protein families database in 2019. Nucleic Acids Res 47:427–432. https://doi.org/10.1093/nar/gky995

Fagan WF, Siemann E, Mitter C, Denno RF, Huberty AF, Woods HA, Elser JJ (2002) Nitrogen in insects: implications for trophic complexity and species diversification. Am Nat 160:784–802. https://doi.org/10.1086/343879

Fisher MC, Henk DA, Briggs CJ, Brownstein JS, Madoff LC, McCraw SL, Gurr SJ (2012) Emerging fungal threats to animal, plant and ecosystem health. Nature 484:186–194. https://doi.org/10.1038/nature10947

Franceschetti M, Maqbool A, Jiménez-Dalmaroni MJ, Pennington HG, Kamoun S, Banfield MJ (2017) Effectors of filamentous plant pathogens: commonalities amid diversity. Microbiol Mol Biol Rev 81:1–17. https://doi.org/10.1128/mmbr.00066-16

Gaderer R, Lamdan NL, Frischmann A, Sulyok M, Krska R, Horwitz BA, Seidl-Seiboth V (2015) Sm2, a paralog of the *Trichoderma* cerato-platanin elicitor Sm1, is also highly important for plant protection conferred by the fungal-root interaction of *Trichoderma* with maize. BMC Microbiol 15:2. https://doi.org/10.1186/s12866-014-0333-0

Gao Q, Jin K, Ying SH, Zhang Y, Xiao G, Shang Y et al (2011) Genome sequencing and comparative transcriptomics of the model entomopathogenic fungi Metarhizium anisopliae and *M. acridum*. PLoS Genet 7:e1001264. https://doi.org/10.1371/journal.pgen.1001264

Ghareeb H, Drechsler F, Löfke C, Teichmann T, Schirawski J (2015) SUPPRESSOR OF APICAL DOMINANCE1 of *Sporisorium reilianum* modulates inflorescence branching architecture in maize and arabidopsis. Plant Physiol 169:2789–2804. https://doi.org/10.1104/pp.15.01347

Godfrey D, Böhlenius H, Pedersen C, Zhang Z, Emmersen J, Thordal-Christensen H (2010) Powdery mildew fungal effector candidates share N-terminal Y/F/WxC-motif. BMC Genomics 11:317. https://doi.org/10.1186/1471-2164-11-317

Gorzelak MA, Asay AK, Pickles BJ, Simard SW (2015) Inter-plant communication through mycorrhizal networks mediates complex adaptive behaviour in plant communities. AoB Plants 7:plv050. https://doi.org/10.1093/aobpla/plv022

Gough J, Karplus K, Hughey R, Chothia C (2001) Assignment of homology to genome sequences using a library of hidden Markov models that represent all proteins of known structure. J Mol Biol 313:903–919. https://doi.org/10.1006/jmbi.2001.5080

Guzmán-Guzmán P, Alemán-Duarte MI, Delaye L, Herrera-Estrella A, Olmedo-Monfil V (2017) Identification of effector-like proteins in *Trichoderma* spp. and role of a hydrophobin in the plant-fungus interaction and mycoparasitism. BMC Genet 18:1–20. https://doi.org/10.1186/s12863-017-0481-y

Hacquard S, Kracher B, Hiruma K, Münch PC, Garrido-Oter R, Thon MR et al (2016) Survival trade-offs in plant roots during colonization by closely related beneficial and pathogenic fungi. Nat Commun 7:11362. https://doi.org/10.1038/ncomms11362

Hassing B, Winter D, Becker Y, Mesarich CH, Eaton CJ, Scott B (2019) Analysis of *Epichloë festucae* small secreted proteins in the interaction with *Lolium perenne*. PLoS One 14:e0209463. https://doi.org/10.1371/journal.pone.0209463

Huang A, Lu M, Ling E, Li P, Wang C (2020) A M35 family metalloprotease is required for fungal virulence against insects by inactivating host prophenoloxidases and beyond. Virulence 11:222–237. https://doi.org/10.1080/21505594.2020.1731126

Irieda H, Inoue Y, Mori M, Yamada K, Oshikawa Y, Saitoh H et al (2019) Conserved fungal effector suppresses PAMP-triggered immunity by targeting plant immune kinases. Proc Natl Acad Sci U S A 116:496–505. https://doi.org/10.1073/pnas.1807297116

Jashni MK, Dols IHM, Iida Y, Boeren S, Beenen HG, Mehrabi R et al (2015a) Synergistic action of a metalloprotease and a serine protease from *Fusarium oxysporum* f. sp. *lycopersici* cleaves chitin-binding tomato chitinases, reduces their antifungal activity, and enhances fungal virulence. Mol Plant-Microbe Interact 28:996–1008. https://doi.org/10.1094/MPMI-04-15-0074-R

Jashni MK, Mehrabi R, Collemare J, Mesarich CH, de Wit PJGM (2015b) The battle in the apoplast: further insights into the roles of proteases and their inhibitors in plant–pathogen interactions. Front Plant Sci 6:1–7. https://doi.org/10.3389/fpls.2015.00584

Khang CH, Park SY, Lee YH, Valent B, Kang S (2008) Genome organization and evolution of the AVR-pita avirulence gene family in the *Magnaporthe grisea* species complex. Mol Plant-Microbe Interact 21:658–670. https://doi.org/10.1094/MPMI-21-5-0658

Kloppholz S, Kuhn H, Requena N (2011) A secreted fungal effector of glomus intraradices promotes symbiotic biotrophy. Curr Biol 21:1204–1209. https://doi.org/10.1016/j.cub.2011.06.044

Kombrink A, Rovenich H, Shi-Kunne X, Rojas-Padilla E, van den Berg GCM, Domazakis E et al (2017) *Verticillium dahliae* LysM effectors differentially contribute to virulence on plant hosts. Mol Plant Pathol 18:596–608. https://doi.org/10.1111/mpp.12520

Kou Y, Tan YH, Ramanujam R, Naqvi NI (2017) Structure–function analyses of the Pth11 receptor reveal an important role for CFEM motif and redox regulation in rice blast. New Phytol 214: 330–342. https://doi.org/10.1111/nph.14347

Li S, Peng X, Wang Y, Hua K, Xing F, Zheng Y et al (2019) The effector AGLIP1 in *Rhizoctonia solani* AG1 IA triggers cell death in plants and promotes disease development through inhibiting PAMP-triggered immunity in *Arabidopsis thaliana*. Front Microbiol 10:2228. https://doi.org/10.3389/fmicb.2019.02228

Liu L, Xu L, Jia Q, Pan R, Oelmüller R, Zhang W, Wu C (2019) Arms race: diverse effector proteins with conserved motifs. Plant Signal Behav 14:1557008. https://doi.org/10.1080/15592324.2018.1557008

Lo Presti L, Lanver D, Schweizer G, Tanaka S, Liang L, Tollot M et al (2015) Fungal effectors and plant susceptibility. Annu Rev Plant Biol 66:513–545. https://doi.org/10.1146/annurev-arplant-043014-114623

Lyu X, Shen C, Fu Y, Xie J, Jiang D, Li G, Cheng J (2016) A small secreted virulence-related protein is essential for the necrotrophic interactions of *Sclerotinia sclerotiorum* with its host plants. PLoS Pathog 12:e1005435. https://doi.org/10.1371/journal.ppat.1005435

Martin F, Aerts A, Ahrén D, Brun A, Danchin EGJ, Duchaussoy F et al (2008) The genome of *Laccaria bicolor* provides insights into mycorrhizal symbiosis. Nature 452:88–92. https://doi.org/10.1038/nature06556

Marton K, Flajsman M, Radisek S, Kosmelj K, Jakse J, Javornik B, Berne S (2018) Comprehensive analysis of *Verticillium nonalfalfae in silico* secretome uncovers putative effector proteins expressed during hop invasion. PLoS One 13:1–28. https://doi.org/10.1371/journal.pone.0198971

Mei L, Chen M, Shang Y, Tang G, Tao Y, Zeng L et al (2020) Population genomics and evolution of a fungal pathogen after releasing exotic strains to control insect pests for 20 years. ISME J 14: 1422–1434. https://doi.org/10.1038/s41396-020-0620-8

Moonjely S, Bidochka MJ (2019) Generalist and specialist *Metarhizium* insect pathogens retain ancestral ability to colonize plant roots. Fungal Ecol 41:209–217. https://doi.org/10.1016/j.funeco.2019.06.004

Moonjely S, Keyhani NO, Bidochka MJ (2018) Hydrophobins contribute to root colonization and stress responses in the rhizosphere-competent insect pathogenic fungus *Beauveria bassiana*. Microbiology 164:517–528. https://doi.org/10.1099/mic.0.000644

Moonjely S, Zhang X, Fang W, Bidochka MJ (2019) *Metarhizium robertsii* ammonium permeases (MepC and Mep2) contribute to rhizoplane colonization and modulates the transfer of insect derived nitrogen to plants. PLoS One 13:e0198971. https://doi.org/10.1371/journal.pone.0223718

Neu E, Debener T (2019) Prediction of the *Diplocarpon rosae* secretome reveals candidate genes for effectors and virulence factors. Fungal Biol 123:231–239. https://doi.org/10.1016/j.funbio.2018.12.003

Newman MA, Sundelin T, Nielsen JT, Erbs G (2013) MAMP (microbe-associated molecular pattern) triggered immunity in plants. Front Plant Sci 4:139. https://doi.org/10.3389/fpls.2013.00139

Pattemore JA, Hane JK, Williams AH, Wilson BAL, Stodart BJ, Ash GJ (2014) The genome sequence of the biocontrol fungus *Metarhizium anisopliae* and comparative genomics of *Metarhizium* species. BMC Genomics 15:660. https://doi.org/10.1186/1471-2164-15-660

Pedruzzi I, Rivoire C, Auchincloss AH, Coudert E, Keller G, De Castro E et al (2015) HAMAP in 2015: updates to the protein family classification and annotation system. Nucleic Acids Res 43: 1064–1070. https://doi.org/10.1093/nar/gku1002

Plett JM, Gibon J, Kohler A, Duffy K, Hoegger PJ, Velagapudi R et al (2012) Phylogenetic, genomic organization and expression analysis of hydrophobin genes in the ectomycorrhizal basidiomycete *Laccaria bicolor*. Fungal Genet Biol 49:199–209. https://doi.org/10.1016/j.fgb.2012.01.002

Plett JM, Daguerre Y, Wittulsky S, Vayssieres A, Deveau A, Melton SJ et al (2014) Effector MiSSP7 of the mutualistic fungus *Laccaria bicolor* stabilizes the *Populus* JAZ6 protein and represses jasmonic acid (JA) responsive genes. Proc Natl Acad Sci U S A 111:8299–8304. https://doi.org/10.1073/pnas.1322671111

Quarantin A, Hadeler B, Kröger C, Schäfer W, Favaron F, Sella L, Martínez-Rocha AL (2019) Different hydrophobins of *fusarium graminearum* are involved in hyphal growth, attachment, water-air interface penetration and plant infection. Front Microbiol 10:751. https://doi.org/10.3389/fmicb.2019.00751

Rafiqi M, Jelonek L, Akum NF, Zhang F, Kogel KH (2013) Effector candidates in the secretome of *Piriformospora indica*, a ubiquitous plant-associated fungus. Front Plant Sci 4:228. https://doi.org/10.3389/fpls.2013.00228

Ramírez-Valdespino CA, Casas-Flores S, Olmedo-Monfil V (2019) *Trichoderma* as a model to study effector-like molecules. Front Microbiol 10:1030. https://doi.org/10.3389/fmicb.2019.01030

Romero-Contreras YJ, Ramírez-Valdespino CA, Guzmán-Guzmán P, Macías-Segoviano JI, Villagómez-Castro JC, Olmedo-Monfil V (2019) Tal6 From *Trichoderma atroviride* is a LysM effector involved in mycoparasitism and plant association. Front Microbiol 10:2231. https://doi.org/10.3389/fmicb.2019.02231

Salas-Marina MA, Isordia-Jasso MI, Islas-Osuna MA, Delgado-Sánchez P, Jiménez-Bremont JF, Rodríguez-Kessler M et al (2015) The Epl1 and Sm1 proteins from *Trichoderma atroviride* and *Trichoderma virens* differentially modulate systemic disease resistance against different life style pathogens in *Solanum lycopersicum*. Front Plant Sci 6:77. https://doi.org/10.3389/fpls.2015.00077

Schweizer G, Münch K, Mannhaupt G, Schirawski J, Kahmann R, Dutheil JY (2018) Positively selected effector genes and their contribution to virulence in the smut fungus *Sporisorium reilianum*. Genome Biol Evol 10:629–645. https://doi.org/10.1093/gbe/evy023

Sędzielewska Toro K, Brachmann A (2016) The effector candidate repertoire of the arbuscular mycorrhizal fungus *Rhizophagus clarus*. BMC Genomics 17:101. https://doi.org/10.1186/s12864-016-2422-y

Seidl V, Marchetti M, Schandl R, Allmaier G, Kubicek CP (2006) Epl1, the major secreted protein of *Hypocrea atroviridis* on glucose, is a member of a strongly conserved protein family comprising plant defense response elicitors. FEBS J 273:4346–4359. https://doi.org/10.1111/j.1742-4658.2006.05435.x

Selin C, de Kievit TR, Belmonte MF, Fernando WGD (2016) Elucidating the role of effectors in plant-fungal interactions: progress and challenges. Front Microbiol 7:600. https://doi.org/10.3389/fmicb.2016.00600

Sigrist CJA, De Castro E, Cerutti L, Cuche BA, Hulo N, Bridge A et al (2013) New and continuing developments at PROSITE. Nucleic Acids Res 41:344–347. https://doi.org/10.1093/nar/gks1067

Sonah H, Deshmukh RK, Bélanger RR (2016) Computational prediction of effector proteins in fungi: opportunities and challenges. Front Plant Sci 7:126. https://doi.org/10.3389/fpls.2016.00126

Spatafora JW, Sung GH, Sung JM, Hywel-Jones NL, White JF (2007) Phylogenetic evidence for an animal pathogen origin of ergot and the grass endophytes. Mol Ecol 16:1701–1711. https://doi.org/10.1111/j.1365-294X.2007.03225.x

Sperschneider J, Dodds PN, Gardiner DM, Singh KB, Taylor JM (2018) Improved prediction of fungal effector proteins from secretomes with EffectorP 2.0. Mol Plant Pathol 19:2094–2110. https://doi.org/10.1111/mpp.12682

Taylor A, Vágány V, Jackson AC, Harrison RJ, Rainoni A, Clarkson JP (2016) Identification of pathogenicity-related genes in *Fusarium oxysporum* f. sp. *cepae*. Mol Plant Pathol 17:1032–1347. https://doi.org/10.1111/mpp.12346

Thatcher LF, Gardiner DM, Kazan K, Manners JM (2012) A highly conserved effector in *Fusarium oxysporum* is required for full virulence on *Arabidopsis*. Mol Plant-Microbe Interact 25:180–190. https://doi.org/10.1094/MPMI-08-11-0212

Thatcher LF, Williams AH, Garg G, Buck SAG, Singh KB (2016) Transcriptome analysis of the fungal pathogen *Fusarium oxysporum* f. sp. *medicaginis* during colonisation of resistant and susceptible *Medicago truncatula* hosts identifies differential pathogenicity profiles and novel candidate effectors. BMC Genomics 17:860. https://doi.org/10.1186/s12864-016-3192-2

Tini F, Beccari G, Benfield AH, Gardiner DM, Covarelli L (2020) Role of the XylA gene, encoding a cell wall degrading enzyme, during common wheat, durum wheat and barley colonization by *Fusarium graminearum*. Fungal Genet Biol 136:103318. https://doi.org/10.1016/j.fgb.2019.103318

Trdá L, Boutrot F, Ciaverie J, Brulé D, Dorey S, Poinssot B (2015) Perception of pathogenic or beneficial bacteria and their evasion of host immunity: pattern recognition receptors in the frontline. Front Plant Sci 6:219. https://doi.org/10.3389/fpls.2015.00219

Tsuzuki S, Handa Y, Takeda N, Kawaguchi M (2016) Strigolactone-induced putative secreted protein 1 is required for the establishment of symbiosis by the arbuscular mycorrhizal fungus *Rhizophagus irregularis*. Mol Plant-Microbe Interact 29:277–286. https://doi.org/10.1094/MPMI-10-15-0234-R

Uhse S, Djamei A (2018) Effectors of plant-colonizing fungi and beyond. PLoS Pathog 14: e1006992. https://doi.org/10.1371/journal.ppat.1006992

Van Den Burg HA, Harrison SJ, Joosten MHAJ, Vervoort J, De Wit PJGM (2006) *Cladosporium fulvum* Avr4 protects fungal cell walls against hydrolysis by plant chitinases accumulating during infection. Mol Plant-Microbe Interact 19:1420–1430. https://doi.org/10.1094/MPMI-19-1420

Volk H, Marton K, Flajšman M, Radišek S, Tian H, Hein I et al (2019) Chitin-binding protein of *Verticillium nonalfalfae* disguises fungus from plant chitinases and suppresses chitin-triggered host immunity. Mol Plant-Microbe Interact 32:1378–1390. https://doi.org/10.1094/MPMI-03-19-0079-R

Voß S, Betz R, Heidt S, Corradi N, Requena N (2018) RiCRN1, a crinkler effector from the arbuscular mycorrhizal fungus *Rhizophagus irregularis*, functions in arbuscule development. Front Microbiol 9:2068. https://doi.org/10.3389/fmicb.2018.02068

Wawra S, Fesel P, Widmer H, Timm M, Seibel J, Leson L et al (2016) The fungal-specific β-glucan-binding lectin FGB1 alters cell-wall composition and suppresses glucan-triggered immunity in plants. Nat Commun 7:13,188. https://doi.org/10.1038/ncomms13188

van de Wouw AP, Cozijnsen AJ, Hane JK, Brunner PC, McDonald BA, Oliver RP, Howlett BJ (2010) Evolution of linked avirulence effectors in *Leptosphaeria maculans* is affected by genomic environment and exposure to resistance genes in host plants. PLoS Pathog 6: e1001180. https://doi.org/10.1371/journal.ppat.1001180

Yamamoto N, Wang Y, Lin R, Liang Y, Liu Y, Zhu J et al (2019) Integrative transcriptome analysis discloses the molecular basis of a heterogeneous fungal phytopathogen complex, *Rhizoctonia solani* AG-1 subgroups. Sci Rep 9:19626. https://doi.org/10.1038/s41598-019-55734-2

Zeng T, Rodriguez-Moreno L, Mansurkhodzaev A, Wang P, van den Berg W, Gasciolli V et al (2020) A lysin motif effector subverts chitin-triggered immunity to facilitate arbuscular mycorrhizal symbiosis. New Phytol 225:448–460. https://doi.org/10.1111/nph.16245

Zhang L, Ni H, Du X, Wang S, Ma XW, Nürnberger T et al (2017) The *Verticillium*-specific protein VdSCP7 localizes to the plant nucleus and modulates immunity to fungal infections. New Phytol 215:368–381. https://doi.org/10.1111/nph.14537

Zhang Y, Gao Y, Liang Y, Dong Y, Yang X, Qiu D (2019) *Verticillium dahliae* PevD1, an Alt a 1-like protein, targets cotton PR5-like protein and promotes fungal infection. J Exp Bot 70:613–626. https://doi.org/10.1093/jxb/ery351

Zhu X, Soliman A, Islam MR, Adam LR, Daayf F (2017) *Verticillium dahliae*'s isochorismatase hydrolase is a virulence factor that contributes to interference with potato's salicylate and jasmonate defense signaling. Front Plant Sci 8:399. https://doi.org/10.3389/fpls.2017.00399

Zimmermann G (2007) Review on safety of the entomopathogenic fungi *Beauveria bassiana* and *Beauveria brongniartii*. Biocontrol Sci Tech 6:553–596. https://doi.org/10.1080/09583150701309006

The manufacturer's authorised representative in the EU is Springer Nature Customer Service Centre GmbH, Europaplatz 3, 69115 Heidelberg, Germany. If you have any concerns regarding our products, please contact ProductSafety@springernature.com

Printed and bound by CPI Group (UK) Ltd, Croydon, CR0 4YY
30/09/2026
02225241-0007